NANOTECHNOLOGY SCIENCE AND TECHNOLOGY

NANOSTRUCTURES

PROPERTIES, PRODUCTION METHODS AND APPLICATIONS

NANOTECHNOLOGY SCIENCE AND TECHNOLOGY

Additional books in this series can be found on Nova's website under the Series tab.

Additional e-books in this series can be found on Nova's website under the e-book tab.

NANOTECHNOLOGY SCIENCE AND TECHNOLOGY

NANOSTRUCTURES

PROPERTIES, PRODUCTION METHODS AND APPLICATIONS

YU DONG
EDITOR

nova publishers
New York

Library of Congress Cataloging-in-Publication Data

ISBN: 978-1-62618-081-9

Published by Nova Science Publishers, Inc. † New York

CONTENTS

PREFACE

Richard P. Feynman, a Nobel Laureate in Physics in 1965, had his famous saying of "There is plenty of room at the bottom.", which unraveled the scientific exploration of nanoscience and nanotechnology. As materials are scaled down from macro-, micro- to nanoscaled level with the advanced fabrication techniques such as nanolithography, nanomoulding, electrospinning, etc., nanostructures, normally possessing the characteristic of at least one dimension smaller than 100 nm, have drawn the overwhelming attention to scientists, material engineers and industrial innovators owing to their widespread usages leading to excellent physical, mechanical and multifunctional properties and a variety of marvelous applications from energy storage, cosmetics, sensing devices, nanofiltration, drug delivery to semiconductors.

The present book targets the audiences at different levels but is not limited to engineering and material researchers, technological developers, scientific patent consultants, policymakers in science and technology, and of course postgraduate students and junior researchers in the relevant fields, in order to facilitate the mutual exchange of spark ideas and in-depth discussions on various nanostructure topics in relation to their properties, production methods and applications.

The brief book structure, including ten chapters, is given to readers for an overview, which is categorized as "nanostructures" (Chapters 1-4), "nanoparticles/nanomaterials" (Chapters 5-8) and "nanocomposites" (Chapters 9 and 10). Chapter 1 presents the preparation and characterization of nanostructures based on polymer or glass surfaces under plasma and laser treatments to evaluate the chemical structure, electrical properties, optical properties, layer ablation, surface properties and material crystallinity. Chapter 2 broadly reviewed the synthesis, characterization and application of different CuO nanostructures in various dimensions and CuO nanoparticles with the guidelines for the future investigations. Chapter 3 offers a brief review on the recent progress in a special application of radiation techniques for synthesis and modification of nanostructured materials. Available experimental data were comprehensively examined which applied to radiation-induced formation of ion track membranes as nanoporous filters and templates for the synthesis of metallic nanostructures, preparation of radiation-cured polymeric nanocomposites and nanogels, as well as production of nanopowders of different substances. The main directions in radiation-induced modification of carbon nanostructures were also reviewed. Chapter 4 gives the recent research progress on hierarchical colloidal photonic crystals covered from preparation methods to structural properties and applications. It also shows the importance of a hierarchy of multi-scaled interconnected structures with very high surface-to-volume ratio to significantly enhance the tailored material properties and alter the band-gap features of photonic crystals.

Chapter 5 provides a critical review on a recent general solid state method to obtain metallic, metal oxide and phosphate nanoparticles and crystals by pyrolysis at 800 °C using organometallic derivatives of cyclo and polyphosphazene precursors containing diverse organometallic fragments linked to polymeric or oligomeric phosphazenes. Chapter 6 investigates the physical characteristics of array TiO_2 nanotubes based on electrochemical anodization growth such as mechanical characteristics, contact angle, and photoluminescence (PL) and X-ray diffraction (XRD) spectra for the application of photodetectors or sensors. Chapter 7 describes the study of filtration and separation of nanoparticles, influence of morphological structures, surface charge state and temperature on filtration with the support of theoretical prediction on capture or separation efficiencies. Additionally, the processing, size classification, agglomeration, de-agglomeration, charging and sintering of nanoparticles in the gas phase were also elaborated with the supplementary review of developing nanofilter media. Chapter 8 presents the production, mechanical characterization and mechanics modeling of continuous nanofibers via electrospinning. The unique mechanical responses of electrospun nanofibers such as the size effect, failure mechanisms, surface instability, etc. were explicitly discussed. The recently developed continuum mechanics models were correlated with the nanofiber experimental observations in terms of wave propagation, surface rippling and wetting, adhesion and contact, hydroelastic response and so on, which are believed to lay out a rational foundation for manufacturing, characterization and applications of electrospun nanofibers.

Chapter 9 concentrates on melt compounding, mechanical and thermal characterization, and clay dispersability of raw Muscovite and organomodified Muscovite reinforced polypropylene (PP) nanocomposites. Chapter 10, as the last chapter, demonstrates a holistic development of a unique systematic approach to determine the preferred nanocomposite formulations and optimized manufacturing process, experimentally characterize PP/clay nanocomposites, numerically predict their mechanical behavior in comparison to conventional composites theoretical models, and understand the deformation mechanisms due to the clay dispersion. Under this comprehensive investigation, the processing-structure-property relationship of tailored PP/clay nanocomposites has been well established.

I expect that *Nanostructures: properties, production methods and applications* will become a very useful technical reference or literature to inspire the future outstanding research work in nanostructures in order to accomplish more breakthroughs in nanotechnology and ultimately improve the livelihood of mankind. Nanostructures physically have extremely midget sizes which are not visible with our naked eyes, but their impacts can be huge and unprecedented.

I would like to take this opportunity to express my sincere gratitude to all chapter contributors for their great endeavors on completing this masterpiece work on nanostructures after rigorous scholarly peer-review process. I also wish to acknowledge Carra Feagaiga and Bahar Kaya from NOVA Science Publishers who offered invaluable advice at every stage of editing this book. Finally, I am indebted to my wife Marie Ma and lovely daughters Joyce Dong and Emma Dong for their patience and strong support.

I sincerely hope that readers find the contents of this work useful for their scientific research and appreciate any comments and feedback to improve this book quality in the future endeavors.

Yu Dong
Department of Mechanical Engineering
School of Civil & Mechanical Engineering
Curtin University, Perth, Australia

NANOSTRUCTURES

In: Nanostructures
Editor: Yu Dong

ISBN: 978-1-62618-081-9

Chapter 1

NANOSTRUCTURING OF SOLID SURFACES

***V. Švorčík*[1,*]*, P. Slepička*[1]*, J. Siegel*[1]*, A. Řezníčková*[1]*, O. Lyutakov*[1]*, O. Kvítek*[1]*, T. Hubáček*[1]*, N. Slepičková Kasálková*[1] *and Z. Kolská*[2]

[1]Department of Solid State Engineering, Institute of Chemical Technology, Prague, Czech Republic

[2]Department of Chemistry, J.E. Purkyne University, Usti nad Labem, Czech Republic

ABSTRACT

In particular, over the last decade, our group dedicated itself to the study of preparing nanostructures and the nanostructuring of solid surfaces,especially using plasma treatment, laser irradiation and thermal modification. Up to this day,much effort has been put into producing smart materials with extraordinary properties usable in a broad range of technological applications. In the last two decades, it has been demonstrated that the production of materials with completely new properties does not onlydepend on its chemical composition, but also on dimensions associated withbuilding blocks of conventionally known materials. Primarily, our group dealt with structures which could find applications in tissue engineering (cell adhesion and proliferation improvement), electronics (adhesion improvement between polymer or glass substrate and metal coatings) and optoelectronics (optical lattice).

In this chapter we are going to focus on the preparation and characterization of nanostructures prepared on polymer or glass surfaces. These materials were modified by plasma and laser treatments and were subsequently subjected to various analytical methods to reveal: chemical structure (FTIR, UV-Vis, XPS, AAS, RBS and Raman spectroscopy), electrical properties (sheet resistance, Van der Pauw method), optical properties (ellipsometry), ablation of layer (gravimetry), surface properties (AFM, FIB-SEM, goniometry, electrokinetic analysis), and crystallinity (XRD).

In this work,information is summarized from the following areas:

(i) Irradiation of polymers (PET) with a linearly polarized laser light (KrF), resultingin the formation of coherent ripple patterns with a lateral periodicity in the order of laser light wavelength and a structure depth of several 10 nm.

[*]Corresponding author's email address: vaclav.svorcik@vscht.cz.

(ii) Plasma ablation and liquid etching (e.g., in water, methanol) of selected polymer foils, determination of ablated and etched polymer masses, polymer wettability, surface morphology and chemical properties of plasma-modified and etched samples. A synergetic effect of thermal treatment and plasma irradiation was employed for cell adhesion and proliferation improvement in the case of biocompatible polymers (PLLA and PMP).
(iii) Deposition of thin gold layers, resulting in continuous metal coverage, copying the nanostructured polymer surface, versus the formation of gold nanowires on the ridge of the laser-induced ripples.
(iv) New methods for achieving achemical bond between Au nanoparticles and plasma-modified materials (polymer, glass). Au grafting increases both the biocompatibility of the substrate surface and consecutive adhesion of deposited gold.
(v) The possibility of surface patterning of arbitrary polymer films was proposed and experimentally verified. The main approach of our experiments consists of dopingthin films with optical chromophores on polymers,particularly porphyrin molecules, and scanning doped materials with a focused laser beam. Doped polymer films were shown to mechanically respond on laser scanning, the response depending on light wavelength.
(vi) Carbon structures were deposited by flash evaporation, sputtering and laser deposition onto substrates. Consequently, the adhesion and proliferation of cells on the substrate with a deposited carbon layer was studied by the *in vitro* method.
(vii) Physical and chemical properties of the modified polymer surfaces mentioned above, such as polarity, wettability, electrical charge and conductivity, roughness and morphology, rigidity or elasticity, as well as the presence of various chemical functional groups that strongly influence cell adhesion and proliferation, were studied.

1. Introduction

The rapid development of nano-science and nanotechnology in the 20th and 21st centuries creates a lot of questions about the structure and surface properties of nano-materials. Polymers have wide-ranging applications in food packaging, decorative products, manufacturing medical devices, insulatingoptic and electronic devices and tissue engineering. Polymeric materials are chosen for their excellent bulk properties, but the inert nature of their surface may be a disadvantage in several applications(e.g., when a metal coating has to be applied or when an enhanced bioactivity is required). For some applications, the polymer surface should be tailored to match specific requirements. Several techniques are used for the modification of polymer surface properties including mechanical abrasion, treatment with chemicals, irradiation with UV radiation, exposition to plasma discharge or ion implantation.

In this study the effects on the surface properties of plasma, laser and ion beam treatments, and surface grafting of "common synthetic" polymers are reviewed and discussed. Results collected by our group over the last 10 years are presented. Nanotechnology can be defined as the creation of functional materials and substrates at the range of up to 100 nm [1]. The chemical and physical properties of material surfacesin such a scale can be significantly different from those of bulk materials of the same composition [2-4]. Plasma treatment, as well as laser exposure, can also guarantee suitable control of the surface properties at the nanometer scale and the preparation of nanostructured surfaces both on polymer or glass substrates. These modified surfaces can be consequently used for grafting chemicals and/or biocompatible experiments and can be further used in nanotechnology for the construction of new products and devices. Nanomaterials exhibit unique mechanical, chemical, physical, thermal, electrical, optical and biological properties. Nanostructured materials offer usage in a wide variety of electronical applications, such as transparent electrodes, solar energy cells, FET transistors, photonic devices, biosensors, targeted drug delivery or tissue engineering [5, 6].

Amanipulation in the nanometer scale and preparing of defined nanostuctures (e.g., periodic nanostructures with defined chemical composition or roughness) proposes the nanomaterials to a wide range of biomedical and electronic interests and applications. The chemical composition, thickness, crystallographic orientation or electrical properties can be measured more precisely by the atomic force microscopy (AFM), scanning electron microscopy (SEM), transmission electron microscopy (TEM), X-ray diffraction (XRD), or by precise equipment capable of accurately measuring the mass-loss or layer resistance. Other additional information about surface properties can be obtained by many other methods, such as goniometry, optical methods (XPS, IR, Raman spectroscopy, RBS), ellipsometry, electrokinetic analysis, nanoindentation or biocompatibility. Without the never-ending improvement in this analytical sphere and new technological achievements,the preparation and innovation in the field of nanomaterial engineering would be almost impossible. Details of various thin-layer preparation techniques (sputtering, evaporation), surface layer modification (plasma and laser treatments) and surface layer activation (organic groups and Au grafting) are discussed with their classifications. Also, results from research associated with the properties of these materials and theircharacterization and applications, mostly withadhesion in the metal layer preparation process and biocompatible studies, are presented.

The main goal of this article is to summarize information about properties, characterization techniques and results of nanostructured materials and their contribution to optics, electronics and tissue engineering sciences.

2. Production Methods of Nanostructured Solid Surfaces

As mentioned above, there are several methods of material modification. In this chapter we are going to focus on the preparation of nanostructures on polymer or glass surfaces with "our" methods: (i) plasma treatment, (ii) ion implantation, (iii) laser treatment, (iv) metal deposition and (v) grafting of different chemicals on modified surfaces.

2.1. Plasma Treatment

Plasma treatment, which is an electric gas discharge, is usually performed in a vacuum chamber in a pressure ranging from 1-100 Pa. An electric field, applied to the gas, accelerates free electrons to energies, which are sufficient for the ionization, fragmentation and excitation of the gas molecules. In this way, highly reactive gaseous products are produced, which are able to react chemically with exposed surfaces.

The processes of the plasma-polymer interaction and the resulting surface properties of the modified polymer are dependent on several factors such as the processing gas and its properties, the chemical structure of materials and the properties of the plasma discharge (density and energy of electrons and ions). The nature of the interaction of the plasma constituents with polymers is also affected by the plasma reactor configuration. Depending on the adjustment of all these factors, different outcomes of plasma-polymer interactions may result. The most important from the practical-applications point of view are the:

(i) *reactive removal of surface layers,* leading to etching or cleaning of the polymer surface,
(ii) *activation or more general modification* of the polymer surface,
(iii) *deposition or plasma polymerization* (the process when a coating with a thickness in the sub-micrometer range is deposited onto the exposed surface).

In general, however, all processes (desorption, activation and deposition) mentioned above may simultaneously contribute to the alteration of the polymer surface.

The basic instrumentation (see Figure 2.1) comprises a vacuum chamber with a pumping system, gas introduction system and an excitation source (direct current or radio-frequency). The size of the chamber is mostly determined by the size and shape of samples. The chamber construction materials should not interact with the processing gas and its dissociative products. Because of electrical potential formed in the chamber, the positive ions from the plasma are accelerated toward the substrate's surface. The ions attain energyfrom several eV up to a few keV.

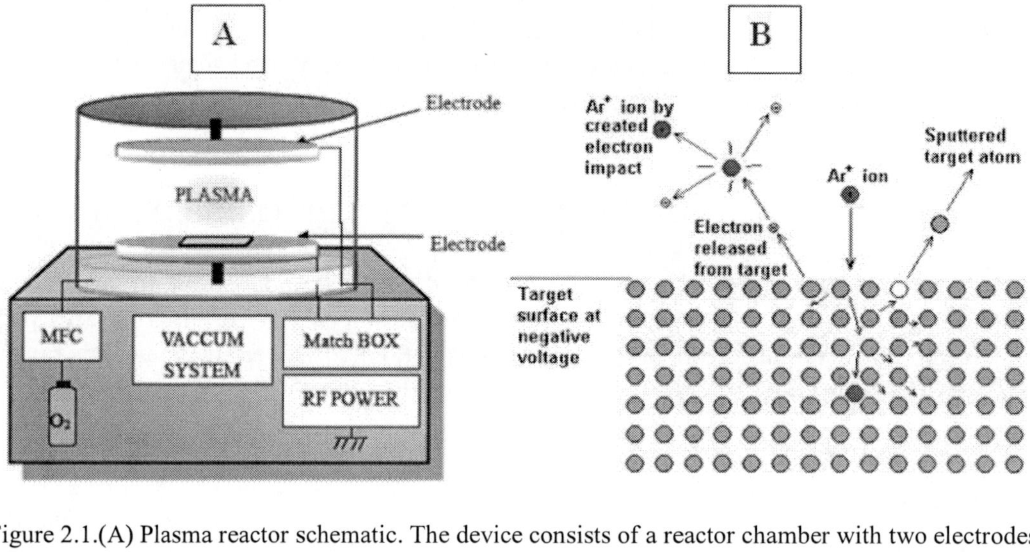

Figure 2.1.(A) Plasma reactor schematic. The device consists of a reactor chamber with two electrodes, an RF power source feeding plasma discharge, vacuum system and system of gas supply (MFC), and the substrate is placed on bottom electrode; (B) the Ar^+ ions hit the polymer surface and create a collision cascade in the surface layer by which a number of atoms are set in movement. As a result,the ionization of atoms and molecular bond cleavage take place and some of the liberated atoms are ejected (sputtering process) [7].

2.2. Ion Implantation

Ion implantation is performed in a high vacuum (see Figure 2.2) and, contrary to plasma treatments, there is no complex interaction of reactive gaseous species from an ambient atmosphere with a polymer surface. Instead, the polymer modification is caused solely by an interaction of energetic ions with the polymer matrix.

The basic instrumentation consists of an ion implanter, commonly a small dedicated accelerator with associated ion optics;a beam scanning system; and a target chamber with a suitable sample holder.

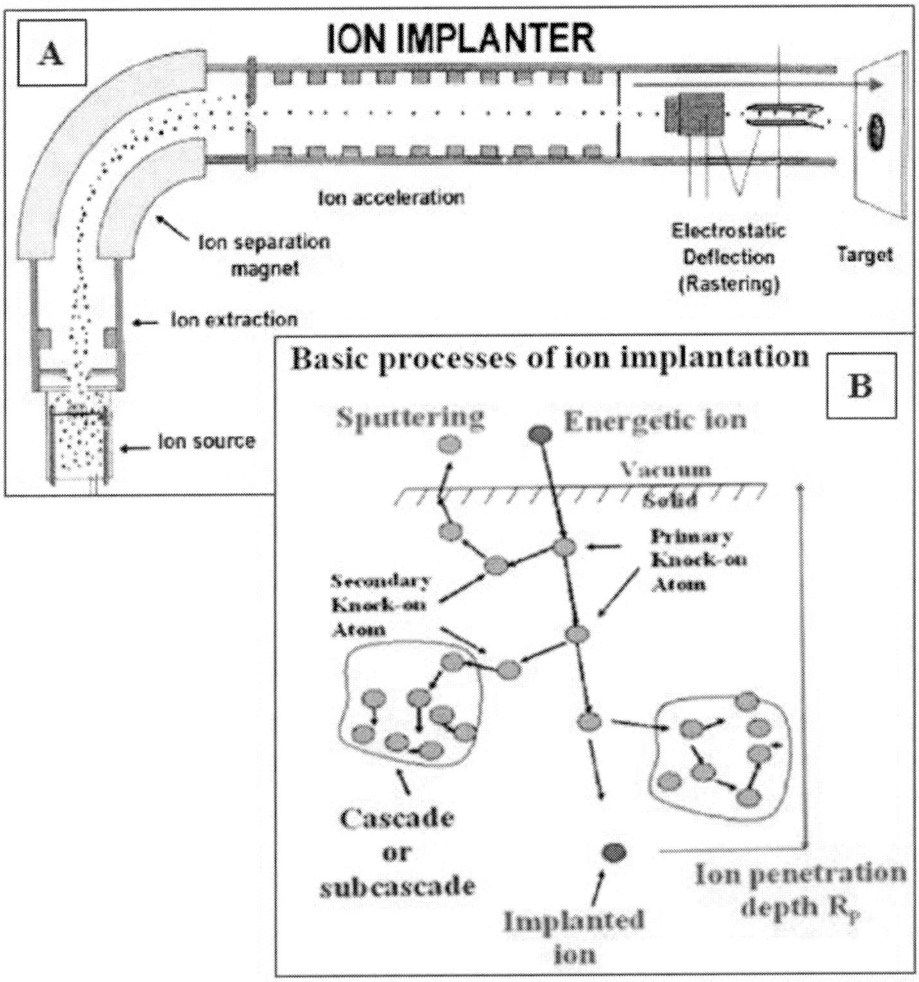

Figure 2.2. Principles of ion implantation: (A) basic components of an implanter are the ion source with the system of ion extraction, ion separation magnet separating the ions according their charge/mass ratio, ion accelerating part, electrostatic system for scanning the ion beam over the sample surface and target chamber with sample holder and beam intensity monitor; (B) basic processes of ion interaction with matter. The thickness of the modified surface layer is determined by the ion projected range R_P, measured in the direction of the sample surface normal [7].

The ions are usually accelerated to energies ranging from a few keV up to several MeV. The ions are mass separated and sent toward the sample. A beam scanning system ensures homogenous irradiation of the sample surface. The indispensable part is a beam-monitoring system measuring the total fluence of the ions. The ions entering the polymer substrate interact with atoms, which are ionized and exited.

As a result, secondary electrons are created, initiating additional ionization along the ion trajectory. The ionization and excitation may lead to chemical bond cleavage and production of highly reactive species, free radicals, ions and molecular fragments, which subsequently interact with each other, and at last, stable degradation products are created. This complex sequence of processes can deliberately be divided into two basic phases: the initial physical phase, in which the ion energy is dissipated to electrons and atoms, and the chemical phase comprised of theinteraction of the reactive species and production of the final stable products.

Two kinds of degradation products are created:low-mass species, created by macromolecular chain scission, and two- to three-dimensional structures produced by macromolecule cross-linking (mediated by free radical reactions).

Relative portions of these two basic products depend on the polymer's initial composition and on the density of dissipated energy. A specific phenomenon is associated with thedesorption of light weight, volatile products in the course of the ion implantation. Since the desorbed species are hydrogen enriched, the polymer surface is progressively carbonized. The thickness of the polymer surface layer modified by implanted ions is determined by the ion's projected range, which is an increasing function of the ion energy and varies, depending on the ion mass and energy and the substrate composition, from several nanometers to tens of micrometers. Typical ion fluences vary from 10^{9}-10^{18} cm^{-2}. The properties of the modified polymer could also be influenced by implanted atoms, which remain embedded in the polymer matrix. In polymers, this effect can usually be neglected because of a rather low concentration of implanted atoms.

2.3. Laser Treatment

Lasers can be broadly classified into four categories: gas discharge lasers; semiconductor diode lasers; optically pumped lasers; and "other," a category which includes chemical lasers, gas-dynamics lasers, x-ray lasers and others. Advantages associated with a laser beam include it being a monochromatic, coherent, and highly collimated light source with very low divergence.

The term excimer or "excited dimer" refers to a molecular complex of two atoms which is stable (bound) only in an electronically excited state. These lasers produce intense output in the ultraviolet and deep ultraviolet. Excimer lasers generate laser light in ultraviolet to near-ultraviolet spectra. The lasers in this family are: XeF (351 nm), XeCl (308 nm), KrF (248 nm), KrCl (222 nm), ArF (193 nm), and F_2 (157 nm). Since these lasers have very short wavelengths, the photons have high energy, resulting in a reduced interaction time between laser radiation and the material that is being processed; therefore, the heat-affected zone is minimized. In laser material processing two regimes of laser-material interactions can be found, includinga heat treatment of the surface with a laser light intensity below the ablation threshold, which causes a temperature rise with or without surface melting (Figure 2.3A, [8]).

The second regime is characterized by a liquid-phase explosion, vaporization and plasma formation (Figure 2.3B).

The illumination of polymers by a polarized UV laser beam induces self-organized, ripple-structure formations ("ripples") within a narrow fluence range well below the ablation threshold. Ripples originate from interference of the incident (k_i) and reflected (θ)/refracted (φ) laser light with scattered (diffracted) light near the interface, together with positive feedback [8]. The period of the ripples depends on the laser wavelength (λ), the angle of incidence associated with the radiation (θ) andthe effective refractive index of the material, and their direction is related to the direction of the laser beam polarization.

These structures develop on the original material surface having little roughness, a result of treatment with one laser beam with a uniform intensity distribution. The interference between the incoming and the surface scattered waves plays an important role in the structure formation.

The interference causes an inhomogeneous intensity distribution, which together with a feedback mechanism results in the enhancement of the modulation depth. Also, additive mechanisms can influence the surface modification(e.g., thermal and non-thermal scissoring of polymer chains, amorphization of crystalline domains, local surface melting, photo-oxidation and material transport).

2.4. Deposition of Structures

Sputtering

Sputtering of a target material by energetic ions is assumed to result from cascades of atomic collisions [9]. There are three main types of sputtering that are dominant in thin film technology: DC sputtering, RF sputtering and magnetron sputtering (Figure 2.4A). DC sputtering can be achieved by applying large (~2,000) DC voltages to the target (cathode).

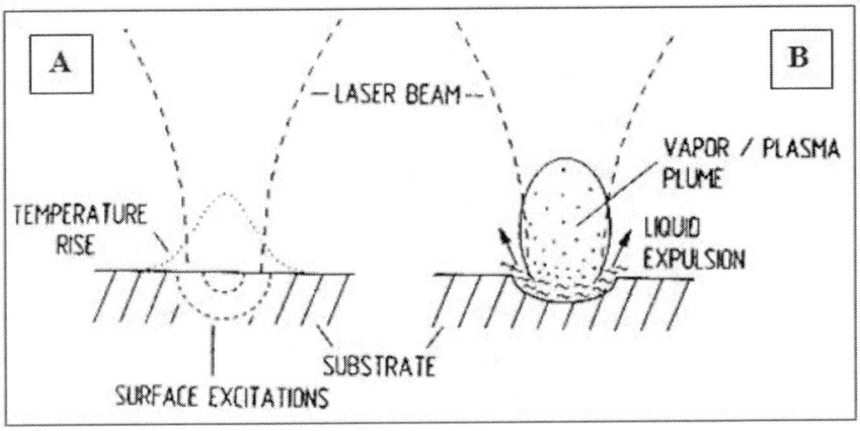

Figure 2.3.Two regimes of laser-material interactions employed in conventional laser processing. (A) Heat treatment of the surface with laser light intensity below the vaporization threshold when the absorbed laser light causes a temperature rise with or without surface melting. The second regime is characterized by liquid-phase expulsion, vaporization, and plasma formation (B). The ablation process takes place in this regime [8].

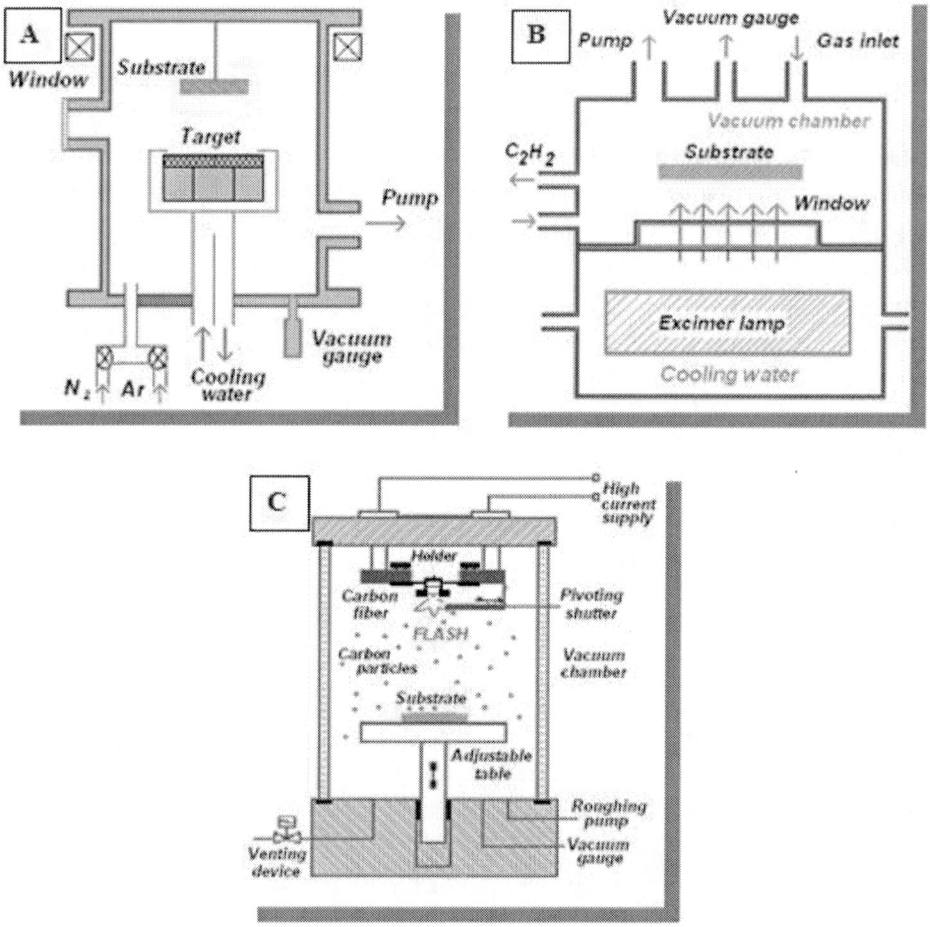

Figure 2.4. Deposition of structures on the substrate schematic by: (A) magnetron sputtering, (B) UV induced chemical vapor deposition (CVD) and (C) evaporation (e.g.,from carbon fibers) [9].

A plasma discharge is established and the inert gas ions (e.g., Ar^+) are attracted and impact the target sputtering of target atoms.

In DC sputtering, the target must be electrically conductive, otherwise the target surface will charge up with the collection of Ar^+ ions and repel other argon ions, stopping the process. Radio frequency (RF) sputtering allows for the sputtering of targets that are electrical insulators (SiO_2, etc). The target attracts argon ions during one half of the cycle and electrons during the other half cycle. The electrons are more mobile and build up a negative charge called self bias that aids in attracting the Ar ions, which makes the sputtering possible. Magnetron sputtering (Figure 2.4A) is based on the magnetic field's influence on electron mobility, so the electron path is increasing and thus the probability of gas ionization is as well. The magnitude of the magnetic field can be about several hundreds of Tesla, and it is chosen so the path of electrons (low mass) is influenced and the path of ions (high mass) remains practically unchanged.

Chemical Vapor Deposition (CVD)

Chemical vapor deposition (schematically Figure 2.4B) relies on the chemical reactivity between ionized gas within the deposition chamber. Ionized gas is injected through control valves into the deposition chamber where a chemical reaction occurs between the precursor and substrate. Ionized reactive gas is forced to interact with the substrate and deposit films through chemical reactions.

The type of reaction is dependent on chosen materials. The precursor comes into contact with the substrate, and absorption of materials on the surface occurs. The gas simultaneously diffuses through the boundary layer, and the decomposition of precursor molecules occurs and converts into a solid film. Plasma Enhanced Chemical Vapor Deposition (PECVD) uses electrical energy to generate a glow discharge (plasma), in which the energy is transferred into a gas mixture.

This transforms the gas mixture into reactive radicals, ions, neutral atoms and molecules, and other highly excited species. These atomic and molecular fragments interact with a substrate and, depending on the nature of these interactions, either etching or deposition processes can occur at the substrate. Since the formation of the reactive and energetic species in the gas phase occurs by collision in the gas phase, the substrate can be maintained at a low temperature.

Evaporation

The vacuum thermal evaporation technique consists of heating until evaporation of the material to be deposited. The material vapor finally condenses in the form of a thin film on the cold substrate surface and on the vacuum chamber walls. Usually, low pressures are used, about 10^{-4} – 10^{-5} Pa, to avoid reactions between the vapor and atmosphere. At these low pressures, the mean free path of vapor atoms is the same order as the vacuum chamber dimensions, so these particles travel in almost straight lines from the evaporation source towards the substrate.

A special evaporation technique is called “flash evaporation”, and there is also a derived method,the“wire flash explosion” (Figure 2.4C). It formerbased on the principle of almost immediately vaporizingthe target material. In the case of the standard flash evaporation, the material is in contact with the heated area of the evaporator, the material being in the form of powder, where the prepared layer has almost the same composition as the target material.

The “wire explosion“ technique is based on the quick evaporation of the wire being exposed to strong electric impulses (up to 10^6 A·cm^{-2}, 10 kV), placed between two electrodes. If the energy is quite enough and impulse time sufficiently short (of about 100μs), ahigh evaporation velocity can be achieved (up to 10^5 nm·s^{-1}).

The mechanism is based on the principle of high temperature plasma creation in the place of a wired area, caused by electric power. From the plasma plume, the particles (mostly ionized) are targeted to the substrate material.

2.5. Grafting of Modified Surfaces

This is the process in which a polymer surface is chemically modified by grafting (see Figure 2.5) with organic or inorganic species (particles or groups), or by the generation of active sites that can lead to the initiation of creatinga chemical bond. Modification of polymer surfaces can be rapidly achieved by plasma treatment due to the possibility of the formation of various active species on the surface of modified polymers as well as other solid substrates (e.g., glass). Plasma treatment also results in the deposition of a macromolecular structure, graft polymerization, functionalization, etching, roughening, and cross-linking. Consequently, the realized grafting process is most often performed from the solution. Surface grafting can been carried out on fiber, film, and plastic to improve the surface properties, such as wettability, antistatic properties, adhesion and biocompatibility,without affecting their bulk properties.

A wide range of polymers are usedas substrates: (polyethylene (PE), polyethyleneterephthalate (PET), polystyrene (PS andpolytetrafluoroethylene (PTFE) with application in tissue engineering. By grafting from the solution, the added surface reactants can be divided into amino-acids (glycin, alanin, RGD), organic polymers (polyethyleneglycol (PEG)) or other chemical groups or metal nanoparticles (-SH, -NH_2, Au, Ag). Above mentioned surface reactants are predominantly used for increasing surface biocompatibility or adhesion to consecutively prepared continuous metal layers.

The main goal of these modifications is to prepare nanostructured materials and their contribution to electronic and tissue engineering sciences (Figure 2.5).

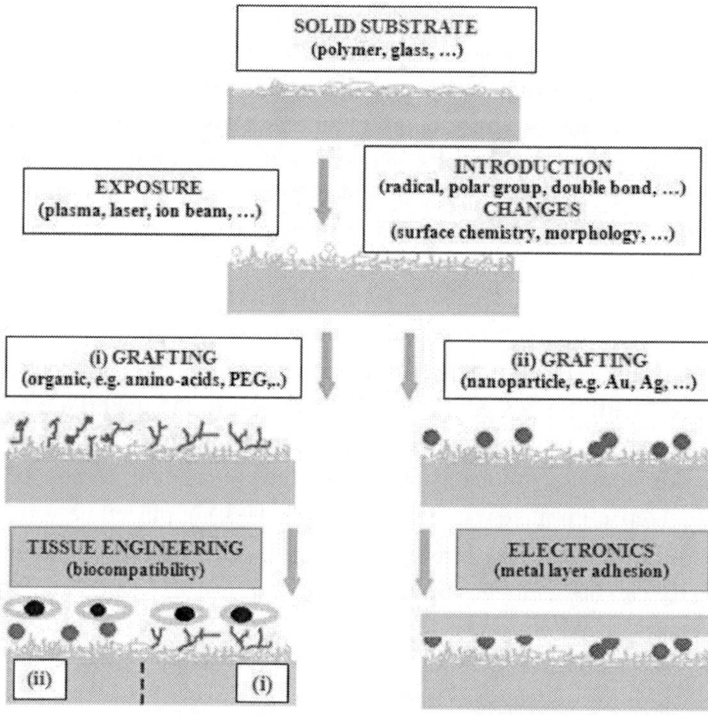

Figure 2.5. Treatment and grafting of schematically modified surfaces. A pristine polymer is modified using "differencial" methods. Thus the reactive species (e.g., free radicals, oxygen containing groups and double bonds) are introduced on the pristine surface. The activated surface is further grafted with (ii) organics (e.g., amino-acids, PEG directly) or (ii) with inorganics (e.g., gold nanoparticles) using a two-step process when the modified surface is initially grafted (e.g., with the thiol group) and then with gold nanoparticles or nanotubes. Those grafted "surfaces" are attractive for tissue engineering and electronics.

3. METHOD OF NANOSTRUCTURED SOLID SURFACE CHARACTERIZATION

Several techniques can be used for "our" characterization of the physical and chemical properties of pristine substrates and substrates modified by various chemical and physical treatments. The techniques can deliberately be divided in two groups:those determining the material properties of application importance (mechanical strength and hardness, surface morphology, wettability, electrical conductivity, dielectric properties and bioactivity) and those suitable for the characterization of the material structure and composition on the microscopic scale, which in turn affect the macroscopic properties of the polymers mentioned above. The latter techniques include, but are not limited to, different sorts of optical and electron microscopies, X-ray photoelectron spectroscopy (XPS), X-ray and neutron diffraction (XRD), infrared and Raman spectroscopy (IR or FTIR), atomic force microscopy (AFM) and techniques based on the elastic scattering of charged particles (RBS). In what follows, selected techniques are briefly described with respect to their usage for material characterization. Most of the diagnostic methods listed below are notoriously known and their principles are described in standard, easily available surveys [10].

Spectroscopy

In *X-ray Phootelectron Specroscopy (XPS)*the sample is irradiated by monoenergetic soft X-rays, which induce a photoionization of the atoms, and photoeletrons emitted from the sample surface layers, several nanometers thick, are analyzed by an electrostatic spectrometer. The photoelectron energy spectrum consists of peaks corresponding to electron energy states of the sample. By the XPS method it is possible to determine the elemental composition of a near-surface layer of the sample and identify the chemical state of the elements. Chemical state identification is based on measured energy shifts related to electron density in the atoms of interest. The XPS method is indispensable for the determination of chemical and structural changes in polymers exposed to plasma discharge or ion irradiation. Apparent disadvantages theXPS method in some instances are associated witha sampling depth that is too shallow and its rather high cost of instrumentation.

*InfraRed spectroscopy (IR).*Infrared spectroscopy is an effective method for the characterization of solids(e.g., of polymers)regarding their chemical structure. IR spectra of the sample examined are obtained by two basic types of IR spectrometers: dispersive or Fourier transformed (FTIR). Infrared spectra are usually presented as a dependence of absorption (in percent transmission) on wave length or wave number (in cm^{-1}). Any molecule or its part may be viewed as a mechanical system exhibiting certain natural frequencies at which it can be excited. The absorption bands in IR spectra correspond to these frequencies, which are characteristic for different chemical structures.

*Raman spectroscopy.*Raman spectroscopy is an analytical technique that yields information about the molecular structure of the sample materials based on an observation of scattered light spectra. Small fractions of light scattered by molecules exhibit frequency shifts from the incident light, which is related to the frequency of molecular vibrational states. It is similar to FTIR spectroscopy, but it has a few advantages.For example, Raman can be used to study solids, liquids, powders, gels, slurries and aqueous solutions. Raman does not require any special sample preparation, meaning that many studies may be performed *in situ*. Raman spectroscopy is based on the detection of scattered light(i.e., the Raman effect). The radiation may be scattered elastically, that is without any change in its wavelength, and this is known as Rayleigh scattering. Conversely, the radiation may be scattered inelastically, resulting in the Raman effect.

Both Raman and Infrared spectroscopy provide qualitative and quantitative information about chemical species through the interaction of radiation with molecular vibrations. Raman spectroscopy complements infrared spectroscopy, particularly for the study of non-polar bonds and certain functional groups. It is often used as an additional technique for the elucidation of the molecular structure and symmetry of a compound. Raman spectroscopy also provides facile access to the low frequency region (less than 400 cm^{-1}Raman shift), meaning that the area more difficult for infrared spectroscopy.

*UV-Vis-NIR spectroscopy.*UV-Vis spectroscopy probes the electronic transitions of molecules as they absorb light in the UV and visible regions. Any species with an extended system of alternating double and single bonds will absorb UV light, and anything with color absorbs visible light, making UV-Vis spectroscopy applicable to a wide range of polymeric materials. While UV-Vis and UV-Vis-NIR spectroscopy are well established analytical techniques being effectively applied across a broad spectrum of well-known industries, high-performance UV-Vis spectroscopy has expanded into new applications, especially in the life sciences. Even in the traditional application areas, such as materials characterization, optics, coatings, glass, pharmaceutical and color control, UV-Vis spectroscopy has become more sophisticated, providing higher quality results faster and easier than ever.

Ion beam analytical techniques. Analytical techniques based on the use of energetically charged particles (protons, alpha particles, heavy ions), such as Rutherford Back Scattering (RBS), Elastic Recoil Detection Analysis (ERDA) and Nuclear Reaction Analysis (NRA), are very powerful due to the broad spectrum of elements, which can be determined simultaneously with high sensitivity and excellent depth resolution down to few nanometers. The energy spectra of scattered or recoiled particles bring information on the sample composition and the depth structure of the sample as well. For all these methods the analyzed depths are limited to several micrometers by the range of primary particles and typical detection limits vary from10^{11}-10^{15}at·cm^{-2}. Thus, the time during which the samples are exposed to particle irradiation should be as short as possible, and the radiation damages caused by the analyzing beam should be taken into account when evaluating the experimental data. These methods are insensitive to the chemical state of elements in the sample. General disadvantages of these methods arethe high costs associated withthe basic equipment, accelerator and data acquisition technique.

Gravimetry

The *thickness*of the surface layer removed by the treatment ablation or etching in solutions is possible to study using a weighing machine(e.g., microbalance). The thickness of the removed layers is calculated from the weight of samples before and after treatment.

Goniometry

This measurement gives us information about surface polarity and wettability (contact angle goniometry), and it is based onthe measurement of the contact angle between a drop of a liquid and substrate surface (glass or polymers). The contact angle is a direct measurement of the substrate surface energy and, consequently, of its wettability. Higher surface energy results in a lower contact angle.

Microscopy

*Atomic force microscopy (AFM).*The atomic force microscopy (AFM) provides quantitative topographic images of solid surfaces at a nanometer scale. It uses a sharp tip scanned over the sample surface to sense attractive or repulsive forces between the tip and the surface. The microscopic image of the surface is obtained as a surface, representing the locus of points of constant force between the tip and the sample. It is well suited to examining surfaces containing features and defects less than 1 μm in size.

*The Scanning Electron Microscope (SEM)*is a standard imaging technique based on electron back-scattering from the sample surface. It analyzes the surfaces of solid objects, producing images with resolutionsthatare about an order of magnitude better than those of optical microscopy (typically 10 nm). The SEM avoids the problem of thin samples (TEM), but the SEM observation requires the deposition of a thin conductive metal film on the sample surface to prevent sample charging.

An*FIB (focused ion beam)* cut is prepared with an adapted scanning electron microscope *(FIB-SEM)*. The FIB cut is made by a Ga ion beam, for example, after the cutsurface is polished.

The polishing procedure is performed to clean and flatten the investigated surfaces. The direct measurement of the layer thickness is done with another scanning electron microscope. The SEM images were taken under constant angle.

*Transmission Electron Microscopy (TEM)*is a standard laboratory technique. TEM is an indispensable tool for the high-resolution observation of very fine structures on the polymer surface created by a modification by plasma discharge or ion implantation. The resolution of TEM is about one order of magnitude better than that of SEM. It corresponds to 1 nm. High resolution transmission microscopes (HRTEM) also exist, with a resolution down to 0.1 nm, capable to resolve individual atomic lattice planes. The TEM observation requires very thin specimens (100 nm - 1 μm), the preparation of which may be difficult in some specific cases.

Electrokinetic Potential (Zeta Potential)

Electrokinetic potential (zeta potential) is a characteristic parameter for the description of solid surface chemistry. It is an important physico-chemical property of polymers, glass and other planar solid samples, giving information about chemistry, polarity, swelling and other surface characteristics. Zeta potential can be determined using electrokinetic measurements and for the planar samples, as glass or polymers, the streaming potential or streaming current approaches are recommended. Streaming current or streaming potential is generated by pressure-driven flow through a measuring cell with sample and zeta potential is calculated by two equations, the Helmholtz-Smoluchowski and Fairbrother-Mastins [11-12]. It is a very sensitive method indicating even slight changes of surface properties before and after sample modification. As an electrolyte we most frequently used the 0.001 mol·dm^{-3} water solution of KCl.

Nanoindentation

*Surface mechanical properties (Nanoindentation).*The nanoindentation technique has an important role in the characterization of mechanical properties associated with materials on submicron and nanometric scales.

The ultra-low load indentation systems (nanoindentors) carry out continuous records of load and displacement as an indentor is pushed into a surface. The indentation curve can be used to determine mechanical properties even when the indents are too small to be imaged conveniently. An analysis of this data gives a number of useful parameters and information forthe complete assessment of elastic, plastic and fracture properties of surface sat the scale of real asperity contacts. The hardness, elastic modulus and other properties are evaluated from the indentation curve.

The indentation methods can be used for the measurement of adhesion between the coating and the substrate when the indents create cracks and delamination on the coating or on the interface onthe polished section. The static indentation methods for the measurement of adhesion are based on measuring stresses necessary for breaking the bond between the coating and the substrate during static pressing of the indentor. The scratch test characterizes the surface mechanical properties of thin films and coatings(e.g., adhesion, fracture and deformation).

XRD Diffraction

*X-Ray Diffraction (XRD)*is a well known technique based on the scattering of X-rays. It can be applied to semi-crystalline or crystalline polymers and structural changes induced by plasma treatment or ion implantation. Occasionally observed shifts in the X-ray peak positions might indicate a distortion of the crystal structure due to increasing strain. The width of the X-ray reflection peaks yields information on the size of the crystals under investigation.

Magnetic Resonance

*Electron paramagnetic resonance (EPR)*uses exhibiting spin magnetic moments for the detection and identification of unpaired free electrons (radicals). This method applies an electron paramagnetic resonance corresponding tothe transition of electrons between states with different spin orientations, initiated by the absorption ofthe*rf* electromagnetic signal in the sample. It is indispensable for the detection of free radicals produced by a modification of polymers.

Electrical Measurement

Measurement of*electrical sheet resistance*can be arranged as atwo- or four-points method. The simplest technique for the measurement of sheet resistance of polymer samples is the two-point method, a prerequisite of which isthe creation of two ohnmic contacts on the sample surface.

The metallic contacts are commonly produced bythe sputtering or vacuum evaporation of suitable metal through a mask onto a polymer surface. The easiest way is to use a special conductive adhesive. The conductivity measurement proceeds in the standard manner.*Volume electrical resistivity* of the surface layers can possibly be measured by the Van der Pauw method as well. *Concentration and mobility of free charge* ***carriers*** are determined from the measured resistance and the Hall constant.

Ellipsometry

Ellipsomery is an optical technique for the investigation of the *dielectric properties* (complex refractive index or dielectric constant) of thin films. It has applications in many different fields, from *semiconductor physics to electronics and biology*. As an optical technique, spectroscopic ellipsomery is a non-destructive and contactless approach. Upon the analysis of the change of polarization of light, which is reflected off a sample, ellipsomery can yield information about layers that are thinner than the wavelength of the probing light itself, even down to a single atomic layer. Ellipsomery can probe the complex refractive index or dielectric function tensor, which gives access to fundamental physical parameters and is related to a variety of sample properties, including morphology, crystal quality, chemical composition, or electrical conductivity.

It is commonly used to characterize film thickness for single layers or complex multilayer stacks ranging from a few angstroms or tenths of a nanometer to several micrometers with excellent accuracy. The name "ellipsomery" stems from the fact that the most general state of polarization is elliptic. The technique has been known for almost a century, and has many standard applications today.

Biocompatibility

The word "*biocompatibility*" has been mentioned for the first time in peer-review journals and meetings in 1970 [13].Biocompatibility is related to the behavior of biomaterials in various contexts.

The term may refer to specific properties of a material without specifying where or how the material is used (for example, that it elicits little or no immune response in a given organism, or is able to integrate with a particular cell type or tissue). The ambiguity of the term reflects the ongoing development of insights into how biomaterials interact with the human body and eventually how those interactions determine the clinical success of a medical device (such as pacemaker, hip replacement or stent). Modern medical devices and prostheses are often made of more than one material, so it might not always be sufficient to talk about the *biocompatibility* of a specific material.

Five definitions of biocompatibility [14]:

(i) an ability of a material to perform with an appropriate host response in a specific application, Williams' definition,
(ii) a necessity to have no toxic or injurious effects on biological systems,
(iii) a comparison of the tissue response produced through the close association of the implanted candidate material to its implant site within the host animal so that tissue response recognized and established as suitable with control materials,
(iv) an ability of a biomaterial to perform its desired function with respect to a medical therapy, without eliciting any undesirable local or systemic effects in the recipient or beneficiary of that therapy, but generating the most appropriate beneficial cellular or tissue response in that specific situation, and optimizing the clinically relevant performance of that therapy,
(v) a capability of a prosthesis implanted in the body to exist in harmony with tissue without causing deleterious changes.

4. Plasma and Laser Modification of Polymers (Ablation and Etching)

Common surface modification techniques include treatments by flames, coronas, plasmas, photons, electron beams, ion beams, X-rays, and γ-rays. These treatments have been applied to achieve the following purposes: (i) to produce special functional groups at the surface for specific interactions with other functional groups, (ii) to increase surface energy, (ii) to increase hydrophobicity or hydrophilicity, (iv) to improve chemical inertness, (v) to introduce surface cross-linking, (vi) to remove weak boundary layers or contaminants, (vii) to modify surface morphology - increase or decrease surface crystallinity and roughness, (viii) to increase surface electrical conductivity, (ix) to increase surface lubricity and (x) to improve dyeability [15].

The interest in plasma modification of polymer surfaces has gradually increased in the plastics industry since it can improve wettability, adhesion, biocompatibility and printability, among other things. Moreover, a plasma treatment can be carried out within only a few nanometers of the surface without altering the bulk properties of the polymer [15].

Surface etching consists of material removal by liquid from the treated sample surface [16]. Water and methanol are mainly applied due to the following reasons.

The water is chosen since in potential biological applications the polymer samples are exposed to a physiological solution, and the methanol is chosen for its potential applications in electronics as a cleaning agent.In this chapter, a pulsed F_2 laser and argon plasma discharge were used for the modification of the polymer surfaces of polyethylene, polypropylene, polytetrafluoroethylene, polystyrene and polyethylenterephthalate. The amount of ablated and etched layers was measured by gravimetry. The wettability of polymers before and after the treatment was determined from the contact angle measurement. Chemical structures of the modified polymers were studied by *X*-ray photoelectron spectroscopy (XPS) and by Electron Paramagnetic Resonance (EPR). Surface morphology and roughness of the pristine and laser treated polymers were determined by atomic force microscopy (AFM).

4.1. Surface Properties of Polymers Treated with an F_2 Laser

One of the basic parameters affecting the interaction of the laser radiation with the polymer surface above an ablation threshold is the number of applied laser pulses. The contact angles and the surface ablation were examined as a function of thenumber of F_2 laser pulses.

Dependence of the contact angle on the pulse number is shown in Figure 4.1. The laser treatment results in a decline of the contact angle and corresponding increase of polymer wettability in all cases exceptthe PP sample. It is known [17-18] that the changes in the contact angle (wettability) are caused by thecreation of polar groups, exhibiting enhanced hydrophilicity, at the polymer surface. Changes of the contact angles (wettability) depend not only on the morphology and the roughness of the polymer surface, but also on its chemistry. After an initial rapid decrease of the contact angle, the saturation is achieved and the contact angle remains unchanged for more than 2,000 laser pulses. A rapid initial decrease of the contact angle is observed on PE and PET, and a much slower decrease is observed on PTFE and PS. The increase of the contact angle in the case of PP (see Figure 4.1) is probably due to the fact that PP is ablated easier than other polymers. In comparison with plasma treatment [19], the laser treatment does not affect the wettability of polymers so dramatically (see section 4.2).

The thickness of the ablated layer as a function of the number of the laser pulses for all polymers is shown in Figure 4.2.

The thickness was determined from the weight-loss using known polymer bulk density. As could be expected, the thickness of the ablated layer is an increasing function of the number of laser pulses. The thickness of the ablated polymer layer after the same laser treatment increases in order: PE = PET < PS < PTFE << PP. The higher ablation rate observed on PTFE could be explained by its higher density and the presence of highly reactive fluorine radicals, which may more easilyfacilitate thelow-molecular degradation of products (LMWOS-low molecular weight oxidized structures) [17,20]. The highest ablation rate of PP is surprising since its behavior is expected to be similar to that of PE. PP and PE are expected to have similar behaviors because of similar structures. Both polymers are built up of linear polymer chains and are non-polar. They have a different crystallinity and their laser treatment results in a dramatic decrease of R_a.

AFM imagesof pristine and laser-treated polymers (250 pulses) are shown in Figure 4.3. It is seen that the pristine polymers exhibit quite a different surface morphology and roughness (the reason why different z-scales are chosen).

The differences are connected with different techniques of polymer foil preparation and different fractions of thecrystalline phase [21]. Laser treatment significantly changes thepolymer surface morphology,and the degrees of the changes dependon the polymer type. For all polymers the laser ablation leads to a “polishing” of the surface structure and the polymer surface becomes “smoother”. Surprisingly enough, the most pronounced changes in the surface morphology are observed on PP. Dramatic changes in the surface morphology and the roughness of PTFE are due to its high ablation rate. It can be supposed that thePTFE amorphous phase is ablated faster than crystalline one [21]. A faster ablation of theamorphous phase was also reported for high density PE. [20] Thepreferential ablation of PE results in thecreation of lamellar structures reflecting thearrangement of molecular chains on the PE surface [20]. In other polymers with lower crystalline fractions, the tiny, sharp formations created by the plasma treatment at the polymer surface may represent low-molecular-weight oxidized structures (LMWOS) [20, 22].

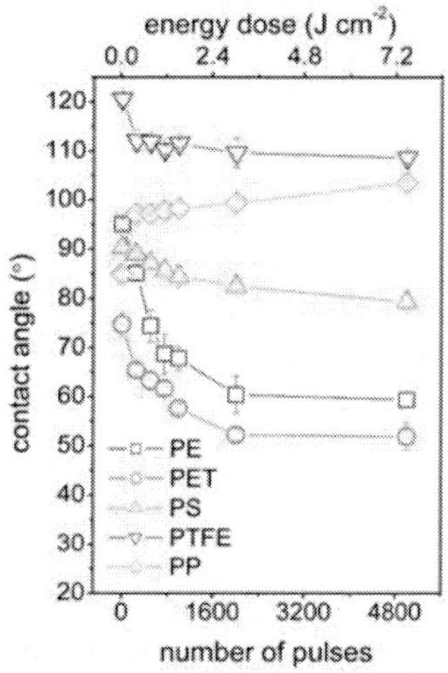

Figure 4.1. Dependence of the contact angle on the number of the laser pulses for PE, PET, PS, PTFE and PP. The measurement was accomplished 10 minutes after the laser exposure [18].

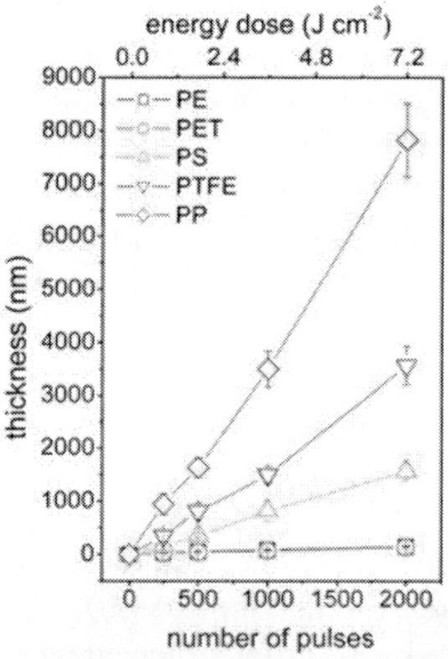

Figure 4.2. The dependence of the thickness of the ablated layer at PE, PET, PS, PTFE, PP on the number of the laser pulses [18].

For all polymers with only exception of PS and PET, the surface roughness decreases immediately after the beginning of the laser treatment. Significant changes in the roughness are due to polymer ablation. After 250 laser pulses the surface becomes much smoother comparedto that of pristine polymers and then, for higher pulse numbers, the morphology remains unchanged. In the case of PS and PET, the situation is different, the laser treatment results in an increase of roughness.

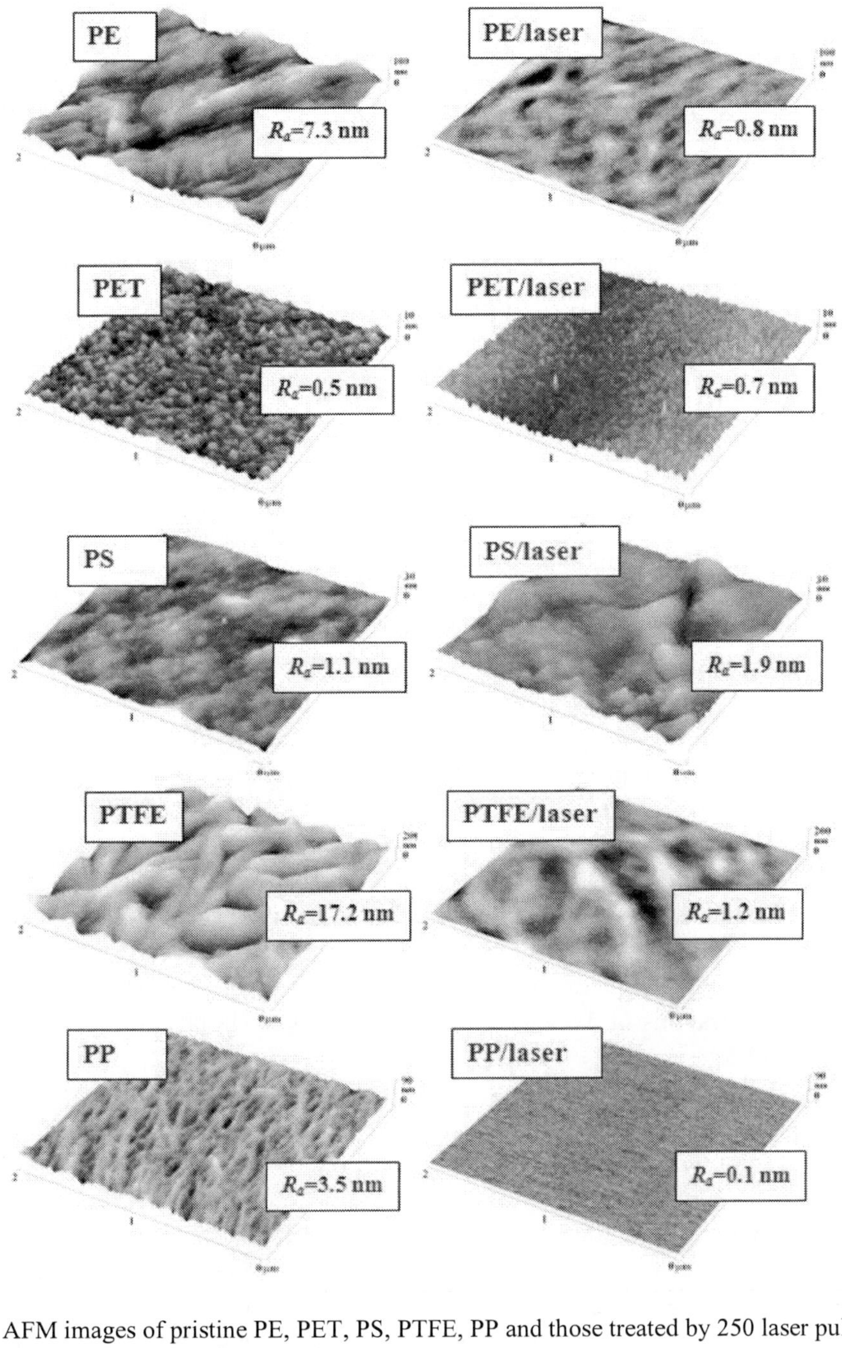

Figure 4.3. AFM images of pristine PE, PET, PS, PTFE, PP and those treated by 250 laser pulses. R_a is the measured surface roughness in nm [18].

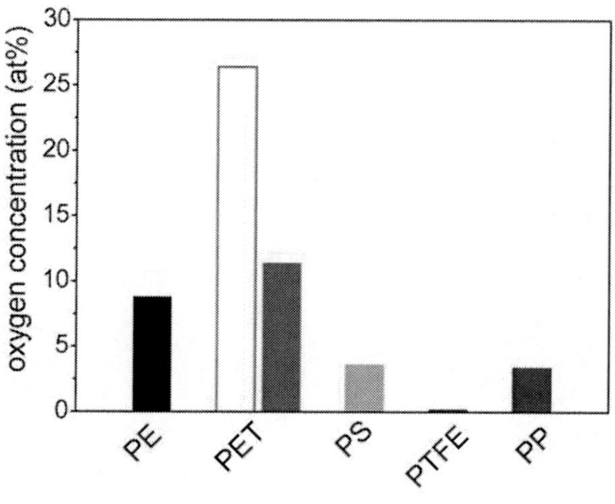

Figure 4.4. Oxygen concentrations measured by XPS on pristine PET (empty column) and on the laser treated PE, PET, PS, PTFE and PP (250 pulses, full columns). Laser-modified samples were stored for 5 weeks under the standard laboratory conditions (RT, air) [26].

The ablation thresholds of PTFE, PE, and PET for 157 nm F_2 laser irradiation are reported to be 54, 67, and 29 mJ·cm^{-2}, respectively [23,24]. The laser fluences employed in the current work are below these values. However, we have also shown that F_2 irradiation in this fluence range can result in changes of the surface topography or in surface polishing, at least for PTFE. [25] In this case, the polishing process is probably due to a transient temperature rise of several hundred degrees at the irradiated area, accompanied by a surface melting and decomposition of PTFE. Due to lower scattering, the laser-polished PTFE foils show enhanced reflectivity and transparency [25]. This initial finding became one of the main motivations for more a detailed investigation of the effects of low-fluence irradiation with an F_2 laser light on the surface topography and chemical composition of other polymers.

Irradiation with the laser light changes the chemical composition of the polymer surface which may lead to theincorporation or depletion of oxygen. Presence and concentration of oxygen in 8-10 surface atomic layers on the polymer surface was determined by the XPS method. The oxygen comes either from remnants in the irradiation chamber (for instance humidity or CO_2 coverage of the pristine polymer surface) or it is due to theformation of radicals in the irradiated polymer, which can react with oxygen after the treatment during storage of the samples in air.

The results of XPS measurements are summarized in Figure 4.4. For the sake of clarity, only atomic concentrations of oxygen determined from the O1s peak of the XPS spectra are quoted.

The laser treatment results in an increase of theoxygen concentration within the surface layer of all polymers with only exception of PET, where a decrease in the oxygen concentration is observed. The observed increase in the oxygen concentration is much lower compared to that reported for the same polymers treated in plasma discharge [26].A smaller increase in the oxygen concentration is probably due to dramatic ablation, especially on PS, PTFE and PP (see Figure 4.2) since the oxygen-containing degradation products created by the laser treatment are ablated too. The observed decrease in the oxygen concentration on PET after the laser treatment may be caused by several factors. In addition tothe ablation degradation reaction, which can be described as photochemical scission of the chains,the preferential release of CO and CO_2 groups (Norrish type II reaction) [27] may also play a role. Oxygen depletion on the PET modified withthe F_2 laser has been recently reported [28] as well. It was further found that the oxygen concentration measured by XPS is lower than thatdeclared for pristine PET. [28, 29] Lower oxygen concentration on the surface may be due to a reorientation of the polar groups toward the polymer bulk [30].

Surface properties of PP, PE, PS, PTFE and PET polymers irradiated with the F_2 laser light were studied by different techniques and compared with those of pristine polymers. As expected, the ablation of the polymer surfaces is an increasing function of the number of laser pulses applied. The ablated polymer thickness after laser treatment increases in order PE = PET < PS < PTFE << PP. With the only exception of PP, the contact angle of all polymers is a decreasing function of the number of laser pulses. The saturation is achieved above 2,000 pulses,and for the still-increasing pulse number the contact angle remains practically unchanged.

The surface wettability of the laser-treated polymers increases. Laser irradiation of all polymers, with theexception of PS, causes an eradication of initial surface structures and a decrease of the surface roughness. XPS measurements show that the concentration of oxygen (polar groups) increases on PE, PP and PS surfacesafter the laser treatment. Laser treatment of PET, however, leads to a decrease in the oxygen content owing to polymer degradation and rapid surface ablation. A “minimum” oxygen concentration after the laser treatment is observed on PTFE.

4.2. Surface Properties of Polymers Treated by an Argon Plasma Discharge

The thicknesses of the ablated layers on the plasma-modified PE, PET, PS, PP and PTFE, measured by gravimetry, are shown in Figure 4.5. The thickness was determined from the weight-loss using thepolymer bulk density. As could be expected, the thickness of the ablated layer is an increasing function of the plasma exposure time. A minimum ablation rate is observed on PS and PET. A higher ablation rate exhibits on PP. The much higher ablation rate observed on PTFE could be explained by thehigher density of PTFE and the presence of highly reactive fluorine radicals in the plasma discharge, which may facilitate theeasier production of low-molecular degradation products (LMWOS-low molecular weight oxidized structures) [20]. Evidence for the creation of LMWOS by plasma treatment is the etching of the plasma-treated polymers in non-polar and polar media demonstrated in several previous experiments [17,31]. The creation of LMWOS was also proved in our previous work [17] on PE exposed to direct Ar plasma under the experimental conditions identical to those of the present work.In Figure 4.6, dependence of the contact angle on the plasma exposure time for plasma-treated polymers is shown. The measurement was accomplished 5 minutes after the exposure to the plasma discharge. Plasma treatment results in a dramatic decline of the contact angle and a corresponding increase of polymer wettability.

The wettability increase is caused by thecreation of polar groups, exhibiting enhanced hydrophilicity, at the polymer surface [17,18]. The highest contact angle is observed on the plasma-treated PTFE. It is well known that the contact angle is mostly affected by the chemical structure and morphology of the polymer surface layer, especially by the presence of oxidized degradation products [32]. Aging of the plasma-treated polymers is usually accompanied by an increase of the contact angle which is due to a reorientation of the molecular segments, produced by the plasma treatment, thanks to their interaction with the ambient atmosphere [33,34]. A major role that this process may play isassociated with thereorientation of polar groups toward the polymer bulk [33]. Aging of the plasma-treated PET and PE was also proved by theXPS technique in our previous studies [35] and [17] performed under the experimental conditions identical with those of the present work.

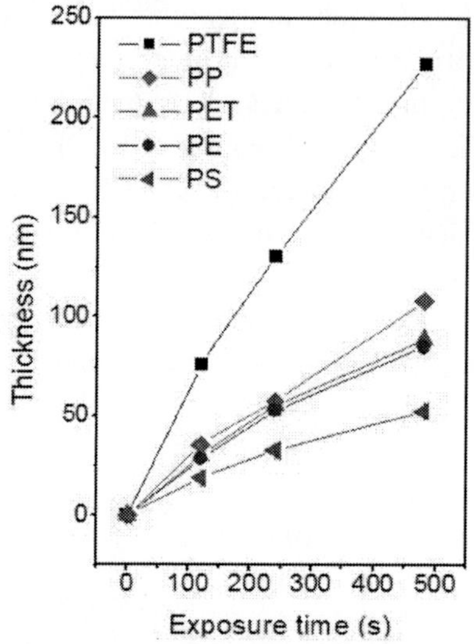

Figure 4.5. Dependence of the thickness of the ablated layer on the plasma exposure time for plasma-treated PP, PET, PS, PE and PTFE [26].

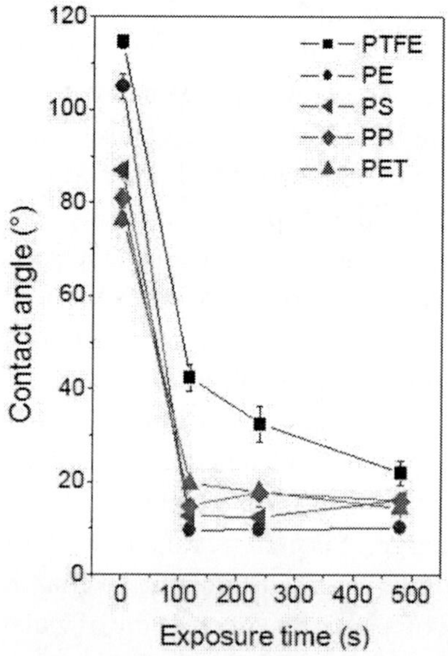

Figure 4.6. Dependence of the contact angle on the plasma exposure time for plasma treated PP, PET, PS, PE and PTFE. The measurement was accomplished 5 minutes after the plasma treatment [26].

With increasing times of the plasma treatment, the total oxygen content in the polymer surface layer increases [36]. Creation of the oxidized structures has been proven before on plasma-treated polyolefines [17,21].

In RBS spectra the presence of oxygen in Ar plasma-treated PE was observed up to the depth of about 100 nm with the oxygen concentration decreasing from the surface to the bulk [17,21]. Obviously, the oxygen from theambient residual atmosphere in the plasma reaction chamber interacts with the activated polymer surface and, in this way, different oxidized structures are created on the polymer surface (carbonyl, carboxyl, ester and hydroxyl groups) [17,21].

The oxygen containing structures are produced during sample exposure to both inert [17,21] and oxidizing plasmas [32,37]. In the above mentioned works the effects of the plasma treatment on PS, PP and PE polymers have been studied.In this work the presence of oxygen on the very surface of the plasma-treated polymers (in 8-10 surface atomic layers) was examined by the XPS technique.

The results of the XPS measurements are summarized in Figure 4.7. Only atomic concentrations of oxygen determined from the O1s peak of the XPS spectra are quoted. After the plasma treatment, the highest oxygen concentration was found on PS, lower on PP and PE and lowest one on PTFE. Also in PET, a low increase of the oxygen content is observedin the molecule of which contains oxygen. Low oxygen content in PTFE may be due tothe high ablation rate of the polymer (see Figure 4.5), and it can be related tothe high contact angle of PTFE (see Figure 4.6) and its low wettability.

Due to the plasma treatment, unpaired spins (i.e., free radicals) are generated on the polymer molecular chain. Not only C-H but also C-C, C-F and C-O (at PET) and "new" C-O (at all exposed polymers) bonds are likely to break.

The bond breakage leads to fragmentation of the polymer chain and to theablation of polymer surface. The free spins- free radical concentration determined bythe EPR method is shown in Figure 4.8. It is evident that after the plasma treatment the highest concentration of the free radicals was found on PS; a slightly lower concentration was found at PTFE. This can be explained by the higher ablation rate on PTFE corresponding to its known low radiation resistance [20]. By a faster ablation, a part of the created free radical isremoved from the PTFE surface during the plasma treatment.

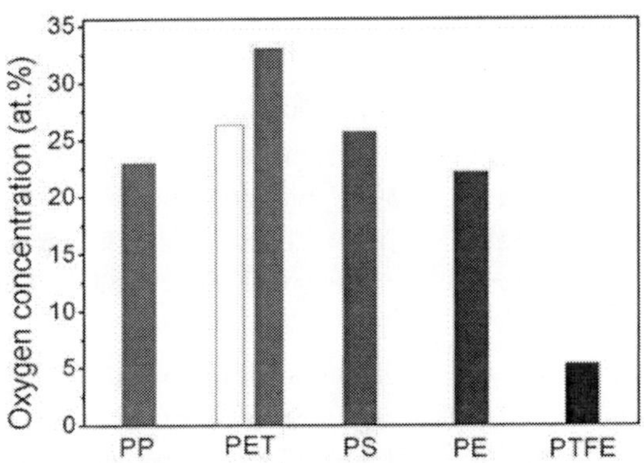

Figure 4.7. Oxygen concentrations measured by XPS on pristine PET (empty column) and PP, PET, PS, PE, PTFE plasma treated for 120 seconds (full columns). Before the measurement the plasma-modified samples were stored on air (RT) for 2 weeks [26].

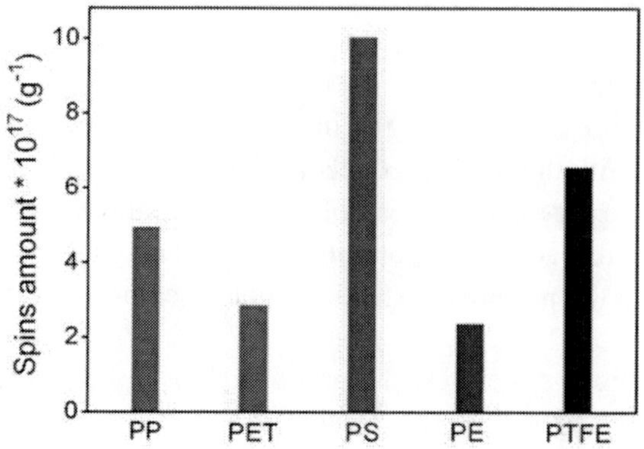

Figure 4.8.Spin amount (i.e., free radical concentration) determined for PP, PET, PS, PE and PTFE plasma treated for 120 seconds. The measurement was accomplished 2 hours after the exposure to the plasma discharge [26].

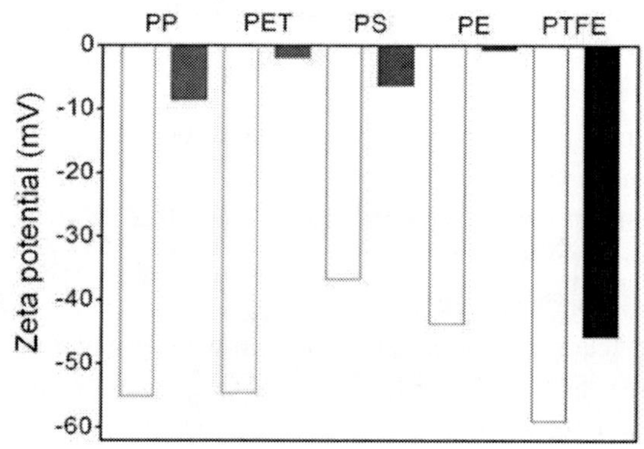

Figure 4.9. Zeta potential determined by SurPASS for pristine PP, PET, PS, PE and PTFE (empty columns) and those treated in plasma discharge for 120 seconds (full columns) [26].

The radical concentrations on PP, PET and PE are significantly lower. An exceptionally high radical concentration on the plasma-treated PS is probably due to the stabilization of the radicals by a mesomeric effect of the benzene rings, aprocess which is common in organic chemistry [20].

It is known that the radical concentration decreases during the aging of the plasma treated polymers as a result of radical recombination and oxidation [17]. If the oxygen in PS (and in PP and PET as well) is bound to LMWOS, its content should decrease after the sample rinsing. After the rinsing the unpaired spins (free radicals), observed in the EPR spectra, can initiate a cross-linking in the polymer surface layer. This process has actually been observed in a couple of previous papers [38,39].

The effects of the plasma treatment on the surface properties of different polymers were studied by various diagnostic techniques. Plasma treatment leads to the ablation of the polymer surface layers, the ablation rate being different for different polymers. As expected, the thickness of the ablated layer is an increasing function of the plasma exposure time. The lowest ablation rate exhibits PS; the highest ablation rate observed on PTFE is probably due to the creation of highly reactive, fluorine-containing species. EPR data shows different concentrations of free radicals created on the treated surface of all polymers. Plasma treatment results in a dramatic decline of the contact angle. The decline is similar on different polymers with only an exception of PTFE with the smallest decline observed. Despite the fact that the plasma treatment proceeds in inert Ar plasma, the oxidized degradation products are produced on the polymer surface which contribute to an increased hydrophilicity (wettability) of the plasma-treated polymers. The highest concentration of the oxidized structures was found to be in PS, lower in PP and PE and much lower in PTFE. Pristine polymers investigated exhibit very different surface morphology and roughness. The plasma treatment leads to significant changes in the polymer surface structure. The changes are different on different polymers with themost pronounced change being observed on PTFE.

Also, electrokinetic analyses wwere carried out for pristine polymers and polymers treated in plasma for 120 seconds (Figure 4.9, pristine polymers are represented by empty columns, the plasma-treated by full ones). The zetapotential of the pristine polymers increases in the order PTFE < PP ≈ PET < PE < PS. Its value is affected by several factors, mainly by thechemical composition and polarity of the polymer surface and also by surface morphology. It is therefore a bit surprising that the zetapotentials of polar PET and non- polar PP are comparable. A possible explanation can be found in our previous study [35], in which we found that in pristine PET most of the oxygen-containing molecular segments are oriented toward the polymer bulk and first atomic layer if effectively depleted of oxygen. Due to this phenomenon, the pristine PET, with reduced surface concentration of oxygen and PP behave in the same manner as can also be seen from similar values of the contact angle (Figure 4.6). Plasma treatment results in a dramatic increase of the zetapotential and its values increase in the order PTFE < PP < PS < PET < PE, which corresponds to wettability changes (see Figure 4.6) and increase of oxygen groups (see Figure 4.7).

4.3. Water and Methanol Etching of Polymer Surfaces

Surface etching consists ofthe removal of material from the treated sample surface [16] in liquid surroundings. The thickness of the PTFE surface layers ablated by the plasma treatment and etched in water or methanol as a function of the plasma exposure time is shown in Figure 4.10. The thickness of the removed material is an increasing function of the plasma exposure time. Additional material removal by etching in water and methanol is an increasing function of the plasma exposure time as well. As a result of the plasma treatment, the C-F and C-C bonds break and a fragmentation of the PTFE macromolecular chains takes place. The lower mass fragments, produced in this way, are easily ablated or etched [40]. In comparison with other polymers, the PTFE exhibits a very high ablation rate which is related to thewell known low radiation stability of PTFE. The higher rate of the material removal in water is due to thehigher polarity of water molecules.

In Figure 4.11 the dependence of the contact angle on the plasma exposure time for the plasma-treated and etched PTFE is shown. Plasma treatment causes a decrease of the contact angle (see also Figure 4.6). Subsequent water or methanol etching leads to an increase of the contact angle, which is due to a removal of a part of themodified material from the sample surface.

The contact angles of the etched samples are close to that of the pristine PTFE. Considering the mass removal during the etching (see Figure 4.10), we can conclude that two processes running simultaneously may be responsible for the contact angle increase. The first process isthe removal of formerly created polar groups from the polymer surface by water and methanol etching, and the second one is the rearrangement and reorientation of polymer segments in the course of the sample aging [41].

The chemical structure of the modified PTFE surface was investigated by the XPS method. By the XPS, only a thin surface layer (typically 8-10 atomic layers thick) is examined, but this layer affects surface properties of PTFE significantly. In contrast with the previous results obtained on polyethylene [17], no nitrogen was observed in the surface layer of the plasma treated PTFE. From the profile and area of individual peaks in the XPS spectra, it is possible to determine the concentration of different chemical groups present on the PTFE surface. The deconvolution of C(1s) is shown in Figure 4.12 for illustration. As in previous studies [42], a carbon contamination fromthe ambient atmosphere was detected. A concentration depth profile of oxygen could, in principle, be determined by theRBS method, which has been applied in our previous studies; but in this case, the RBS can not be used because ofthe extreme radiation sensitivity of PTFE [17].

Nevertheless, it could be supposed that the oxygen is not distributed homogenously [17], and this is one of the reasons why the samples prepared under different conditions exhibit different concentrations of oxidized groups.

The XPS measurements show that the plasma treatment leads to a formation of oxidized structures on the PTFE surface, which could partly be removed by subsequent plasma etching. Oxidized groups (-C-O-C-, -C=O and -O-C=O) may be created directly in the reactor atmosphere during the plasma treatment or subsequently during storage of the modified samples under standard laboratory conditions [17].

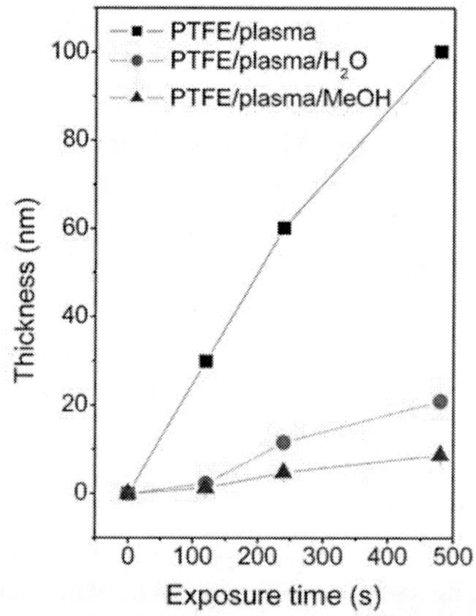

Figure 4.10. Dependence of thicknesses of the ablated polymer layers and layers etched in water or methanol on the plasma exposure time measured by gravimetry [43].

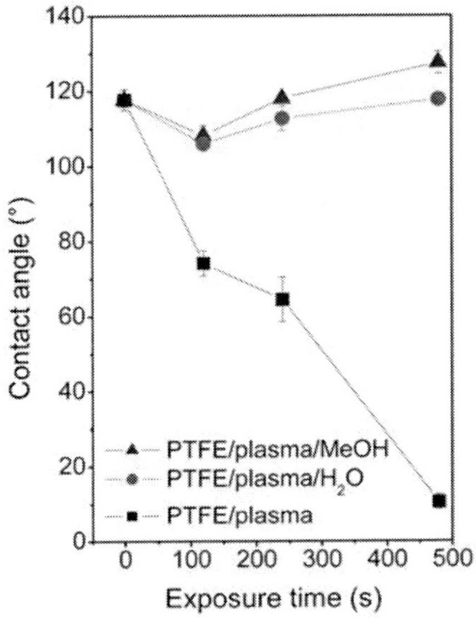

Figure 4.11. Dependence of the contact angle on plasma exposure time for the plasma-treated and etched (water or methanol) PTFE [43].

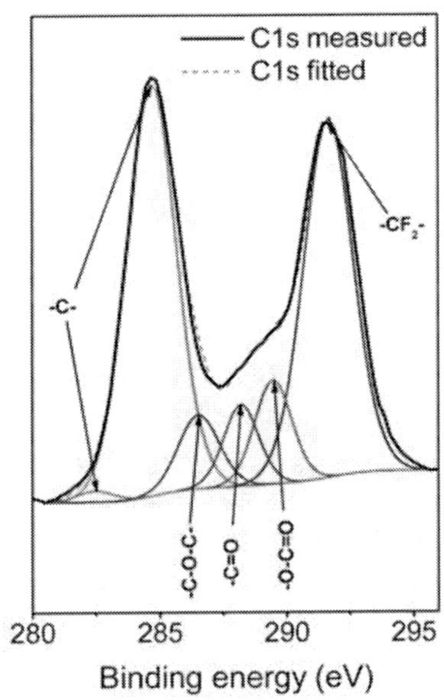

Figure 4.12. Deconvolution of C(1s) peak from the PTFE, plasma treated for 240 seconds (solid line). Measured spectrum (solid line) is compared with that fitted (dashed line). The position and size of particular components are also shown [43].

Present results prove that the plasma treatment of PTFE results in thecleavage of themolecular chain and production of polar groups [43], the presence of which leads to a decrease of the contact angle observed after the plasma modification and etching (see Figure 4. 6 or 4.11).

The polar groups are known to enhance solubility of the modified polymers in polar solvents [17]. This effect was also observed in the present experiments on etching of the plasma treated PTFE in water and methanol (see Figure 4.11).Plasma treatments of the PTFE result in an immediate, dramatic decline of the contact angle and a corresponding increase of PTFE wettability. Aging of the plasma-treated PTFE is accompanied by an increase of the contact angle which is due to a reorientation of the molecular polar segments (carbonyl, ester and ether oxygen groups), produced by the treatment. Plasma treatment results in an ablation of the PTFE surface layer. Subsequent etching in water or methanol leads to another removal of the plasma-modified material from the PTFE surface. The etching leads to an increase of the contact angle due to the removal of polar groups.

5. Irradiation of Polymers with a Linearly Polarized Laser

Laser micro- and nano-nanofabrication of various passive and active structures and devices in polymer solid media have attracted a lot of research efforts in the recent years. There are several optical processes and techniques that may be utilized for producing structures on a flat surface or in the polymer bulk: surface relief grating, localized photochemical reactions or surface ablation, interference lithography, holography, laser scanning, etc. These techniques utilize the change of thepolymer structure, chemistry or surface relief under the photon absorption. Interference patterns of laser lights are used to prepare a periodical structure in a relatively simple way. The process can be performed on different polymers, including those commercially available or specially synthesized in a laboratory. In the following, one of these perspective techniques – "Single-laser beam nanopatterning of polymers" – is described in more detail.

The trend to higher integration in current microelectronics, micromechanics, and micro-optics requires the preparation of well-defined thin metal layers of complex shapes on different substrates. Nano-patterning of polymer materials can be employed forthe preparation of semiconducting or insulating regions on flexible polymeric substrates(e.g., for theproduction of polymeric field effect transistors (PFETs) [44]), and regular nanostructures on polymer surfaces may allow the realization of polymer-based electronic memories [45]. Surface texturing with laser lights is used also for the fabrication of biomedical surfaces [46] or the preparation of superhydrophobic polymer surfaces [47]. Biosensing currentlyrepresents one of the important applications of the metal-polymer structure. A PET- based flexible polymer glucose biosensor with a polymer-gold interlayer was reported in Ref. [48]. A silver nanoparticle-functionalized PET surface was used for thedetection of protein–protein interactions in a biotin sensor device [49]. Nano-structured dielectric materials like polymer or glass are also regarded as promising substrates for the lithography-less preparation of metal nano-wires [28,50].

In addition to the latterapplication, laser lights may also be employed forthe micro- or nano-fabrication of semiconductors [51], polymers [52] or thin films [53]. Today, common methods for thepreparation of sub-micrometer features on various substrates depend mostly on electron [54] or DUV (deep ultraviolet) laser lithography [55-57]. For some applications, thedirect exposure of samples to polarized excimer laser lights may be an alternative, allowing mask-less and photoresist-less formations of coherent periodic structures [58], particularly on polymers [28,59]. Here, the formation of the nano-pattern at the surface originates probably from interference between the incoming laser light and light scattered at material corrugations and inhomogeneities [8,60]. However, the mechanisms involved are complex and, especially for polymers, there is no conclusive explanation of the structure formation process yet.

In this section, we attempt to comprise a long-term experimental work of our working group dealing with thelaser-assisted nanopatterning of polymers. When coated with metal nanolayers with the thickness of several tens nanometers, these nano-structured polymer foils appear as promising substrates forthe lithography-less production of metal nanowires.

The following experiments were performed on polyethyleneterephtalate (PET) foil which was irradiated either by thelinearly polarized light of KrF excimer laser or by linearly polarized light of an F2 excimer laser.In a certain range of irradiation parameters, the irradiation resulted in the formation of coherent ripple patterns with a lateral periodicity in the order of the wavelength of the laser light and with a corrugation height of several 10 nm. Resulting structures were consecutively coated by gold. The gold deposition was accomplished by sputtering.

The irradiated polymers and deposited layers were analyzed by atomic force microscopy (AFM), focused ion beam (FIB) cuts, scanning electron microscopy (SEM), and angular resolved X-ray induced photoelectron spectroscopy (ARXPS). Gold sputtering on theKrF laser irradiated PET and led to the formation of separated ''nano-wires'' at the ridges of the nano-patterns and not to a continuous metal layer, as we obtained in case of gold sputtering onto F2 irradiated PET.

The results of the XPS analysis indicated that the KrF irradiation caused degradation on the ridge of the ripples, whereas no noticeable degradation occurred for theF2 laser treatment. We attribute the different growth mechanisms of the deposited gold layers mainly to the difference in surface chemical composition of laser irradiated PET with the two different lasers employed. We also demonstrated that the structure period could be controlled by variation of the incidence angle of the laser beam. Moreover, we have shownthat the modification of the PET surface with linearly polarized light from pulsed KrF laser has a significant effect on the properties of subsequently deposited gold nanolayers, and the choice of the deposition technique (evaporation and/or sputtering) is crucial owing to the quality of prepared coatings. Subsequent deposition (either by evaporation or sputtering) of a 200nm-thick gold layer caused a decrease of the surface roughness. While by evaporation a continuous metal coverage is formed, copying thenano-structured polymer surface, in the case of sputtering a nanowire-like structure of the gold coating can be observed. In conclusion, it will be shown that the width of the nano-wires can be tailored by the width of the ripples formed by preceding laser irradiations.

5.1. Single Laser Beam Nanopattering of a Polymer

This chapter will give an overviewof an unconventional method for thenano-structuring of polymeric materials and for thelithography-less preparation of metal nano-wires. Such a nano-structured composite material may find its application especially in optics, like a metamaterial with a negative refractive index or difraction grids.

Several studies have shown that the illumination of polymers by a polarized UV laser beam induces self-organized ripple structure formations within a narrow fluence range well below the ablation threshold [61-65]. The properties of these periodic structures have been frequently studied since the first report about them [51]. The smallest features on polymers were produced by our group by a 157 nm F_2 laser irradiation of polyethyleneterephthalate [59]. The period of the ripples depends on the laser wavelength and on the angle of incidence of the laser beam, and their direction is related to the direction of the laser beam polarization [8]. The spacing of the structures can be described by the relation:

$$\Lambda = \lambda / (n - \sin\theta), \quad (1)$$

where Λ is the lateral periodicity of structures, λ is the wavelength of the excitation laser light, n is the effective refractive index of the material and θ is the angle of incidence of the laser beam [8].These structures develop on the original material surface having a small roughness, as a result of treatment with one laser beam with a uniform intensity distribution. It is known that the interference between the incoming and the surface scattered waves plays an important role in the structure formation [60].

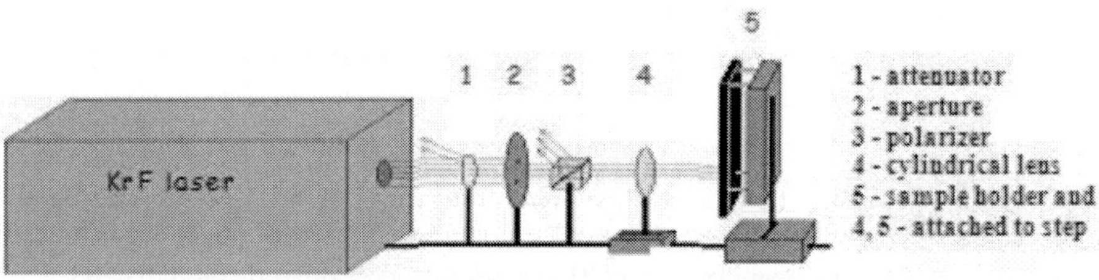

Figure 5.1. Experimental set-up for KrF laser irradiation.

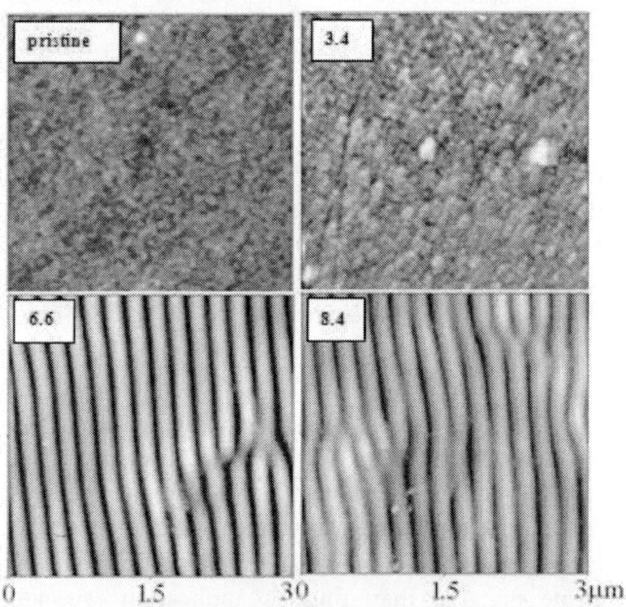

Figure 5.2. AFM images of the PET irradiated at different KrF laser fluences; the numbers in the inset refer to the laser fluences in mJ·cm^{-2} employed for the irradiation of the PET foils, while pristine stands for un-irradiated pristine PET [70].

The interference causes an inhomogeneous intensity distribution, which together with a feedback mechanism results in anenhancement of the modulation depth [63].

However, the mechanisms involved are complex and, especially for polymers; there is no conclusive explanation of the structure formation process, yet. Nevertheless, different processes have been reported as responsible for ripple formation [35,59,66-68] such as thermal and non-thermal scissoring of polymer chains, amorphization of crystalline domains, local surface melting and photo-oxidation and material transport.

5.2. Threshold Fluence and Periodic Structure Formation

As it was already mentioned above, many studies have shown [61-65] that the illumination of polymers by a polarized UV laser beam induces a self-organized ripple structure formation within a narrow fluence range well below the ablation threshold. In these processes, certain fluence thresholds always exist below which the periodic ripple structure is not created on the polymer surface. For the irradiation of the PET samples, the light from a KrF excimer laser (see Figure 5.1, Lambda Physik Compex, wavelength 248 nm, pulse duration of 20 to 40 ns) was used in combination with a Glan-Thompson polarizer prism [28]. After, the light passed through laser beam was divided into the two mutually perpendicularly polarized beams (σ, π) inclined by 110°. For the modification,only the π beam was used, beam σ pointed out of the sample. For a homogeneous illumination of the samples, only the central part of the beam profile defined by an aperture with a length of 10 mm and a width of 0.5 mm was used. The samples were mounted onto a translation stage and scanned in a fix position. The PET surfaces were irradiated with 6,000 pulses per area at different laser fluences between 2.0 and 8.4 mJ·cm^{-2} and different angles of incidence of the laser beam (0°, 22.5° and 45°). The experiments were performed in an ambient air atmosphere. Figure 5.2 displays AFM images of pristine PET and PET irradiated at different laser fluences and under the incidence angle of 0°. The sample irradiated with 3.4 mJ·cm^{-2} fluence exhibits a rougher surface than the pristine PET. There is a noticeable nodulation, although no ripple formation is visible. At higher laser fluences, periodic ripple structures have developed over the irradiated area. At the laser fluence of 6.6 mJ·cm^{-2}, a regular and uniform coverage of the PET surface with ripples is reached.

A similar development of the surface morphology has been observed during F2 excimer laser irradiation of thePET foil (see Figure 5.3), published also in ref. [59]. During these experiments we used an F2 laser (Lambda Physik LPF 202) that operates at a wavelength of 157 nm, with pulse duration of 15 ns, and we applied a repetition rate of 5 Hz.

The light was polarized linearly with a Rochon prism and for the homogeneous illumination of the samples we used an aperture of 10 x 3.5 mm^2. The samples were mounted onto a translation stage and scanned at a speed of 6.25 mm·s^{-1} and they were irradiated with 56,00 pulses per area. We carried out the experiments in a flow box purged with nitrogen at a pressure of 110 kPa. In this case, the ripple threshold was found at 3.8 mJ·cm^{-2}, whereas the most regular and uniform filling of irradiated area with ripples was achieved during theirradiation at the fluence of 4.2 mJ·cm^{-2}.

The measured period of the structures was about 130 nm. As it has already been mentioned, applied fluences in the case ofthe KrF laser (248 nm) as well as in the case of theF2 laser (157 nm) irradiations were well below the ablation thresholds of PET which are 29.6mJ·cm^{-2} [59] and 40 mJ·cm^{-2} [52], respectively.

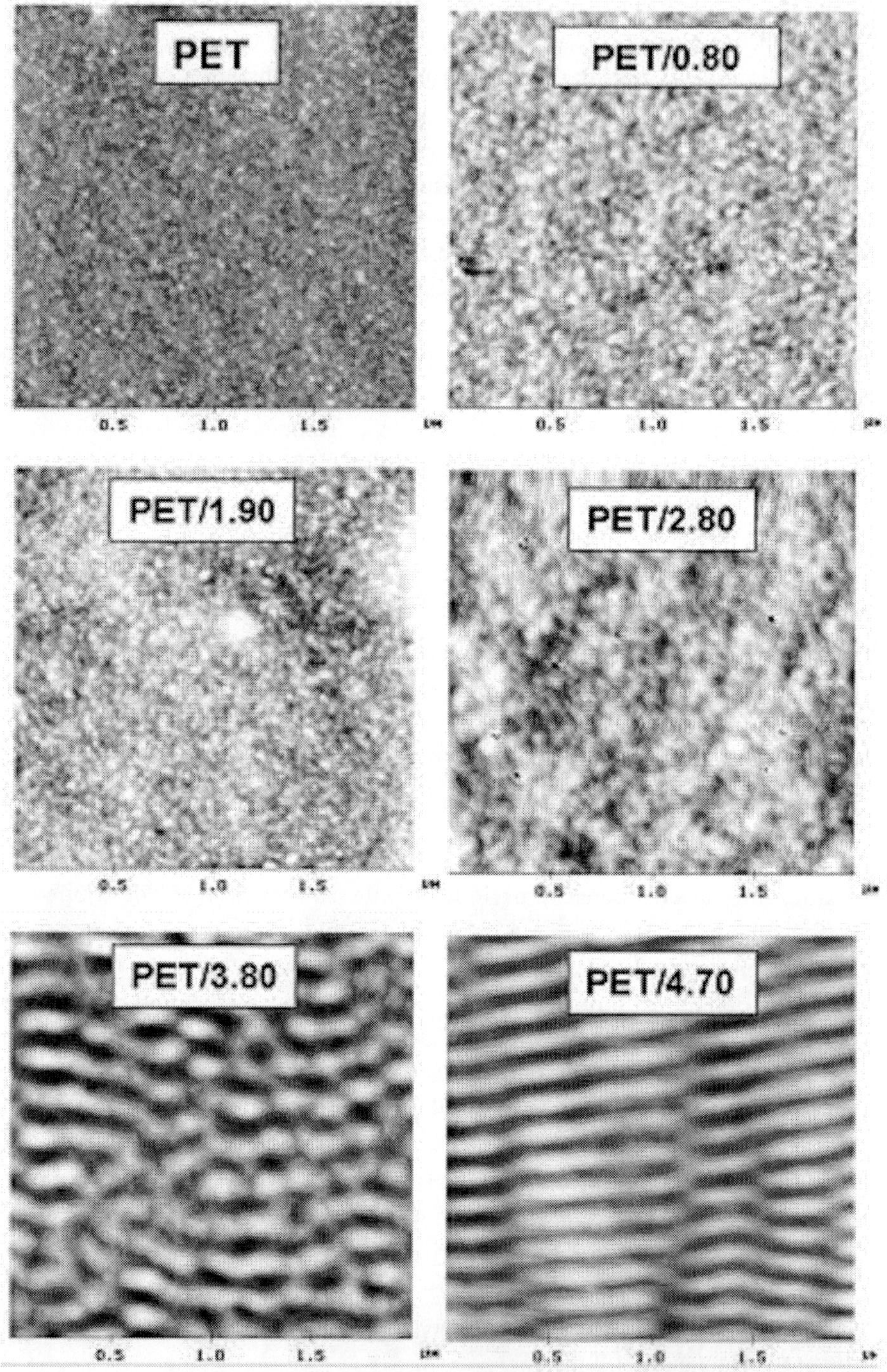

Figure 5.3. Surface morphology (AFM images) of pristine and F_2 laser-modified PET at the indicated fluences (in mJ·cm^{-2}). The vertical scale is 50 nm [59].

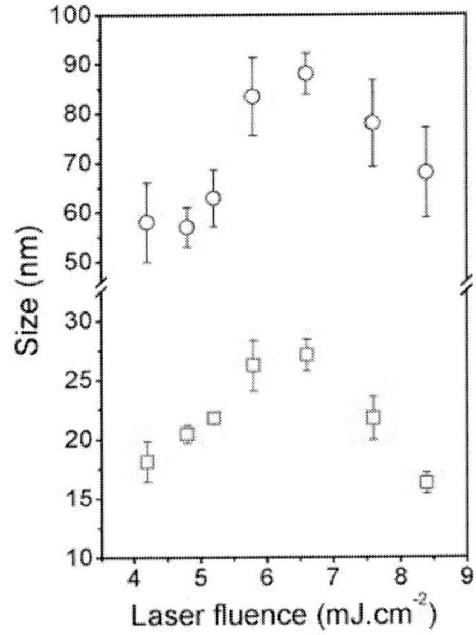

Figure 5.4. Dependence of the ripple height (○) and roughness (□) on the KrF laser fluence employed for the PET irradiation [70].

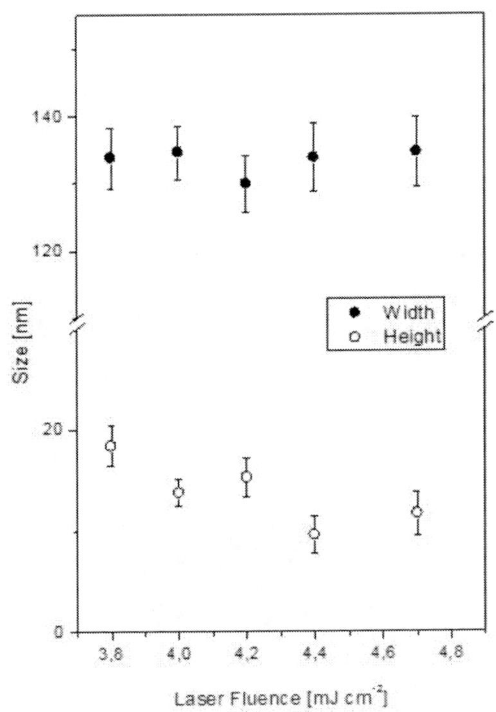

Figure 5.5. Dependence of the width and the height of the ripple-like periodic structures formed on theirradiated PET on laser fluence (F2 laser) [59].

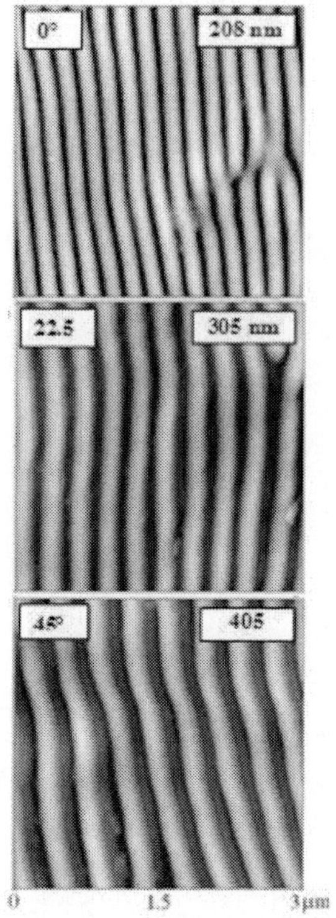

Figure 5.6. AFM images of the PET irradiated at a KrF laser fluence of 6.6 mJ·cm^{-2} under adifferent incidence angle of the laser beam (0°, 22.5° and 45°). The numbers in the insets in the upper left corner refer to the angle of incidence of the laser beam and in the insets in the upper right corner to the ripple period in nm [70].

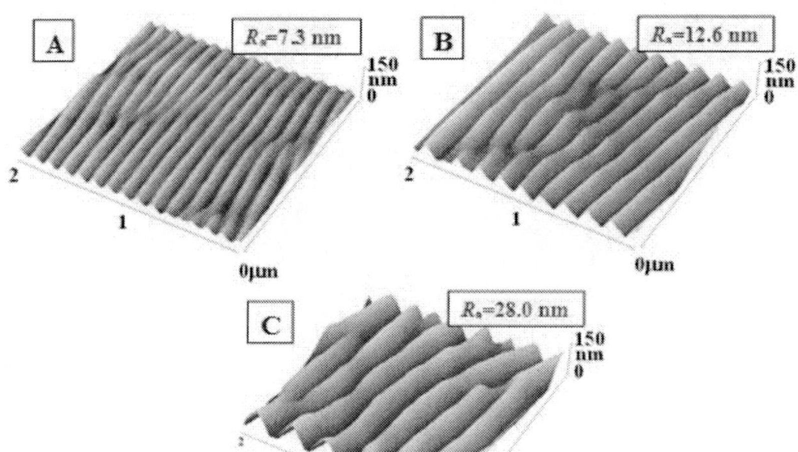

Figure 5.7. AFM images of PET samples irradiated with F2 157 nm laser (4.40 mJ·cm^{-2}) under different angles of thelaser beam; (A) angle of incidence 0° (perpendicular to the sample surface), (B) 22.5° and (C) 45°. R_a is average surface roughness in nm [79].

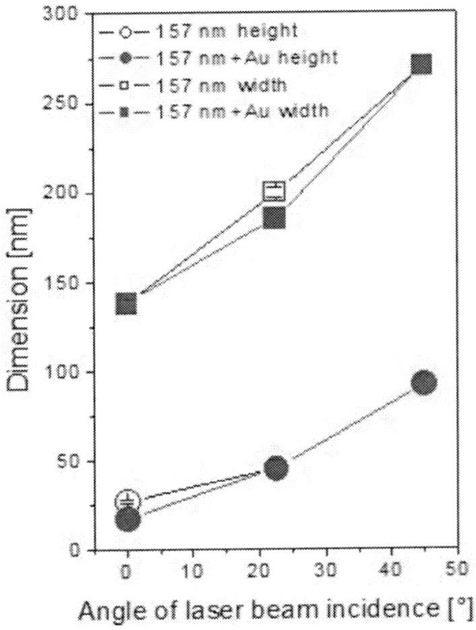

Figure 5.8. Dependence of structure width and height on theangle of laser beam incidence towards the PET surface for a PET modified with a 157 nm laser (4.40 mJ·cm^{-2}) (-○- 157 nm height, -ϒ- 157 nmwidth) and the same structures sputtered with 100 nm of Au (-●-157 nm +Au height, -■- 157 nm +Au width) [79].

Further experiments performed on the laser-irradiated PET proved that the height and the roughness of the ripples depend strongly on the applied laser fluence (see Figure 5.4). This fact is illustrated in Figure 5.5, which shows the height and the roughness of the ripples as a function of the applied laser fluence in thecase of theKrF laser treatment. The dependence follows from an AFM measurement shown in Figure 5.2. There is a good correlation between height and surface roughness of the ripples overthe whole laser fluence range examined (see Figure 5.5).

Both parameters reach the maximum value at the fluence of 6.6 mJ·cm^{-2}, which corresponds to the ripple height of about 90 nm. From the technological point of view this is the most promising structure. On the other hand, Figure 5.5 displays the dependence of the nanostructure width and height resulting from theF2 laser treatment. One can see that whereas ripple height strongly depends on the laser fluence (it monotonously decreases with applied fluence) the width is nearly independent. This finding is in a good accordance with eq. 1.

Since the periodicity of the formed nano-structures (ripples) could be altered by changing the irradiation parameters (according to eq. (1)), other experiments proving the validity of this formula were done. Figure 5.6 shows AFM images of the PET irradiated under different incidence angles of the laser beam. For larger angles of incidence, the spacing between two neighboring ripples is wider. For the incidence angle of 0° and 22.5°, the observed spacing of the ripples is in good agreement with the value calculated by eq. (1) with an effective index of refraction $n \approx$ 1.2. The agreement for the incidence angle of 45° is less convincing. The discrepancy may be due to changes of the polymer refractive index induced by the UV laser irradiation as reported earlier in [69]. Surface morphology of the PET samples irradiated by the 157 nm laser light (fluence 4.4 mJ·cm^{-2}) under different angles of incidence can be seen from AFM images shown in Figure 5.7.

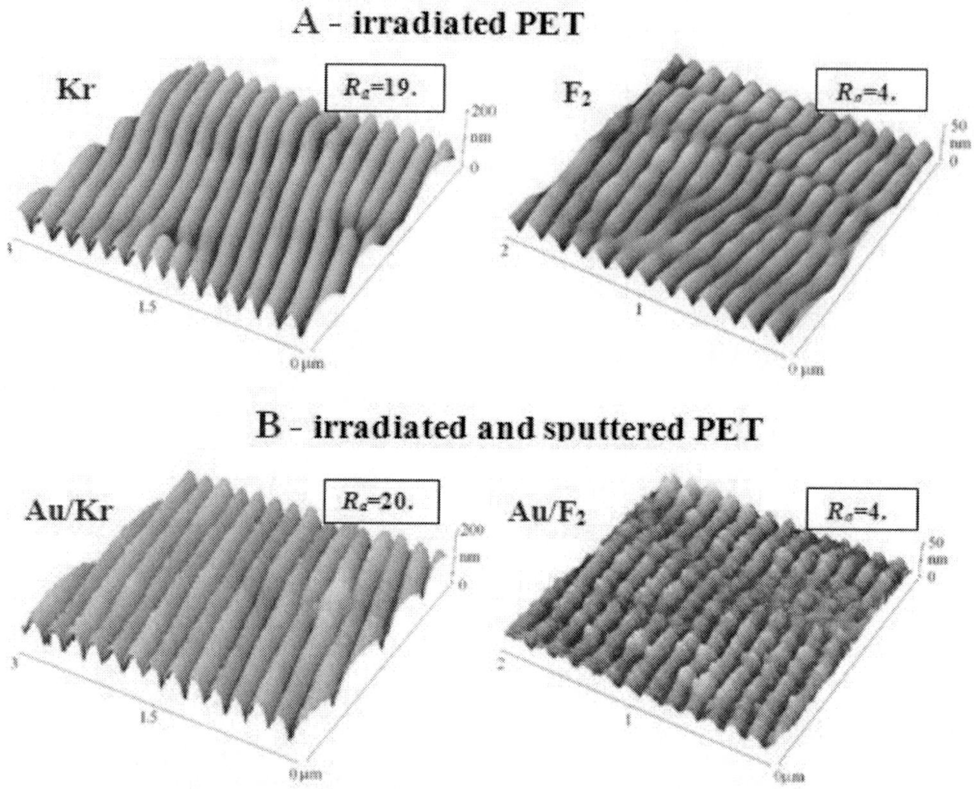

Figure 5.9. AFM images of PET samples: (A) irradiated by KrF (6.6 mJ·cm^{-2}) and F_2 (4.4 mJ·cm^{-2}) lasers and (B) irradiated and than sputtered with 200 nm (KrF) and 50 nm (F_2) Au layer. R_a is the average surface roughness in nm [28].

In Figure 5.8 the dependence of the width and height of the observed structures on the angle of the laser beam incidence is shown. From Figures 5.7 and 5.8 it is evident, that the periodicity and the height of the ripples increases with an increasing angle of incidence of the laser beam. The average surface roughness also increases with an increasing angle of incidence, and it achieves its maximum for the PET irradiated under the maximum angle (45°) of incidence. Theoretical values calculated from the eq.1 (n = 1.21) are 130 nm for the 0°, 190 nm for 22.5° and 310 nm for the 45° incidence angle.

Since the mechanism of the ripple pattern formation is very complex (some details are described in [8]) and especially for larger incidence anglesvariations of the measured values are expected [8], the correspondence between measured and expected values (Figure 5.8) is satisfactory. Present experiments confirmed the possibility of tailoring the width of ripples by changing the incidence angle of the laser beam for both KrF and F2 lasers.

5.3. Metal Coating and Nano-wire Formations

For the successful construction of optically active materials, it is essential to perform a selective metal coating of the previously formed periodic structures. There are many conventional ways how to reallize this technological step, however, only a few of them are relatively simple and inexpensive [28,50,70]. Metal deposition on polymers can be carried out by sputtering, vacuum evaporation, and also by various electrochemical procedures [71-73]. The thicknesses of the prepared layers are from a few to hundreds of nanometers. Here, an effective process of selective metallization by the sputtering method will be presented.

In Figure 5.9, we compare the morphology of F_2 (laser fluence of 4.4 mJ·cm^{-2}) and KrF (6.6 mJ·cm^{-2}) irradiated samples. The ripples formed under KrF laser exposure have larger widths and heights in comparison to that formed by the F_2 laser. The periodicity of ripples, Λ, was about 208 nm in the case of KrF laser irradiation, while in the case of F_2 laser irradiation we obtained a periodicity of about 140 nm. The heights of the ripple structure (top-bottom) were about 100 nm and about 15 nm for the KrF laser and the F_2 laser irradiation respectively.

The sputtering of gold layers onto the ripple structures was accomplished, and the surface morphology was examined by AFM. For easier comparison of the growth mechanism of the gold layer, the gold-to-ripple height ratio was kept constant. F_2 irradiated samples were sputtered with a 50-nm thick Au layer, whereas KrF irradiated samples were covered with a 200-nm thick Au layer. Figure 5.9 shows the morphology of sputtered gold layers on the laser-induced ripple structures for both laser light wavelengths. For both wavelengths, the ripple structure of the PET surface is transferred into the relief of the gold layer. The roughness values,R_a, were nearly identical for the surface, with and without a gold coating, in both cases. However, for the F_2 laser-irradiated samples it seems that the gold coated surface has a more pronounced granular structure than the uncoated ripple structure. Here one has to keep in mind that the AFM images show a convolution of the surface morphology with the geometry of the AFM tip. Therefore, small features and features with a high aspect ratio may not be feasible in the AFM images. Supplementary to the AFM analysis, the samples were cut by FIB and the FIB cuts were investigated with the SEM (see Figure 5.10). The gold sputtered onto the KrF laser induced ripples is deposited in the form of "nano-wires", which grow on the ridges of the ripples.

The FIB cuts reveal that the gold layer is not continuousand there are gaps between the individual wires. Additionally, grains are visible along the wires, but the FIB cut images suggest that these grains are interconnected. The morphology of the gold layer deposited on the F_2 laser-induced ripples is distinctly different. The gold is also deposited in the valleys of the ripple structure (see Figure 5.10) and the gold coating is continuous. The Au coatingpartially smoothes out the initial ripple profile. In general, the Au layers show cracks which, however,are not directly related to the ripple pattern.

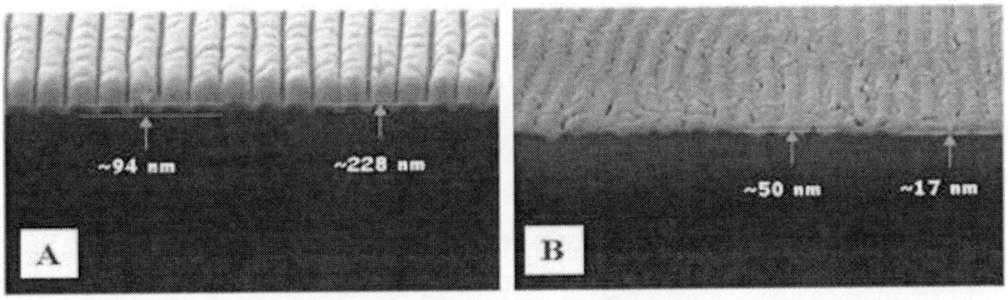

Figure 5.10. SEM images of FIB cuts of lasers irradiated and Au sputtered samples: (A) PET irradiated by KrF laser (6.6 mJ·cm^{-2}) and sputtered with 200 nm of Au layer, (B) PET irradiated by F_2 laser (4.4 mJ·cm^{-2}) and sputtered with 50 nm of an Au layer [28].

5.4. Chemical Composition of Nano-Structured PET

The chemical composition of the polymer surface can affect the growth morphology of the gold layers on polymer surfaces [74]. Therefore, we analyzed KrF and F_2 irradiated PET with the ARXPS method [35]. By this method it was shown that the oxygen content in first ten atomic layers of pristine PET is depth dependent [35]. This phenomenon is explained by the reorientation of the polar groups in the surface layers leading to lower oxygen content in near surface layers in comparison with the theoretical value for pristine PET (29 at %) [29].

The dependence of the oxygen concentration (at %) derived by ARXPS on the angle of incidence of the XPS primary beam is shown in Figure 5.11 for pristine PET and PET irradiated with KrF and F_2 lasers. For pristine PET the oxygen content decreases from 25.5 at % for an incidence angle of 0° to 19.5 at % at an incidence angle of 80°. For F_2 laser irradiation, the oxygen content is about 18 at % and is slightly decreasing to about 16 at %; for KrF laser the O content increases from 28.5 to 33 at % when going from the surface to the bulk.The ARXPS analysis of the irradiated PET showed that the ridges and the valleys of the F_2 laser-induced ripples have nearly similar chemical composition, at least regarding the oxygen content. The oxygen content in F_2 irradiated PET is lower than in pristine PET, whereasin the case of KrF laser irradiation the oxygen content is higher and towards the surface the oxygen content increases. It may, therefore, be concluded that in case of KrF the ridges of the ripple structure contain more oxygen groups than the areas at the valley, which are shielded at higher angles of incidence of the primary X-ray beam. The oxygen on the ridges is even higher than that in the pristine PET. This could be due to a thermal degradation (connected with oxidation) of the polymer at the ripple ridges during the irradiation process.

The fact that the F_2 laser-treated surface shows a lower oxygen content than that of pristine PET may be connected with a reorientation of the polar groups toward the polymer bulk [29,35] or by a photochemical scission of the polymer chains in combination with a preferential release of CO and CO_2 groups (Norrish type II reaction[27]).

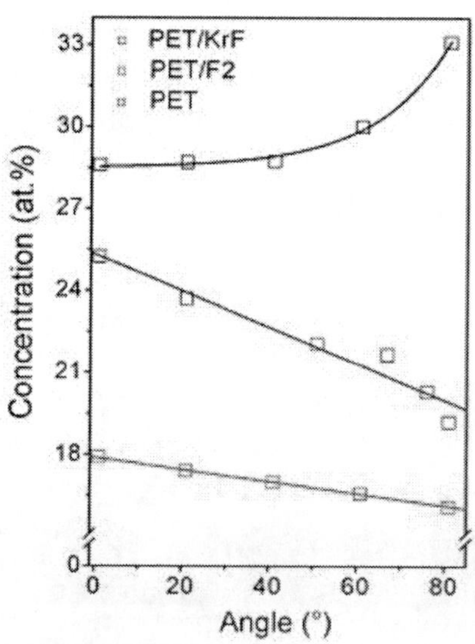

Figure 5.11. Dependence ofthe oxygen concentration (at %) on the angle of incidence of the XPS primary beam for pristine PET (PET) and KrF (PET/KrF) and F_2 lasers (PET/F_2) irradiated PET [28].

KrF laser irradiation of PET foils can induce the formation of nano-ripple structures with a periodicity of about 200 nm and a height of about 100 nm. Gold sputtering onto these PET surfaces leads to the formation of gold "nano-wires" at the ridges of the ripple structures.

The gold layers seem to be non-continuous, with gaps between neighboring wires. The gold layer morphology is distinctly different for gold sputtered onto F_2-laser induced ripples, which are narrower and considerablyshallower than those induced by the KrF laser irradiation. Here the layers are continuous, showing only irregularly distributed cracks. From the XPS analysis it is possible to conclude that in the case of KrF irradiation, the exposure leads to degradation (oxidation), especially at the ridges of the ripple structures, with an increased occurrence of oxygen containing groups.

The altered chemical composition probably results in improved metal deposition on the tops of the structures rather than in the valleys between. Together with shadowing effects, this may be the reason for the formation of gold wires with a height considerably higher then the ripple structure itself.

5.5. Angle Dependent Irradiation and Sputtered Vs. Evaporated Coatings for KrF Lasers

In the previous paragraph it was shown that by the combination of an excimer laser nano-fabrication of PET and an Au coating by the sputtering technique it is possible, at certain experimental conditions, to prepare gold nano-wires. This procedure is very effective, since it does not require rather complex, conventional techniques that utilize procedures such aslithograpy and mask processing.

This paragraph will demonstrate that modification of the PET surface with linearly polarized light from a pulsed KrF laser has a significant effect on the properties of subsequently deposited gold nanolayers and that the choice of the gold deposition technique is crucial for the quality of the gold coatings. Here, the difference between the sputtering and evaporation deposition techniques will especially be discussed.

Figure 5.12 shows the AFM morphology of laser-induced ripples on PET foils and the ripples subsequently coated by gold layers prepared by sputtering (sputt.) and by evaporation (evap.). The ripples on the PET surface were prepared by KrF laser irradiation under incidence angles of 0° and 22.5°, respectively. The wider ripples obtained under the angle of incidence of 22.5° also led to wider structures in the gold coatings. The evaporated and sputtered layers have almost the same topography.

The main difference is in the roughness of the nano-structures (see Figure 5.13), which is slightly lower in the case of evaporated layers. It should be mentioned that the AFM images represent a convolution of the surface morphology with the geometry of the AFM tip. The dependence of the surface roughness on the laser fluence for PET without a coating and for gold-coated PET is shown in Figure 5.13. There are two distinct intervals of the roughness values depending on the laser fluence.

The first interval, up to the fluence of 4.0 mJ·cm^{-2}, is characterized by a small increase of the surface roughness (compared to pristine PET), which correlates with small changes in the surface topography in Figure 5.3 for the fluence of 3.4 mJ·cm^{-2} at which no ripple structure is created. At fluences above a threshold about of 4.2 mJ·cm^{-2}, periodic ripple structures are formed (second range) and the surface roughness increases rapidly between the two "intervals". In the first interval, gold sputtering causes only a slight increase of the surface roughness, and after the gold's evaporation the roughness remains practically unchanged for all fluences in this interval. In the second interval, gold deposition results in a decrease of the surface roughness (in comparison to samples only exposed to laser radiation without a gold coating) for both sputtering and evaporation techniques.

The roughness of the evaporated layer seems to show less variations for different fluences compared to the roughness of the sputtered layers. FIB cuts of the laser-irradiated and gold-coated PET samples were investigated by SEM (see Figure 5.14. After sputtering, the gold is deposited in the form of "nano-wires", which grow mainly at the ridges of the ripples. The FIB cuts reveal that there could be gaps between the individual wires and that the metal layer may be discontinuous.

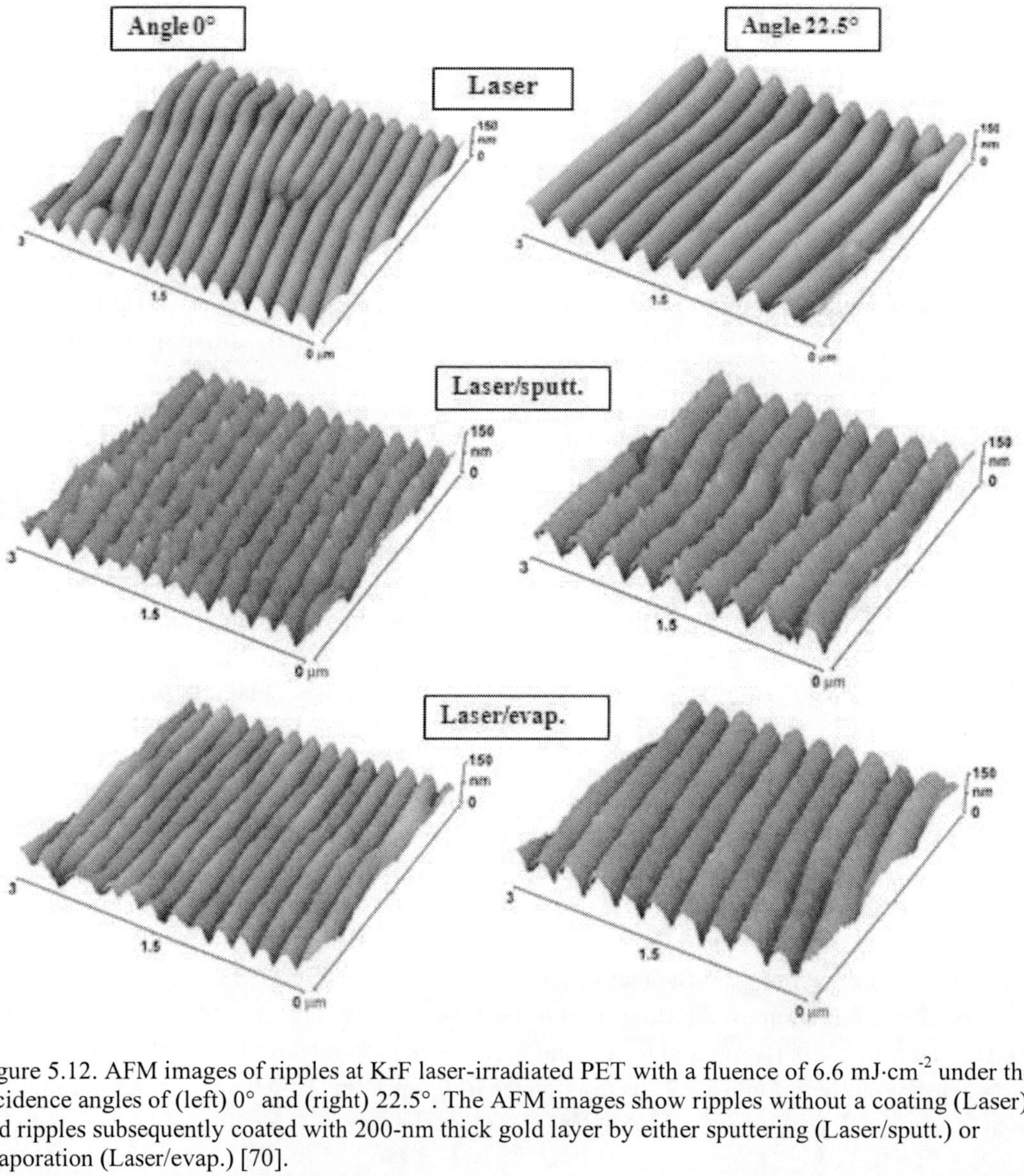

Figure 5.12. AFM images of ripples at KrF laser-irradiated PET with a fluence of 6.6 $mJ\cdot cm^{-2}$ under the incidence angles of (left) 0° and (right) 22.5°. The AFM images show ripples without a coating (Laser) and ripples subsequently coated with 200-nm thick gold layer by either sputtering (Laser/sputt.) or evaporation (Laser/evap.) [70].

The width of the gold nano-wires directly correlates with the width of the ripples formed before the gold deposition. Additionally, a granularity is visible along the wires, but the FIB cut images suggest that the grains may be interconnected. The morphology of the gold layers deposited by evaporation is distinctly different.

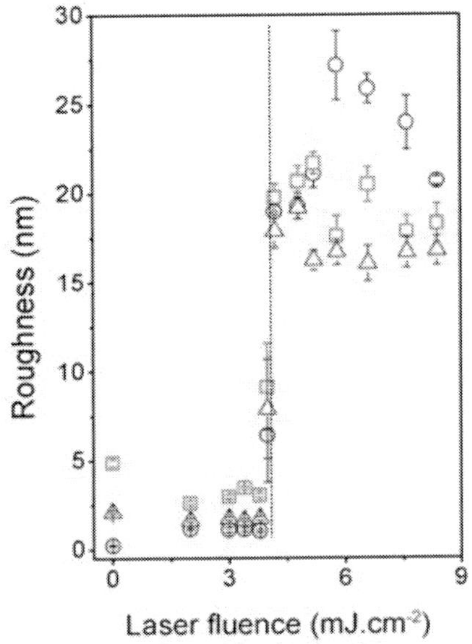

Figure 5.13. Dependence of the surface roughness on the KrF laser fluence for irradiated PET withoutcoating (○) and for irradiated and gold-coated PET (sputtering (□) and evaporation (Δ)) [70].

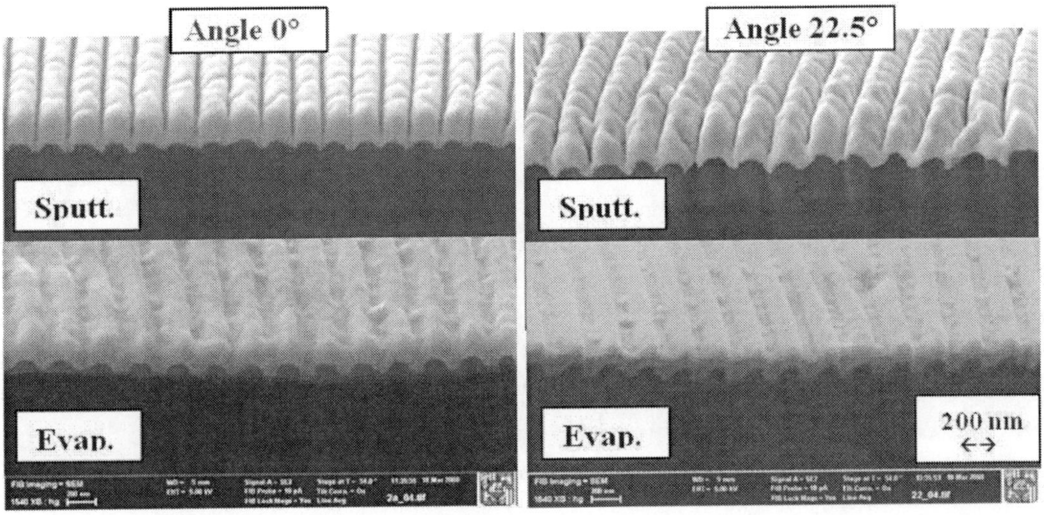

Figure 5.14. FIB-SEM images of the gold coatings on PET samples irradiated by a KrF laser under incidence angles 0° and 22.5° (fluence 6.6 mJ·cm^{-2}). The gold deposition was performed either by sputtering (Sputt.) orevaporation (Evap.) [70].

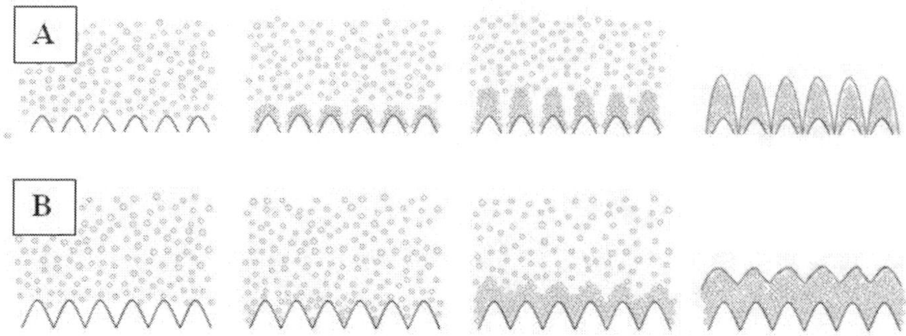

Figure 5.15. (A) Scheme of the growth mechanism of the gold layer on the nano-structured PET surface during the sputtering process and (B) the gold growth mechanism on the nano-structured PET surface during the evaporation process.

The gold is deposited in the valleys of the ripple structure as well. A comparison with Figure 5.14 reveals that there is a difference between AFM and FIB-SEM analyses(i.e., the gaps between the sputtered gold nano-wires are not observed on the AFM images). We attribute this difference to the fact that small objects with high aspect ratio may not be observed in the AFM images due to the above-mentioned convolution of the surface topology with the AFM tip geometry. The nano-wire structure of the sputtered gold layer can also be observed after the gold deposition onto PET samples irradiated by the laser under an incidence angle of 45°. Again, the gold evaporation leads to a continuous coverage copying the nano-structured polymer surface. The reason for the different observed gold morphologies after sputtering (nano-wires) and evaporation (homogeneous gold coverage) is still unclear. The different particle energies in both processes are one possible explanation. For sputtering, the particle energy may be considerably higher because of a sample charging effect, while the evaporated materials should be slower (i.e., colder) and closer to a thermodynamic equilibrium.

The electrical charge of the sputtered particles can have direct influence on layer formation too, while the evaporated material should be mainly neutral. Other reasons may be the different deposition rates, which were a factor of two lower for sputtering than for evaporation, and possible differences of the substrate and gold layer temperature during the deposition by the two different techniques.

The abovedescribed results lead to a conclusion that the growth mechanism of gold nano-structures depends strongly on the deposition technique. Figure 5.15A shows the proposed growth mechanism in the coating of nano-structured PET by the sputtering technique. Separate Au atoms nucleate preferentially on the top of the ripple ridges. On the other hand, during evaporation (see Figure 5.15B) no preferential nucleation is observed and thus the gold layer is copying the nano-structured PET surface, forming homogeneous Au coverage.

5.6. Angle Dependent Irradiation and Sputtered Vs. Evaporated Coatings for the F2 Laser

In this section the attention is focused on the effect of the pattern dimensions (width, height) on the properties of subsequently deposited gold layers of a constant thickness (about

100 nm) on F2 laser-processed PET foil. We have chosen the thickness 100 nm as the height of the ripple pattern modified under the highest angle 45°, approx. (see Figure 5.8).

Figure 5.16 shows the morphology of the gold-coated ripple patterns created on the PET surface by laser irradiation under incidence angles 0°, 22.5°and 45°. On the sample patterned under the angle 0° and coated with gold, the patterns become less pronounced and the gold exhibits a pronounced granular structure. The gold coating results in a slight decrease of the surface roughness [59]. The surface morphology of the samples irradiated under larger angles of laser beam incidence and then coated with gold is similar to that of uncoated samples. The ripple width and height on the gold-coated samples increase with increasing incidence anglesas well (see Figure 5.8). The structure of the gold layer can be seen from FIB-SEM images shown in Figure 5.17. It is known from the previous studies on polymers treated by the KrF laser light (see previous chapter) that certain shapes, chemical compositions of ripple patterns, together with an appropriate deposition technique,are necessary for the creation of gold nano-wires in gold coverage [28]. Since the form of the ripple pattern created by the 157 nm laser light at a 45° incidence angle is similar to that of created by the KrF laser [28], one could expect a similar behavior (growth mechanism and morphology) of the gold layer in both cases. Present results show that this is not the case; on the contrary, no nano-wires are observed and the deposited gold forms continuous coverage, copying the ripple pattern (Figure 5.17C).

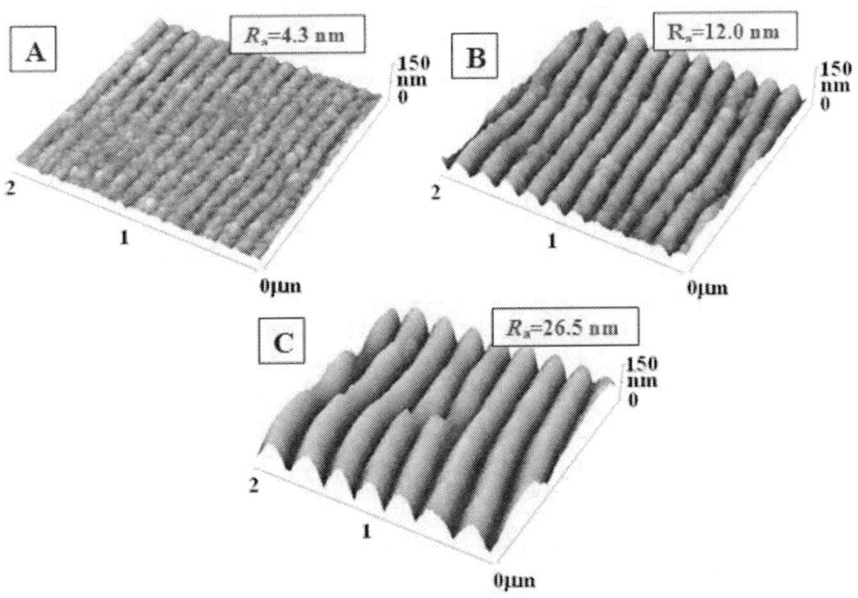

Figure 5.16. AFM images of the PET samples irradiated with 157 nm laser (4.40 mJ·cm^{-2}) under different angles of incidence (A) angle of incidence 0°, (B) 22.5° and (C) 45°) and sputtered with 100 nm of gold. R_a is the average surface roughness in nm [79].

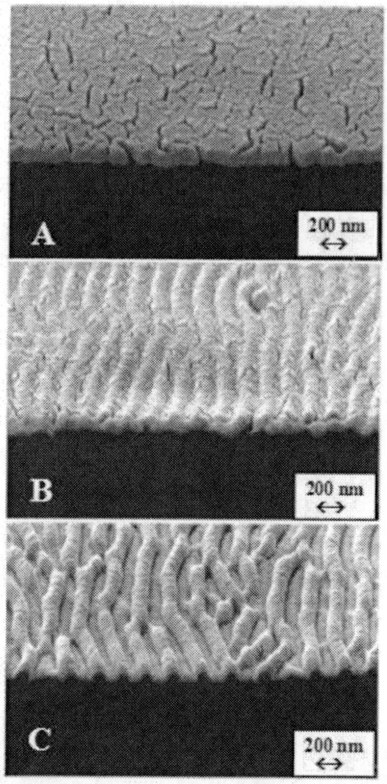

Figure 5.17. FIB-SEM images of F2 laser irradiated (4.40 mJ·cm^{-2}) by different angles of laser incidence (A) angle of incidence 0°, (B) 22.5°and (C) 45°, and gold-sputtered (thickness 100 nm) samples [79].

The FIB-SEM measurements revealed gaps between the ripples, in this case, the presence of which could affect the growth and resulting morphology of the gold layer (see Figure 5.17C). The situation is different on the PET laser treated under the incidence angles of 0° (Figure 5.17A) and 22.5° (Figure 5.17B). For the PET modified under the incidence angle of 22.5° (Figure 5.17B) the AFM measurement shows that the ripple structure remains visible after the gold deposition and the roughness is unchanged.

However, the FIB-SEM image (Figure 5.17B) shows that the deposited gold fills the spaces between the ripples and the initial ripple pattern is partially smeared out.

The change in the surface morphology is much more pronounced in the case of the PET laser treated at the incidence angle of 0°. In this case the gold fills up the spaces between ripples, the surface becomes smoother and the surface roughness decreases significantly after the gold deposition (Figures 5.16A and 5.17A), the ripple pattern being less pronounced (Figure 5.17A). The gold deposition smoothes the initial ripple profile and this observation is in good correspondence with our previous results [59,75]. There is an obvious difference between the surface morphology visible on FIB-SEM and AFM images of the PET modified under the 0° incidence angle (Figures 5.16A and 5.17A).

The difference is due to a much higher resolution of the AFM method, which makes it possible to observe fine periodic structures of the ripple pattern which is invisible by theFIB-SEM method.

6. Deposition of Thin Gold Layers Resulting In Continuous Metal Coverage

The ever-increasing role of polymers in biotechnology and nanotechnology is connected with their advantageous physical, mechanical and electrical properties and easy processing [76]. There are several optical processes and techniques that may be utilized for producing structures on a flat surface or in the polymer bulk: surface relief grating, localized photochemical reactions or surface ablation, interference lithography [77], holography [78], laser scanning [79] and, last but not least, plasma irradiation [36]. These techniques utilize the changes of polymer structuring, chemistry or surface relief under photon absorption. The processes can be performed on different polymers, including those commercially available or specially synthesized in a laboratory [80]. The modification process of a polymer's surface chemistry involves two basic categories: (1) modification by direct a chemical reaction with a certain solution (wet modification); (2) covalent bond creation of appropriate macromolecular chains on the material's surface (grafting). The examples of chemical modification of surfaces are: alcaline or acid etching, oxidation (e.g. via O_3) and other chemical mechanisms [81]. Principles based on physical methods of modifying polymer surfaces can be also categorized into two main groups: (i) modification of the surface layer; (ii) deposition of a thin layer of material different from the bulk one. Because of low reactivity of the pristine polymer surface, the first type of modification requires the creation and usage of high-energetic species, such as radicals, ions or molecules in excited states. The second type of modification (thin layer deposition) includes the plasma techniques, such as sputtering or plasma polymerization, induced evaporation by heating or the electron beam. Metallic plastics are nowadays widely used in products ranging fromcar reflectors, compact discs, electric shielding to foils for food packing and biosensors [82-84]. The polymer metallization is accomplished using sputtering, evaporation and Chemical Vapor Deposition (CVD) techniques [85-87]. The formation of the metallic ad-layers is governed by complex nucleation processes, which are still not completely understood [88]. Several techniques have been developed for measuring of the thickness, surface continuity and morphology of the metal layers deposited on polymers [89]. As was mentioned above, polyolefins represent a wide range of candidates for plasma and laser treatments as well as consecutive studies involving metal layer growth, gold nanoparticles grafting and tissue engineering applications. Selected results from plasma modification; and the mettalization or thermal annealing of polypropylene (PP), polystyrene (PS), polymethylpentene (PMP) and polyl *L*-actic acid (PLLA) will be presented in the next chapters.

6.1. Plasma Treatment of Polypropylene

We exposed PP foils to plasma treatment and their responses for different plasma-powerexposure times on surface properties (roughness, morphology, chemical structure) were determined. The effect of the polymer foil aging on surface wettability and chemistry was described. A wettability study on the plasma-exposed PP samples revealed the rapid decrease of the contact angle which was observed immediately after 20 seconds of modification for both plasma discharge powers. The decline in the contact angle indicates acorresponding increase in the surface wettability. For consecutive applications (e.g., biocompatibility) it is necessary to obtain the surface no longer exhibiting fluctuations in surface wettability (aging study, [90]). A sharp increase of the contact angle of plasma-treated PP is observed in the first few hours. Regardless of the power of the plasma discharge, the full surface relaxation and contact angle restoration are achieved after about 70 hours of aging. Enhanced surface hydrophylicity of the plasma-treated polymers is due to the presence of oxygen-containing groups [21, 40, 91], and the subsequent relaxation is connected with their reorientation and possibly also with the rearrangement of degraded macromolecules on the polymer surface [29, 74]. The most pronounced changes in the surface wettability were determined for the treatment with a plasma power of 8W.

6.2. Surface Chemistry Before Metallization

The RBS method was chosen for the PP surface oxygen determination, since it allows a study of the surface chemistry up to a few hundreds of nanometers. The oxygen concentration depth profile in the polymer surface layer is introduced in Figure 6.1. The oxygen containing groups are present not only on the very surface of the sample, but the presence of oxygen was determined up to the depth of about 60 nm. With an increasing time of the plasma treatment, the surface and total oxygen concentration increases. Different oxidized structures are created on the PP surface, creation of carbonyl, carboxyl and ester groups on the plasma treated polyolefine has been proven earlier in similar experiments [40, 91]. The oxygen concentration on the surface of the plasma-modified PP sample was determined using the XPS method. The oxygen concentration was measured at the different angles of photoelectron beam detection [74]. The studied samples were plasma-modified PP after 1 and 170 hours of aging [90]. The amount of oxygen-containing groups decreased with increasing aging time due to a rearrangement of the degraded macromolecules and oxygen-containing groups produced by the plasma treatment.

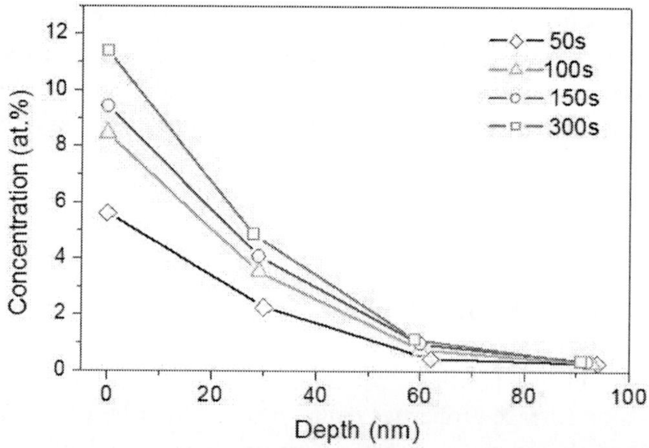

Figure 6.1. Oxygen concentration depth profiles from the PP modified by the plasma discharge at 8.3 W power and for exposure times (50, 100, 150, 300 s). The profiles were obtained fromthe RBS measurement [36].

6.3. Au Nanolayers on Plasma Treated Polymer

Especially in automotive industry, the metallization process and its precise understanding is necessary for successful process applications. Therefore, the influence of plasma irradiation and different surface morphologies on consequently sputtered gold nanolayers was studied. First of all, the dependence of Au layer thickness on sputttering time, both for pristine and modified PP (power 8.3W, exposure time 240 seconds), was determined. These irradiation conditions were chosen because the most significant changes in the contact angle and in the surface roughness weremeasured in our previous experiments [90]. In a range of sputtering time from 10-500 seconds we observed that for both polymers the dependence of thickness on sputtering time is almost linear and that the preceding exposure with energetic particles does not have significant influence on the thickness of consecutively sputtered gold nanolayers. The sheet resistance (R_s) of Au layers deposited on the plasma modified PP as a function of the deposition time is shown in Figure 6.2. The electrically continuous layer (R_s is aprox. constant) is formed earlier for Au sputtered on PP modified at 3.1 W. It can be concluded that within an average layer thickness of 10 nm (60 seconds sputtering time) the Au layer becomes electrically continuous for both discharge powers.

Similar studies were performed on polystyrene (PS) foils. For the sake of clarity, only some selected results will be presented. We observed that with increasing aging time from the plasma treatment of PS foils the contact angle increases, and after about 50 hours of aging the contact angle reaches the saturated value. XPS revealed the presence of oxygen-containing groups on the PS surface modified by the plasma discharge.

The aging of the plasma-modified PS leads to a significant change in the oxygen concentration and its depth profile, the changes being dependent on the plasma discharge power.

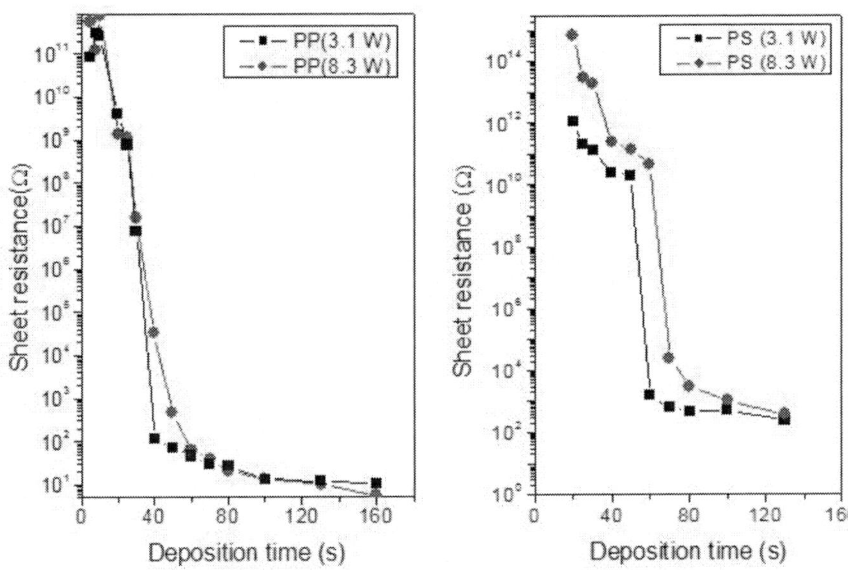

Figure 6.2. Dependence of the sheet resistance on the Au sputtering time for Au layers deposited onto PP [90] and PS plasma modified at the discharge powers of 3.1 and 8.3 W (exposure time 240 seconds).

The effect of plasma exposure of thePS substrate on the continuity of consecutively sputtered gold layers was observed, the decrease of the sheet resistance being more rapid on Au-sputtered samples modified at a lower plasma discharge power (Figure 6.2).

6.4. Nanoindentation of Au Nanolayers

It was shown earlier that the plasma modification of PET and PE does not enhance the adhesion of sputtered Au layers [35,41]. Scratch tests were performed on the samples prepared by gold sputtering on pristine and plasma-treated PS. The sputtering on the plasma-treated PS was performed either immediately after the plasma exposure or after 168 hours aging time (PS surface provably aged). The results are illustrated on SEM images in Figure 6.3.

The scratch paths in the gold layer are qualitatively different on pristine and plasma-treated PS films. The scratch in the gold layer deposited on pristine PS is accompanied by conformal-type buckling cracks. On the gold layer deposited onto the plasma-modified PS immediately after the plasma treatment one can see buckling cracks in addition topeeled off flakes with an orientation opposite to that of the cracks. On the gold layer deposited onto the PS sample subjected to 168 hours of aging, the scratch includes conformal- type buckling cracks of an elliptical shape. The different mechanical properties of the gold layers deposited on different substrates (pristine vs. modified) are obviously due to different structures and physico-chemical properties of the substrate which are affected by the plasma treatment and aging time.

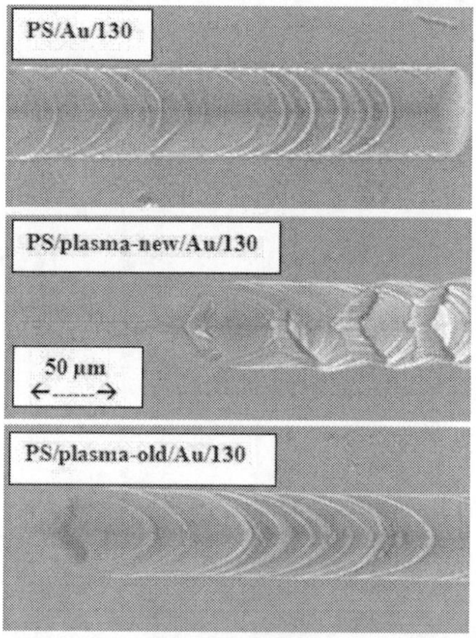

Figure 6.3. SEM images of the scratch paths produced on the PS surfaces by movement of the tip gradually loaded 50 mN. The surface morphology of the scratch of pristine PS sputtered with Au 130 s (PS/Au/130), PS modified (240 s with 8.3 W) and than sputtered with gold 130 s immediately after the exposure (PS/plasma-new/Au/130) and after 168 hours of aging (PS/plasma-old/Au/130).

7. Interaction of Biocompatible Polymers with a Plasma Discharge

Many advances have been made in the development of techniques for the modification of chemical and physical properties of polymer surfaces without affecting polymer bulk properties [35,74]. Polymers arewidely used materials with advantages of easy manufacturing, low density and optimal physical, mechanical, thermal and electrical properties [76]. In general, special surface properties with regard to chemical composition, hydrophilicity, roughness, crystallinity, conductivity, lubricity and cross-linking density are required for successful applications. Surface modification techniques which can transform polymers into highly valuable finished products have become an important part of plastic technologies [92]. Surface modification of polymers with low-pressureplasma has gained great scientific and industrial importance, and it is often used to improve the adhesion of coatings, wettability, printability, biocompatibility and other surface-related properties of polymers [93-95]. Polymers are often modified using a rare gas (He, Ne, Ar) or reactive gas (O, F) plasma. The low temperature and the high chemical activity of plasma species play animportant role in modification process [96].

Cell adhesion is closely connected with surface properties of materials, such as wettability, morphology, surface polarity, surface free energy, electrical properties, roughness, or chemical composition [97, 98]. The polymer surface can be modified in several ways and its properties can be changed with the aim of cell-polymer interaction improvement. Wettability of material is usually evaluated by the contact angle measurement [99]. Hydrophobicity or hydrophilicity of the substrate surface is of great importance in cell-substrate interaction.

The role of polymer-based materials in healing therapy, implantable medical devices, synthetic organs, prosthetic materials, eye surgery medicine, stomatology, tissue engineering and target drug delivering is of great importance [100]. Cell affinity of biomaterials involves two main aspects: cell adhesion and, consequently, cell growth. The anchorage of cells in the first stage of cell-material interaction is the most important for the ability of consequent cell proliferation and differentiation within the contact ofan implant. Cell adhesion is of great importance influencing the cell growth, migration and differentiation of cell functions [97, 98].

As was mentioned above, one of the progressive applications of biocompatible polymers is tissue engineering and thus the controlling and modifying of their biocompatibility. In the field of biological and tissue engineering,polystyrene (PS)is well known for its good biocompatible properties. This polymer is very often used as a comparative material for the determination of changes in a material's biocompatibility (increase of cell growth, adhesion and proliferation).Nonetheless, it is necessary to search for new innovative materials, polymers known as biocompatible, to study their physico-chemical properties in a wide range of parameters, resulting in cell adhesion, growth and proliferation improvement.

As new innovative materials, the polyl *L*-actic acid (PLLA) and polymethylpentene (PMP) were chosen.

The next chapter will describe the influence of plasma irradiation on the surface chemistry, morphology and chemical composition of these biocompatible polymers. The samples were used consequently for the biocompatibility studies after their surface wettability was stable; therefore, it is necessary to operate with an"aged" surface.

7.1. Wettability of Biopolymers

A surface energy determination was performed on the PLLA and PMP samples modified under the same conditions (3 and 8W, 240s) as a function of the aging time and the results are shown in Figure 7.1A and 7.1B for PLLA and PMP respectively. The results of the surface energy measurements are in good correspondence with the results of the contact angle measurements [101]. During the aging process a decrease of the surface energy is observed, the decrease being strongly influenced by the direction and the concentration of polar groups on the polymer surface [35].

The PLLA aging leads to an initial rapid decrease of the surface energy which is followed by the saturation. For the samples treated at 3 W power, the saturated value of the surface energy is close to that of pristine PLLA (doted line), but for the discharge power of 8W the saturated value on approx. 25 $mJ{\cdot}m^{-2}$ is significantly lower than that of pristine PLLA. Also, a similar dependence of the surface energy on the aging time is observed on PMP. After the initial rapid decrease the surface energy reaches saturated values which are close for both discharge powers, the value of surface energy being slightly lower for the discharge power of 8W. The aged PMP exhibits similar surface free energy as the pristine PMP (Figure 7.1), but its contact angle is significantly lower in comparison with the pristine polymer [101].

As a comparative study of plasma treatment influence,the surface morphology of studied polymers was also determined which can, together with the wettability changes,influence the biocompatibility of foils. Atomic force microscopy was used for the surface morphology and roughness determination of pristine and plasma-modified PLLA and PMP.

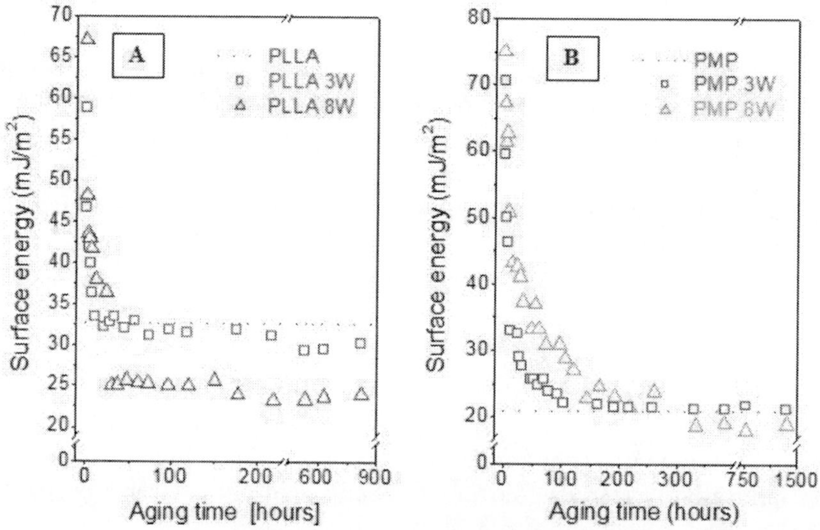

Figure 7.1. Dependence of surface energy of (A) PLLA and (B) PMP surface plasma modified (with 3 and 8W) on aging time. Value for the pristine polymer is shown by a doted line [101].

The modification of PLLA at the discharge powers 3 and 8W and for exposure time 0-240 seconds leads to the increase of the average surface roughness [101]. When compared, pristine PLLA exhibits lower surface roughness in comparison with the modified samples. During plasma treatment, the ablation of the polymer surface takes place. The changes in surface morphology and its differences when compared PMP and PLLA may result from the different ablation rate of crystalline and amorphous phases [17, 40], resulting in the appearance of the crystalline structures on the modified PLLA surface, the surface roughness being significantly increased. For the sake of brevity only the surface morphology of PLLA modified with 8W is shown. The granular structure of PMP is preserved during the plasma modification at both discharge powers 3 and 8W.

7.2. Ablation as a Plasma Treatment Resultand Thermal Annealing of Biocompatible Polymers

As a result of plasma species interacting with a biodegradable polymer surface, the ablation of a polymer material takes place. As a result of ablation, the mass loss and therefore the thickness of the ablated layerof PLLA and PMP was studied by gravimetry. The samples were modified at discharge powers of 3 and 8W. The dependence of the thickness of ablated layer on the exposure time is shown in Figure 7.2A (PLLA) and 7.2B (PMP).

From Figure 7.2 it is evident that the thickness of the ablated layer depends on the polymer type, plasma exposure time and discharge power. With increasing exposure time and the discharge power, the thickness of the ablated layer increases.

It can be concluded that PMP is more resistant in comparison with PLLA. The higher ablation rate of PLLA, mostly of its amorphous phase, also causes the surface morphology changes discussed above. The surface morphology of polymers after the thermal treatment was studied by AFM.

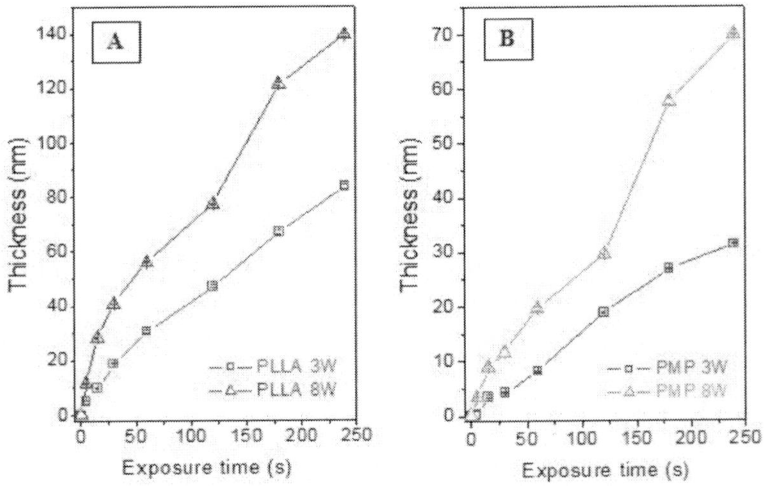

Figure 7.2. Dependence of the thickness of the ablated layer of (A) PLLA and (B) PMP on the Ar plasma exposure time (power 3 and 8 W) [101].

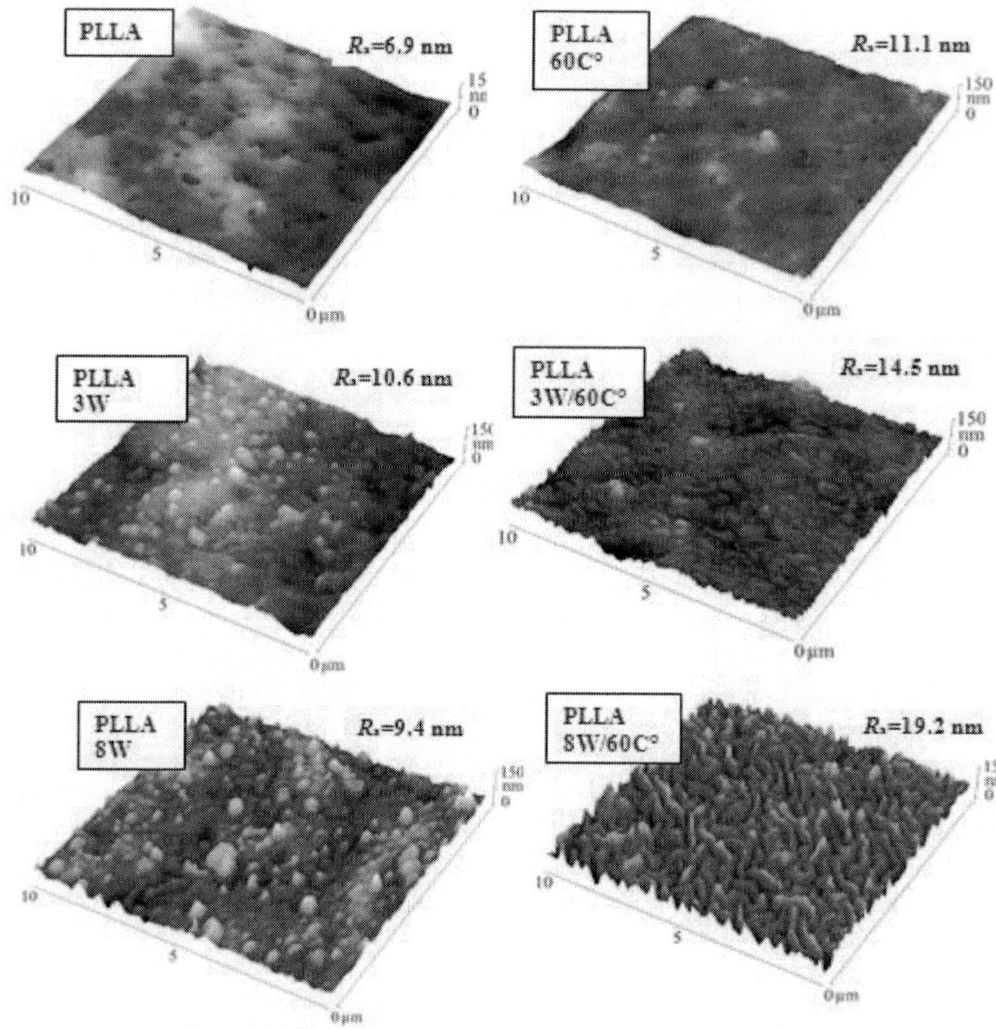

Figure 7.3. AFM images: pristine PLLA (PLLA), pristine PLLA thermally stressed with 60°C (PLLA/60C°), PLLA irradiated with plasma (power 3W, time 240s (3W)), PLLA irradiated with plasma (3W) and thermally treated with 60°C (3W/60 C°), PLLA irradiated with plasma (power 8W, time 240s (8W) and PLLA irradiated with plasma and thermally treated with 60°C (8W/60 C°) [101].

The AFM scans for pristine, plasma modified and thermally treated PLLA are shown in Figure 7.3. It is obvious that the thermal annealing of PLLA causes a significant increase of surface roughness, the increase being most pronounced for plasma-modified samples. This phenomenon can be explained by the fact that the plasma exposure causes ablation of the PLLA surface layer (the higher ablation rate inthe amorphous phase[40]). The remaining crystallites on the PLLA surface melt during the thermal heating and re-crystallize during the cooling process again. It can be concluded that the plasma modification and consequent thermal treatment at the temperature close to the glass transition temperature of PLLA causes significant morphological changes. The changes are caused by the plasma ablation of PLLA, melting and re-crystallization processes. The most pronounced changes in the surface morphology were observed for the PLLA modified at 8W power for 240seconds.

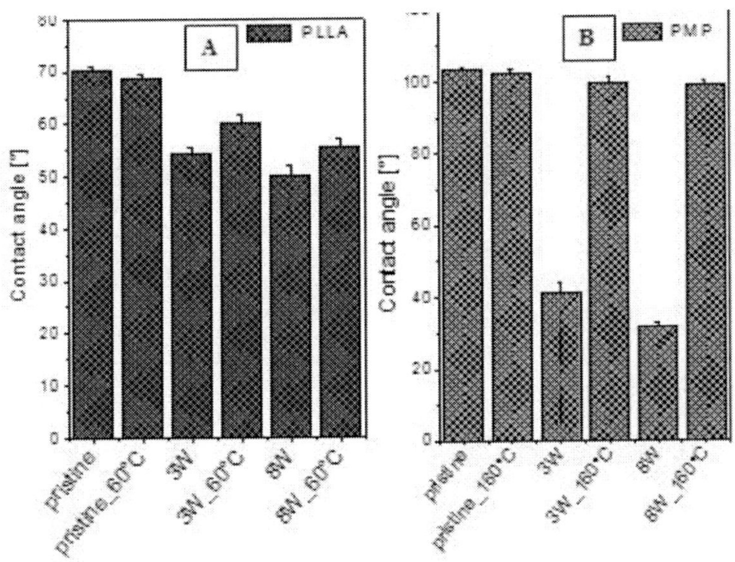

Figure 7.4. Contact angles of (A) PLLA and (B) PMP: pristine (pristine), pristine thermally stressed with 60°C resp.160°C (pristine_60C°), polymers irradiated with plasma (power 3W, time 240s (3W), polymers irradiated with plasma (3W, 240s) and thermally treated with 60°C resp.160° (3W_60C°), polymers irradiated with plasma by 8W and 240s (8W)) and polymers irradiated with plasma (8W, 240s) and thermally treated with 60°C resp. 160°C (8W_60C°) [101].

The influence of plasma treatment and thermal annealing of the surface wettability was determined using goniometry and the results are shown inFigure 7.4. From Figure 7.4 it follows that for both PMP and PLLA the contact angle of pristine samples and pristine samples consequently thermally annealed are almost the same. It is obvious that the thermal treatment has no significant effect on the wettability of pristine samples. The plasma exposure causes a significant increase of surface wettability (decrease of the contact angle). During the aging process the wettability decreases again (see Figure 7.1). When analyzing the Figure 7.4 we can conclude that the thermal treatment has a significant effect on the wettability of PMP (Figure 7.4B) and less for PLLA (Figure 7.4A). The wettability studies revealed that the treatment at an elevated temperature causes an increase of the contact angle, both for PLLA and PMP; the increase is more significant forPMP. The thermal treatment can be successfully used for an acceleration of the aging process.

7.3. Chemical Structure of Modified Polymers

The influence of thermal and plasma treatment on the chemical"structure" of the polymer surface was studied with the XPS method. The PLLA and PMP samples, pristine, plasma modified (8W, 240s), plasma modified and thermally treated (60 resp.160°C) and aged (260 hours) were analyzed. The higher plasma discharge power was chosen with respect to the previous studies in order to achieve more pronounced chemical changes. The higher plasma discharge power was chosen with respect to the previous studies in order to achieve more pronounced chemical changes. The chemical structure of pristine PLLA consists of oxygen in the basic polymer chain. Plasma treatment does not cause an additional oxidation of PLLA.

On the contrary, the thermal treatment leads to an increase in oxygen concentration on the PLLA surface. The chemical structure and composition of pristine, plasma-treated, aged and annealed PMP was studied by the XPS method and typical XPS spectra are shown in Figure 7.5. While practically no oxygen is observed on the surface of pristine PMP, the plasma-treated PMP exhibits significant oxygen contamination with asurface concentration of 28.6%. By subsequent annealing, the oxygen concentration is reduced to 19.3%. Elevated temperature releases conformations of the polymer chain, and the observed concentration decrease may be due to the reorientation of polar, oxygen-containing structures toward the sample bulk [35, 40]. Aging for 260 hours leads to a slight decrease of the oxygen surface concentration to 22.4% in comparison to plasma-treated PMP.

8. Possibility of Surface Patterning of Arbitrary Polymer Films

Polymers are perfectly fit to bridge the gap between so-called top-down and bottom-up nanotechnology due to their size in between the atomic and macroscopic scales. Furthermore, polymers are versatile materials for nanotechnology due to their processability, low cost and tunable properties. Laser micro- and nano-fabrication of various passive and active structures and devices in polymer solid media have attracted a lot of research efforts in recent years.

There are several optical processes and techniques that may be utilized for producing structures on a flat surface or in the polymer bulk: surface relief grating, surface ablation, interference lithography, holography, laser scanning and so on. These techniques utilize the change of polymer structure, chemistry or surface relief under photon absorption.

The process can be performed on different polymers, including those commercially available or specially synthesized in laboratory. The traditional polymer patterning can include the optical lithography and direct laser writing. Tendencies to improve the resolution of these techniques and the quality of prepared structures have ledto the appearance of immersion lithography, deep ultraviolet lithography, and near-field optical lithography [102-104]. In traditional lithography, polymers are used as resists. Exposure of the polymer resist to light radiation through a mask induces reactions in the polymer that make it either more soluble (positive resist) or less soluble (negative resist). Photolithography is an expensive technique since the substrate size is limited by the size of the exposure tools. Furthermore, the maximum resolution that can be obtained with traditional lithography is limited by the wavelength of applied light [105]. Immersion lithography and extreme ultraviolet lithography (EUV) have been proposed for improving the resolution of traditional lithography. In the case of immersion lithography an immersion fluid with a refractive index larger than one is introduced with the aim to construct projection optics with numerical apertures (NAs)>1. It was demonstrated that 45 nm patterns can be achieved by using a chemically amplified resist [102]. In EUV lithography photons with a wavelength of 13.4 nm are used to pattern photoresists [103]. It was demonstrated that 14.5 nm patterns were achieved by using anEUV lithography. Several mechanisms have been proposed to explain the origin of the driving force responsible for surface relief production, due to the inherent shape mechanisms, pressure gradients resulting from isomerization, and the supposition of interactions between the azo-structure dipoles and the electric field of the light interference pattern.

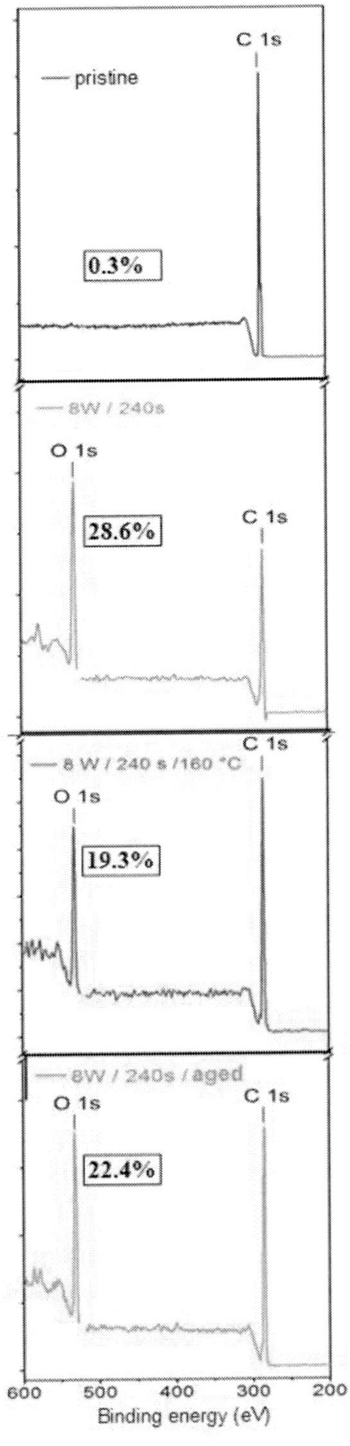

Figure 7.5. The XPS spectra of pristine PMP (pristine), PMP plasmatreated at the power of 8 W for 240 s – measured immediately after the plasma treatment (8W/240s), PMP plasma treated at the power of 8 W for 240 s – measured immediately after the treatment and annealed for 30 min at 160°C (8W/240s/160°C), PMP plasma treatedat the power of 8 W for 240 s and aged for 260 h (8W/240s/aged). The numbers in frames represent the measured oxygen concentrations.

The more plausible option seems to be a mechanism based on mutual dipole interactions and aggregations [106], and a mechanism based on a variation of the material susceptibility at the film surface [107] because these mechanisms include both polarization and intensity dependences observed in experiment.

Typical experimental set-up for SRG formation represents laser beams, arising from one source, splitand intersected at the polymer surface [108]. The grating spacing was usually controlled by changing the angle between the two writing beams and was found to be consistent with the theoretically calculated spacing for the interference pattern. Under the optimum conditions, a surface modulation depth greater than 500 nm could be produced [109].

Another application of NLO polymers can be attributed to the so-called photorefractive effect. The laser beam, when transmitted through materials, can cause a reversible change in the refractive index of the material. This particular light-induced change in the refractive index is now known as the photorefractive effect [110]. In the polymers the photorefractivity is a combined effect of a linear electro-optic response and photoconductivity simultaneously present in a material [111]. It arises from a light-induced generation and migration of charges in the polymer, which give the rise to internal space charge fields, and change the refractive index of material.

Several experimental sets up for interference lithography have been proposed. In a simpler way, multiple beams forming the interference pattern can be obtained by two independent optical elements and steps: splitting the laser output into two beams either by a dielectric beam splitter or by a grating, then superposing them at the exposure area by another specially designed prism [112]. The more widely used technique is the holographic lithography based on the interference of four laser beams. In theoretical reports, four beam holographic lithography configurations were used to create structures which emulate particular cases of triply periodic minimal surfaces, shown to produce large 3D structure [113, 114]. Other proposed configurations include the use of multiple exposures and phase shifts between exposures and a hybrid five-beam configuration with three beams forming a two-dimensional pattern and two other beams forming a one-dimensional pattern in the out-of-plane direction [115].

8.1. Scanning by One Laser Beam

In our experiments the possibility of surface patterning of arbitrary polymer films was proposed and experimentally verified. The main approach of our experiments consists of point-by-point and line-by-line scanning of polymers by focused laser beams with a simultaneous mechanical movement of the sample. Polymer films were shown to respond mechanically on laser scanning. Additional simultaneously continual mechanical movement results in the formation of a regular grating pattern.

Particular experimental arrangement consists of surface writing, which was performed by scanning a0.25 mW laser emitting at a 405-nm wavelength. The laser beam was focused to a spot with a diameter of approx. 0.5 μm. The scanning was performed line by line. Mechanical movement was applied in the direction parallel to scanned lines. The pattern formation was monitored by a confocal microscope,evaluation of the reflected laser light or in a usual optical regime (Figure 8.1).

Polymerdoping procedures were performed with the aim to increase the absorption properties of materials at acertain wavelength. Doping was performed either in the thin surface layers or in the polymer bulk. In the case of surface doping, two techniques wereapplied.

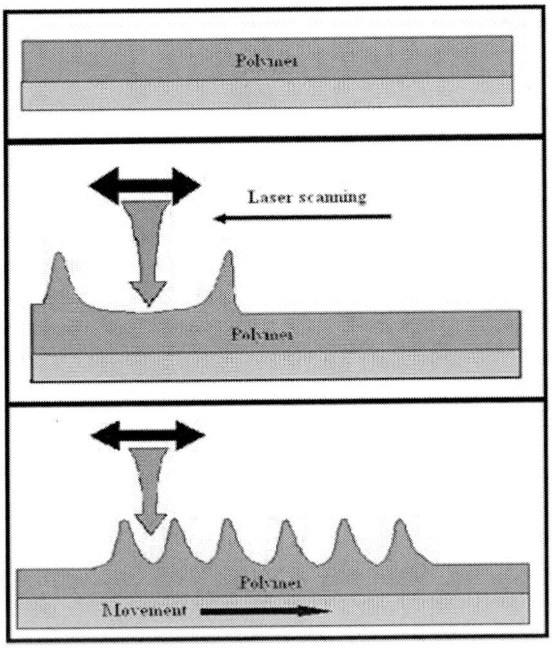

Figure 8.1. Principle of the periodic structures createdby a single laser beam.

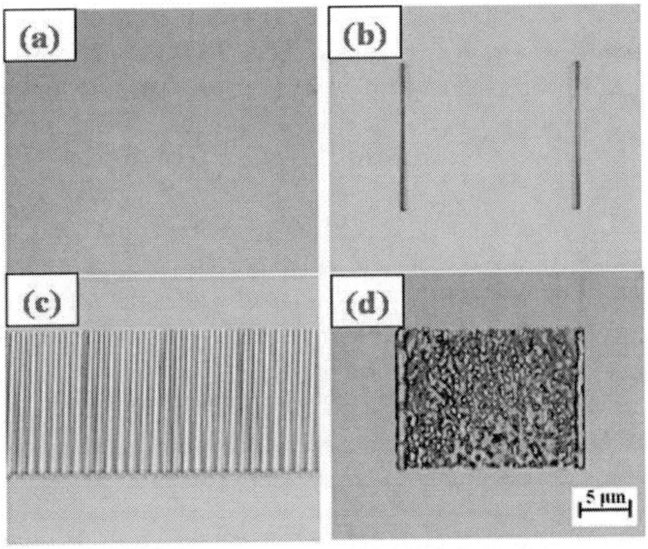

Figure 8.2. Various stages of the surface patterning as a function of the laser intensity and the velocity of the mechanical movement. Laser scanning was applied across the area at the center of the image. Part (A) corresponds to laser intensity of 0.03 mW, (B) 0.12 mW, (C) 0.12 mW and simultaneous mechanical movement, and (D) 0.25 mW [116].

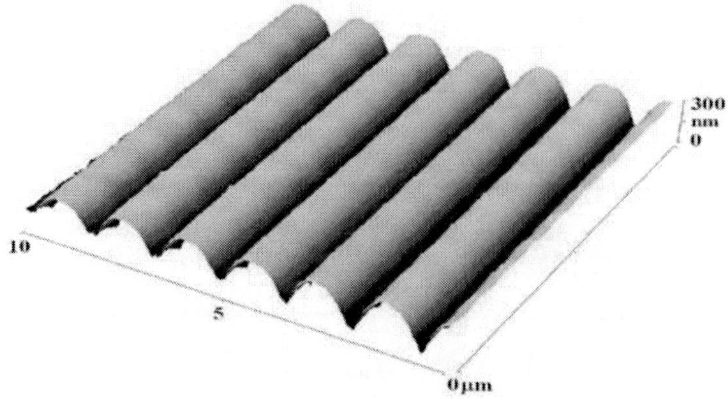

Figure 8.3. AFM image of polymer grating prepared by a single beam laser scanning.

The first techniqueconsists of a deep coating process: the polymer film was immersed in asolution of chromophore in suitable solvent for several seconds, and then dried at ambient conditions. The second technique consists of a vacuum evaporation of chromophore molecules onto the polymer surface. Both techniques exhibit specific advantages and restrictions. Deep coating is simpler, but leads to surface damages. Vacuum evaporation of chromophores does not change the surface morphology, but a chemical destruction of chromophores during the doping process can occur. Bulk doping of polymers was performed in a simple way by mixing polymer and chromophore solutions and preparing, by the spin-coating method, the composite films on a suitable substrate. This method can be applied only for polymers with good solubility. Experiments were performed on different polymers, including poly(methylmethacrylate) (PMMA) with different molecular weights, polystyrene, Su-8 (resist) and so on. The fluid temperature and structure of polymers seems to be important parameters for the patterning process. Another parameter, affecting the patterning is the initial thickness of the polymer films, especially in the case of doped polymers or polymers containing covalently linked chromophore molecules. After preparation and doping, the polymer films wereexposed to periodic dotty irradiation using a focused laser beam (see above) scanned over a defined area. The laser energy was absorbed by chromophore molecules and in this way a local heating of polymer film was induced. The initially featureless surface of polymers undergoes surface distortion at the boundaries of the exposed area. An addition of concurrent mechanical movement leads to the formation of a regular, periodic pattern on the polymer surface. The periodicity of the structure (optical grating) can be changed in the range from tenths to tens of micrometers by selecting a suitable scanning rate of the laser beam, its intensity and sample speed (photo in enclosure) (see Figure 8.2 [116]). The benefit of the proposed approach is the possibility to prepare microstructures with optical quality in an extremely simple way. No additional etching is necessary. It must be noted that the proposed process requires relative lowlaser intensity. Creation of the linear structures is created if the laser intensity exceeds some threshold value. The threshold depends mainly on the dopant concentration, its absorption coefficient and the properties of the polymer matrix. For PMMA films, being doped with 0.1% meso-tetraphenylporphine 0.1 mW laser power, which scans a polymer area 50×50 microns, was found to be enough for structure creation. Irradiation with the smaller intensity does not lead to surface patterning, regardless of the time of irradiation. At the other hand, when the laser intensity is too high, polymer degradation occurs and a structure with poorly expressed periodicity is formed (see Figure 8.2d [116]). Surface morphology of the samples prepared in this way was examined by atomic force microscopy. It confirms that the simple sinusoidal pattern is obtained. The resulting profile consists of a periodic array of lines on which the polymer mass was pulled up and above initial film surface (see Figure 8.3, [116]). Pattern parameters were found to be a function of the experimental conditions and polymers properties. The periodicity primarily depends on the velocity of the sample movement. The grating constant of several tenths to several tens of micrometers was achieved. Simple linear dependence between the velocity of the sample movement and pattern periodicity was observed in the experiments. In this experiment, a pattern periodicity down to about 300 nm was obtained, which was limited by achievable velocity of sample movement. It was also shown that the pattern form depends on the porphyrin concentration and the time of immersion. With asimilar procedure, a single lattice of a pillar array with predetermined geometrical parameters, location, and orientation can be obtained. Additionally, it was found that the pattern amplitude depends mostly on the concentration of dopant and the laser intensity. The grating width of the pattern components from several nanometers to one micrometer was obtained by changing the patterning procedure. Temperature treatment under a polymer glass transition temperature of the prepared structures for several hours does not lead to surface smoothing and leaves the grating structure unchanged. Considering this thermal behavior and the laser intensity threshold, it can be concluded that the polymer molten state is a prerequisite of the pattern formation. A possible explanation of the pattern formation is in the combination of the laser scanning and polymer flow induced by temperature and surface tension gradients (so-called Marangoni effect). First, the polymer surface is fused and then a periodical heating of the polymer surface introduces the surface tension gradient slipped in the direction of laser scanning. The gradient in the surface tension causes the polymer to flow away from a region of low surface tension. It is obvious that this phenomenon must be more pronounced in the direction of "quick" laser scanning, when the temperature gradient has not had enough time to be relaxed. On the boundary of the exposed area the surface tension gradient reaches itsgreatest value.

Figure 8.4. SEM image of a structure prepared by a two-step procedure. After first stage of modification by laser scanning the sample was rotated by 90° and the previously generated structures were over-scanned by an additional modification step.

The material flow stops after an untouched polymer outside of the exposed area is achieved. As a result, the polymer surface is locally pushed up. Additional mechanical movement of the sample results in the drift of the exposed boundary area and in the grating formation.

Not onlylinear patternsareprepared in this way. After the first stage of the modification by laser scanning the sample can be turned by 90° and the previously generated linear structure can be over-scanned in an additional modification step. As a result, the initial linear pattern is changed into asystem of dots, with good quadratic symmetry (see Figure 8.4).

8.2. Metal Coating of a Patterned Polymer Surface

Patterning of the polymer surface seems to be useful for further applications of prepared structures in electronics and photonics. Patterned polymers opens a wide way forthe preaparation of perspective hybrid structures. These structures can be serve as, for example, media with a negative refractive index. Optics has traditionally dealt with dielectric materials. Here, the incident electric-field vector of the light excites microscopic electric dipoles that re-emit electromagnetic waves just like a dipole radio antenna. Other dipoles are excited by this re-emission and so on. This successive excitation and re-emission clearly modifies the phase velocity of light and determines the optical properties of the material. One of the "little" revolutions in optics is the fact that magnetic dipoles can be excited by the magnetic-field component of the light, which can be cast into the effective magnetic permeability μ of the material [117-121].

The interplay of dielectric constantsε and μ has given rise to interesting new aspects of electromagnetism that have been reviewed several times [122-125]. In general, both ε and μ are tensors (i.e., the dipole direction is not necessarily identical to the exciting vector direction [126]). An understanding of this fact leads to a theoretical prediction associated with the appearance ofnegative refractive index materials (NIMs).

Although discovered only 10 years ago, NIMs have been the target of intense study, drawing researchers from physics, engineering, materials science, optics, and chemistry. These artificial "metamaterials" are fascinating because, as it was shown above, they allow for the design of substances with optical properties that simply do not occur in nature [122, 127, 128]. The development of NIMs at microwave frequencies [117-119, 129] has progressed to the point where scientists and engineers are now vigorously pursuing microwave applications. In contrast, research on NIMs that operate at higher frequencies [130-132] is at an early stage, with issues of material fabrication and characterization still being sorted out.

Metamaterials are typically formed by a combination of dielectrics and metals forming binary composites. A comprehensive review about basic structures with a negative refractive index was published in [125]. A frequently used structure that exhibits a negative refractive index is a split-ring resonator (SRR). By using an array of SRRs and thin wires in alternating layers, several groups [117, 119] showed a negative refractive index at giga hertz frequencies. Although the choice of the metal constituting the structure is not critical in the microwave regime, it is crucial in the optical and the visible regimes because the metamaterial losses are dominated by metal losses. An alternative is silver utilization. Silver exhibits the lowest losses at optical frequencies, and indeed, going from gold [130-132] to silver drastically reduced the losses at similar frequencies [133].

Furthermore, the use of silver has enabled the preparation of the first negative-index metamaterials at the red end of the visible spectrum [134] (wavelength 780 nm). Another group has also reported a negative refractive index [135, 136], but this has been questioned recently [137].

One of the possible practical realizations of metamaterials hasbeen proposed as a combination of a patterned polymer surface with a further partial metal coating. In our work we futher developed this technique, with the aim to prepare metamaterials operating at infrared and visible optical frequences. The main idea is consisted in the application associated with the evaporation of metal under the angle to a structured polymer surface. This leads to the appearance of a shade effect during metal deposition. The proposed schema is given on Figure 8.5. The resulted structure must present the system of well separated metallic nanowires (see Figure 8.6). According to theoretical prediction given in the work, this systems can exibit the negative refractive index in certain wavelength range.

The mean wavelength range is closely dependent on the geometrical parameters of the patterned polymers' surface. Generally, the periodicity of surface grating must be at least one order of magnitude smaller than the operating wavelength.

The latter is why operating at visible frequencies of periodicity with approximately a value of 20 nm is needed. This condition islessly strong in the case of near-infrared and infrared frequencies. In these cases the periodicities up to 100-300 nm will be usable. It is evident that the effect of metal deposition on the "shaded" areas must be avoided. For this reason,ametal deposition by temperature evaporation under a high vacuum was chosen in our work. The futher analyses performed by SEM and AFM methods (material hardness regime) confirmed the absence of metal in these areas and a partial metallization of the polymer surface. Additionally, attention must be focused on the amplitude-periodicity ratio. For better utilization ofthe shade effect, this ratio must take the value as great as possible. In our experimets,avalue of ½, was achieved which seems to be sufficient for our goals.

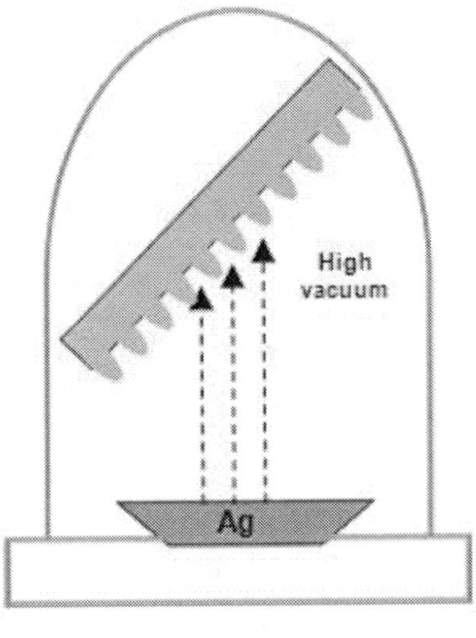

Figure 8.5. Schema of vacuum evaporation of silver on a patterned polymer surface, oriented under the angle to the targetplain.

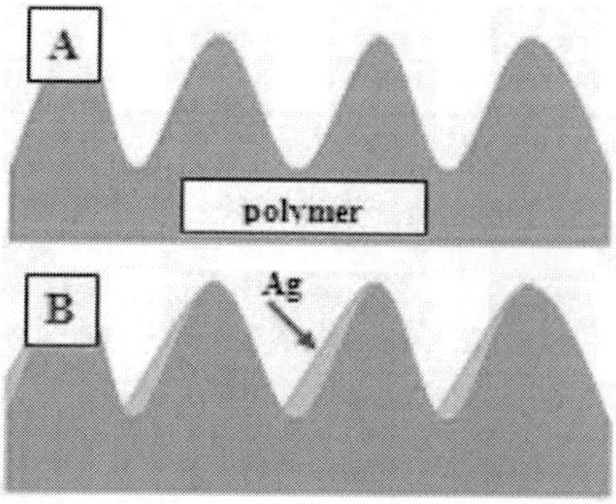

Figure 8.6. Hybrid structure for metamaterials preparation (A) before and (B) after a metal (Ag) coating.

8.3. Application in Optics

Structures prepared on a polymer surface or in bulk were extensively used in photonics, electronics, medicine and in some other fields. Applications in photonics include waveguide couplers, wavelength filters, lasers, sensors and so on. Surface gratings were extensively used as input and output couplers in integrated optics [138]. The recently developed single-step photofabrication process of SRG has offered a quick formation of grating couplers with desired periods. The utilization of this one-step fabrication of surface gratings as slab waveguide couplers was demonstrated [139]. Tunable wavelength filters are useful devices for the purpose of selecting one wavelength carrying desired information among the wavelength division multiplexor signals [140]. In the work [141] a tunable wavelength filter with a Bragg grating on a low-loss polymer waveguidewas demonstrated. The possibility of incorporating different chromophores into polymers seems to be a perspective for the preparation optically active materials [142, 143]. Simultaneously, these chromophores will be responsible for polymer patterning by a interference lithography technique. It means that by practically one step, an optically active material can be prepared and patterned [144, 145]. In literature several sensors prepared by SRG or interference lithography techniques on the planar channel or fiber waveguides have been reported [146, 147]. Polymer fiber gratings are many times more sensitive to these influences than silica fiber gratings. Prepared polymer structures can be used in medicine as biosensors, for example. Additionally, techniques of optically active polymers modification will find their application in an optical data storage area [148,149]. The need for data storage is explosively increasing, triggered by the development of multimedia and electronic communication networks, which is expected to cross 1,020 bits/storage media. Recently, the focus has shifted from two-dimensional storage to three-dimensional ones [150]. For this reason the use of two-photon processes for 3D optical data storage seems to be perspective [151,152]. The basic components of a two-photon memory fabrication are a medium which exhibits a change in its optical properties after two-photon absorption [153]. At higher light intensity the two photons-induced photochemical process shifts the optical properties of the written region. The data storage will be in the form of stacked compact disk arrangement(i.e., in multiple layers with storage capacity exceeding terabits per cubic centimeter).

8.4. Application in Electronics

Patterned polymers can also find their application in electronics. The attention will be focused on the preparation of a hybrid structure, where the patterned polymer surface can serve as substrate for the next deposition of metal or semiconductor layers. Vacuum-deposited metal thin films play an important role in electronics. Especially of interest are metalized polymers, which can be used for different goals, as a diffusion barrier, as dielectric layers in microelectronics and so on. During recent years a new application for metalized polymer films has emerged in the fast developing field of organic electronics. Here, thin metal films are served as electrodes, to apply to the electrical field on the active layer of the electronic device - organic transistors, light emitting diodes and solar cells.

An important concern in metal coatings is associated withthe interactions occurring at the metal-polymer interface. Related to this is aquestion associated with the growth kinetics of the metal film on the polymer surface and how the polymer influences the metal film morphology in the initial growth stage. A number of parameters are involved in the growth process, such as surface morphology and free energies; applied temperature and polymer wettability; and other experimental conditions, determining size, shape, and structure of the metal clusters. In turn, polymer film morphology and chemistry influences the characteristics of the interface and consequently the response of the metal-polymer system. By varying the polymer surface morphology or chemistry it is possible to create different types of metal coatings. An application of the nanopatterned polymer surface can lead to a creation of systems of nanowires or ensembles of periodically ordered metal nanoparticles. Additionally, the nanopatterning of a polymers surface can resolve one of the most important problems in polymers-semiconductor technology – adhesion of a metal or semiconductor onthe polymer surface. For the use of polymeric substrates in plastic electronics, adhesion between the substrate and the metallic wires is crucial for the reliability of these devices. However, promising substrates like poly(ethyleneterephthalate) and poly(ethylene naphthalene) are not very adhesive. A patterned polymer surface exhibits a greaterdegree of surface roughness, which can lead to metal adhesion improvement.

Moreover,withhybrid technology, several classes of polymers with high conductivity can be used in electronics as semiconductor materials. These polymers are known as melanins and include polyacetylene, polypyrrole, polyaniline and their copolymers. Compare with traditional conductive or semiconductor materials conductive polymers are lighter, more flexible, and less expensive. Additionally, polymers can be more easily fabricated and prepared in the form of nanostructures. It opens a wide way for the utilization of these materials in several new applications, unattainable on the base of traditional materials. New applications of conductive polymers in nanotechnology will include smart windows and electronic paper.

9. Termal Preparing Au Nanoparticle Modified Surfaces

Gold in the form of thin films is currentlyused in a vast range of applications, such as micro- and nanoelectromechanical systems (MEMS and NEMS) [154], sensors [155] for bioengineering or devices for surface-enhanced Raman scattering [156]. In recent studies, however, it has been shown that a controlled formation of the surface morphology of Au films may bring new possibilities for advanced applications. Therefore, many approaches to obtain a discontinuous layer of Au structures on the substrate surface were introduced. Our group studied two very different methods to form Au structures on glass and polymer substrates.

9.1. Thermal Treatment of Sputtered Au Structures

Previous studies have shown that the size and shape of gold structures affect their thermodynamic stability. For spherical gold particles, the melting temperature decreases abruptly from 1,100 K to 600 K when the particle diameter changes from 5 to 2 nm [157,158]. Actually, the measured melting point of Au nano-structures, such as Au nano-rods, decreases faster compared to theoretical prediction, especially for larger particles [159]. As supposed, thin metal films also show a decrease of thermal stability, which is very important in electronics applications. Elevated temperature causes a permanent alteration of surface morphology, leading to an increased electric resistance of the component. Although this effect is undesirable in certain systems, the material obtains new properties that differ dramatically from that of a former continuous layer.

Au Structures on a Glass Substrate

Experiments were performed on gold layers sputtered on a 2.4 × 2.4 cm microscopic glass. The deposition conditions were: DC Ar plasma, gas purity 99.995%, discharge power of 7.5W, sputtering time 5–500 s. Under these experimental conditions, the homogeneous distribution of gold over the glass surface is achieved [89]. Post-deposition annealing of Au-covered glass was carried out in air at 300°C (±3°C) for 1 hour using a thermostat Binder oven. The heating rate was $5°C{\cdot}min^{-1}$ and the annealed samples were left to cool in air to room temperature (RT).

Photographs of the sample surface for different sputtering times and for sputtered and annealed samples are shown in Figure 9.1. The as-deposited samples become darker with increasing sputtering time, the darkening being related to an increasing thickness of the gold structures. Also, a gradual change of color of the structures from blue to green occurs. After annealing, all structures exhibit a red color regardless of the sputtering time. These changes in the layer color indicate pronounced alteration of the gold structure caused by the annealing. This could be in accordance with previously presented results: the small gold sample about 10 nm absorbs green light and thus appears red [160].

This effect was also confirmed by UV-Vis spectroscopy. For the sake of clarity, only some of UV-Vis spectra from as-sputtered and annealed samples are shown in Figures 9.2A and 9.2B, respectively. The absorbance of gold structures increases with increasing sputtering time and structure thickness, as expected. From comparison of the spectra obtained from the sputtered and annealed samples respectively, it is seen that the annealed structures have qualitatively different shapes and lower absorbance. Both phenomena point at structural changes due to annealing. The observed shift of the 530 nm absorption peak (corresponding to surface plasmon resonance) with increasing sputtering time towards longer wavelengths is probably related to the interconnection and mutual interaction of gold nano-particles in the structure. From present UV-Vis spectra it can be seen that the SPR peak disappears for as-sputtered samples prepared with a deposition time of 30 seconds, whereas for annealed samples the SPR peak is still strong for sputtering timesas long as 150 seconds. Aualitative difference between absorbance spectra of the as-sputtered structures and annealed ones may indicate a transition from structures, comprising discrete gold islands to continuous gold coverage. Values of the optical band gap (E_g^{opt}) were calculated from the UV-Vis spectra as a function of the sputtering time for sputtered and annealed samples using the well-known Tauc's model [161]. This dependence is shown in Figure 9.3A.

Also, the gold structure thickness vs. sputtering time measured by the AFM method on the scratch step is presented. Structures prepared with a time of deposition below 150 seconds exhibit non-zero E_g^{opt}. A rather dramatic change in the E_g^{opt} is observed after annealing, where the values of the E_g^{opt} are much higher in comparison to those of the sputtered sample. For samples sputtered for times around 300 seconds a non-zero E_g^{opt} is observed. Regarding the behavior of ultra-thin metal structures (<10 nm) the surface-size and quantum-size effects must be considered [160,162,163]. This quantum-size effect in small structures leads to a semi-conducting character, which is accompanied by non-zero E_g (band gap) or E_g^{opt}. The dependence of the electrical sheet resistance (R_s) of gold structures on the sputtering time before and after annealing is shown in Figure 9.3B.

For the sputtered samples, sheet resistance decreases rapidly for the sputtering times above 50 seconds, as a continuous gold coverage is formed. From the measured R_s and effective layer thickness, the layer resistivity R (Ω·cm) was calculated, which appears to be about one order of magnitude higher than that reported for metallic bulk gold ($R_{Au}= 2.5 \times 10^{-6}$Ω·cm) [164]. Resistivity of thin gold structures increases due to the size effect according to the Matthiessen rule [165]. After annealing, a dramatic change in the R_s is observed. The annealed structures are electrically discontinuous up to the sputtering time of 250 seconds when the continuous coverage is created and a percolation limit is overcome. However, even for longer sputtering times up to 500 seconds the sheet resistance changes slowly and it does not achieve a saturation which is observed on the sputtered layers.

This phenomenon results from structural changes of the Au layer during post-deposition thermal treatment. A rapid decrease of sheet resistance does not occur, as observed for as-sputtered layers during deposition times between 25 to 50 seconds. Transition from an electricallydiscontinuous to an electrically continuous layer in the case of annealed samples is more gradual and occurs between the deposition times of 200-500 seconds. AFM images, illustrating surface morphology and roughness (R_a) of gold-coated glass before and after annealing, are shown in Figure 9.4.

For the sake of comparison only, an identical vertical scale was used. It is clear that surface morphology of the as-sputtered structures does not depend significantly on the sputtering times. A monotonous decrease of surface roughness with deposition time is determined by the stage of the layer growth. Duringthe initiatory stage of metal growth, a layer is formed over isolated islands. After that, during ongoing deposition, interconnections between clusters are formed and deposited layers become homogeneous and uniform.

Decrease of surface roughness is direct evidence of the formation of a thicker layer during the sputtering process on a flat substrate.

After annealing, however, the surface morphology changes dramatically. Similar changes in the morphology of the thin gold structures have also been observed on the samples annealed at 200°C for 20 hours [166] and at 450°C for 2 hours [157]. Annealing leads to the formation of spherolytic and hummock-like structures from the gold layers.

The formation may be connected with an enhanced diffusion of gold particles at elevated temperatures and their aggregation into larger structures. It is well known that the melting point of gold nano-particles decreases rapidly with decreasing particle size [157, 158, 160]. Migration of the gold nano-particles and formation of larger structures may be caused by a lower thermodynamic stability of the gold nano-particles and the low gold wettability of glass.

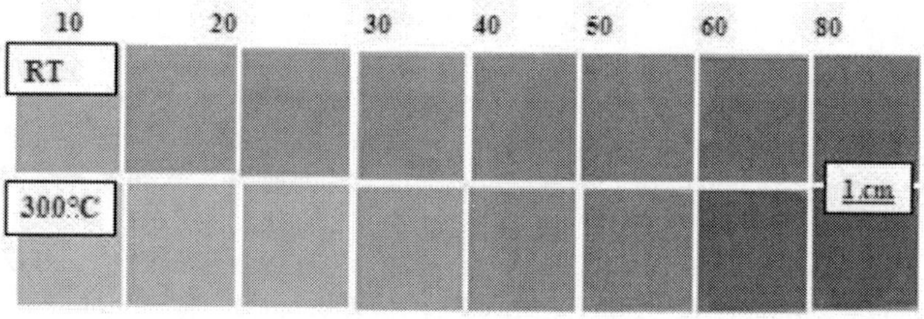

Figure 9.1. Images (photograph cuts) of the glass samples with gold structures sputtered for increasing times. The as-sputtered (RT) and annealed samples (300°C) are shown for comparison [167].

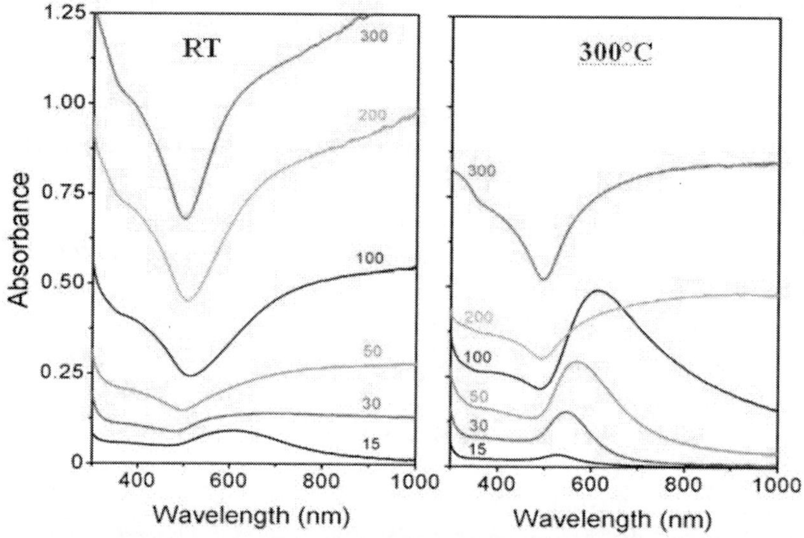

Figure 9.2. UV-Vis spectra of gold structures sputtered on glass before (RT) and after annealing (300°C). The numbers are sputtering times in s [167].

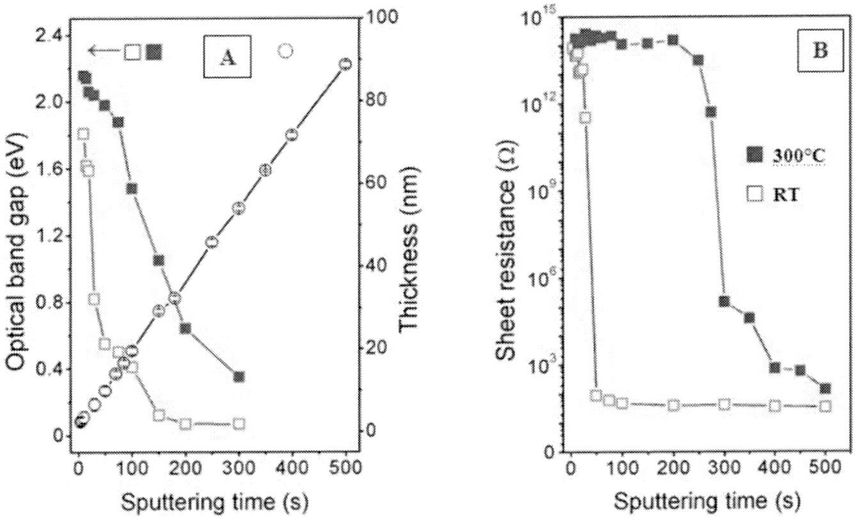

Figure 9.3. Dependences on the sputtering time of (A) the optical band gap of gold structures before and after annealing, (B) the electrical sheet resistance (*Rs)* of gold structures before (RT) and after annealing (300°C) [167].

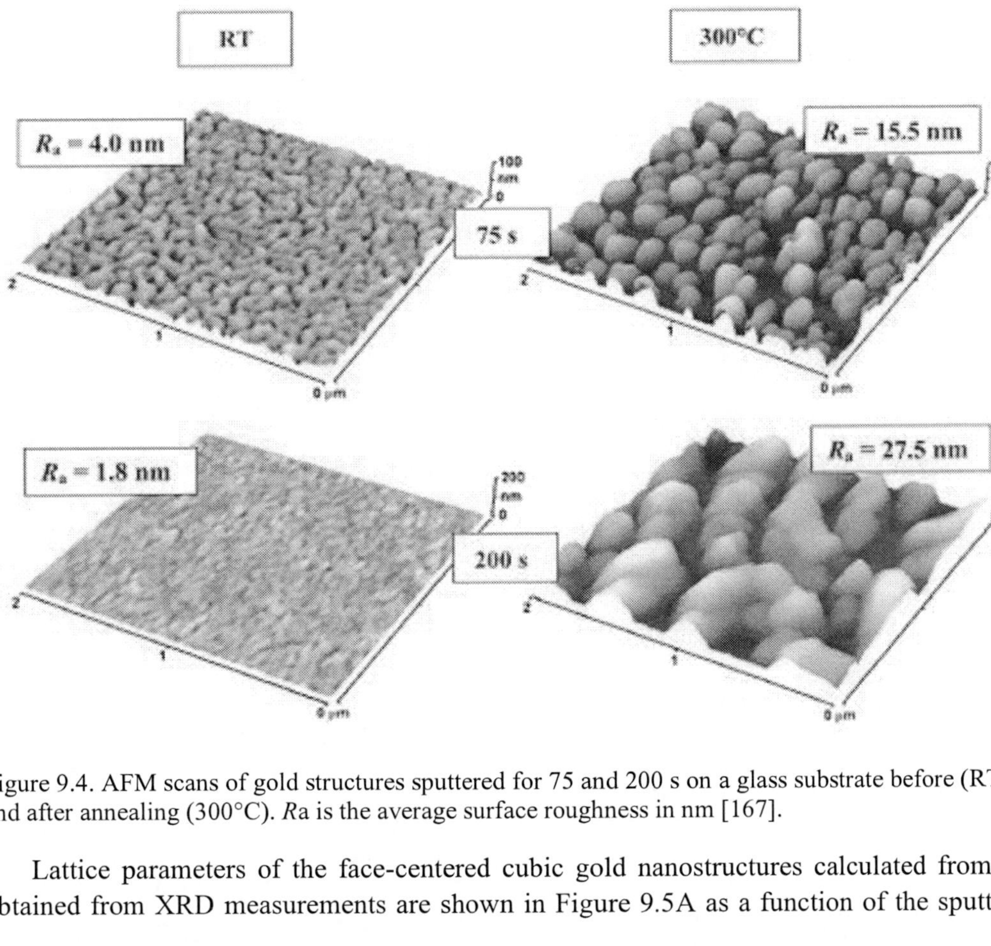

Figure 9.4. AFM scans of gold structures sputtered for 75 and 200 s on a glass substrate before (RT) and after annealing (300°C). *R*a is the average surface roughness in nm [167].

Lattice parameters of the face-centered cubic gold nanostructures calculated from data obtained from XRD measurements are shown in Figure 9.5A as a function of the sputtering time (i.e., effective layer thickness). Lattice parameters were calculated using the Rietveld procedure (full pattern fitting). For this purpose the five strongest diffraction lines were taken into account. A dramatic difference is found in the dependences of the lattice parameter on the sputtering time between as-sputtered and annealed samples. For the as-sputtered samples the lattice parameter varies rapidly with the increasing sputtering time(i.e., with the increasing mean size of the gold crystallites) [163, 167, 168].

The maximum lattice parameter is observed after 100 seconds of sputtering(i.e., for a layer thickness of about 18 nm). For both the thinner and thicker layers the lattice parameter declines significantly. The same trend in the lattice parameter vs. sputtering time dependence was also reported for silver structures, for which the lattice parameter increases slightly up to the structure size of 12 nm and then decreases [169]. In contrast to the as-sputtered gold structures, the lattice parameters of the annealed ones arenearly independent of the sputtering time and the size of the structures.

The dependence of crystallite size on sputtering time (before and after annealing) calculated from data obtained from XRD measurements is shown in Figure 9.5B.

The dependences are quite different for the as-sputtered and annealed samples. Whereasin the as-sputtered samples the crystallite size is an increasing function of the sputtering time, the crystallite size first increases rapidly up to the sputtering time of 250 seconds, achieves a maximum and then decreasesin the annealed ones. The method of crystallite size evaluation by the XRD technique is based on the determination of coherently diffracting domains with their mean dimensions in a direction perpendicular to the film surface. This is the reason why the size of these crystallites calculated in this way significantly exceeds layer thickness in some cases.

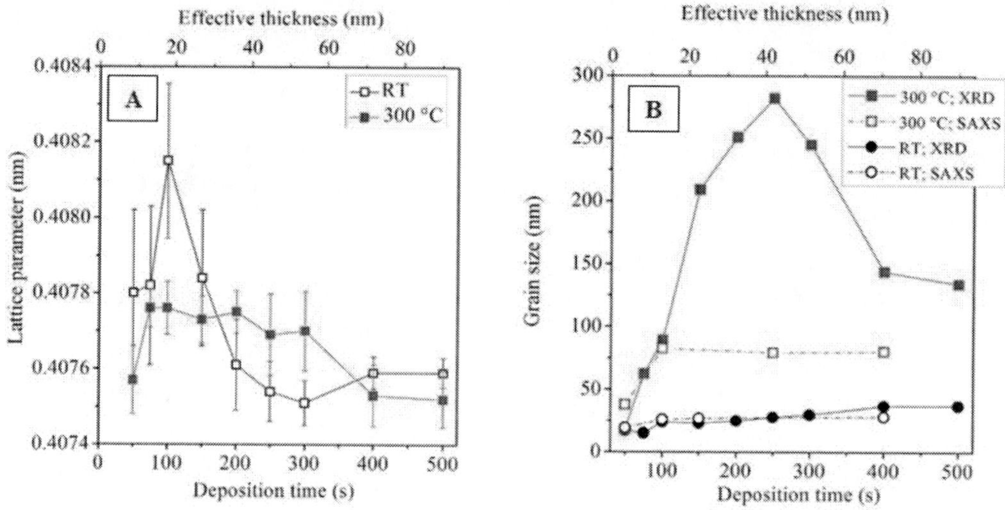

Figure 9.5. (A) Dependence of the gold lattice parameter on the sputtering time (i.e., effective thickness) measured before and after annealing at 300°C, (B) Dependence of the size of the gold crystallites on the sputtering time (i.e., layer effective thickness) measured before and after annealing at 300°C using SAXSess and XRD methods [167].

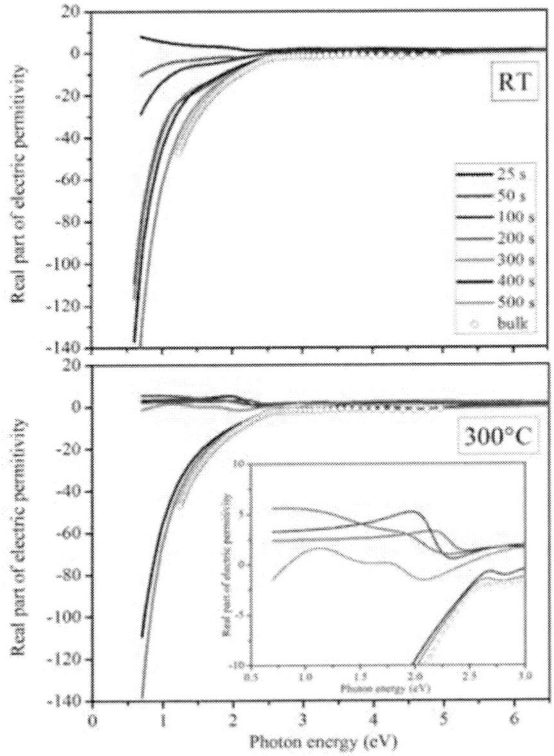

Figure 9.6. Real part of the electric permittivity spectra of as-sputtered (RT) and annealed (300°C) gold structures obtained by spectroscopic ellipsometry for different sputtering times [167].

Figure 9.6 presents the real part of electric permittivity in the studied spectral range. Spectroscopic features in the Drude (IR) region clearly show the tendency of Au films to lose their metallic behavior with decreasing thickness due to gold coalescence, leading to a layer discontinuity [167]. Film discontinuity of the as-deposited thin layers is a natural consequence of the mechanism of the layer growth. The percolation threshold is reached at the layer thickness of about 7 nm [170], corresponding to the deposition time of about 25 seconds in the present case. Figure 9.6 also shows that the strong change in the surface morphology, induced by the annealing, shifts the metal-insulator transition toward greater layer thicknesses (i.e., deposition times). The thickness variation of the IR end of the real part of the electric permittivity spectra of annealed gold layers (positive value reaching the maximum and then passing through zero to negative values with increasing deposition times) is consistent with previous studies of metallic films around the percolation threshold [170].

9.2. Au Structures on a PTFE Substrate

In this chapter we focus on investigation of gold nanostructures and nanolayers sputtered on a polytetrafluoroethylene (PTFE) surface and their changes induced by post-deposition annealing at 100-300°C. Special emphasis is devoted especially to the transition from discontinuous to continuous gold coverage regarding Au layer thickness and the degree of post-deposition annealing. The transition zone was found to be between Au thicknesses 10-15 nm and this threshold remains practically unchanged after the annealing at the temperatures below 200°C. The annealing at 300°C, however, leads to a significant rearrangement of the gold layer and the transition threshold increases to about 70 nm. Significant carbon contamination and the presence of oxidized structures on gold-coated samples are observed in XPS spectra. Gold coating leads to a decrease in the sample surface roughness. Pristine PTFE and gold-coated PTFE annealing at 300°C results in a significant increase of the sample surface roughness. In addition to electrical measurements, another quantity characterizing the structure of the sputtered gold layers, however closely related to the surface conductivity state, is zeta potential determined from electrokinetic analyses.

The dependence of the electrical sheet resistance (R_s) of the gold layer on its thickness before and after annealing (at 100, 200 and 300°C) is shown in Figure 9.7. For the as-sputtered samples the sheet resistance decreases rapidly in the narrow thickness from 10-15 nm when an continuous,electrical gold coverage is formed. The resulting sheet resistance is saturated at a level of ca 200 Ω. From the measured R_s and effective layer thickness, layer resistivity R (Ω·cm) was calculated, which appears to be orders of magnitude higher than that reported for metallic bulk gold ($R_{Au}^{bulk} = 2.5\times10^{-6}$Ω·cm [164], e.g., for 100 nm thick Au layer $R_{Au}^{100nm} = 1\times10^{-3}$Ω·cm). The higher resistivity of thin gold structures is due to the size effect in accord with the Matthiessen rule [165]. Annealing at temperatures below 200°C causes only a mild shift in the resistance curve towards thicker layers. The transition from an electrically discontinuous to electrically continuous layer in the case of low-temperature annealed samples is more gradual and occurs between the effective layer thicknesses from 10 to 20 nm regarding the annealing temperature.

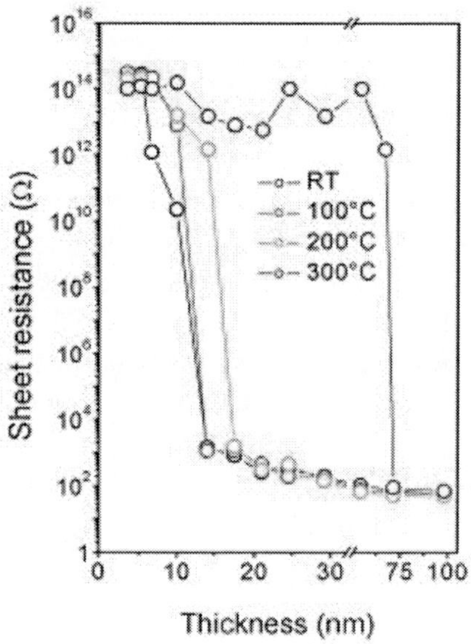

Figure 9.7. Dependence of the sheet resistance (R_s) on Au layer thickness for as-sputtered samples (RT) and the samples annealed at 100, 200, 300°C [171].

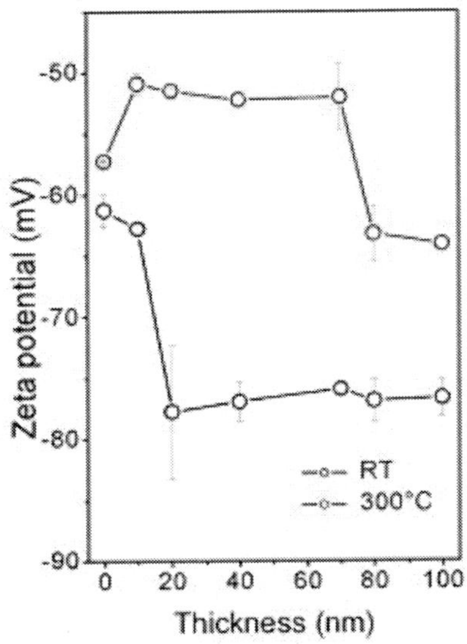

Figure 9.8. Dependence of zeta potential on the Au layer thickness for as-sputtered samples (RT) and the samples annealed at 300°C [171].

After annealing at 300°C, a dramatic change in the resistance curve is observed. The annealed layers are electrically discontinuous up to the Au effective thickness of 70 nm, above which the continuous coverage is created and a percolation limit is overcome. However, for longer sputtering times up to 550 seconds, the sheet resistance changes slowly and it achieves a saturation which is observed on the as-sputtered layers and layers annealed at low temperatures.

As mentioned above, surface electrokinetic potential (zeta potential) is closely related to the surface conductivity state and one could therefore expect a certain degree of correlation comparing sheet resistance and zeta potential values.

The dependence of zeta potential on the gold structure thickness for as-sputtered samples (RT) and annealed samples at 300°C is shown in Figure 9.8. For as-sputtered samples and very thin gold layers, the zeta potential is close to that of pristine PTFE due to the discontinuous gold coverage since the PTFE surface plays a dominant role in thezeta potential value. Then, for thicker layers, where the gold coverage prevails over the original substrate surface, the zeta potential decreases rapidly;for the thicknesses above 20 nm it remains nearly unchanged, indicating a total coverage of theoriginal substrate by gold. For annealed samples the dependence on the layer thickness is quite different. It is seen that the annealing leads to a significant increase of the zeta potential for very thin layers. This increase may be due to a thermal degradation of the PTFE accompanied by production of excessive polar groups on the polymer surface, which plays animportant role when the gold coverage is discontinuous. Moreover, the surface roughness increases at this moment as well [171]. Then, for medium thicknesses, ranging from 20 to 70 nm, the zeta potential remains unchanged and finally it decreases again for higher thicknesses due to the formation of continuous gold coverage. It appears that the results of the electrokinetic analysis and measurement of the sheet resistance (Figures 9.7 and 9.8) are highly correlated.

The rapid decrease in the sheet resistance occurs at the same layer thickness as the decrease in zeta potential. Both correlated changes are connected with the creation of continuous, conductive gold coverage.

Another interesting fact is that even for the layers with thicknesses above 80 nm the values of the zeta potential measured on as sputtered and annealed samples differ significantly. This can be due to a higher fluorine concentration in the annealed samples and the fact that the C-F bond is more polar and exhibits higher wettability [171]. It should be also noted that the value of the zeta potential may be affected by the surface roughness too. In general, it follows that the thicker the gold coverage the lower the zeta potential is, reflecting the electrokinetic potential of pure metal itself.

10. Au Nanoparticles Grafted on a Plasma-Treated Surface

Nanoparticles are at the center of research interests because of their unique physico-chemical and optoelectric properties. Gold nanoparticles, due to their unique electrical, optical and photophysical properties, are expected to play a role in a variety of potential applications as electronic and sensor materials [172-174].

Because of its well known stability, non-reactivity and bioinertness,gold is used with an advantage in medical diagnostics [175], Raman spectroscopy [176], biological imaging [177] and biosensors [178]. The deposition of Au nanoparticles on a solid surface is a critical step ina process to form a functional material that could be used in certain analytical methods. Metal colloid solutions are used in some methods mostly due to their affordable and simple preparation, but a big drawback is their instability leading to possible aggregation, which causes measurement irreproducibility [179, 180]. Attaching AuNPs to solid surfaces would improve stability of the SERS substrate, retaining its great sensitivity. Binding AuNPs to a solid surface enables miniaturization, multiplexing and compatibility with in-flow assays [181]. The AuNP interlayer is also useful in various applications to improve adhesion of subsequently deposited metal films as adhesion is of great importance in determining the lifetime of a coated component [182, 183].

10.1. Au Nanoparticles Grafted on a Polymer Substrate

The structures, combining gold nanoparticles with polymers, are promising materials for biomedical applications [184, 185]. Since gold is the most inert metal, the adhesion between gold nanoparticles and thepolymer surface is expected to be very poor. However, gold can easily react with thiol (-SH) groups, which can be fixed on the surface modified beforehand in a proper manner (e.g., by plasma discharge, irradiation with ultraviolet light from an excimer lamp or by ion irradiation) [186, 187].

Experiments were performed on surfaces of four thermoplastic polymers (polyethylene, polyethyleneterephthalate, polytetrafluoroethylene and polystyrene), which were modified by plasma discharge and subsequently grafted with dithiol to introduce thiol (-SH) groups. Dithiol is expected to be fixed, via one of its –SH groups, to radicals created by the preceding plasma treatment.

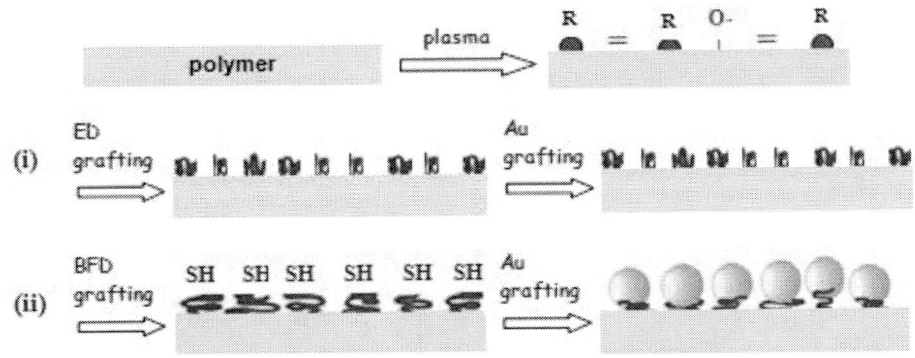

Figure 10.1. Scheme of the plasma treatment of PET, grafting of modified PET by (i) ethanedithiol (ED) and (ii) biphenyldithiol (BFD) + then grafted by gold nanoparticles [26].

The other –SH group is free to interact with gold nanoparticles (see schematic in Figure 10.1). Surface chemistry and radical concentration of four thermoplasts (pristine and modified by the plasma discharge) were studied by X–ray Photoelectron Spectroscopy (XPS). Determination of the surface morphology and the roughness of the plasma-treated polymers, and vizualization of gold nanoparticles was accomplished by Atomic Force Microscopy (AFM).

Chemical Structure of Plasma-Modified and Grafted Surfaces

Plasma treatment leads to the cleavage of chemical bonds (C-H, C-C and C-O) [26]. The bond breaking leads to a fragmentation of the polymer chain, to an ablation of the polymer surface layer and to the creation of free radicals, conjugated double bonds and excessive oxygen-containing groups [26]. An activated polymer surface can be grafted with thiol groups. The binding of the molecules is mediated by free radicals, present on the surface of the plasma-treated polymers.

The binding on new double bonds has not been proven [188]. Cleavage of the molecular chains facilitates solubility of the initially insoluble polymer in common solvents(e.g., water) [189].The progress of the present experiment and the main results of this work are schematically summarized in Figure 10.1 and Figure 10.2. It was shown that the plasma treatment results in thedegradation of thepolymer chain and thecreation of free radicals, double bonds and excessive oxygen groups on the PET surface. The "flexible" molecule of 1,2-ethanedithiol is bonded to the surface radicals, probably by both -SH groups in contrast to the "rigid" molecule of 4,4'-biphenyldithiol, where one of -SH group remains "free" for the subsequent chemical reaction with the gold nanoparticle. The gold nanoparticles are grafted on the PET surface only in the case the pretreatment with 4,4'-biphenyldithiol. [190,191]

FTIR spectroscopy was used for the characterization of thechemical composition of modified PET samples. In Figure 10.3 the differential FTIR spectra of the PET samples (i) plasma-treated and grafted in biphenyldithiol (BFD) and (ii) treated and grafted with BFD and then with Au nanoparticles are shown. The band at 790 cm^{-1} corresponds to the absorption of the S-C group, and the band at 761 cm^{-1} is assigned to the S-Au group. After the grafting of plasma-treated PET with ethanedithiol and Au nanoparticles,apeak at 761 cm^{-1} (S-Au) in FTIR spectra was not detected. This finding supports the conclusion that no Au nanoparticles are bonded to the PET treated in ethanedithiol.

From the results presented in Figure 10.3 it is apparent that Au nanoparticles are grafted only on the PET surface previously activated by biphenyldithiol. This can be explained by the concept that the "flexible" molecule of ethanedithiol is bonded to an activated polymer surface by both -SH groups, while the more "rigid" molecule of biphenyldithiol is grafted only via one of -SH groups and the second one is free for a chemical reaction with theAu nanoparticle.

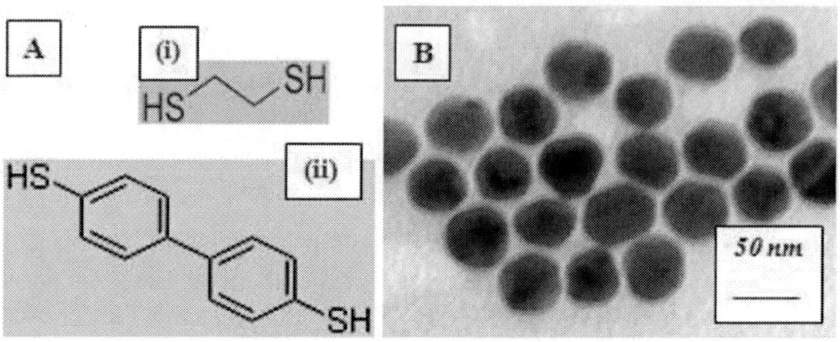

Figure 10.2. (A) Molecular structure of (i) ethanedithiol (ED) and (ii) biphenyldithiol (BFD), (B) TEM images of Au nanoparticles from a Transmission Electron Microscope. For structural characterization we used TEM (JEOL JEM-1010, USA) operated at 80 kV [26].

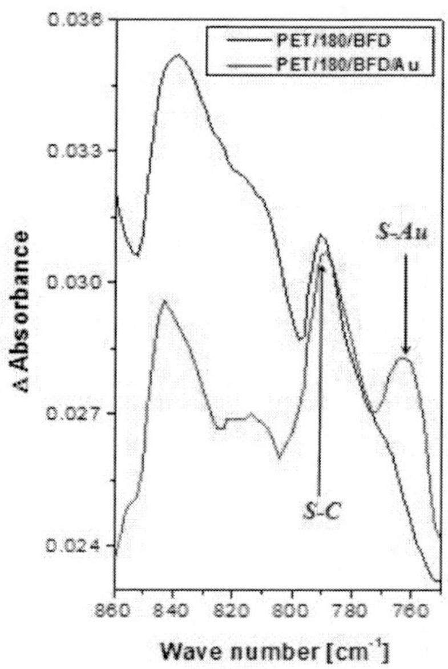

Figure 10.3. Differential FTIR spectra of (i) plasma treated and with biphenyldithiol grafted (PET/plasma/BFD) and (ii) plasma treated; grafted with BFD + then with Au nanoparticles (PET/plasma/BFD/Au) [26].

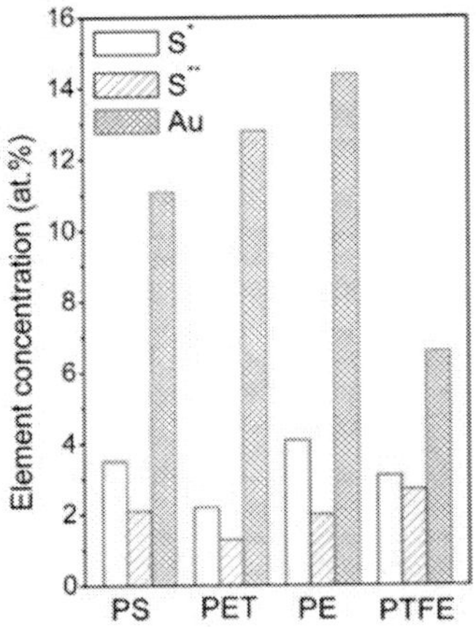

Figure 10.4. Concentration of sulphur and gold in the polymer surface layer, after the dithiol grafting (S^*) and after the dithiol and gold nanoparticles grafting (S^{**}) [26].

The composition of the first 8-10 surface atomic layers of pristine and plasma-treated polymers, dithiol grafted and coated with gold nanoparticles polymers was investigated using the XPS method. Concentrations of sulphur and gold in the surface layer of polymers after the dithiol grafting (S^*) and the dithiol and gold nanoparticles grafting (S^{**}) are shown in Figure 10.4. Binding the gold nanoparticles leads to a decrease of sulphur concentration on all polymers examined. The effect is observable due to the fact that the XPS method is sensitive only to 8-10 surface atomic layers [192]. A fraction of the sulphur atoms on the samples grafted with gold nanoparticles is covered by gold and can not be seen in XPS spectra. With the only exception of PTFE, sulphur concentration halves after coating with the gold nanoparticles, the gold concentration remains relatively high.

This is understandable since the gold is not bound as single atoms but in the form of nanoparticles 50 nm in diameter.

Surface Homogenity of Au Nanoparticles on Polymers

The surface morphology of polymers, plasma treated and then grafted with dithiol and gold nanoparticles, was studied by the AFM method. The results are illustrated in Figure 10.5 where the images from the initially non-polar PE and polar PET are compared. For the sake of completeness, images taken in both tapping and phase modes are shown. The distribution of the gold nanoparticles over the polymer surface is probably affected by the polymer surface morphology which is changed by the plasma treatment. After the plasma treatment the surface roughness of PE ($R_{a\ PE}$ = 8.4 nm) is significantly higher than that of PET ($R_{a\ PET}$ = 1.9 nm) [26]. The gold particles are spread homogenously on the polymer surface, but particle aggregation is observedon the PET. Grafted gold nanoparticles do not create continuous coverage of the polymer surface and it is therefore not surprising that the electrical resistance of the gold coated samples is similar to that of pristine polymers [188]. Figure 10.5 (AFM) also documents that more gold nanoparticles are grafted onto the surface of plasma-treated PE than onto PET. Gold nanoparticles homogenously distributed over the polymer surface may have a positive effect on interactions with living cells (tissue engineering) [189] or they may facilitate theadhesion of additional gold structures to polymeric substrates (electronics) [188].

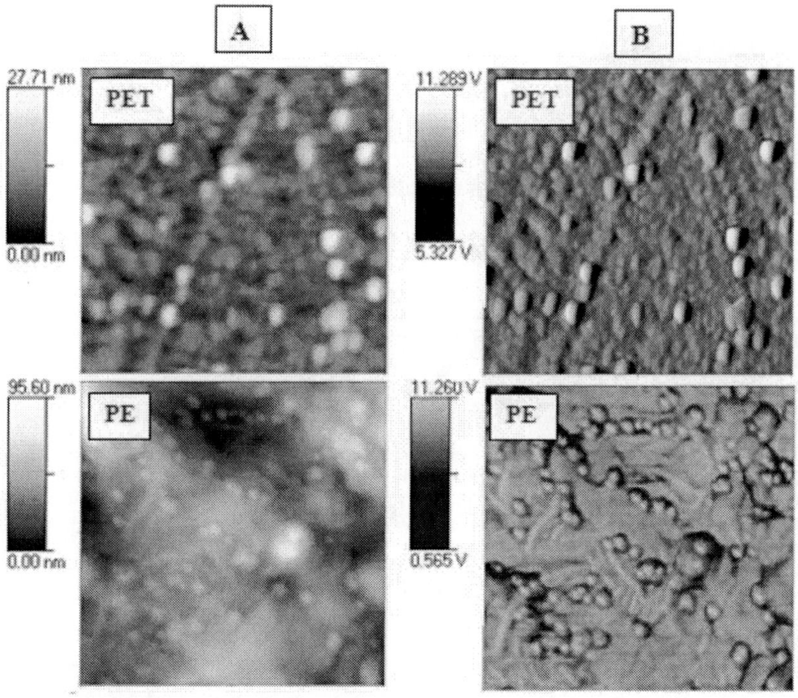

Figure 10.5. AFM images of PET and PE treated by plasma, grafted by dithioland then grafted with Au nanoparticles taken in (A) tapping and (B) phase modes [26].

10.2. Au Nanoparticles Grafted on a Glass Substrate

Many approaches were employed to try and create a glass and Si-based surface to bind AuNPs, most notably thesilanization of glass with various organic compounds [193, 194], fabrication of self assembled monolayers [195] or laser ablation/gas deposition [196]. Our group uses Ar plasma etching to increase the reactivity of thepolymer surface, but plasma treatment also has effect on glass as it is often used to clean its surface. Experiments were carried out with Ar plasma at a power of 10 W on borosilicate microscopic glass 1.5 x 1.5 mm^2. Immediately after plasma etching, samples were immersed in 1,2-ethanedithiol for 24 hours and subsequently in an AuNP colloid solution for 24 hours. Samples were cleaned with methanol after each step of preparation. A surface morphology of samples was investigated by the AFM method (Figure 10.6). Both topographic and phase images were obtained. Images of the sample modified in plasma for 240 seconds show sparse single AuNPs on the glass surface. With increased modification time, more AuNPs are bound to thesurface, but these get aggregated, form clusters and the surface is not covered evenly. UV-Vis spectra were obtained from absorption measurements (Figure 10.7) for samples after plasma modification and for samples with subsequently deposited AuNPs. Increased absorption for modified glass samples at shorter wavelengths is partly caused by a reaction of the modified surface with ambient reactive molecules and partly by a contamination with thematerial from the vacuum chamber used for modification. Samples treated in thethiol solution and subsequently the AuNP colloid show weak absorption peaks at ca 530 nm suggesting a surface plasmon resonance of deposited AuNPs.

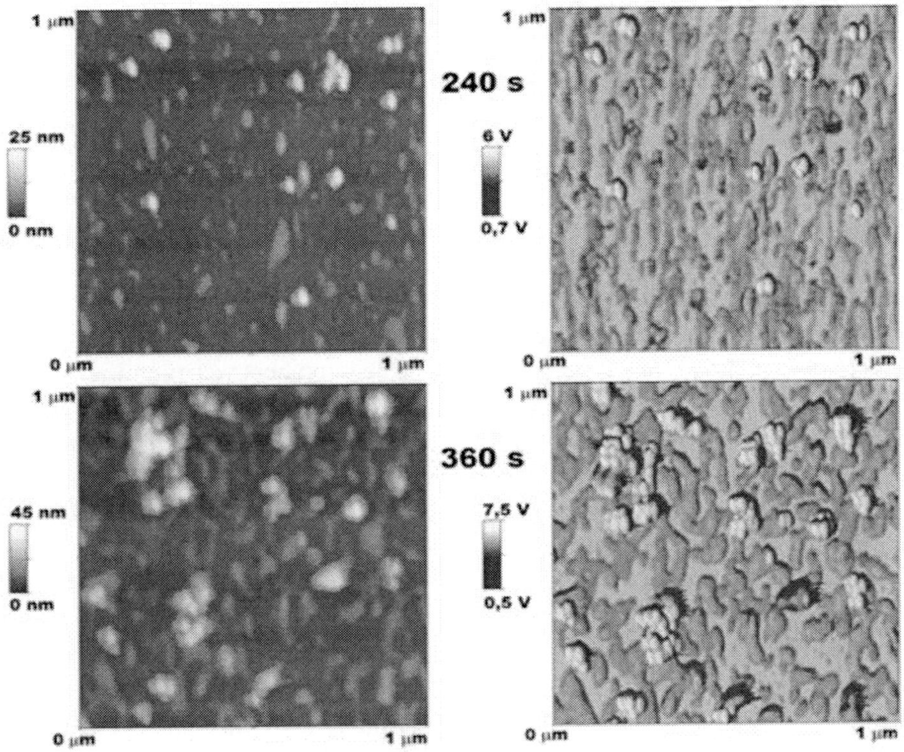

Figure 10.6. AFM images of a glass surface treated by plasma for 240 and 360 s resp. grafted by dithioland then grafted with Au nanoparticles taken in phase (left) and tapping mode (right).

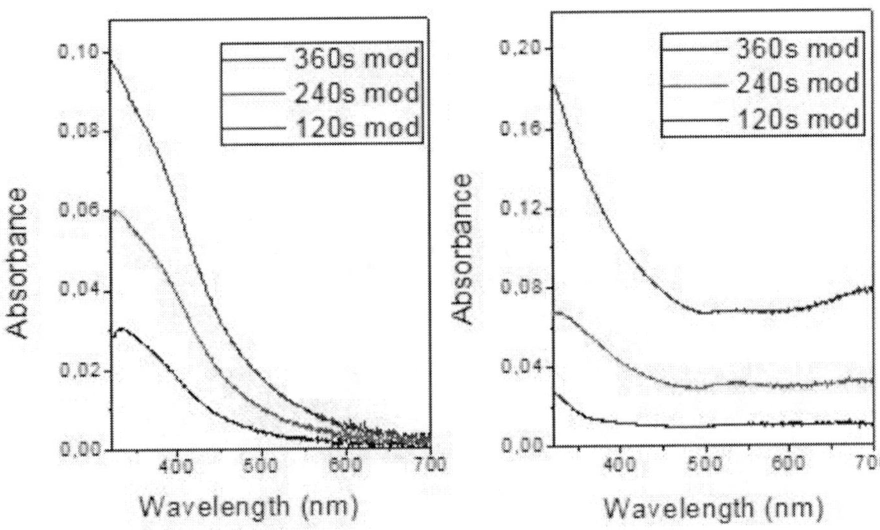

Figure 10.7. UV-Vis spectra of samples modified for 240 and 360 s resp. after modification (left) and after grafting by dithiol and AuNPs (right).

10.3. Some Important Applications of Au-Grafted Polymers

Cells Adhesion and Proliferation

The cell-material interaction was studied and compared on pristine PE, PE exposed to Ar plasma and plasma-activated samples grafted with Au nanoparticles [189]. The number of initially adhered VSMC is similar in all three groups of PE samples(i.e., this parameter is not influenced by treatment with plasma or plasma/Au grafting) [189]. However, both PE modifications enhanced the proliferation activity of VSMC (Figure 10.8). Two days after seeding, the number of VSMC on both types of modified PE samples was higher than that on pristine PE samples. This difference showed increasing tendency in the following days of cultivation [189]. On days 5 and 7 after seeding, the cell numbers were significantly higher on modified PE samples, especially those grafted with Au nanoparticles (Figure 10.8). On the latter samples, the cell population density was approximately 4 times higher than the value on pristine PE. The higher proliferation activity of VSMC on modified PE samples can be explained by the oxidation of the polymer surface after the plasma treatment, resulting in the increase in surface wettability (at least in the samples treated for 50 s).

This increase was well apparent especially in samples grafted with Au nanoparticles, where the samples were wettable after all times of exposure to plasma. It is believed that hydrophilic surfaces adsorb the cell adhesion-mediating molecules (such as vitronectin or fibronectin present in the serum supplement of cell culture media) in a more physiological conformations, supporting the accessibility of specific amino acid sequences (e.g., RGD) in these molecules to cell adhesion receptors (e.g., integrins). Other factors influencing the growth activity of cells is the surface roughness and morphology of the adhesion substrate. Both plasma treatment and Au grafting created nano-scale irregularities on the material surface.

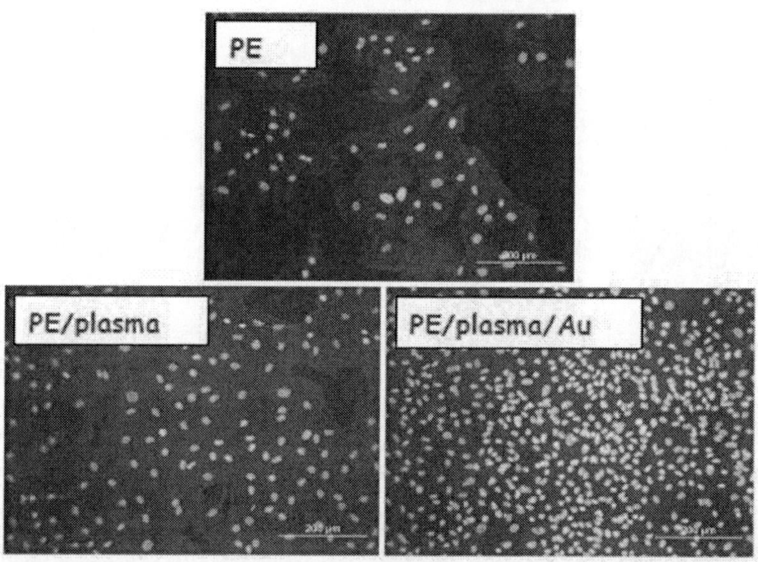

Figure 10.8. The photographs of VSMC prolifered (7th day) on pristine PE (PE), PE treated for 150 s by plasma (PE/plasma) and subsequently grafted with Au nanoparticles (PE/plasma/Au) [189].

The surface nano-roughness acts synergetically with the surface wettability(i.e., by improving the geometrical conformation of adsorbed ECM molecules, and thus a better exposure of active sites on these molecules to the cell adhesion receptors) [189]. Also the shape of the irregularities on the material surface is important for its colonization with cells. After grafting with the Au nanoparticles, the surface irregularities on plasma-treated PE lost their conical shape and sharpness and became rounded and filiform, which supports the better spreading of cells and their subsequent proliferation activity. These results have animportant effect in tissue engineering.

Gold Thin Layers Adhesion

The main goal of our work [188] was to verify whether the pretreatment by the plasma discharge and by subsequent grafting with dithiol may affect theadhesion of an Au layer deposited onto PE by sputtering. For that purpose the mechanical properties of gold layers sputtered on pristine, plasma treated, and plasma treated and dithiol grafted PE were examined using a nanointender.

The results of scratch tests are illustrated in Figure 10.9. The scratch path in the sputtered gold layer for the pristine PE and plasma-treated PE is qualitatively different (Figures 10.9A,B).

In several recent studies, no significant effects of plasma treatment on the adhesion of sputtered gold layer have been observed. Dramatic changes in the shape and the character of the scratch path are observed on the dithiol-grafted PE (see Figure 10.9C). It is seen from Figures 10.9 A-C that under the same indenter load (about 10 mN) the most pronounced adhesive destruction of the gold layer takes place on the pristine PE, where the underlying substrate appears. Lower adhesive destruction is observed on the gold layer deposited on the plasma-treated PE (see Figure 10.9B). On the dithiol-grafted PE (Figure 10.9C) adhesive destruction is much smaller and cohesive destruction becomes dominant.

The cohesive destruction is demonstrated by a radial spreading of scratches in the Au layer. It may therefore be concluded that the adhesion of the sputtered gold layer may greatly be enhanced by the dithiol grafting. The higher adhesion is probably due to the chemical binding of the sputtered Au atoms to –SH groups present on the plasma treated and dithiol grafted PE. These results have an important effect in electronics and optics.

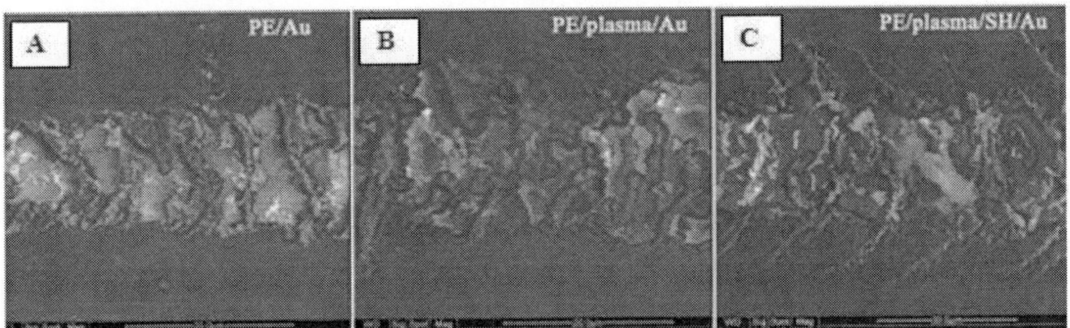

Figure 10.9. SEM images of the scratch path produced in the Au layer, about 50 nm thick, deposited onto pristine PE ((A), PE/Au), plasma-treated PE ((B), PE/plasma/Au) and plasma-treated and dithiol-grafted PE ((C), PE/plasma/SH/Au) [188].

11. THE PREPARATION CARBON STRUCTURES ON A SUBSTARTE

One of the topical problems of tissue engineering is a possibility associated with thecultivation of human tissues on polymeric substrates [197]. The biocompatibility of polymers can be increased, in additionto other techniques, by theaddition of carbon allotropes(e.g., carbon black [198] or carbon nanotubes [199]). The commonly studied is polytetrafluoroethylene (PTFE) [200], which is already used for the preparation of blood vessel prostheses [201]. PTFE is usually classified as a biologically inert biomaterial and it cannot be directly used for similar purposes in clinical practice. Modification of PTFE properties can be achieved by thedeposition of a carbon layer(e.g., amorphous hydrogenated carbon (a-C:H) film [202,203]). The layer protects the polymer surface and facilitates cell adhesion and proliferation [204]. The carbon layer also increases the surface wettability, which has positive effects on tissue growth [205].

Carbon layers with different structures (from amorphous carbon [a-C] to diamond or diamond-like [DLC] layers) are prepared by various techniques such as evaporation [206], flash evaporation [207], sputtering [208], plasma-assisted chemical vapor deposition [209] or ion beam deposition [210]. The structure and properties of carbon layers prepared by various deposition techniques have been studied extensively using different experimental methods [211]. However, little attention has been devoted to the study of carbon layers prepared by the simplest deposition technique – flash evaporation. By flash evaporation carbon layers of different thickness can be produced for routine SEM and TEM electron microscopy. In general, there is a need for these layers to be of a fine grain, with even coatings, and with uniform and controllable layer thicknesses. Flash deposition is distinguished from other techniques (e.g., vacuum evaporation, ion beam) by its short deposition time and low total power input [207].

The thickness of a flash-prepared carbon layer should be controlled, but at present none of the conventional methods in general use allows precise and reproducible deposition and layer thickness control [212]. Flash deposition can be accomplished either by therapid evaporization of a carbon filament [213] or by thepulsed laser vaporization of a carbon target [214]. The former technique is based on therapid evaporization of a carbon filament caused by an electric discharge.

11.1. Homogeneity and Thickness of a Deposited Carbon Layer

In first experiment the carbon layer was deposited at the distance of 2 cm from the filament on the glass sample, 5 cm in diameter. Homogeneity of the deposited layer was examined by a thickness measurement of thecarbon layer at positions situated along two perpendicular directions in 0.5 cm steps starting from the sample center. As expected, the layer thickness depends on the distance from the sample center. The layer thickness measured at 10 positions varies within ±10% over the distance of ±1.5 cm from the sample center.

The dependence of the layer thickness, measured by the scratch technique, on the distance between the substrate and the carbon filaments is shown in Figure 11.1 for arrangements with one and two filaments.

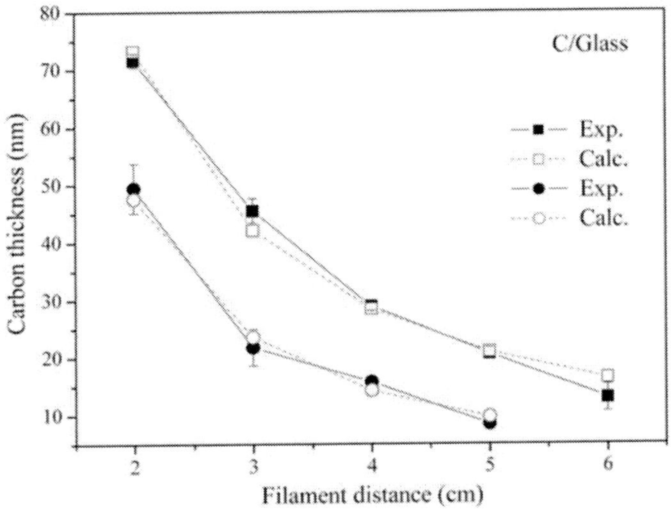

Figure 11.1. Dependence of the thickness of the flash-evaporated carbon layer on the distance of the glass substrate from 1 (●) and 2 (■) filaments (exp.), determined by AFM technique. The values calculated according to formulae presented in the text are also shown (1 (○) and 2 (□) filaments (calc.)) [68].

The thickness of thinner layers prepared at the deposition distances above 6 cm (or 5 cm for 1 filament) can not be determined by the scratch technique. One can see from Figure 11.1 that the layer thickness decreases with deposition distance monotonically, as expected. The observed dependence of the layer thickness (t) on the distance between the carbon source and substrate (d) can be described by the equation $t = a/d^b$ [215]. From the least-squares fit of the experimental data the following values of adjustable parameters were obtained: $a = 158 \pm 13$; $b = 1.7 \pm 0.1$ for deposition from one filament and $a = 189 \pm 18$; $b = 1.4 \pm 0.1$ for a two-filament case. Good coincidence between measured and calculated curves is seen from Figure 11.1. It should be noted that the deposition process is affected by several phenomena.

The filament does not represent a point-like carbon source [215], and during the evaporation process it breaks without fully evaporating along its complete length. The flash evaporation proceeds in a rather low vacuum, and evaporated carbon particles under a large number of inelastic scattering events with molecules of residual gas on theway from the filament to the substrate, by which their energy is reduced, and flight direction is changed randomly. Moreover, the excited carbon atoms may recombine with molecules of residual gas and create molecular fragments(e.g., different oxidized structures, which were observed in deposited layers). Both processes become more significant for larger deposition distances.

11.2. Chemical Composition and Structure of Deposited Layers

Figure 11.2 shows a typical XPS spectrum from a carbon layer deposited onto PTFE. PTFE was chosen for the sake of simplicity since it does not contain oxygen. One can see (Figure 11.2) that oxygen is also observed in addition to carbon. The presence of oxygen is explained by oxygen absorption from the residual atmosphere during thedeposition process. C1s and O1s peaks correspond to about 94.1 and 5.8 at% of carbon and oxygen concentration, respectively. The structure of theO1s band is also shown in the inset of Figure 11.2 with positions of groups –C– (284.7), –C–O– (286.5), –C=O (288.0) and –O–C=O (289.2 eV) [216, 217].

The presence of –C–O–, –C=O and –O–C=O structures in the carbon layer is evident. The RBS study did not confirm the presence of oxygen, probably because of its too low concentration in thesurface layer's interior. A hydrogen depth profile was determined by ERDA (Figure 11.3). One can see that the concentration of carbon and hydrogen decreases with increasing depth while the concentration of fluorine increases. No hydrogen is detected beyond 75 nm, and at deeper theregions concentration ratio of C and F is close to 1.

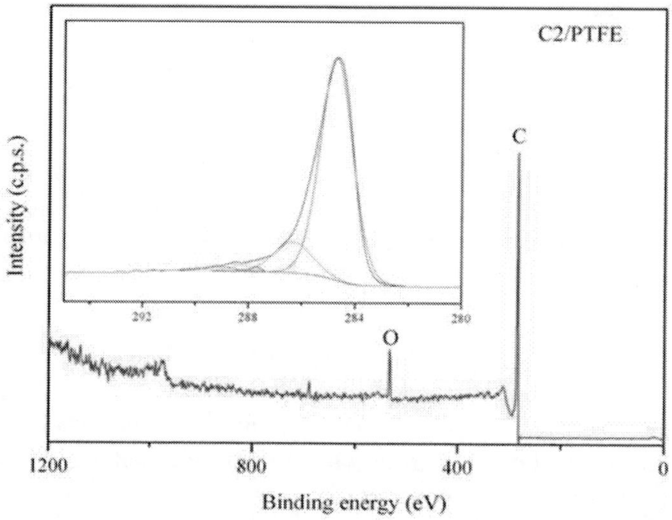

Figure 11.2. XPS spectrum of the carbon layer deposited from 2 cm distance on PTFE; decomposition of the O1s band is shown as well[68].

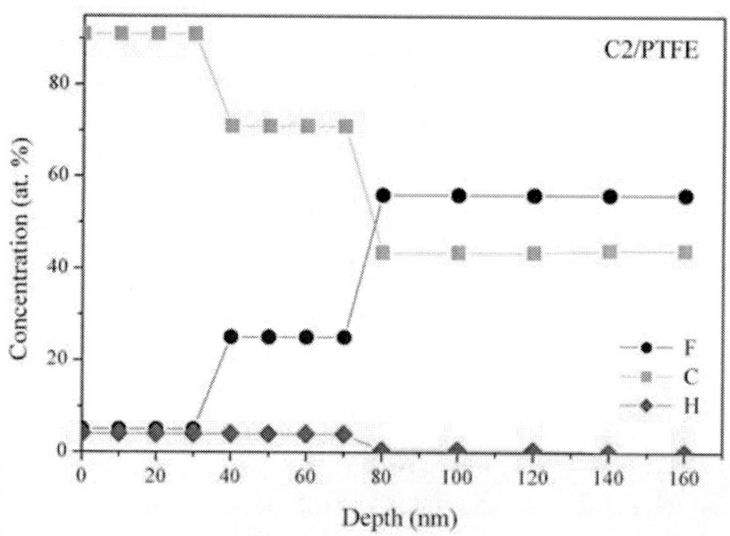

Figure 11.3. Depth concentration profiles of C and F, determined by theRBS method and H, determined by theERDA method on the PTFE sample carbon coated at a deposition distance of 2 cm [68].

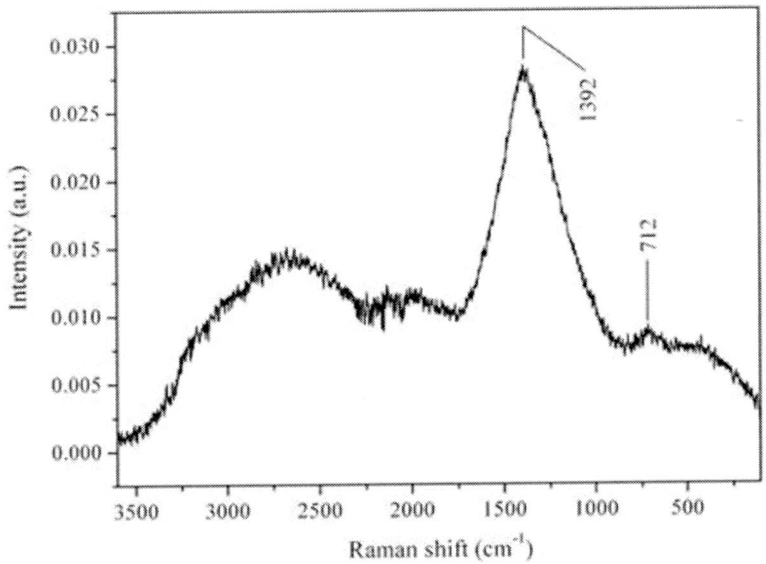

Figure 11.4. Raman spectrum of a carbon layer deposited onto Au/Pt backing at adeposition distance of 2 cm [68].

These observations may indicate that the thickness of thedeposited carbon layer is about 75 nm (in accordance with measurements performed on glass samples, see Figure 11.1) and at larger depths the composition approaches pristine PTFE. The structure of the carbon layer was further characterized by Raman spectroscopy. Because of a layer thickness that was too low, thetechnique described in [218] was applied instead of classical one. The Raman spectrum of the sample deposited at adistance of 2 cm is shown in Figure 11.4. The main band observed at 1,392 cm^{-1} corresponds to amorphous carbon (a-C), and it could be assigned to theD (disorder)-band of carbon materials [202,219]. The band at 712 cm^{-1} is typical for thenon-flat deformation of carbon layers. The band at 1,604 cm^{-1}, corresponding to graphitic carbon, is not present.

11.3. Surface Properties of Carbon Layers

Typical sheet resistances of pristine PET and PTFE are $Rs \approx 1x10^{14}$ Ω [75] and $R_s \approx 1x10^{17}$ Ω [220], respectively. Carbon deposition results in a rapid resistance decrease (comparing to pristine polymers) indicating the formation of a continuous carbon layer on the polymer. The decrease of R_s is more pronounced on PET, probably due to lower surface roughness (this will be discussed later). For the measurement of the sheet resistance R_s as a function of the temperature, the C-PTFE samples were placed in the cryostat cooled by LN_2. Temperature dependence of the sheet resistance, measured in the temperature range from 80-350 K, is shown in Figure 11.5. One can see that over a broad temperature range the sheet resistance decreases rapidly with increasing temperature. The decrease is typical for semiconductors. The semiconductivity of amorphous carbon (a-C) was reported earlier in [221].

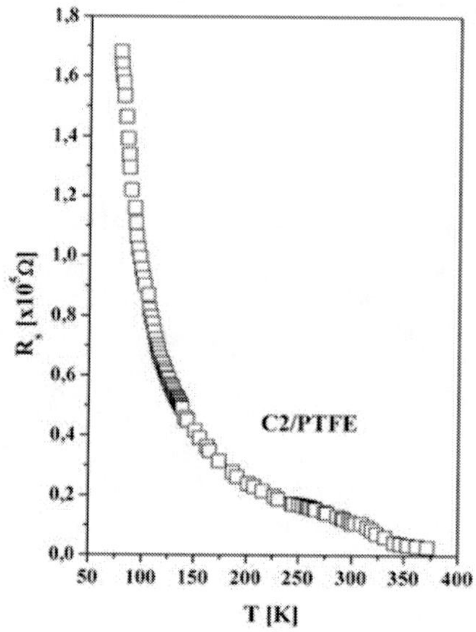

Figure 11.5. Temperature dependence of the sheet resistance (R_s) for carbon layer, 70 nm thick, flash evaporated onto PTFE [258].

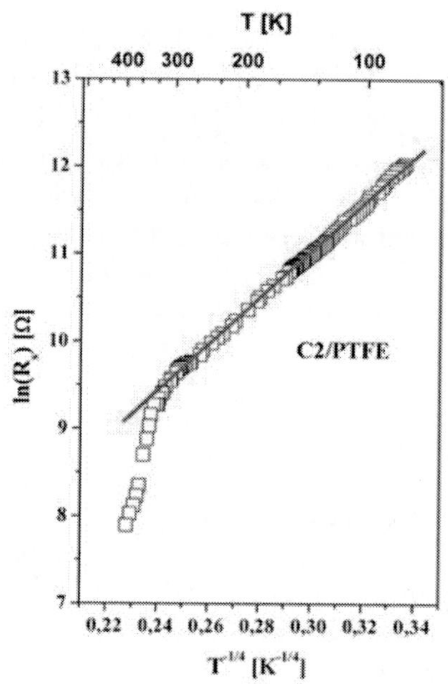

Figure 11.6. Temperature dependence of the sheet resistance R_s in form ln R_s vs. $T^{-1/4}$ for the same sample as in Figure 11.5 [258].

It is believed that the mechanism of the charge transport in carbon layers proceeds according to a variable range hopping (VRH) mechanism, suggested by Mott [222], which depends on the density of the states present near the Fermi level [223]. The VRH model leads to the dependence $R_s(T) \approx \exp((T_0/T)^{1/4})$, where T_0 is an adjustable parameter.

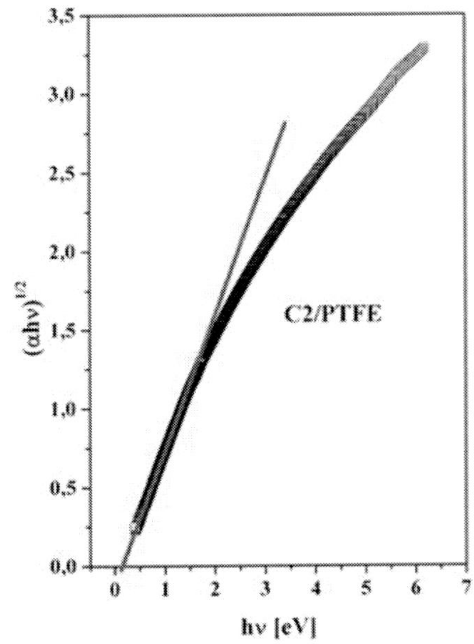

Figure 11.7. Tauc's plot $(\alpha h\upsilon)^{1/2}$ vs. $h\upsilon$ deduced from UV-Vis spectra taken from the same sample as in Figure 11.6 [258].

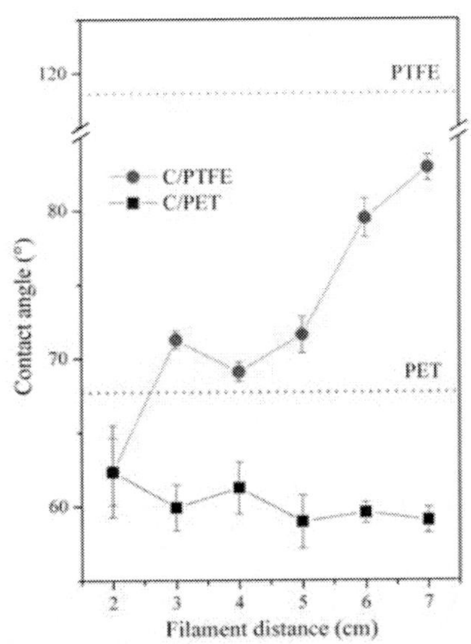

Figure 11.8. Contact angle determined by standard goniometry on carbon-coated PET and PTFE in dependence on the deposition distance [68].

The same dependence as in Figure 11.5 is shown in Figure 11.6 in the ln $R_s(T)$ vs.$T^{1/4}$ representation. One can see that the VRH model describes the charge transport in the temperature range from 80-350 K well. It is supposed that the electron states are localized on carbon clusters sp^2, dispersed randomly within the carbon layer. After application of an external electrical field electron hopping between the clusters takes place.

The electron movement proceeds via phonon-assisted tunneling, and with decreasing temperature the electrons tend to hop to larger distances on sites which are energetically closer than the nearest neighbor.

The values of the optical gap width (E_g^{opt}) of the carbon layers with different structures have been reported. They lie between a minimum ($E_g^{opt} = 0$ eV) for graphite and a maximum ($E_g^{opt} = 5.48$ eV) for diamonds [224]. Experimentally, the optical gap width can be determined from observable phenomena related to the electron transitions between valence and conductance bands. The transitions can be initiated thermally (E_g^{term}) or by photon absorption (E_g^{opt}). Here, the optical gap width (E_g^{opt}) was determined from measured UV-Vis spectra using thewell known Tauc's relation [161]. The dependence ($(\alpha h\nu)^{1/2}$ vs. $h\nu$) is shown in Figure 11.7. By an extrapolation shown in Figure 11.7 the value of the optical gap width,E_g^{opt} = 0,1 eV, was estimated. This value is close to the energy gap of graphite. It is supposed that the estimated E_g^{opt} corresponds to electron transitions between band edges of valence and conductance bands of sp^2 carbon atoms, taking part in theVRH conductivity mechanism. Other, larger values of E_g^{op} may play a role too. It should be noted that the electrical width of the band gap may be larger that optical ones due to additional energy needed for the dissociation of theexciton pair [223].

Surface wettability is dependent on the surface chemical structure [35] and is commonly characterized by contact angle [35, 225]. Contact angles as a function of the distance of PET and PTFE substrates from the filament are shown in Figure 11.8. On PTFE the contact angle increases with an increasing deposition distance (thinner layer thickness), while on PET the contact angle does not change within experimental errors for deposition distances from 2 to 7 cm.

According to XPS analyses, the chemical composition of the deposited carbon layer is the same for both polymers with low concentrations of oxidized, polar structures. It is supposed that the wettability might also be affected by thesurface morphology and roughness of both polymers.

Figure 11.9. SEM image of the scratch path produced in the carbon layer, deposited from 2 cm distance on PET. During the movement the tip was gradually loaded from 0 to 20 mN [68].

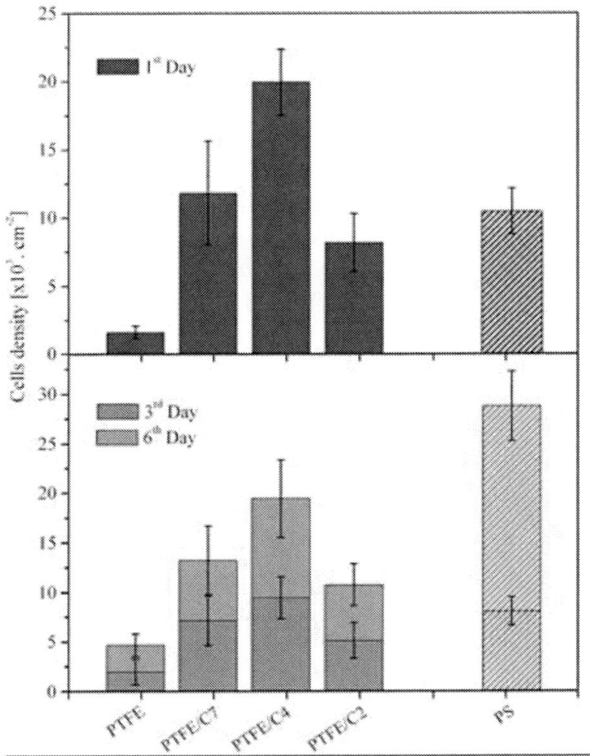

Figure 11.10. The amount of the of HDF cells after adhesion (1st day, the higher figure part) and proliferation (3rd and 6th day, the lowe figure part) on pristine PTFE, PTFE with carbon layers prepared by carbon flash evaporation from the distances 2, 4 and 7 cm (C2, C4 and C7) and on the tissue polystyrene PS.

While the roughness of PET after carbon coating remains unchanged within experimental error, a dramatic change of the roughness is observed on PTFE.With increasing thickness of the carbon layer, the PTFE roughness decreases from 13.3 nm for pristine PTFE to 2.9 nm for thethickest layer (deposition from 2 cm). Significant differences in the surface morphology are found between both polymers before and after carbon deposition. Surface morphology of PET and PTFE, before and after a 2-cm deposition distance is dependent on the polymer surface.

For pristine PET, the surface is composed of tiny, rounded formations, homogenously distributed over the sample surface. Carbon deposition does not result in any significant change in the surface morphology and roughness. Surface of the pristine PTFE is markedly wrinkled and its roughness is higher compared to PET. Carbon deposition results in a dramatic morphology change, and roughness declines indicating a preferential carbon accumulation into holes.

Dependence of the surface hardness and elastic modulus of on the tip displacement into the surface decline with increasing depth, the decline being much slower at the depths of about 75–80 nm. This depth coincides with the carbon layer thickness (see Figures. 11.1 and 11.3). Therefore,it may be concluded that the carbon layer has higher hardness and elastic modulus than the underlying pristine PET. A scratch test on carbon coated polymers was performed using a NanoIntender device. The result obtained for the C2/PET sample (carbon evaporated from of 2 cm) is shown in Figure 11.9 for illustration. The test shows a semi-crystalline structure of the carbon layer. However, this result was not confirmed by Raman (missing 1,604 cm^{-1} band [202]) and XRD (too thin of a carbon layer).

11.4. Cells Adhesion and Proliferation

The cytocompatibility of samples was determined from *in vitro* experiments on the adhesion and proliferation of HDF cells performed on pristine and carbon-coated PTFE. For comparison, the same experiments were performed on polystyrene (PS) tissue as well. The results are presented in Figure 11.10. It can be seen that the carbon rating significantly increases the adhesion (Figure 11.10A) and proliferation (Figure 11.10 B) of HDF cells in comparison with pristine PTFE. Cell proliferation occurs threedays after the seeding is comparable with PStissue. For both cell adhesion and proliferation, the maximum positive effect is seen on the samples carbon coated from adistance 4 cm.

In this case, the carbon layer is about 32 nm thick and it exhibits the higher roughness, good electrical conductivity and a contact angle of about 80°. Low surface roughness and wettability seem to have a negative effect on the cell adhesion and proliferation.

12. Polymer Grafting and Plasma Treatment as a Tool for Cell Colonization Improvements

There are various ways of modifying the polymer surfaces to make them more suitable for cell cultivation. A modified polymer surface exhibits increased chemical reactivity, due to the presence of free radicals and unsaturated bonds, and it can easily be grafted with suitable agents. For this purpose, the surfaces have been exposed to ultraviolet (UV) irradiation [226], beams of various ions (e.g., oxygen, nitrogen, noble gases, or halogens) for biological applications [227] or the plasma discharge. For more pronounced changes in the physico-chemical properties of the modified surface, some of these processes are realized in a gas atmosphere(e.g., in acetylene or ammonia [226]). A common approach to tailoring surface properties of polymer materials is an exposition to the plasma discharge, by which the polymer surface chemistry and morphology can be changed. Plasma treatment has become an important industrial process for the modification of polymer surfaces [228,229]. Plasma constituents, electrons, ions and radicals interact with the polymer surface and modify its chemical and physical properties. The plasma treatment of polymers leads to the creation of new chemical groups [229], the branching and crosslinking of macromolecules, and the formation of low molecular weight oxidized (LMWO) structures [22, 230, 231].

These modifications change the stability, roughness, morphology, mechanical properties and chemical composition of the polymer surface due to the creation of functional chemical groups and surface ablation. The oxygen-containing groups, created by the plasma treatment, render the material surface more wettable, which supports the adsorption of physiological extracellular matrix (ECM) molecules in a flexible and re-organizable form. This form enhances the accessibility of oligopeptidic ligands in these molecules for cell adhesion receptors [186, 187, 232]. Another interesting property of the modified polymers is the formation of conjugated double bonds between the carbon atoms; from this follows an increase of the material's electrical conductivity. It has been shown that this has positive effects on the adhesion, growth and maturation of various cell types, including vascular smooth muscle and endothelial cells [232, 233]. Also, under some specific conditions, plasma etching takes place, leading to surface ablation [15, 234] and to an alteration of the polymer surface morphology [21, 235]. The changes in morphology and wettability of the polymer surface [21] may strongly influence the cell-substrate interaction [236]. Unsaturated chemical bonds and radicals created bythe plasma treatment [17, 21, 35] can be utilized for the grafting of new chemical functional groups, atoms and molecules [225, 237] and, in this way, creating nanostructured surfaces with new functionalities.

The nanostructures of the material surface (irregularities smaller than 100 nm) are considered to the morphology of the cell membrane, as well as the architecture of ECM molecules [238]. It is known that cell adhesion to artificial materials is mediated by ECM molecules spontaneously adsorbed to the material surface from biological fluids. Various materials have been used for constructing nanostructured surfaces (e.g., polymers [239], ceramics, metals [238, 240] and combinations of these materials). However, among metallic materials, gold, including its nano-sized forms, has shown no cytotoxicity *in vitro*or*in vivo* [241]. The non-toxicity of gold is related to its well-known stability, non-reactivity and bioinertness. For example, gold nanoparticles have often been conjugated with antibodies [242]. Especially in the last few years, the interaction of nanoparticles with living cells has been intensively studied. The investigation of this phenomenon brings a lot of promises in the area of cellular therapy and offers great opportunities for tissue engineering or organ replacements. Targeted nanoparticulate drug delivery is one of the most promising techniques for increasing the efficiency of drugs [243] and also of influencing the differentiation of stem cells. The specific molecules can be prepared as "smart darts" and the nanoparticles are then used as a specific carrier, which enables the delivery of "smart drugs" directly to the place where needed.

The ability to integrate metal nanoparticles into biological systems has its greatest impact in biology and biomedicine [244]. Noble metals are used in biomedicine because of their ability to resist corrosion at physiological conditions and because of their advantageous mechanical properties: chemical inertness and relative mechanical softness. Shukla et al. suggested that Au(0) nanoparticles are not cytotoxic, reduce the production of reactive oxygen and nitrite species, and do not elicit secretions of proinflammatory cytokines TNF-α and IL1-β, making them suitable candidates for nanomedicine. Their studies thus underline the noncytotoxic, nonimmunogenic, and biocompatible properties of gold nanoparticles with the potential for applications in nanoimmunology, nanomedicine, and nanobiotechnology [245]. Although important for the final outcome of biomaterials, little is reported about early events between pure noble metals and blood [246]. There are a lot of approaches regarding how to stabilize gold nanoparticles (AuNPs). Thiol-stabilized AuNPs can exhibit desired reactivities due to itsvariety of functionalizations and the strong Au-S bond between the soft acid Au and the soft thiolate base [247]. The formation of self-assembled mono- and multilayer films of small-ligand-stabilized metal nanoparticles opens up applications as separative layers and chemical sensors [248, 249]. Cellular cytotoxicity has shown that gold nanoparticles, gold nanoparticles functionalized with methoxy-PEG-thiol (i.e., gold-PEG-OMe nanoparticles), and coumarin-PEG-thiol functionalized gold nanoparticles were not cytotoxic at the concentrations studied and the biocompatibility of the system was not affected [250].

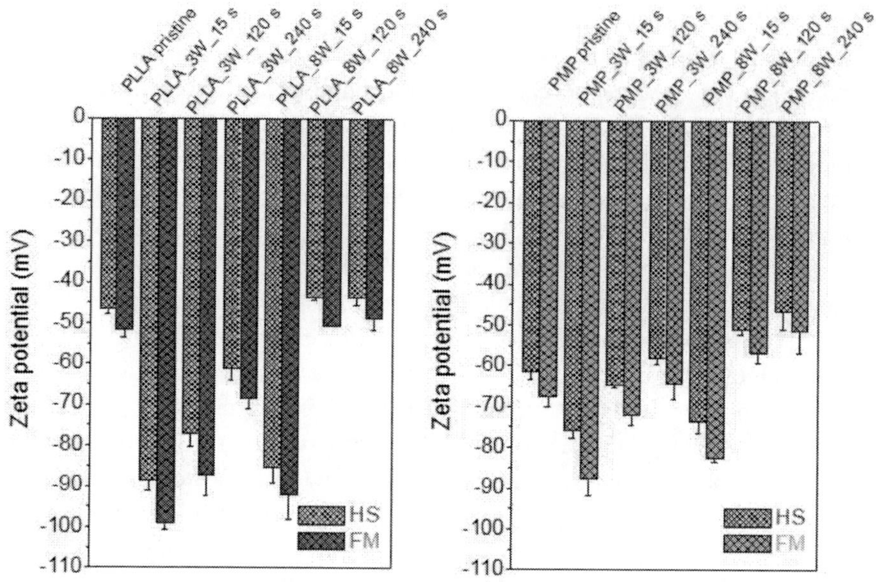

Figure 12.1. Zeta potential values of pristine (A) PLLA and (B) PMP and aged samples modified by Ar plasma (power 3 and 8W) with exposure times 15, 120 and 240 s. The value of the zeta potential was determined using common different approaches and equations (streaming current and Helmholtz-Smoluchowski eq. - HS and streaming potential and Fairbrother-Mastins eq. - FM) [101].

Our scientific group used a plasma discharge for the activation of the polymer surface. Polymer substrates were chosen among all polyethylene (PE), and biopolymers poly(*L*-lactic acid) (PLLA) and poly-4-methyl-1-pentene (PMP). Grafting procedures were performed among others with polyethyleneglycol (PEG) and gold nanoparticles. The hypothesis is that gold nanoparticles may also serve as building blocks for creating artificial bioinspired, nanostructured surfaces for tissue engineering. Physico-chemical changes of plasma exposed or grafted polymer surfaces in connection with biocompatibility enhancement were determined. The adhesion and proliferation of VSMC in cultures derived from the rat aorta, as well as mouse fibroblasts of the line 3T3, were evaluated on the modified or grafted polymers.

12.1. Physico-Chemical Properties of a Grafted Surface

Properties of the polyethylene samples modified by the plasma treatment and grafted with Au nanoparticles were studied using various analytical methods. The presence of Au in the modified surface layer was proved by X-ray photoelectron spectroscopy (XPS), and by Rutherford backscattering spectroscopy (RBS).

The plasma treatment led to an ablation of the PE surface layer [91]. As a result, the surface morphology was changed dramatically and surface roughness increased. Also, a lamellar structure appeared on the PE surface which could be due to the different ablation rates of amorphous and crystalline phases [21]. Observed changes might also have been due to the preferential removal of oxidized low-mass degradation products from the PE surface [29, 251]. The presence of oxygen and nitrogen on the surface of the plasma-treated and gold-grafted PE was examined using the XPS method [21]. The presence of oxygen on the PE surface was proved in the XPS spectrum [252]. Obviously, the oxygen from the ambient residual atmosphere interacted with the plasma-activated PE surface, and created different oxidized structures. The creation of carbonyl, carboxyl and ester groups on the plasma-treated PE was proven earlier in similar experiments [91, 92].

The Au4f peak was clearly seen indicating the presence of Au atoms on the surface of the gold-grafted PE. The presence of gold was also proved by RBS measurement, the Au particles were present not only on the very surface of the PE, but also in the near surface layer of about 100 nm thick [252]. It is well known that chemical structure and surface morphology significantly affect surface wettability [225, 237], which in turn may affect the adhesion and proliferation of living cells. From the wettability studies of plasma-modified and grafted PE, it was observed that short exposure times to the plasma discharge result in a rapid growth of hydrophilicity which can further be increased by the gold grafting. For longer exposure times, however, the hydrophilicity diminishes and finally the samples become hydrophobic[252].

Surface morphology is well known to affect the interaction of cells with polymers [186]. Plasma treatment results in an increase of surface roughness and leads to the exposure of the PE lamellar structure [197]. This phenomenon is a typical demonstration of the plasma-stimulated ablation of PE [253]. After water etching, the morphology changes due to a dissolution of the low molecular weight degradation products [21]. Subsequent PEG grafting ledto a changein the surface roughness and, regardless of the PEG molecular weight, to a decrease in surface roughness. While the grafting with a PEG of molecular weight $M = 300$ $g{\cdot}mol^{-1}$ preserves the PE lamellar structure, the grafting with a PEG of molecular weight $M = 20000$ $g{\cdot}mol^{-1}$ results in a significant change of the morphology and in the creation of branched star-like formations.

As was described above, surface chemistry and wettability are important factors influencing polymer surface biocompatibility. In addition tocontact angle evaluations, a complementary method for biopolymer surface characterization was used: the zeta potential determination [101]. The negative value of the zeta potential can be connected with the higher contact angle of the polymer surface and its lower wettability (hydrophobic character) [254] However, the relationship between the zeta potential and the character of the polymer surface is very complex, involving surface chemistry, surface charge, wettability, surface roughness and morphology [254]. Thezeta potential was determined on the aged samples of PLLA and PMP modified at discharge powers 3 and 8W (Figure 12.1). Pristine PMP has more negative zeta potential compared to that of PLLA; therefore, the pristine PMP appears to be more hydrophobic than PLLA. This conclusion was confirmed by the contact angle measurement [101]. The more pronounced decrease of zeta potential for both polymers was observed after the plasma modification with an exposure time of 15 seconds (both for 3 and 8 W). With increasing exposure time (120 and 240 seconds), the value of the zeta potential is close to that of pristine polymers (PMP), in the case of PLLA it is even higher. The lattercorresponds to results that the contact angle of polymers increases more significantly for shorter exposure times after aging [255],meaning the different changes of chemical and physical properties of polymer surfaces for shorter (15 seconds) and longer exposure times [101].

12.2. Cell Proliferation and Growth

The cell adhesion represents the first stage of the cell–material interaction, and above all the quality of this phase influences the cell ability to proliferate and to differentiate in contact with the substrate. After the cell adhesion procedure on the substrate is finished, the new stage, the so-called lag-stage begins. This stage lasts from 24 to 48 hoursfrom the seeding. In this period the adaptation of the cell on the different environment takes place. If the cell stays alive through this period, the proliferation process can start.

Grafting with Au Nanoparticles

A study of cell proliferation and growth on polyolefin substrates was performed on polyethylene (PE) foils seeded with vascular smooth muscle cells (VSMC) derived from the rat aorta by an explantation method [232]. The cell-material interaction with polyolefins was studied and compared on pristine PE, PE exposed to Ar plasma and plasma-activated samples grafted with Au nanoparticles.

The number of initially adhered VSMC was similar in all three groups of PE samples(i.e., this parameter is not influenced by treatment with plasma or plasma/Au grafting). Both PE modifications (plasma and grafting) enhanced the proliferation activity of VSMC. Two days after seeding, the number of VSMC on both types of modified PE samples was higher than that on pristine PE samples. On days 5 and 7 after seeding, the cell numbers were significantly higher on modified PE samples, especially those grafted with Au nanoparticles (Figure 12.2). The higher proliferation activity of VSMC on modified PE samples can be explained by the oxidation of the polymer surface after the plasma treatment, resulting in anincrease in the surface wettability.

This increase of cell numbers was well apparent, especially in samples grafted with Au nanoparticles, where the samples were wettable after all times of exposure to plasma [252],(Figure 12.3). It is believed that hydrophilic surfaces adsorb the cell adhesion-mediating molecules (such as vitronectin or fibronectin present in the serum supplement of the cell culture media) in a more physiological conformation, which supports the accessibility of specific amino acid sequences (e.g., RGD) in these molecules to cell adhesion receptors (e.g., integrins). Receptor-mediated adhesion and appropriate cell spreading on the adhesion substrate are prerequisites for proliferation capabilities in all types of anchorage-dependent cells, including VSMC. Other factors influencing the growth activity of cells is the surface roughness and morphology of the adhesion substrate. The mouse embryonic fibroblasts (NIH 3T3) were used as a second model cell line on PE [189] The obtained data strongly argued that surfaces modified either by plasma or plasma/Au are significantly more suitable for cell adhesion than the pristine PE, as markedly higher numbers of cells were attached to the modified samples 24 hours after inoculation [252]. Moreover, they had a typically elongated shape without any visible granulation, suggesting their good physiological state. This correlates well with their rapid proliferation on both plasma and plasma/Au-treated surfaces (on day 3 after seeding).

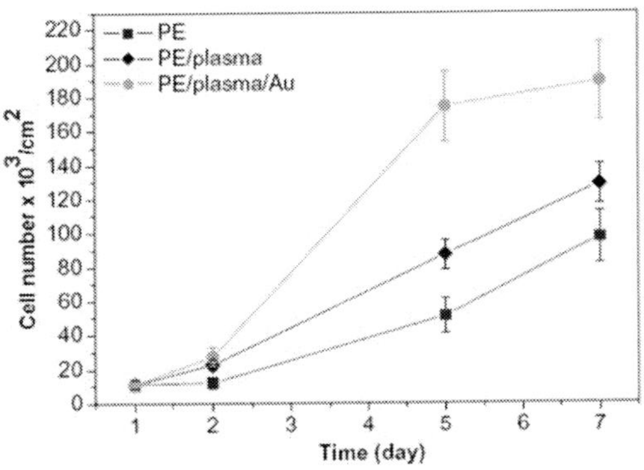

Figure 12.2. Dependence of the number of VSMC cells on the cultivation time for pristine PE (PE), PE treated for 150 s with plasma (PE/plasma) and subsequently grafted with Au nanoparticles (PE/plasma/Au). The plasma power is 2W [189].

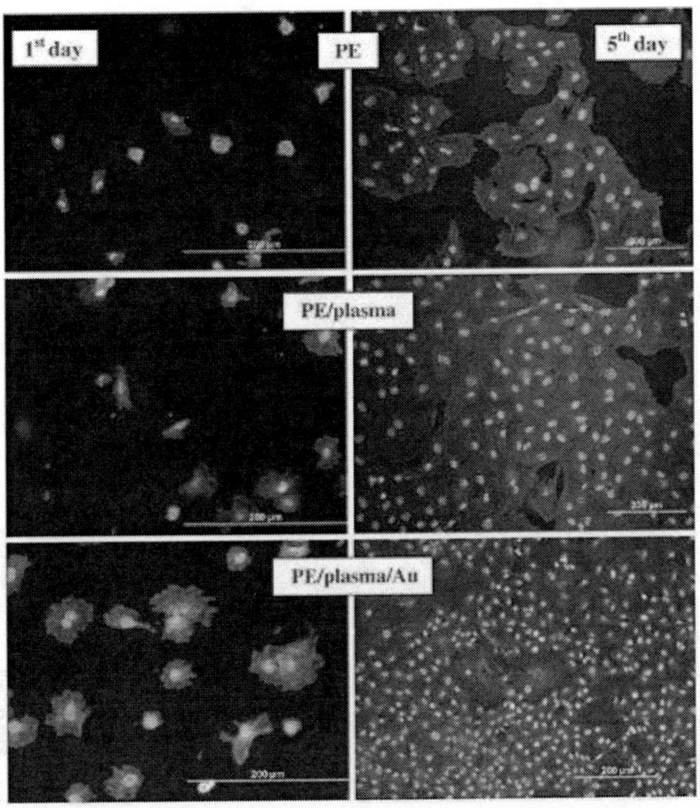

Figure 12.3. The photographs of VSMC adhered (1st day) and grown (5th day) on pristine PE (PE), PE treated for 150 s with plasma (PE/plasma) and subsequently grafted with Au nanoparticles (PE/plasma/Au). The plasma power is 2W [189].

Grafting with PEG

Oriented high-density polyethylene (HDPE) was used for the PEG grafting studies. The substrate was immediately grafted after plasma treatment (plasma-activated PE) by immersion to an aqueous solution (of poly(ethylene glycol) (PEG, molecular weight (*M*) 300 and 20,000 g·mol^{-1}) [197]. The non-bound PEG was removed by an immersion of the samples into distilled water for 24 hours. Adhesion and proliferation of rat VSMCs was studied on pristine PE; PE modified by a plasma discharge for 50, 150 and 400 seconds; and on PE modified by a plasma discharge and subsequently grafted with PEG (*M* = 300 and 20,000 g·mol^{-1}). The anti-adhesive action of PEG is due to its very high hydrophilicity and mobility of itschains, which hamper stable adsorption of the proteins that mediate cell adhesion.

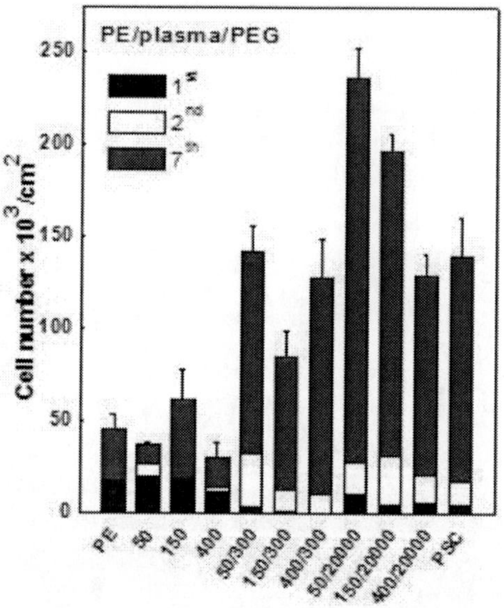

Figure 12.4. Dependence of the number of VSMC cells on the cultivation time for pristine PE (PE); PE plasma modified (2W) for 50, 150 and 400 s (PE/plasma/50-400); and subsequently grafted with PEG (molecular weight M = 300 g·mol-1, PE/plasma/50-400/300) and PEG (molecular weight M = 20,000 g·mol-1, PE/plasma/50-400/20,000) and tissue polystyrene (PSC) [197].

However, at the same time, the anti-adhesive action of PEG is strongly dependent on its concentration on the polymer surface and on the length of its chain [256]. For comparison, the same procedures were applied to tissue culture polystyrene (TCPS). All experiments were performed *in vitro*.

The total number of adhered (after 1 day cultivation) and proliferated (after 2 and 7 days of cultivation) cells is shown in Figure 12.4. During the first 24 hours the cells adhere very well on pristine PE and PE modified by a plasma discharge. The lowest adhesion was observed on the PE grafted with PEG (*M* = 300 g·mol^{-1}). Higher adhesion was found on the PE grafted with PEG (*M* = 20,000 g·mol^{-1}) and on the TCPS. A dramatic increase of proliferating cells was observed on grafted samples 7 days after seeding. Many more cells grew on the samples grafted with PEG of higher molecular weights.

After 7 days on pristine and plasma-exposed PE, the cells were still well spread but their distribution was not homogeneous and cell aggregates appeared. On the contrary, on the plasma-modified and PEG-grafted PE, the dense and homogeneous cell coverage was observed after 7 days of cultivation, in spite of the fact that grafting did not affect initial cell adhesion [197].

Biopolymers

Biopolymers poly-*L*-lactic acid (PLLA) and linear isotactic poly-methylpentene (PMP) were modified with a diode plasma discharge at a power of 3 and 8 W. For cell culture experiments, four pristine polymer samples and the same number of plasma modified polymers (power 3 and 8W, exposure times 15 and 240 seconds) were used. [101]. As the comparative material the tissue polystyrene (TCPS) is introduced. The visual field method was used for the determination of number of adhered cells (after 24 hours) and the number of proliferated cells (72 and 168 hours) from the seeding procedure. The highest increase in the number of proliferated cells on the seventh day from the seeding was observed on the PMP modified for 15 seconds at apower of 8W [101]. The modification of PMP with plasma had a positive effect on VSMC cell adhesion and proliferation. The cell adhesion and proliferation on the pristine PMP was rather poor. In comparison, the pristine PLLA exhibited significantly better proliferation of VSMC cells on the seventh day, which was even comparable to the modified samples. PLLA can be used for biomedical applications(e.g., as appropriate substrate for human osteoblast cell cultivation [257]). The outstanding biocompatibility of even the pristine, non-modified polymer was confirmed.

CONCLUSION

In this chapter we summarized all knowledge and basic information about preparation approaches used by our scientific teamassociated withnanostructures and nanostructured surfaces of polymers. Also, some analytical and diagnostic methods are discussed as an inclusive study of biocompatibility by *in vitro*procedures, which we use for the characterization and description of prepared nanostructures.

The properties of polymer substrates *modified by plasma discharge or by a UV laser light* were investigated. Ablation of the polymer surfaces is an increasing function of the number of laser pulses applied. The contact angle of all polymers is a decreasing function of the number of laser pulses. XPS measurements show that the concentration of oxygen (polar groups) increases after the laser treatment on PE, PP and PS surfaces. As expected, the thickness of the ablated layer is an increasing function of the plasma exposure time. EPR data shows different concentrations of free radicals created on the treated surface of all polymers. Plasma treatment results in a dramatic decline of the contact angle. Despite the fact that the plasma treatment proceeds in inert Ar plasma, the oxidized degradation products are produced on the polymer surface which contribute to the increased hydrophilicity (wettability) of the plasma-treated polymers. The plasma treatment leads to significant changes in the polymer surface structure. Plasma treatment results in an ablation of the PTFE surface layer. Subsequent etching in water or methanol leads to another removal of the plasma-modified material from the PTFE surface. The etching leads to an increase of the contact angle due to a removal of polar groups. The zeta potential of pristine polymers increases in the order PTFE $<$ PP $\approx$ PET $<$ PE $<$ PS. After the plasma treatment, the zeta potential increases dramatically due to increased surface hydrophilicity and wettability of the plasma-treated polymer surface.

*Polymer patterning with laser beams*was illustrated by theauthors' experiments performed on PET films. Modification of the PET surface with linearly polarized light from pulsed KrF and F2 lasers has a significant effect on the properties of subsequently deposited gold nanolayers, and the choice of the deposition technique is crucial owing to the quality of prepared coatings. The subsequent deposition of a 200 nm thick gold layer causes a decrease of the surface roughness. A significant effect was observed during an Au deposition on the KrF laser-modified sample,while by evaporation, a continuous metal coverage is formed, copying a nano-structured polymer surface; in the case of sputtering, a nanowire-like structure of the gold coating can be observed. In contrast to previous results, subsequent gold sputtering onto F2 laser-modified samples leads to the formation of a continuous gold layer copying the ripple pattern created by the preceding laser irradiation. Obviously, the structure of the gold coating on the nano-patterned PET surface depends on several factors, such as the form of the ripples created by the laser irradiation and the chemical composition of the laser-modified PET surface. PET modification with 157 nm laser light performed under different angles of laser beam incidence can be a possibletechnique for preparing continuous gold layers, copying the "ripple character" of the polymer backing. It was shown that the width of the nano-wires can be tailored by the width of the ripples formed by the preceding laser irradiation. Promising techniques for the creation of nano-patterned, regular gold structures (nano-wires) were also shown. In principle, those techniques could be employed for the creation of metal-polymer composites with interesting electrical, mechanical, and optical properties, which could find novel applications in micro- and nano-technology.

In accord with similar previous studies it was found that the *plasma treatment* results in pronounced changes in the surface properties of polyolefins. RBS and XPS measurements showed the presence of oxygen-containing groups on the PP surface modified by the plasma discharge. The thickness of the Au layer sputtered onto pristine and plasma-modified PP is nearly a linear function of the deposition time and it is not affected by the plasma treatment. The different mechanical properties of the gold layers deposited on different substrates (pristine vs. modified) are obviously due to different structures and physico-chemical properties of the substrate which are affected by the plasma treatment and aging time. The properties of PLLA and PMP samples modified by the plasma discharge and thermal treatment were studied by different techniques. Immediately after the modification process the contact angle sharply decreases; with the increasing aging time the surface polarity is spontaneously reduced (as well as surface energy). After the plasma treatment the surface morphology of PLLA is changed dramatically; in the case of PMP the changes in surface morphology are not so pronounced as in the case of PLLA. The amount of ablated material increases with the increasing discharge power and exposure time. The thermal treatment accelerates the aging of the polymer surface and also causes a decline of the concentration of oxygen groups on the PMP surface. The thermal treatment dramatically changes the structure of modified PLLA. In the case of PMP the observed changes are insignificant. The aged samples were consecutively used for VSMC cell adhesion and proliferation.

The *laser light patterning* comprises several techniques, based on photoeffect, surface relief grating, and temperature or electric field gradients. Generally, the application of laser beams for polymer patterning seems to be very appealing because of its simplicity and ability to produce well defined and controllable structures. The main disadvantages of these techniques are that they can only be applied to a limited set of polymers and there are associated problems with the preparation of structures ordered on longer distances. A method,proposed by the authors (one laser beam scanning), seems to be independent of the polymer's nature and be applicable for all classes of these materials. At the end of the chapter some applications of patterned polymers are described. Periodical polymer structures can play a great role in photonic and electronic elements. In photonics, for example, they can be useful as coupling elements, wavelength filters, mirrors and so on. Temperature, pressure and bio-sensors based on polymers seem to be very promising as well. Anew class of materials, metamaterials, can be preparedby the next deposition of metal on the patterned polymer surface.

Thermally induced changes in the surface morphology of Au thin films were generally understood in the past as an undesirable effect. Though some advantageous properties of the material are lost (e. g., electrical conductivity), the formation of a new structure brings new possibilities. Ellipsometric measurements show that the new material obtains a semiconducting character, which is also supported by UV-Vis measurements and the occurrence of optical band gap. Measurements performed on the PTFE substrate suggest the formation of an island structure is caused by a decrease of the thermal stability of a thin film rather than the occurrence of Au-Si eutectic, as the significant increase of sheet electrical resistance takes place at similar temperature. The zeta potential measurements proved to be useful at determining Au layer continuity.

Carbon flash evaporation does not proceed from a "point source". The dominant carbon fraction, the presence of oxidized structures and significant fractions of hydrogen and amorphous carbon (a-C) in the deposited material were observed. The deposition of carbon leads to a significant decline of surface resistance and wettability in comparison with a pristine polymer. The deposited carbon layer has a semi-crystalline character and it exhibits higher microhardness and elastic modulus than the pristine polymer. The charge transport in the temperature range from 80-350 K is well described by Mott's variable range hopping mechanism. The layers exhibit non-linear dependence of the electrical current on the applied voltage which indicates a departure from the ohmic mechanism of the charge transport. The optical gap width was determined to be E_g^{opt} = 0.1 eV from UV-Vis spectra. The cytocompatibility of the samples was tested using human diploid fibroblastoids (LEP). In comparison with pristine PTFE the carbon coated PTFE shows a dramatically increased adhesion and proliferation of LEP on the sample surface.

The properties of *gold nanoparticles* can be successfully *grafted* onto plasma-modified polymers. The "flexible" molecule of dithiol binds to the surface radicals, probably by both -SH groups, in contrast to the "rigid" molecule of 4,4'-biphenyldithiol, where one -SH group remains free for the subsequent chemical reaction with a gold nanoparticle. Grafting from a methanol solution is accompanied with polymer etching. The presence of gold particles on the grafted polymers was proven by AFM method as well. Gold nanoparticles are distributed relatively homogenously over the polymer surface. AuNPs can be succesfully bound by this method to a glass surface as well. AFM measurements, however, suggest that at longer modifications of the sample, the AuNPs aggregate, leads to the formation of clusters.The step of AuNP introduction must therefore be revised at altered deposition conditions to obtain a more homogenously covered structure.

Plasma modification of the PE surface, followed by a grafting of Au-nanoparticles, significantly increased the attractiveness of the PE surface for the adhesion and growth of VSMC cells and 3T3 fibroblasts. Nanoindentation measurements show that the grafting of a plasma-activated PE surface with dithiol significantly increases the adhesion of a sputtered Au nano-layer.

The *plasma treatment* of the polymer, and particularly the presence of *Au nanoparticles,* had a positive influence on spreading, proliferation activity and homogeneity of the distribution of VSMC on the material surface. The positive effect of the plasma treatment of PE on the adhesion of NIH 3T3 fibroblasts was even more pronounced, although their following proliferation was stimulated only moderately. Grafting with PEG leads to the additional dramatic reduction of the contact angle. With a decreasing chain length of PEG, the polarity of the grafted PE decreases. The biological *in vitro* experiments with VSMCs show that PEG grafting does not increase cell adhesion, but it dramatically increases cell proliferation. The pristine and modified *biopolymers* (PLLA, PMP) were used as substrates for the cultivation of VSMC cells. The plasma treatment leads to significant changes in surface wettability, aging and chemistry closely connected with consequent cell proliferation and growth.

The newly constructed surfaces should advantageously combine several features, which will synergetically enhance the attractiveness of the material surface for cell colonization: nanostructure, appropriate chemical composition, wettability, electrical charge and conductivity.

Acknowledgments

The work was supported by the GA CR under the projects Nos. 106/09/0125, 108/10/1106, 108/11/P840, 108/11/P337, 108/12/1168 and GAAS CR under the project KAN200100801. The authors thank Ms. O.Kesselová for her technical assistance.

References

[1] F. Valentini and G. Palleschi, *Anal. Lett.,* 41, 479 (2008).
[2] Y. Xia, Y. Xiong, B. Lim and S. E. Skrabalak, *Angew. Chem. Int. Ed.,* 48, 60 (2009).
[3] C. P. Poole Jr. and F. J. Owens, Introduction to Nanotechnology, Wiley, New York (2003).
[4] P. Reiss, M. Protiere and L. Li, *Small,* 5, 154 (2009).
[5] O. V. Salata, *J. Nanobiotechnology,* 2, 1 (2004).
[6] J. Gaoa and B. Xu, *Nano Today,* 4, 37 (2009).
[7] V. Švorčík and V. Hnatowicz, Polymer Degradation and Stability,Nova, New York, 2007.
[8] D. Bäuerle, *Laser Processing and Chemistry, 3rd edition,* Springer-Verlag, Berlin (2000).
[9] http://www.alacritas-consulting.com/sputter_deposition_for_thin_films.html, 20.2.2007.
[10] R. W. Cahn, Concise Encyclopedia of Materials Characterization, Pergamon Press, London (1993).
[11] B. J. Kirby and E. F. Hasselbrink Jr., *Electrophoresis,* 25, 187 (2004).
[12] T. Luxbacher, H. Bukšek, I. Petrinić and T. Pušić, *Tekstil* 58, 393 (2009).
[13] C. A. Homsy, *J. Macromol. Sci. Chem.,* A4, 615 (1970).
[14] D. F. Williams, *Biomaterials,* 29, 953 (2008).
[15] C. M. Chan, T. M. Ko and H. Hiraoka, *Surf. Sci. Rep.* 24, 3 (1996).

[16] C. Tendero, C. Tixier, P. Tristant, J. Desmaison and P. Leprince, *Spectrochim. Acta., B* 61, 2 (2006).

[17] V. Švorčík, V. Kotál, J. Siegel, P.Sajdl, A. Macková and V. Hnatowicz, *Polym. Degrad. Stabil.*, 92, 1645 (2007).

[18] A. Řezníčková, A. Chaloupka, J. Heitz, Z. Kolská and V. Švorčík, *Surf. Interface Anal.*, DOI 10.1002/sia.3801.

[19] L. D. Laude, N. Boutarek and K. Kolev, *Nucl. Instrum. Meth. B,* 105, 254 (1995).

[20] D. J. Wilson, R. L. Williams and R. C. Pond, *Surf. Interface Anal.*, 31, 385 (2001).

[21] V. Švorčík, K. Kolářová, P. Slepička, A. Macková, M. Novotná and V. Hnatowicz, *Polym. Degrad. Stabil.*, 91, 1219 (2006).

[22] V. Jones, M. Strobel and M. J. Prokosch, *Plasma Process. Polym.* 2, 547 (2005).

[23] A. Costela, *J. Appl. Phys.*, 77, 2343 (1995).

[24] P. E. Dyer, *J. Opt. Soc. Am. B3,* 792 (1986).

[25] T. Gumpenberger, J. Heitz, D. Bäuerle andT. C. Rosenmayer, *Europhys. Lett.*,70, 831(2005).

[26] A. Řezníčková, Z. Kolská, V. Hnatowicz, P. Stopka and V. Švorčík, *Nucl. Instrum. Meth B,* 269, 83 (2011).

[27] G. Wu, M. D. Paz, S. Chiussi, J. Serra, P. Gonzalez, Y. J. Wang and B. Leon, *J. Mater. Sci.-Mater. Med.*, 20, 597 (2009).

[28] J. Siegel, P. Slepička, J. Heitz, Z. Kolská, P. Sajdl and V. Švorčík, *Appl. Surf. Sci.*, 256, 2205 (2010).

[29] K. S. Kim, C. M. Ryu, C. S. Park, G. S. Sur and C. E. Park, *Polymer,* 44, 6287 (2003).

[30] X. L. Zhu, S. B. Liu, B. Y. Man, C. Q. Xie, D. P. Chen, D. Q. Wang, T. C. Ye and M. Liu, *Appl. Surf. Sci.*, 253, 3122 (2007).

[31] K. Buscher, R. Berger, W. Brunger and K. Graf, *Microel. Eng.*, 83, 819 (2006).

[32] E. Bonaccurso and K. Graf, *Langmuir,* 20, 11183 (2004).

[33] V. Švorčík, V. Kotál, O. Bláhová, M. Špírková, P. Sajdl and V. Hnatowicz, *Nucl. Instrum. Meth., B* 244, 365 (2006).

[34] L. J. Gerenser, *J. Adhes. Sci. Technol,* 7, 1019 (1993).

[35] V. Kotál, V. Švorčík,P. Slepička, O. Bláhová, P. Šutta and V. Hnatowicz, *Plasma Proc. Polym.*, 4, 69 (2007).

[36] P. Slepička, A. Vasina, Z. Kolská, T. Luxbacher, P. Malinský, A. Macková, and V. Švorčík, *Nucl. Instrum. Meth. B,* 268, 2111 (2010).

[37] A. Larsson and H. Derandy, *J. Colloid Interf. Sci.*, 246, 214 (2002).

[38] Y. Karade, S. A. Pihan, W. H. Brünger, A. Dietzel, R. Berger and K. Graf, *Langmuir,* 25, 3108 (2009).

[39] Y. Karade, K. Graf, W. H. Brunger, A. Dietzel and R. Berger, *Microel. Eng.*, 84, 797 (2007).

[40] J. Siegel, A. Řezníčková, A. Chaloupka, P. Slepička and V. Švorčík, *Radiat. Eff. Defect. S,* 163, 779 (2008).

[41] V. Švorčík, V. Kotál, P. Slepička, O. Bláhová, P. Šutta and V. Hnatowicz, *Polym. Eng. Sci.*, 46, 1326 (2006).

[42] T. L. Barr, Modern ESCA, CRC Press, Boca Raton (1994).

[43] A. Řezníčková, Z. Kolská, V. Hnatowicz and V. Švorčík *J. Nanopart. Res.*, 13, 2929 (2011).

[44] M. Schrodner, R. I. Stohn, K. Schultheis, S. Sensfuss and H. K. Roth, *Org. Electron.*, 6, 161 (2005).

[45] Q. D. Ling, D. J. Liaw, Ch. Zhu, D. S. H. Chan, E. T. Kang and K. G. Neoh, *Prog. Polym. Sci.*, 33, 917 (2008).

[46] J. J. Ramsden, D. M. Allen, D. J. Stephenson, J. R. Alcock, G. N. Peggs, G. Fuller and G. Goch, *Ann. CIRP,* 56, 687 (2007).

[47] Ch. W. J. Berendsen, M. Škereň, D. Najdek and F. Černý, *Appl. Surf. Sci.*, 255, 9305 (2009).

[48] X. B. Yan, X. J. Chen, B. K. Tay and K. A. Khor, *Electrochem. Commun.*, 9, 1269 (2007).

[49] M. Fan, M. Thompson, M. L. Andrade and A. G. Brolo, *Anal. Chem.*, 82, 6350 (2010).

[50] A. Toma, D. Chiappe, D. Massabó, C. Boragno and F. B. Mongeot, *Appl. Phys. Lett.*, 93, 163104 (2008).

[51] M. Birnbaum, *J. Appl. Phys.*, 36, 3688 (1965).

[52] J. Heitz, E. Arenholz, D. Bäuerle, R. Sauerbrey and H. M. Phillips, *Appl. Phys. A,* 59, 289 (1994).

[53] D. Bäuerle, E. Arenholz, J. Heitz, S. Proyer, E. Stangl and B. Lukyanchuk, *Mater. Sci. Forum,* 173-174, 41 (1995).

[54] E. Herth, P. Tilmant, M. Faucher, M. François, Ch. Boyaval, F. Vaurette, Y. Deblocq, B. Legrand and L. Buchaillot, *Microelectron. Eng.*, 87, 2057 (2010).

[55] S. J. Kim, T. Ahn, M. Ch. Suh, Ch. J. Yu, D. W. Kim and S. D. Lee, *Jpn. J. Appl. Phys.*, 44, 1109 (2005).

[56] Y. Hirai, Y. Inamoto, K. Sugano, T. Tsuchiya and O. Tabata, *J. Micromech. Microeng.*, 17, 199 (2007).

[57] S. Partel, S. Zoppel, P. Hudek, A. Bich, U. Vogler, M. Hornung and R. Voelkel, *Microelectron. Eng.*, 87, 936 (2010).

[58] W. Guo, Y. B. Wang, L. Li, Z. Liu, B. Lukyanchuk and D. J. Whitehead, *J. Laser Micro Nanoeng.*, 2, 212 (2007).

[59] P. Slepička, E. Rebollar, J. Heitz and V. Švorčík, *Appl. Surf. Sci.*, 254, 3585 (2008).

[60] J. E. Sipe, J. F. Young, J. S. Preston and H. M. van Driel, *Phys. Rev. B,* 27, 1141 (1983).

[61] E. Arenholz, J. Heitz, M. Himmelbauer and D. Bäuerle, *Proc. SPIE,* 2777, 90 (1996).

[62] M. Bolle, S. Lazare, M. Le Blanc and A. Wilmes, *Appl. Phys. Lett.*, 60, 674 (1992).

[63] M. Csete and Z. Bor, *Appl. Surf. Sci.*, 133, 5 (1998).

[64] S. Lazare and P. Benet, *J. Appl. Phys.*, 74, 4953 (1993).

[65] M. Csete, R. Eberle, M. Pietralla, O. Marti and Z. Bor, *Appl. Surf. Sci.*, 208-209, 474 (2003).

[66] M. Bolle and S. Lazare, *J. Appl. Phys.*, 73, 3516 (1993).

[67] P. H. Geil, *Europhys. Conf. Abstr.*, 12D, 22 (1988).

[68] V. Švorčík, T. Hubáček, P. Slepička, J. Siegel, Z. Kolská, O. Bláhová, A. Macková and V. Hnatowicz, *Carbon,* 47, 1770 (2009).

[69] D. S. Dunn and A. J. Ouderkirk, *AIP Conf. Proc.*, 191, 375 (1989).

[70] J. Siegel,, J. Heitz and V. Švorčík, *Surf. Coat. Technol.*, 206, 517 (2011).

[71] V. Švorčík, J. Zehentner, V. Rybka, P. Slepička and V. Hanatowicz, *Appl. Phys., A* 75, 541 (2002).

[72] K. Efimenko, V. Rybka, V. Švorčík and V. Hnatowicz, *Appl. Phys., A* 68, 479 (1999).

[73] G. Gustafson, Y. Cao, N. Colaneri and A. J. Heeger, *Nature,* 357, 477 (1992).
[74] V. Švorčík, J. Siegel, P. Slepička, V. Kotál and M. Špirková, *Surf. Interf. Anal.,* 39, 79 (2007).
[75] P. Slepička, V. Švorčík, M. Šlouf, V. Rybka and M. Špirková, *Optoelectron. Adv. Mater. Rapid Com.,2,* 153 (2008).
[76] G. Singh, H. Bhunia, A. Rajor and V. Choudhary, *Polym. Bull.,* 66, 939 (2011).
[77] J. M. Carter, R. C. Fleming, T. A. Savas, M. E. Walsh and T. B. O'Reilly, *MTL Annu. Rep.,* 186, 215 (2003).
[78] Y. J. Liu, H. T. Dai and X. W. Sun, *J. Mat. Chem.* 21, 2982 (2011).
[79] P. Slepička, A. Chaloupka, P. Sajdl, J. Heitz, V. Hnatowicz and V. Švorčík, *Appl. Surf. Sci.,* 257, 6021 (2011).
[80] T. Lippert, *Plasma Process. Polym.,* 2, 525 (2005).
[81] C. Oehr, *Nucl. Instrum. Meth. B,* 208, 40 (2003).
[82] K. L. Mittal, Metallized Plastics: Fundamentals and Applications, Marcel Dekker, New York (1998).
[83] C. P. Wong, Polymers for Electronic and Photonic Applications, Academic Press, Boston (1993).
[84] N. Stutzmann, T. A. Tervoort, K. Bastiaansen and P. Smith, *Nature,* 407, 613 (2000).
[85] K. Akamatsu, N. Tsuboi, Y. Hatakenaha and S. Deki, *J. Phys. Chem. B,* 104, 10168 (2000).
[86] S. A. Wilson, R. P. J. Jourdain and Q. Zhang, *Mater. Sci. Eng. R,* 56, 1 (2007).
[87] K. L. Choy, *Progr. Mater. Sci.,* 48, 57 (2003).
[88] V. Zaporojtschenko, J. Zekonyte, A. Biswas and F. Faupel, *Surf. Sci.,* 532-535, 300 (2003).
[89] V. Švorčík, P. Slepička, J. Švorčíková, M. Špírková, J. Zehentner and V. Hnatowicz, *J. Appl. Polym. Sci.,* 99, 1698 (2006).
[90] A.Vasina, P. Slepička, J. Švorčíková, P. Sajdl, A. Macková and V. Švorčík, *J. Adhes. Sci. Technol.,* 24, 731 (2010).
[91] P. K. Chu, J. Y. Chen, L. P. Wang and N. Juany, *Mater. Sci. Eng., R* 36, 143 (2002).
[92] V. Švorčík and V. Hnatowicz, Polymer Degradation and Stability,Nova, New York, 2007.
[93] Z. Chang and J. A. La Verne, *J. Polym Sci. Polym. Chem.,* 38, 1656 (2000).
[94] L. Calcagno, G. Compagnini and G. Foti, *Nucl. Instrum. Meth., B* 65, 413 (1992).
[95] V. Švorčík, V. Rybka, V. Hnatowicz and K. Smetana, *J.Mater. Sci. Mater. Med.,* 8, 435 (1997).
[96] H. S. Choi, V. V. Rybkin, V. A. Tito, T. G. Sikova and T. A. Ageeva, *Surf. Coat. Technol.,* 200, 4479 (2006).
[97] M. T. Khorasani, H. Mirzadeh and S. Irani, *Radiat. Phys. Chem.,* 77, 280 (2008).
[98] Y. Wan, J. Yang, J. Yang, J. Bei and S. Wang, *Biomaterials,* 24, 3757 (2003).
[99] A. Fujinami, D. Matsunaka and Y. Shibutani, *Polymer,* 50, 716 (2009).
[100] Y. X. Wang, J. L. Robertson, W. B. Spillman and R. O. Claus, *Pharm. Res.,* 21, 1362 (2004).
[101] P. Slepička, S. Trostová, N. Slepičková Kasálková, Z. Kolská, P. Sajdl and V. Švorčík, *Plasma Proc. Polym.,* DOI: 10.1002/ppap.201100126.
[102] D. Bratton, D. Yang, J. Dai and C. K. Ober, *Polym. Adv. Technol.,* 17, 94 (2006).
[103] H. H. Solak, *J. Phys. D: Appl. Phys.,* 39, R171 (2006).

[104] Y. Xia, J. A. Rogers, K. E. Paul and G. M. Whitesides, *Chem. Rev.,* 99, 1823 (1999).
[105] S. Okazaki, *J. Vaccum Sci. Technol. B,* 9, 2829 (1991).
[106] T. Pedersen, P. Johansen, N. Holme, P. Ramanujam and S. Hvilsted, *Phys. Rev. Lett.,* 80, 89 (1998).
[107] J. Kumar, L. Li, X. Jiang, D. Kim, T. Lee and S. Tripathy, *Appl. Phys. Lett.,* 72, 2096 (1998).
[108] F. L. Labarthet, T. Buffeteau and C. Sourisseau, *J. Phys. Chem., B* 102, 2654 (1998).
[109] K. G. Yager and C. J. Barrett, *Curr. Opin. Sol. State Mater. Sci.,* 5, 487 (2001).
[110] D. D. Nolte, Photorefractive effects and materials, Kluwers Academic Publishers, Norwell (1995).
[111] S. R. Marder and J. W. Perry, *Adv. Mater,* 5, 804 (1993).
[112] J. H. Moon, J. Ford, S. Yang, *Polym. Adv. Technol.,* 17, 83 (2006).
[113] M. Maldovan and E. L. Thomas, Periodic Materials and Interference Lithography for Photonics, Phononics and Mechanics, Wiley, New York (2008).
[114] T. Y. M. Chan, O. Toader and S. John, *Phys. Rev. E,* 71, 046605 (2005).
[115] M. J. Escuti and G. P. Crawford, *Opt. Engin.,* 43, 1973 (2004).
[116] O. Lyutakov, I. Hüttel, J. Siegel and V. Švorčík, *Appl. Phys. Lett.,* 95, 173103 (2009).
[117] D. R. Smith, W. J. Padilla, D. C. Vier, S. C. Nemat-Nasser and S. Schultz, *Phys. Rev. Lett.,* 84, 4184 (2000).
[118] R. A. Shelby, D. R. Smith and S. Schultz, *Science,* 292, 77 (2001).
[119] C. G. Parazzoli, R. B. Greegor, K. Li, B. E. C. Koltenbah and M. Tanielian, *Phys. Rev. Lett.,* 90, 107401 (2003).
[120] J. D. Baena, R. Marques, F. Medina and J. Martel, *Phys. Rev. B,* 69, 014402-1 (2004).
[121] S. Linden, C. Enkrich, M. Wegener, J. Zhou, T. Koschny and C. M. Soukoulis, *Science,* 306, 1351 (2004).
[122] D. R. Smith, J. B. Pendry, and M. Wiltshire, *Science,* 305, 788 (2004).
[123] N. Engheta and R. W. Ziolkowski, Metamaterials: Physics and Engineering Explorations, Wiley, New York (2006).
[124] V. M. Shalaev, *Nat. Photon.,* 1, 41 (2007).
[125] C. M. Soukoulis, S. Linden and M. Wegener, *Science,* 315, 47 (2007).
[126] Ch. É. Kriegler, M. S. Rill, S. Linden and M. Wegener, *IEEE Quantum Electron.,*16, 367 (2010).
[127] D. R. Smith and J. B. Pendry, *Phys. Today,* 57, 37 (2004).
[128] C. M. Soukoulis, M. Kafesaki and E. N. Economou, *Adv. Mater.,* 18, 1941 (2006).
[129] M. Bayindir, K. Aydin, E. Ozbay, P. Markos and C. M. Soukoulis, *Appl. Phys. Lett.,* 81, 120 (2002).
[130] S. Zhang, W. J. Fan, N. C. Panoiu, K. J. Malloy, R. M. Osgood and S. R. J. Brueck,*Phys. Rev. Lett.,* 95, 137404 (2005).
[131] G. Dolling, C. Enkrich, M. Wegener, J. F. Zhou and C. M. Soukoulis, *Opt. Lett.,* 30, 3198 (2005).
[132] G. Dolling, C. Enkrich, M. Wegener, C. M. Soukoulis and S. Linden, *Science,* 312, 892 (2006).
[133] G. Dolling, C. Enkrich, M. Wegener, C. M. Soukoulis and S. Linden, *Opt. Lett.,* 31, 1800 (2006).
[134] C. Dolling, M. Wegener, C. M. Soukoulis and S. Linden, *Opt. Lett.,* 32, 53 (2007).

[135] A. N. Grigorenko, A. K. Geim, H. F. Gleeson, Y. Zhang, A. A. Firsov, I. Y. Khrushchev and J. Petrovic, *Nature,* 438, 335 (2005).
[136] A. N. Grigorenko, *Opt. Lett.,* 31, 2483 (2006).
[137] A. V. Kildishev, W. Cai, U. K. Chettiar and V. M. Shalaev, *New J. Phys.,* 10, 115029 (2008).
[138] R. G. Hunsperger, Integrated Optics: Theory and technology, *Springer,* New York (2002).
[139] C. U. Bang, A. Shishido and T. Ikeda, *Macromol. Rapid Commun.,* 28, 1040 (2007).
[140] E. Murphy, Integrated Optical Circuits and Components: Design and Applications, CRC Press, Boca Raton (1999).
[141] S. H. Nam, J. W. Kang and J. J. Kim, *Opt. Commun.,* 266, 332 (2006).
[142] G. S. He, L. X. Yuan, N. Cheng, J. D. Bhawalkar, P. N. Prasad, L. L. Brott, S. J. Clarson and B. A. Reinhardt, *J. Opt. Soc. Amer. B - Opt. Phys.,* 14, 1079 (1997).
[143] K. Ju, J. S. Lim, C. Lee, D. H. Choi and D. W. Kim, *Mol. Cryst. Liquid Cryst.,* 491, 152 (2008).
[144] T. Kavc, G. Langer, W. Kern,G. Kranzelbinder, E. Toussaere, G. A. Turnbull, I. D. W. Samuel, K. F. Iskra, T. Neger and A. Pogantsch, *Chem. Mater.,* 14, 4178 (2002).
[145] G. Kranzelbinder, E. Toussaere, D. Josse and J. Zyss, *Synth. Met.,* 121, 1617 (2001).
[146] G. D. Peng and P. L. Chu, *Fiber Integ. Opt.,* 19, 277 (2000).
[147] Y. Zhang, D. J. Feng, Z. G. Liu, Z. Y. Guo, X. Y. Dong, K. S. Chiang, and B. C. B. Chu, *IEEE Photon. Techn. Lett.,* 13, 618 (2001).
[148] C. Lu and R. H. Lipson, *Laser. Photon. Rev.,* 4, 568 (2010).
[149] G. S. Buckley and C. M. Roland, *Polym. Eng. Sci.,* 37, 138 (1997).
[150] H. E. Pudavar, M. P. Joshi, P. N. Prasad and B. A. Reinhardt, *Appl. Phys. Lett.,* 74, 1338 (1999).
[151] A. S. Dvornikov, C. M. Taylor, Y. C. Liang and P. M. Rentzepis, *J. Photochem. Photobiol., A.* 112, 39 (1998).
[152] H. Strickler and W. W. Webb, *Opt. Lett.,* 16, 1780 (1991).
[153] C. Gu, Y. Xu,Y. S. Liu, J. J. Pan, F. Q. Zhou and H. He, *Opt. Mater.,* 23, 219 (2003).
[154] F. Liu, P. Rugheimer, E. Mateeva, D. E. Sabate and M. G. Lagally, *Nature,* 416, 498 (2002).
[155] V. J. Lumelsky, M. S. Shuk and S. Wagner, *IEEE Sens.,* 1, 41 (2001).
[156] S. Lal, N. K. Grady, J. Kundu, J. B. Lassiter and N. J. Halas, *Chem. Soc. Rev.,* 37, 898 (2008).
[157] C. Kan, X. Zhu and G. J. Wang, *J. Phys. Chem., B* 110, 4651 (2006).
[158] H. B. Liu, J. A. Ascencio, M. Perez-Alvarez and M. Yacaman, *J. Surf. Sci.,* 491, 288 (2001).
[159] M. B. Mohamed, Z. L. Wang and M. A. El-Sayed, *J. Phys. Chem. A,* 103, 10255 (1999).
[160] E. Roduner, *Chem. Soc. Rev.* 35, 583 (2006).
[161] J. Tauc, Amorphous and Liquid SemiconductorsSpringer, Heidelberg (1974).
[162] Z. Kolská, J. Siegel and V. Švorčík, *Coll. Czech. Chem. Commun.,* 75, 517 (2010).
[163] W. Fisher, H. Greiger, P. Rudolf and P. Wissmann, *Appl. Phys.,* 13, 245 (1977).
[164] C. D. Hodgman, Handbook of Chemistry and Physics,Chemical Rubber, Cleveland (1975).
[165] K. Chopra, Thin Film Phenomena, Wiley, New York (1969).

[166] I. Doron-Mor, Z. Barkay, N. Filip-Granit, A. Vaskevisch and I. Rubinstein, *Chem. Mater.*, 16, 3476 (2004).
[167] V. Švorčík, O. Kvítek, O. Lyutakov, J. Siegel and Z. Kolská, *Appl. Phys. A,* 102, 747 (2011).
[168] K. Hauplt, M. Lang and P. Wissmann, *Surf. Interface Anal.*, 9, 27 (1986).
[169] I. Shyjumon, M. Gopinadhan, O. Ivanova, M. Quaas, H. Wulff, C. A. Helm and R. Hippler, *Eur. Phys. J.*, D37, 409 (2006).
[170] M. Hovel, B. Gompf and M. Dressel, *Phys. Rev., B* 81, 035402 (2010).
[171] V. Švorčík, A. Řezníčková, Z. Kolská and P. Slepička, *e-Polymers,* 133,1 (2010).
[172] M. Haruta, *Gold Bull.*, 37, 27 (2004).
[173] M. C. Daniel and D. Astruc, *Chem. Rev.*, 104, 293 (2004).
[174] D. V. Goia, *J. Mater. Chem.*, 14, 451 (2004).
[175] M. M. C. Cheng, G. Cuda, Y. L. Bunimovich, J. R. Heath, C. A. Mirkin, A. J. Nijdam, R. Terracciano, T. Thundat and M. Ferrari, *Curr. Opin. Chem. Biol.*, 10, 11 (2006).
[176] K. Hering, D. Cialla, K. Ackermann, T. Dörfer, R. Müller, R. Mattheis, W. Fritzsche and R. Rösch, *J. Popp, Anal. Bioanal. Chem.*, 390, 113 (2008).
[177] J. F. Hainfeld, D. N. Slatkin and H. M. Smilowitz, *Phys. Med. Biol.*, 49, 309 (2004).
[178] A. K. Sinha, S. Seelan, S. Tsubota and M. Haruta, *Top. Catal.*, 29, 95 (2004).
[179] T. de Beer, N. H. Velthorst, U. A.Th. Brinkman and C. Gooijer,*J. Chromatography A,* 971,1 (2002).
[180] A. Otto, I. Mrozek, H. Grabhorn and W. Akemann, *J. Phys.: Condens. Matter.*, 4,1143 (1992).
[181] B. Sepúlveda, P. C. Angelomé, L. M. Lechuga and L. M. Liz-Marzán, *Nano Today,* 4, 244 (2009).
[182] L. Guzman, A. Miotello, R. Checchetto, M. Adami, *Surf. Coat. Technol.*, 158-159, 558 (2002).
[183] H. K. Pulker, Coatings on Glass, Elsevier, Amsterdam (1984).
[184] R. J. Tseng and J. X. Huang, *J. Ouyang, Nano Lett.*, 5, 1077 (2005).
[185] T. Ishida, K. Kuroda, W. Minagawa and M. Haruta, *J. Colloid Interf. Sci.*, 323, 105 (2008).
[186] J. Heitz, V. Švorčík, L Bačáková, K. Ročková, E. Ratajová, T. Gumpenberger, D. Bauerle, B. Dvořánková, H. Kahr, I. Graz and C. Romanin, *J. Biomed. Mater. Res.*, 67A, 130 (2003).
[187] R. Mikulíková, S. Moritz, T. Gumpenberger, M. Olbrich, C. Romanin, L. Bačáková V. Švorčík and J. Heitz, *Biomaterials,* 26, 5572 (2005).
[188] V. Švorčík, A. Chaloupka, K. Záruba, V. Král, O. Bláhová, A. Macková and V. Hnatowicz, *Nucl. Instrum. Meth. B,* 267, 2484 (2009).
[189] V. Švorčík, N. Kasálková, P. Slepička, K. Záruba, V. Král, L. Bačáková, M. Pařízek, T. Ruml, H. Gbelcová, S. Rimpelová and A. Macková, *Nucl. Instrum. Meth. B,* 267, 1904 (2009).
[190] V. Švorčík, Z. Kolská, O. Kvítek, J. Siegel, A. Řezníčková, P. Řezanka and K. Záruba, *Nanoscale Res. Lett.*, accepted.
[191] V. Švorčík, A. Řezníčková, P. Sajdl, Z. Kolská, Z. Makajová and P. Slepička, *J. Mater. Sci.*, 46, 7917 (2011).
[192] Y. L. Zhou, X. Zheng, L. Z. Wu, J. F. Zhi and M. T. Xu, *J. Mater. Sci.*, 46, 1139 (2011).

[193] N. Hajduková, M. Procházka, J. Štěpánek and M. Špírková, *Colloid Surf. A,* 301,264 (2007).

[194] O. Seitz, M. M. Chehimi, E. Cabet-Deliry, S. Truong, N. Felidj, C. Perruchot, S. J. Greaves and J. F. Watts, *Colloid Surf. A,* 218, 225 (2003).

[195] E. M. S. Azzam, A. Bashir, O. Shekhah, A. R. E. Alawady, A. Birkner, C. Grunwald and C. Wöll, *Thin Solid Films,* 518, 387 (2009).

[196] Y. Kawakami, T. Seto, T. Yoshida and E. Ozawa, *Appl. Surf. Sci.,* 197-198, 587 (2002).

[197] N. Kasálková, Z. Makajová, M. Pařízek, P. Slepička, K. Kolářová, L. Bačáková, V. Hnatowicz and V. Švorčík, *J. Adhes. Sci. Technol.,* 24, 743 (2010).

[198] J. E. Puskas, E. A. Foreman-Orlowski, G. T. Lim, S. E. Porosky, M. M.Evancho-Chapman, S. P. Schmidt, M. El Fray, M. Pátek, P. Prowans and K. Lovejoy, *Biomaterials,* 31, 2477 (2010).

[199] I. Armentano, L. Marinucci, M. Dottori, S. Balloni, E. Fortunati, M. Pennacchi, E. Becchetti, P. Locci and J. M. Kenny, *J. Biomat. Sci. E,* 22, 541.

[200] F. D. Matl, A. Obermeier, S. Repmann, W. Friess, A. Stemberger and K. D. Kuehn, *Antimicrob. Agents Ch.,* 52, 1957 (2008).

[201] [201] M. P. Krafft and J. G. Riess, *J. Polym. Sci. Pol. Chem.,* 45, 1185 (2007).

[202] O. Kubová, V. Švorčík, J. Heitz, S. Moritz, C. Romanin, P. Matějka and A. Macková, *Thin Solid Films,* 515, 6765 (2007).

[203] H. Matsuo, K. Kanasugi, Y. Ohgoe, K. K. Hirakuri and Y. Fukui. *Diam. Relat. Mater.,* 18, 1031 (2009).

[204] G. Dearnaley and J. H. Arps, *Surf. Coat. Tech.,* 200, 2518 (2005).

[205] Y. Sun, Y. L. Yao, X. Y. Zhu, D. M. Zhang, G. Chen and K. Zhong. *New Carb. Mat.,22,* 40 (2007).

[206] J. Besold, R. Thielsch, N. Matz, C. Frenzel, R. Born and A. Mobius, *Thin Solid Films,* 293, 96 (1997).

[207] K. R. Peters, *J. Microsc. Oxford,* 133, 17 (1984).

[208] Z. G. Lu and C. Y. Chung. *Diam. Relat. Mater.,* 17, 1871 (2008).

[209] D. M Mattox, Handbook of physical vapour deposition (PVD) processing, Noyes-Park Ridge, New York (1998).

[210] X. D. Zhu, F. Ding, H. Haramoto and K. Narumi, *Appl. Surf. Sci.,* 253, 1480 (2006).

[211] J. Robertson, *Surf. Coat. Technol.,* 50, 185 (1992).

[212] K. R. Peters, *Scann Electron. Microsc.,* 16, 143 (1980).

[213] D. M. Mattox, A. V. Mullendore and F. N. Rebarchik, *J. Vac. Sci. Technol.,* 4, 123 (1967).

[214] J. A. Greer and M. D. Tabat. *J. Vac. Sci. Technol. A,* 13, 1175 (1995).

[215] J. A. McLaughlin and P. D. Maguire. *Diam. Relat. Mater.,* 17, 873 (2008).

[216] E. Uchida, H. Iwata and Y. Ikada, *Polymer,* 41, 3609 (2000).

[217] A. M. Ektessabi and K. Yamaguchi, *Thin Solid Films,* 377-378, 793 (2000).

[218] K. Záruba, P. Matějka, R. Volf, K. Volka, V. Král and J. L. Sessler, *Langmuir,* 18,6896 (2002).

[219] P. Yang, S. C. H. Kwok, R. K. Y. Fu, Y. X. Leng, J. Wang, G. J. Wan, N. Huang, Y. Leng and P. K. Chu, *Surf. Coat. Technol.,* 177-178, 747 (2004).

[220] http://www.goodfellow.com/catalogue PTFE, HTMLGoodfellow, 13.10.2011.

[221] D. Pradhan and M. Sharon, *Appl. Surf. Sci.* 253, 7004 (2007).

[222] N. Mott, Metal Insulator Transitions, Taylor and Francis, London (1990).

[223] M. Lonfat, B. Marsen and K. Sattler. *Chem. Phys. Lett.,* 313, 539 (1999).

[224] J. Singh, Physics of Semiconductors and Their Heterostructures, McGraw-Hill, New York (1993).

[225] K. Ročková, V. Švorčík, L. Bačáková, B. Dvořánková and J. Heitz, *Nucl. Instrum. Meth. B,* 225, 275 (2004).

[226] V. Švorčík, K. Ročková, E. Ratajová, J. Heitz, N. Huber, D. Bäuerle, L. Bačáková, B. Dvořánková and V. Hnatowicz, *Nucl. Instrum. Meth. Phys. Res. B,* 217, 307 (2004).

[227] K. Walachová, V. Švorčík, L. Bačáková and V. Hnatowicz, *Biomaterials,* 23, 2989 (2002).

[228] J. M. Grace and L. J. Gerenser, *J. Dispers. Sci. Technol.,* 24, 305 (2003).

[229] [229] Y. Choi, J. Kim, K. Paek, W. Ju and Y. S. Hwang, *Surf. Coat. Technol.,* 193, 319 (2005).

[230] M. Kuzuya, S. Kondo, M. Sugito and T. Yamashiro, *Macromolecules,* 31, 3225 (1998).

[231] N. Inagaki, S. Tasaka and S. Shimada, *J. Appl. Polym. Sci.,* 79, 808 (1999).

[232] L. Bačáková, V. Mareš, M. G. Bottone, C. Pellicciari, V. Lisá and V. Švorčík, *J. Biomed. Mater. Res.,* 49, 369 (2000).

[233] L. Bačáková, K. Walachová, V. Švorčík and V. Hnatowitz, *J. Biomater. Sci. Polym. Ed.,* 12, 817 (2001).

[234] R. Li, L. Ye and Y. W. Mai, *Composites,* 28A, 73 (1997).

[235] M. Tahara, N. K. Cuong and Y. Nakashima, *Surf. Coat. Technol.,* 174, 826 (2003).

[236] V. Švorčík, K. Ročková, B. Dvořánková, V. Hnatowicz, R. Öchsner and H. Ryssel, *J. Mater. Sci.,* 37, 1183 (2002).

[237] V. Švorčík, V. Hnatowicz, P. Stopka, L. Bačáková, J. Heitz and H. Ryssel, *Rad. Phys. Chem.,* 60, 89 (2001).

[238] T. J. Webster and T. A. Smith, *J. Biomed. Mater. Res. A,* 74, 677 (2005).

[239] D. C. Miller, A. Thapa, K. M. Haberstroh and T. J. Webster, *Biomaterials,* 25, 53 (2004).

[240] T. J. Webster, C. Ergun, R. H. Doremus and R. Bizios, *J. Biomed. Mater. Res.,* 51, 475 (2000).

[241] C. J. Gannon, C. R. Patra, R. Bhattacharya, P. Mukherjee and S. Curley, *J. Nanobiotechnol.,*6, 2 (2008).

[242] A. C. Curry, M. Crow and A. Wax, *J. Biomed. Opt.,* 13, 014022 (2008).

[243] G. Wang and H. Uludag, *Expert Opin. Drug DeliVery5*, 499 (2008).

[244] P. K. Jain, X. Huang, I. H. El-Sayed and M. A. El-Sayed, *Acc. Chem. Res.,* 41, 1578 (2008).

[245] R. Shukla, V. Bansal, M. Chaudhary, A. Basu, R. R. Bhonde and M. Sastry, *Langmuir,* 21, 10644 (2005).

[246] M. Hulander, J. Hong, M. Andersson, F. Gerveń, M. Ohrlander, P. Tengvall and H. Elwing, *Appl. Mat. Interf.,* 1, 1053 (2009).

[247] M. Giersig and P. Mulvaney, *Langmuir,* 9, 3408 (1993).

[248] G. A. DeVries, M . Brunnbauer, Y. Hu, A. M. Jackson, B. Long, B. T. Neltner, O. Uzun, B. H. Wunsch and F. Stellacci, *Science,* 315, 358 (2007).

[249] R. A. Sperling, P. Rivera Gil, F. Zhang, M. Zanella and W. J. Parak, *Chem. Soc. Rev.,* 37, 1896 (2008).

[250] D. Shenoy, W. Fu, J. Li, C. Crasto, G. Jones, Ch. DiMarzio, S. Sridhar and M. Amiji, *Int. J. Nanomedicine,* 1, 51 (2006).

[251] S. Guimond and M. R. Wertheimer, *J. Appl. Polym. Sci.*, 94, 1291 (2004).

[252] V. Švorčík, A. Chaloupka, P. Řezanka, P. Slepička, Z. Kolská, N. Kasálková, T. Hubáček and J. Siegel, *Rad. Phys. Chem.*, 79, 315 (2010).

[253] L. Guo, N. Kawazoe, Y. Fan, Y. Ito, J. Tanaka, T. Tateishi, X. Zhang and G. Chen, *Biomaterials,* 29, 23 (2008).

[254] C. Werner, U. König, A. Augsburg, Ch. Arnhold, H. Körber, R. Zimmermann and H. - J. Jacobasch, *Colloids Surf. A,* 159, 519 (1999).

[255] G. Borcia, C. A. Anderson and N. M. D. Brown, *Appl. Surf. Sci.*, 225, 186 (2004).

[256] E. Filová, L. Bačáková, V. Lisá, D. Kubies, L. Machová, M. Lapčíková and F. Rypáček, *Eng. Biomater.*, 7, 19 (2004).

[257] H. C. Liu, I. C. Lee, J. H. Wang, S. H. Yang and T. H. Young, *Biomaterials,* 25, 4047 (2004).

[258] T. Hubáček, O. Lyutakov, V. Rybka and V. Švorčík, *J. Mater. Sci.,* 45, 279 (2010).

In: Nanostructures
Editor: Yu Dong

ISBN: 978-1-62618-081-9

Chapter 2

SYNTHESIS, CHARACTERIZATION, AND APPLICATION OF NANO CUPRIC OXIDE

Qiaobao Zhang[1], Daguo Xu[1], Xiang Zhou[1], Zhiqiang Qiao[2], Guangcheng Yang[2], Fude Nie[2] and Kaili Zhang[1,*]

[1]Department of Mechanical and Biomedical Engineering,
City University of Hong Kong, Kowloon, Hong Kong
[2]Institute of Chemical Materials, China Academy of Engineering Physics,
Mianyang, Sichuan, People's Republic of China

ABSTRACT

Nanomaterials are receiving increasing interests because of their remarkable properties and applications. Among various nanomaterials, cupric oxide (CuO) nanostructures have been widely studied due to their promising applications in catalysis, superconductors, electrodes for batteries, field emissions, energetic materials, superhydrophobic surfaces, gas sensors, bio-sensors, initiators, field effect transistors, solar cells, and others.

This chapter deals with the synthesis, characterization, and applications of various CuO nanostructures including CuO 3D/2D nanostructures, CuO 1D nanostructures, and CuO nanoparticles. After a general introduction to CuO nanostructures, CuO 3D/2D nanostructures,1D nanostructures, and nanoparticles are addressed separately.Finally, conclusions and some guidelines for future investigations are provided.

INTRODUCTION

Nanomaterials are being widely studied due to their interesting properties and promising applications. As a p-type semiconductor with a narrow band gap (1.2eV), nanoscale cupric

* Corresponding author: Kaili Zhang, Department of Mechanical and Biomedical Engineering, City University of Hong Kong, 83 Tat Chee Avenue, Hong Kong, E-mail: kaizhang@cityu.edu.hk; Tel: (852)34427845; Fax: (852)34420172.

oxide (CuO) has been extensively investigated because ofits promising applications in various areas including catalysis, superconductors, anode electrodes for batteries, field emissions, energetic materials, superhydrophobic surfaces, gas sensors, bio-sensors, initiators, field effect transistors, solar cells, and others.

The synthesis, characterization, and applications of various CuO nanostructures including three-dimensional (3D), two-dimensional (2D), one-dimensional (1D) CuO nanostructures, and CuO nanoparticles will be covered in this chapter. Firstly, the synthesis, properties, and applications of various CuO 3D/2D nanostructures such as walls, honeycombs, platelets, spindles, flowers, etc. will be reviewed. Secondly, CuO 1D nanostructuresincluding nanowires, nanorods, and nanobelts have also been widely studiedowing to their unique properties and extensive applications.

The synthesis methods, properties and applications of CuO 1D nanostructures will be presented. Among different CuO nanostructures, CuO nanoparticles have been paid special attention because of their various applications. The third part of this chapter will give a review of the recent progresses in CuO nanoparticles including the synthesis methods, properties and applications.

Finally, conclusions and guidelines for future investigations will be provided.

CuO 3D/2D Nanostructures

It is widely accepted that theunique properties of nanomaterials are not only closely related to their sizes but also to their shapes [1-5]. Therefore, controlling the morphologies of nanomaterials is one of the most important issues and effective ways to obtain desirable properties [5-8].

Nanostructures with various shapes such as wires, tubes, honeycombs, platelets, spindles, flowers have beensynthesized using a variety of methodologies [9-11]. Recently, many efforts have been directed toward the fabrication of CuOnanostructuresdue to their promising applications.

Among various CuO nanostructures, CuO 3D/2D nanostructures such as walls, honeycombs, platelets, spindles, flowers, etc. have been extensively investigated due to their large specific surface areas and potentiality for fabricating nanodevices [12-13].

Different methods have been employed to synthesize CuO 3D/2D nanostructures [14-15]. Generally, they can be divided into two major categories. One is the direct synthesis of CuO 3D/2D nanostructures with or without the assistance of surfactants and templates including hydrothermal method, solution-based chemical precipitation method and microwave-assisted synthesis method.

The other is to use copper hydroxide [$Cu(OH)_2$] or basic copper salts as sacrificial precursors/templates, such as ultrasonic assisted synthesis method and solid-state thermal conversion of precursor method [7].

This section will mostly focus on these five methods for synthesizing CuO 3D/2D nanostructures with various morphologies and their related properties and applications.

1. Synthesis And Characterization of CuO 3D/2D Nanostructures

1.1. Hydrothermal Synthesis Method

Hydrothermal synthesis has been widely used to generate different kinds of nanomaterials because its reaction system is simple/green and the post-treatment is convenient. Furthermore, the hydrothermal method possesses some other advantages: (i) many inorganic salts can be well dissolved in water, making it very flexible to adjust the source of the metal ions in accordance with the requirements; (ii) water is low-cost, non-toxic and environment friendly; (iii) small coordinating molecules can be easily applied to modulate the growth of the final nanocrystals; (iv) the strong polarity of water is favorable for the orientation growth of nanocrystals [16].

3D branch-like and 2D flake-like CuO nanostructureshave been synthesized by a hydrothermal process in the presence of sodium citrate (SC) by simply controlling the reaction conditions [17].

Results show that when the molar ratio of sodium citrate to$CuSO_4 \cdot 5H_2O$ (represented by SC/Cu^{2+}) is more than 1, 3D branch-like CuO nanostructures are attained. In the absence of citrate, 2D flake-like CuO nanostructures are achieved, suggesting that the citrate acts as a modifying and reducing agent and plays a critical role in the determination of the morphology and in the control of the phase of the CuO nanostructures during the hydrothermal process. Moreover, the bandgap energies of the 2D flake-like, and 3D branch-like CuO nanostructures are estimated to be 1.60and 1.40 eV, respectively, which are both larger than that reported (Eg =1.2 eV) for bulk CuO. This indicates that the bandgap can be tuned by controlling the morphology of the CuO nanostructures.

Yang et al. [18-19] also used the hydrothermal method to synthesize flower-like, boat-like, plate-like and ellipsoid-like CuO nanostructures with $Cu(NO_3)_2 \cdot 3H_2O$ as the copper source, NaOH as the precipitator, PEG and $NH_3 \cdot H_2O$ as the morphology modulating materials. The morphologies and structures of the obtained CuO were found to be largely affected by reactants, hydrothermal temperature and time.

As can be seen from Figure 1(a), flower-like CuO nanostructures were formed when the time of hydrothermal treatment was 0.5 h. When the hydrothermal treatment time was increased to 2 h, well-defined boat-like structures (Figure 1(b)) were achieved in a high yield (98%).

While without addition of $NH_3 \cdot H_2O$, ellipsoids of uniform shape and size (Figure 1(c)) were obtained on a large scale. When $Cu(NO_3)_2$, NaOH, and $NH_3 \cdot H_2O$ were hydrothermally treated at 150°C without addition of PEG, 2D nanoplates of polygonal shapes were obtained (Figure 1(d)).

These results indicate that flower-like and boat-like CuO nanostructures can be successively fabricated by simply increasing the hydrothermal time in the presence of both PEG and $NH_3 \cdot H_2O$. Plate-like and ellipsoid-like CuO nanostructures were produced by modulating the use of PEG and $NH_3 \cdot H_2O$. The formation mechanisms were given in Figure 2, which show that both PEG and $NH_3 \cdot H_2O$ play important rolesin the formation of the CuO nanostructure with different morphologies and structures.

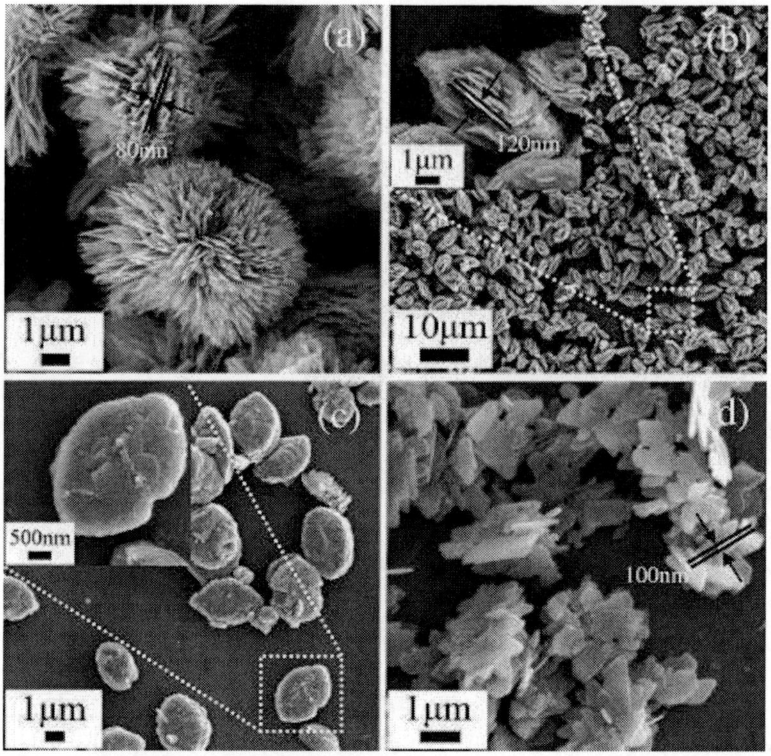

Figure 1. Typical SEM images of CuO nanostructures: (a) flower-like CuO, (b) boat-like CuO, (c) ellipsoid-like CuO, and (d) plate-like CuO [19].

A facile anion-controlled mild hydrothermal method was demonstrated for the construction of CuO nano-architectures on copper foils in [20]. With introducing WO_4^{2-} and MoO_4^{2-} into the reaction system, honeycomb- and flower-like CuO assemblies built from nanowires and nanoribbons were successfully synthesized. On the other hand, by adding certain amount of anions with smaller volume than that of WO_4^{2-} and MoO_4^{2-}, no special CuO architectures were formed.

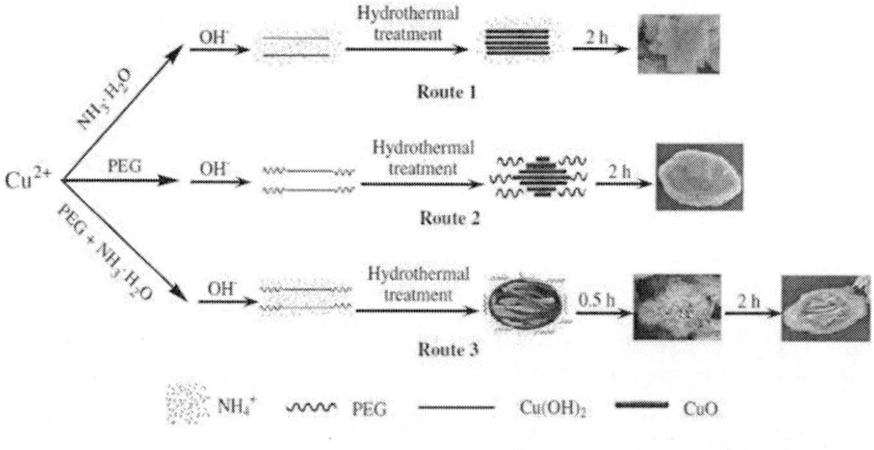

Figure 2.Schematic illustration of the growth of different CuO nanostructures under varied conditions [18].

It can be deduced that different inorganic anions will result in dissimilar morphologies of the product, and anions with larger volumes play key roles in the formation of diverse and novel assemblies of the CuO nanostructures.

The obtained CuO nanostructures in [20] offer exciting opportunities for potential applications in catalysis, electrochemistry, superconductivity, and super-hydrophobic coating.

Urchin-like CuO (a sphere with the diameter of 1 μm) nanostructures, which consist of closely packed nanorods with a diameter of 10 nm, were also fabricated by a PEG-assisted hydrothermal route at a relatively low temperature of 100 °C. No Cu(II)Cu(I) reduction by PEG was observed. This was attributed to high PEG concentration by the authors [21].

Single-crystalline CuO nanoplatelets (Figure 3(a)) were prepared on a large scale by a facile hydrothermal treatment of cupric dodecylsulfate ($Cu(DS)_2$) and NaOH aqueous solution at 120 °C. The $Cu(DS)_2$ was used as both a structure directing agent and a Cu^{2+} resource agent [22]. The average thickness, width and length of the as-prepared CuO nanoplatelets are about 50 nm, 120–300 nm and 480–700 nm, respectively.

The bandgap of the CuO nanoplatelets is found to be about 2.60 eV, which is much bigger than the reported value (E_g = 1.85 eV)for bulk CuO. In [23], well-aligned arrays of CuO nanoplatelets (Figure 3(b)) were also prepared by a hydrothermal method at 140 °C from CuCl and concentrated ammonia. These nanoplatelets are 75 nm in thickness, 200 nm in width, and 1 um in length with [010] growth direction (Figures 3(c-d)).

It was found that the well-aligned arrays of CuO nanoplatelets while acting as anode materials for lithium-ion batteries can exhibit a high initial discharge capacity of 788 mAh/g. This electrode exhibits rather good electrochemical performances with a specific capacity of about 300 mAh/g after 50 cycles with a slow capacity fading rate and a relatively high Coulombic efficiency.Moreover, CuO nanoplatelets (with 10 nm thickness and 1μm width) were also obtained by a hydrothermal oxidation in a solution containing H_2O_2 and NaOH [24]. The formation mechanism can be described as follows:

$$Cu+H_2O_2 \rightarrow Cu(OH)_2 \quad (1)$$

$$Cu(OH)_2 + 2OH^- \rightarrow Cu(OH)_4^{2-} \quad (2)$$

$$2Cu(OH)_4^{2-} \rightarrow Cu_2(OH)_6^{2-} + 2OH^- \quad (3)$$

$$Cu(OH)_4^{2-} + Cu_{n-1}(OH)_{2n}^{2-} \rightarrow Cu(OH)_{2n+2}^{2-} + 2OH^- \quad (4)$$

$$Cu(OH)_{2n+2}^{2-} \rightarrow n\,CuO + nH_2O + 2OH^- \quad (5)$$

Additionally, sheaf-like CuO consisting of nanoplatelets were also synthesized by a hydrothermal method at 100 °C from $Cu(NO_3)_2 \cdot 3H_2O$ and NaOH in the presence of an ionic liquid1-butyl-3-Me imidazolium tetrafluoroborate ($[BMIM]BF_4$) which served as a cosolvent and modifier in the reaction system [25].Furthermore, the rod-like and dandelion-like CuO nanomaterials were prepared by the decomposition of a copper complex $[Cu(pbbt)Cl_2]_2$ CH_3OH(pbbt 1,1′-(1,3-propylene)-bis-1H-benzotriazole) in the presence of suitable surfactants and alkalies under hydrothermal conditions [26]. The average diameter of the dandelion-like CuO microspheres is 2 μm. It was found that the sensitivity of the CuO nanorods was better than that of dandelion-like CuO particles.

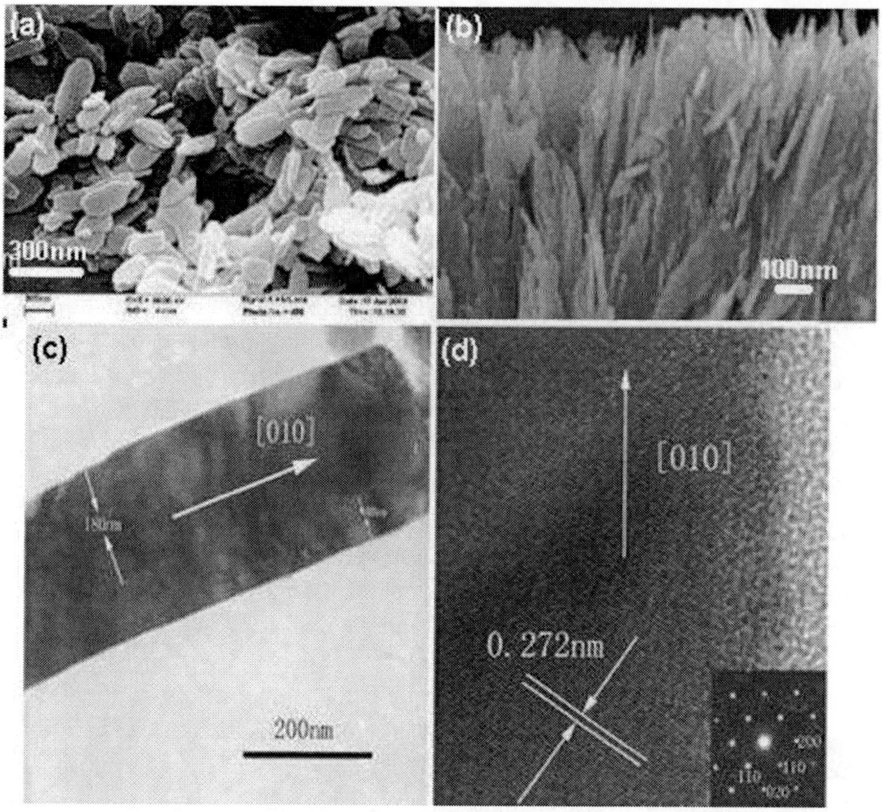

Figure 3. SEM images of (a) single-crystalline CuOnanoplatelets [22], (b) well-aligned arrays of CuO nanoplatelets [23],and (c-d) TEM/HRTEM images of the well-aligned arrays of CuO nanoplatelets [23].

1.2. Solution-Based Chemical Precipitation Method

Yang et al. reported a simple solution route to prepare various CuO nanostructures including nanoribbons and their spherical or flower-like hierarchical assemblies with $Cu(NO_3)_2{\cdot}3H_2O$ and NaOH as the starting materials [27]. The authors found that the morphologies of CuO were largely affected by the molar ratio of NaOH to $Cu(NO_3)_2{\cdot}3H_2O$. When the molar ratio was increased to 200, large-scale and uniform sphere-like or flower-like CuO nanostructures were achieved. These uniform flower-like assemblies (Figure 4(a)) are made up of dozens of loosely compacted thin CuO nanorods of 1.5–2 mm in length and 500–800 nm in width with cuspate tips. Sphere-like CuO nanostructures (Figure 4(b)) are made up of hundreds of compactly self-assembled nanorods with the breadths of 50–400 nm and lengths of 1–2 mm. The formation mechanism of the CuO spherical and flower-like assemblies is illustrated in Figure 5.

In Ref. [28], various CuO nanostructures including spindly-, belt-, and spherical-like CuO nanocrystals have also been prepared by a facile solution-phase method. The method is based on a convenient reaction between $Cu(NO_3)_2$ and NaOH in aqueous solution without using any surfactants.

Figure 4. SEM images of(a) flower-like and (b) sphere-like CuO nanostructures [27].

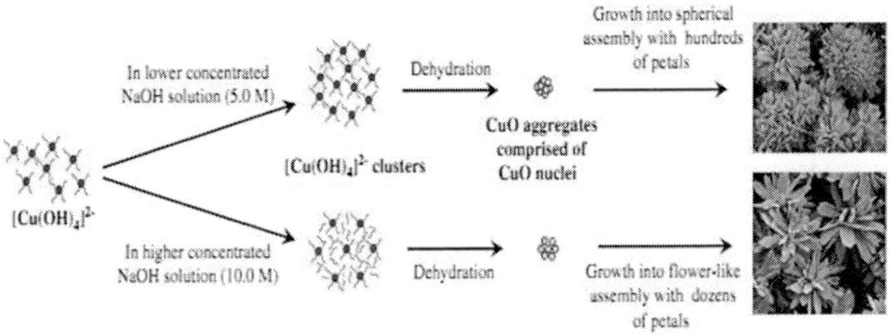

Figure 5. Illustration of the formation mechanism of spherical and flower-like CuO assemblies from $[Cu(OH)_4]^{2-}$ [27.

It was found that the obtained different shapes of CuO nanocrystals can be significantly affected by modulating the adding temperatures of precipitator and they exhibit different Raman properties.Additionally, Singh et al. reported the synthesis of different CuO nanostructures (rectangles, seed-like, belt-like, and sheet-like) in a solution phase with a high yield at low cost by simple reduction of aqueous solution of copper nitrate ($Cu(NO_3)_2$0.2 M) with different alkaline solutions of sodium hydroxide (NaOH 0.1, 0.25, 0.50, 0.75, and 1.0 M) [14]. The obtained different morphologies of CuO nanostructures were synthesized by thermal dehydration of the corresponding different $Cu(OH)_2$ nanomaterials, which were prepared by the reduction of $Cu(NO_3)_2$ solution with different molar concentrations of NaOH solution. The tentative formation mechanism of the CuO nanostructures was given in Figure 6. Various $Cu(OH)_2$ nanostructures are first obtained by addition of sodium hydroxide solution to copper nitrate solution.And then, these $Cu(OH)_2$ nanostructures are transformed into the corresponding nanostructures of CuO after annealing at 80 °C for half an hour.

Furthermore, this chemical precipitation method can also be used to synthesize more complex CuO nanostructures in large-quantity at low-temperature and short time.Monodispersed flower-shaped CuO nanostructures (Figure 7) have been successfully fabricated by a solution process at 100 °C simply by reacting copper nitrate, NaOH, and hexamethylenetetramine (HMTA) for 3 h without the use of any complex apparatus and reagents [29].

The authors observed that the large-quantity flower-shaped morphologies were strongly dependent on the concentration of HMTA, presence or absence of NaOH and HMTA, and the reaction time.

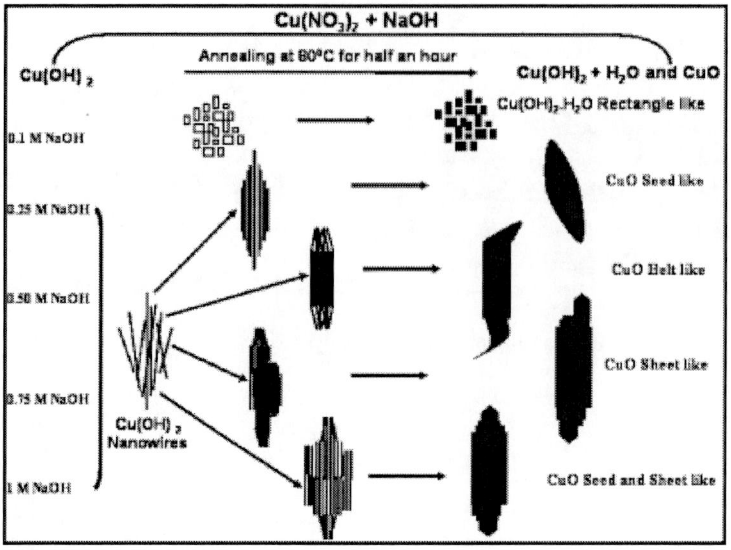

Figure 6.Schematic illustration of the growth of different $Cu(OH)_2$ and CuO nanoarchitectures under different conditions [14].

Figure 7. SEM images of (a) individualpetal-like CuOnanostructures and (b) flower-like CuOnanostructures [29].

The possible growth mechanism for the formation of flower-shaped CuO products is illustrated in Figure 8.As shown in Figure 7, when the copper nitrate was mixed only with HMTA (without NaOH), no flower-like morphology and only individual or few arranged petal-like structures were obtained (Figure 7(a)).However, by using NaOH with HMTA the individual petals started to aggregate and flower-like morphologies were obtained (Figure 7(b)). This indicates that NaOH as a strong electrolyte can neutralize the surface charges of the CuO and affect the aggregation, which plays an important role in the aggregation of individual petals, resulting in the formation of flower-like morphology of CuO nanostructures.

Uniform urchin-like CuO nanostructures (Figure 9) (with the diameter of around 2 μm) composed of aggregates of the single-crystal strips with the yield of 95% were also synthesized by facile reflux method with $Cu(NO_3)_2$ and urea as the starting materials [30]. The proposed growth mechanism is as follows.

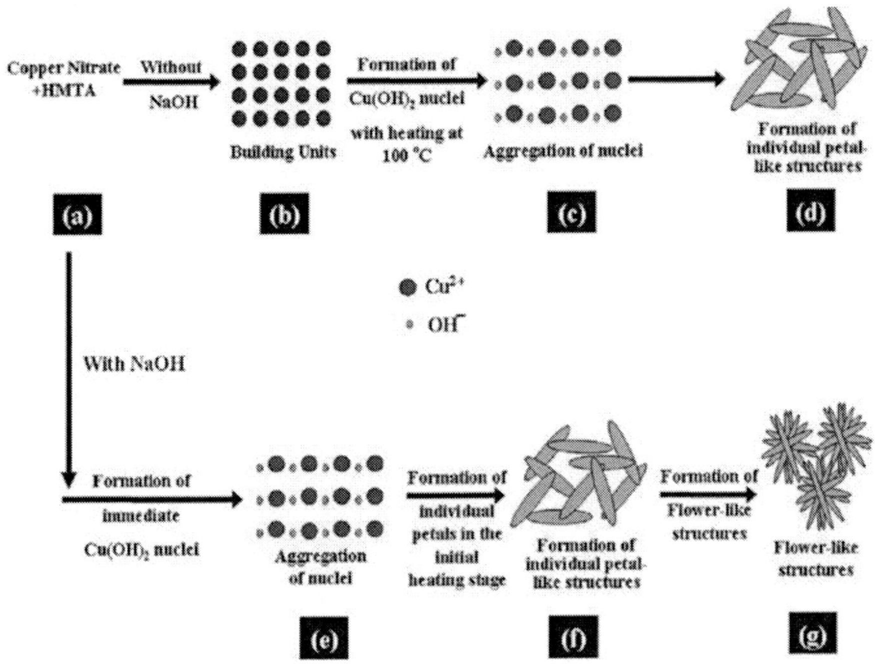

Figure 8.Schematic growth mechanisms for the formation of flower-shaped CuO nanostructures [29].

Hydroxyl groups are generated from the hydrolysis of urea at a certain temperature. Copper hydrate ions react with hydroxyl in the presence of nitrate ions to form a copper nitrate hydroxide precipitation. Then the copper nitrate hydroxide is converted to hydroxide when heating with a small amount of ammonium.Hydroxide is changed into oxide through dehydration. It was found that both copper sources and precipitators play crucial roles in forming the urchin-like CuO nanostructures. Better olefin epoxidation catalytic activity and higher yield of norbornene (89.5%) wereobserved for the urchin-like CuO nanostructures compared with conventional CuO (64.3%) and commercial samples (78.2%).

1.3. Solid-State Thermal Conversion of Precursor Method

Various CuO structures can be generated through the thermal conversion of the precursors and the morphological features of the precursors can be well-preserved in the final products given that the heat treatment is appropriately employed. In recent years, $Cu(OH)_2$ and basic copper salts have become favorable precursor candidates to exploit unusual morphologies of CuO because of their unique and well-known layered structures.2D CuO nanoleaves were successfully synthesized through the sequence of $Cu^{2+} \rightarrow Cu(OA)_2 \rightarrow Cu(OH)_2 \rightarrow Cu(OH)^{2-}_4 \rightarrow$CuO in aqueous solution at room temperature by employing $CuSO_4{\cdot}5H_2O$ and NaOH as the starting reactants[11]. The possible mechanism for $Cu(OH)_2$to convert into CuO nanostructures was proposed by the authors under a kinetics-controlled process. It was found that the velocity of the transformation of $Cu(OH)_2$ to CuO was greatlyaffected by the starting amounts of OH^- ions. By simply changing the pH in the reaction system, the authors could control the transformation velocity, which shows promising potential for preparing for other oxides with low cost at room temperature.

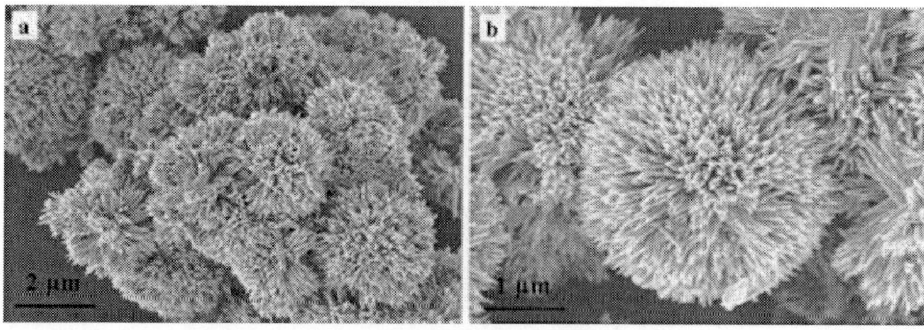

Figure 9. SEM images of urchin-like CuO nanostructures [30].

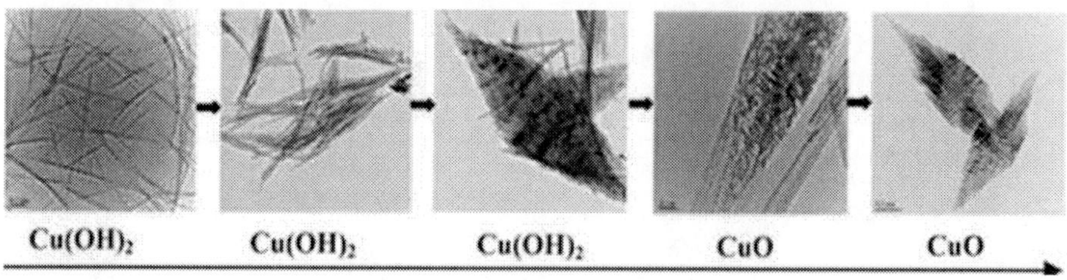

Figure 10. SEM images of the transition process of CuOnanoleaves [31].

In another studyin [31], $Cu(OH)_2$ nanoleaves were converted into to 2D CuO nanoleaveson a large scale by a reconstructive transformation. The processstarted from the nucleation of CuO followed by a two-step oriented attachment of the CuO particles to 1D CuO nanoribbons and then to 2D CuO nanoleaves, in which hierarchical-oriented attachment was involved. The transition process monitored by SEM is shown in Figure 10.

Cross-bedded CuO microstructures were achieved by thermal decomposition of the $Cu_2(OH)_3Br$ precursor obtained via hydrothermalsynthesis in hexadecyl trimethyl ammonium bromide (CTAB) using $Cu(OAc)_2{\cdot}3H_2O$ as the reacting agent at 100°C [32]. The effect of CTAB concentration was studied as shown in Figure 11. The authors show that the gas sensitivity of the as-prepared cross-bedded CuO targeting at ethanol and acetone is two times higher than that of 100 nm CuO particles.

Jia et al. [33]synthesized sisal-like CuO nanostructures (Figure 12(c)) by thermal conversion of sisal-like $Cu(OH)_2$ (Figure 12(b)). For preparing for thesisal-like CuOnanostructures, $Cu_2(OH)_3Cl$ (Figure 12(a)) was firstsynthesized by a simple and template-free hydrothermal method with copper chloride and urea as the starting materials. Then the transformation of the $Cu_2(OH)_3Cl$ to the novel sisal-like $Cu(OH)_2$ was facilely achieved by a NaOH treatment. Using $Cu(OH)_2$ as a sacrificial precursor, CuO nanostructures with well-preserved sisal-like structure were obtained through a thermal treatment. The overall transformation mechanism is illustrated in Figure 13. On the other hand, Lei's group also reported fishbone-like CuO (Figure 14(c-d)) achieved by thermal decomposition of the precursor of fishbone-like $Cu(OH)_2$ (Figure 14(b)) obtained via chemical reaction between $Cu_4SO_4(OH)_6$ and NaOH[34]. The $Cu_4SO_4(OH)_6$ (Figure 14(a)) was also firstly synthesized by a simple hydrothermal reaction with a yield of ~90% and then converted to fishbone-like $Cu(OH)_2$ by NaOH treatment.

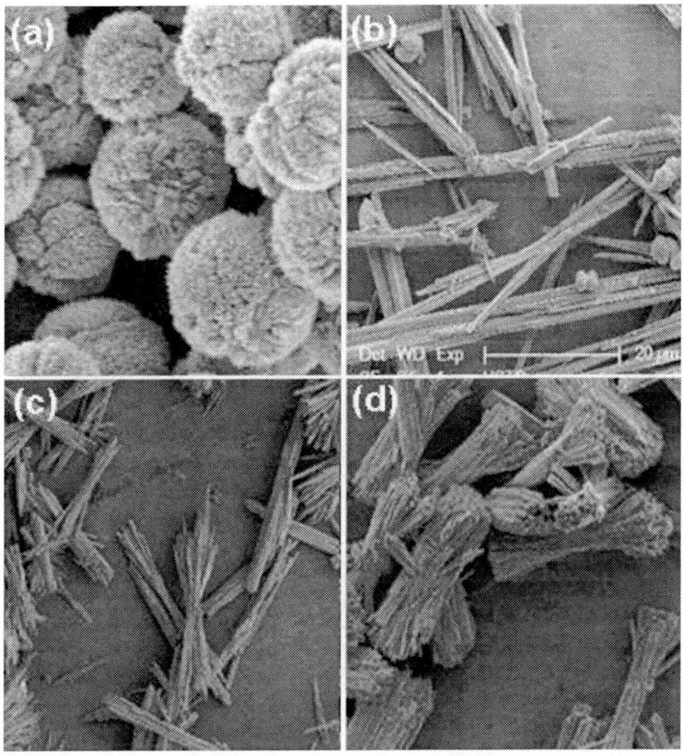

Figure 11. SEM images of the precursors prepared by CTAB with different concentrations:(a) 0.001, (b) 0.005, (d) 0.01, and (d) 0.05 M [32].

Figure 12. SEM images of the as-prepared samples: (a) octahedral $Cu_2(OH)_3Cl$, (b) sisal-like $Cu(OH)_2$, (c) sisal-like CuO, and (d) sisal-like CuO with a higher magnification [33].

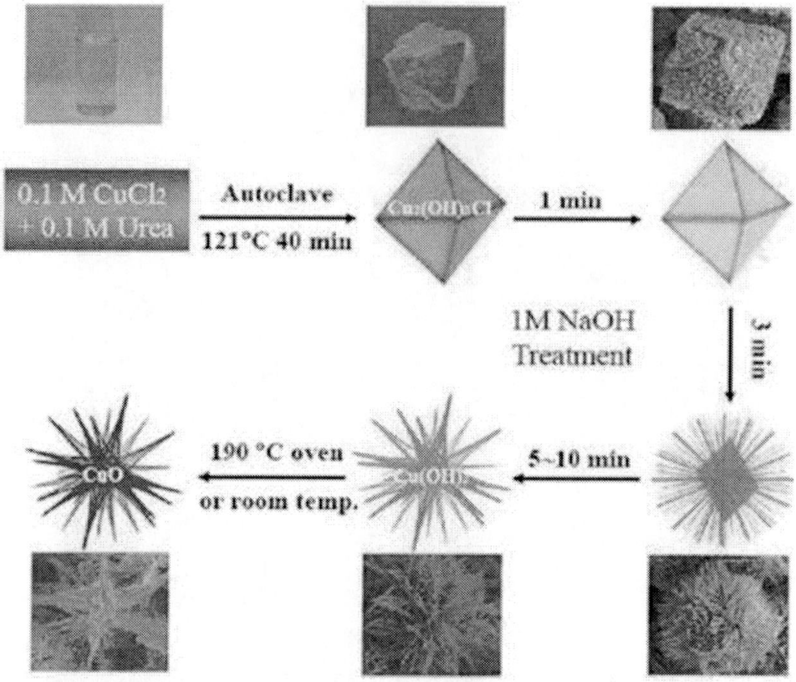

Figure 13. Schematic of the proposed mechanism for thetransformations [33].

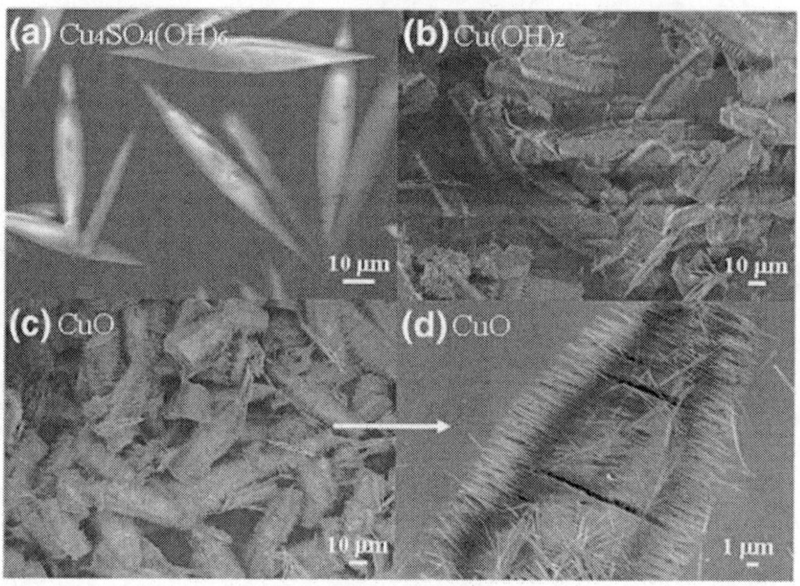

Figure 14. Typical SEM images of the as-prepared samples. (a) the $Cu_4SO_4(OH)_6$; (b) the fishbone-like $Cu(OH)_2$; (c) the fishbone-like CuO; and (d) high magnification of CuO [34].

Finally, fishbone-like CuO was formed by heating treatment of fishbone-like $Cu(OH)_2$. The authors show that the whole synthetic route involves no surfactant, template, orammonia, and CuO with three-dimensional 3D superstructures are obtained from a mild hydrothermal reaction followed by facile NaOH and thermal treatments.This work provides an economic and practical way for large-scale production of CuO with unusual superstructures. In addition, CuO nanostructures with different morphologies such as spheres (Figure 15(a)) and peanut-shaped superstructures (Figure 15(b)) were produced through thermal conversion of the corresponding malachite ($Cu_2(OH)_2CO_3$) [35-36].

Figure 15. SEM images of (a) hierarchical sphere-like and (b)peanut-shaped CuO nanostructures [35-36].

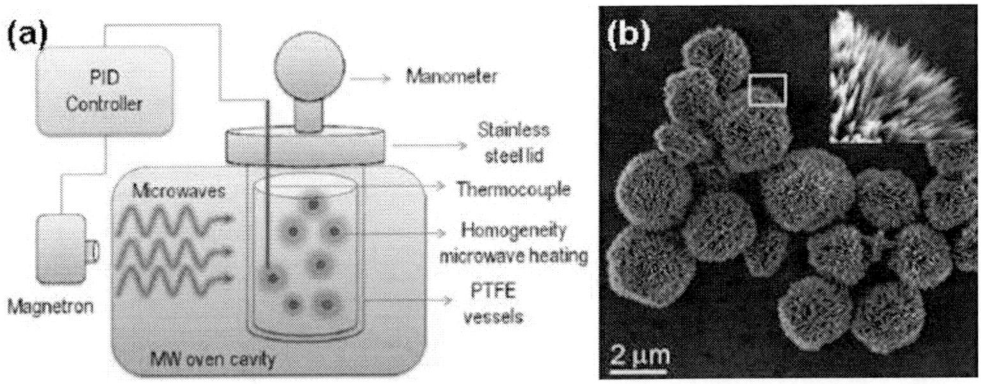

Figure 16. (a) Schematic diagram of the experimental setup and (b) SEM image of the as-prepared sea urchin-like CuO [37].

1.4. Microwave-Assisted Synthesis Method

The microwave-assisted method is a green approach to synthesize materials, and it has shown several advantages: (i) low energy consumption, (ii) rapid heating process, (iii) fast kinetics of crystallization, and (iv) production of new phases [37]. CuO with sea urchin-like morphology (Figure 16(b)) was successfully synthesized by means of microwave-assisted hydrothermal (MAH) synthesis method (the equipment is shownin Figure 16(a)) in the presence of polyethylene glycol as organic additive and using copper (II) chloride dehydrate and ammonium hydroxide as the striating reagents [37]. The CuO architectures were found to grow via a mesoscale self-assembly of the nanoparticles with the formation of mesocrystals as intermediates under the MAH condition. The growth process of the CuO architectures was from hollow nanoparticles to triangular points to 3D cross-like to 3D flower-like to3D sea urchin-like structures as can be seen in Figure 17.

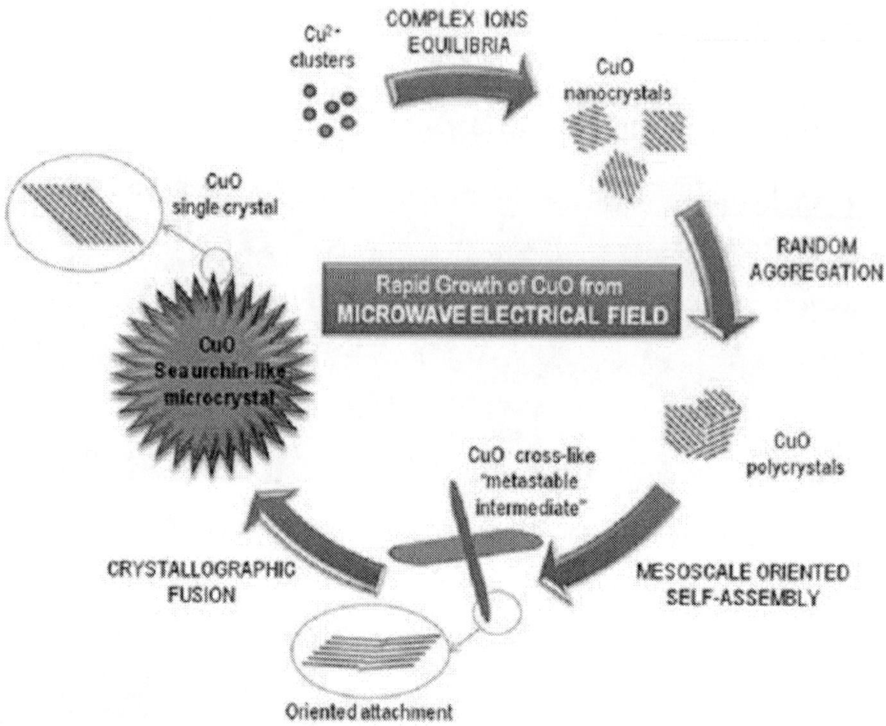

Figure 17. Schematic illustrationof the crystal growth mechanism of theCuO urchin-like microcrystals via a mesoscale self-assembly undermicrowave electrical field [37].

Uniform Polycrystalline urchin-like CuO microspheres with the mean particle diameter of 0.7–1.9 μm and specific surface areaof 170 m^2/g were also obtained usingthe MAH synthesis method by Keysonet al. [38].It was suggested that the formation and growth of the urchin-nanostructures was mainly affected by the addition of polyethylene glycol (PEG). This kind of CuO nanostructures may find potential applications in gas sensors, batteries, bactericides, and catalysis.

In addition, hierarchical cocoon-shaped CuO hollow nanostructures (Figure 18)have been obtained on a large scale by a facile green microwave-assisted aqueous process without using any template and additive. The CuO hollow nanostructures were prepared by the reaction of $Cu(CH_3COO)_2{\cdot}H_2O$ with urea($CO(NH_2)_2$) [39]. The formation of the hollow cocoon-shaped CuO was illustrated in Figure 19.The optical properties of these cocoon shaped CuO were found to be strongly dependent on the morphology and size. The band gap of the hollow CuO cocoons was estimated to be 2.03 eV, which showed blue shift compared with that of the bulk CuO. This kind of CuO nanostructures may become advanced materials for potential applications in biosensors, optical materials, catalyst and so on.Moreover, flower-like and leaf-likeCuO single-crystal nanostructures have been successfully synthesized using ionic liquid 1-octyl-3-methylimidazolium trifluoroacetate ([Omim]TA) under the microwave-assisted approach [40].

The effects of various growth parameters such as reaction temperature and amounts of ionic liquid were studied (Figure 20). It was found that both the reaction temperature and ionic liquid [Omim]TA play important roles in the formation of different morphologies of CuO crystals.

Figure 18.(a) SEM and (b)TEM images of the as-prepared urchin-like CuO microspheres [39].

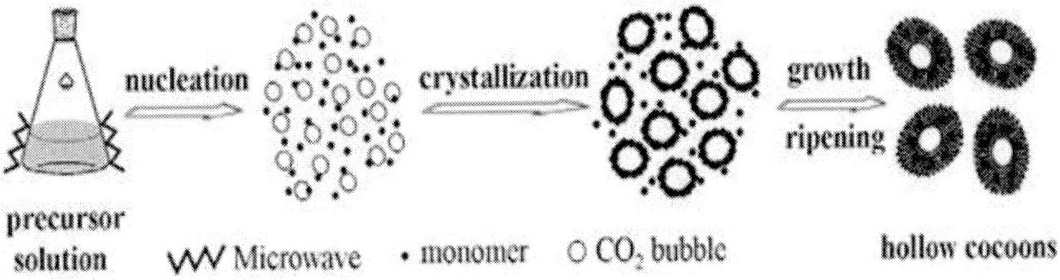

Figure 19.Formation processes of the hollow cocoon-shaped CuO nanostructures [39].

Without adding [Omim]TA, only ribbon-like CuO crystals were obtained. In the presence of 2mL [Omim]TA, flower-like CuO nanocrystals were prepared.With the increase in the amount of [Omim]TA, the morphology of CuO crystal changed from flower-like to leaf-like.While with the increase in reaction temperature from 80 to 100°C in the presence of 3mL[Omim]TA, the leaf-like CuO nanosheets evolved to needle-like CuOnanowhiskers.

The band gaps of the flower-like and leaf-like CuO nanocrystals with ionic liquids were found to be higher than the reported value (Eg = 1.2eV)for bulk CuO, which have promising applications in various fields such as photonics and catalysis. In Ref. [41], leaf-like, chrysanthemum-like and rod-like CuOnanostructures have also been synthesized by microwave-assisted approach using an ionic liquid 1-n-butyl-3-methyl imidazolium tetrafluoroborate ([BMIM]BF_4).

The authors showed that the concentration of [BMIM] BF_4 and reaction time could largely affect the shape of CuO nanostructures. With the prolongation of reaction time, leaf-like (Figure 21(a)) CuO would transform to chrysanthemum-like CuO (Figure 21(b)), which was composed of leaf-like CuO. However, the rod CuO was split from leaf-like CuO by increasing the concentration of [BMIM]BF_4, which serves as an excellent microwave-absorbing agent during this process. The band gap of as-prepared CuO nanoleaves was estimated to be 2.51 eV from the corresponding absorption spectroscopy, which was larger than the reported value for the bulk CuO due to the quantum confinement effects.

Additionally, single-crystalline CuO nanoleaves with an average thickness of ~10 nm and lateral sizes of hundreds of nanometers to several micrometers were synthesized by microwave heating of an aqueous solution containing copper salt and sodium hydroxide [42-43]. The bandgap of the CuO nanoleaves was estimated to be ~2.13 eV, which showed significant blue-shift compared with that of the bulk CuO.

Figure 20. SEM images of various CuO structures prepared in a reflux system for 10 min under microwave irradiation at different temperatures and different amounts of ionicliquid: (a) 80 °C, without any ionic liquid [Omim]TA, (b) 80 °C, 2 mL ionic liquid [Omim]TA, (c) 80 °C, 3 mL ionic liquid [Omim]TA, and (d) 100 °C, 3mL ionic liquid [Omim]TA [40].

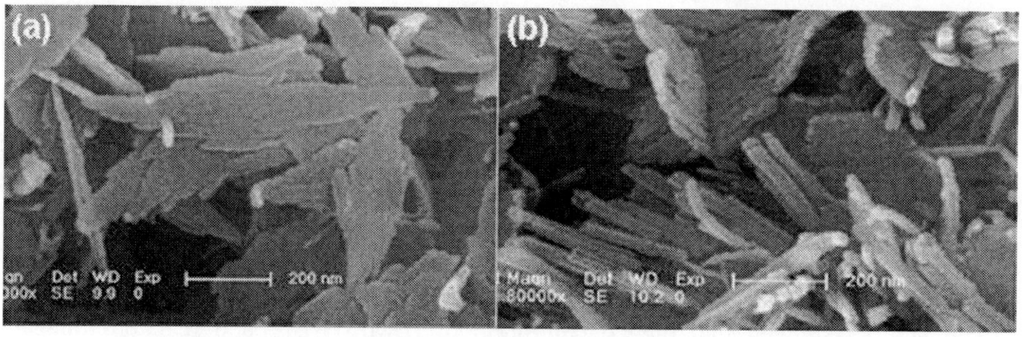

Figure 21. SEM images of (a) leaf-like and (b) chrysanthemum-like CuO nanostructures [41].

1.5. Ultrasonic-Assisted Synthesis Method

The ultrasonic-assisted synthesis method, as a new approach, has attracted much attention for preparing for inorganic nanoparticles. Different from other traditional chemical methods, the sonochemistry route is based on acoustic cavitations. Bubbles will form, grow, and collapse in the liquid during the acoustic cavitations. Ni et al.[44] firstly reported the successful synthesis of $Cu(OH)_2$ nanowires and further conversion to 3DCuO microstructures in a simple aqueous solution containing $CuCl_2{\cdot}2H_2O$ and NaOH under ultrasound irradiation for 5 min without the assistance of any surfactant or template. The effects of growth parameters, such as reaction time and with or without ultrasound irradiation were studied. It was found that the ultrasonic waves mainly played two aspects of roles besides dispersion: shorten the conversion time from $Cu(OH)_2$ to CuO and induce the formation of 3D CuO microstructures. The possible growth and conversion mechanisms for the formation of these 3D CuO microstructures were proposed.First, 1D $Cu(OH)_2$was formed (Equation7) from$[Cu(OH)_4]^{2-}$ ions(Equation6) irradiated by ultrasonic waves. And then 3D CuO microstructures were formed from 1D $Cu(OH)_2$under the continuous ultrasound conditions. The simple and controllable method presented by the authors can be applied as a general route for synthesizing other oxides (Cu_2O, ZnO, and CdO) with special morphologies and properties [44].

$$Cu^{2+} + 4OH- \longrightarrow [Cu(OH)_4]^{2-} \quad (6)$$

$$[Cu(OH)4]^{2-}(aq) \longrightarrow Cu(OH)_2 + 2OH^- \quad (7)$$

2. Applications of CuO 3D/2D Nanostructures

2.1. Lithium Ion Battery

Nanostructured materials have attracted great interest in both fundamental and applied research areas due to their outstanding physical and chemical properties. In recent years, utilizing nanostructured materials as electrodes in lithium-ion batteries has shown enormous promise with respect to both energy density and power density of the battery.

Hierarchical CuO nanoparticles with various morphologies such as leaf, shuttle, flower, dandelion, and caddice clew were obtained by simple self-assemblyand then they were used as anode materials for lithium-ion batteries [45]. The authors noted that various morphologies of CuO particles resulted in different electrochemical performances of electrodes. The leaf-shaped and shuttle-shaped CuO particles exhibited high initial coulombic efficiency but relatively low discharge capacities due to their small specific surface, which limited the large formation of inactive SEI film and reduced the accommodation of Li^+ at the same time. For the dandelion-like and caddice clew-like CuO, although the initial coulombic efficiencies were a little bit lower, the discharge capacities and cycling properties were much better than the leaf-shaped and shuttle-shaped CuO, especially at high rates. It was suggested that the high specific surface and porosity of hierarchical CuO resulted in the large contact area for CuO/electrolyte and shorter diffusion length of Li^+.

Guan et al. [46] synthesized hierarchical CuO hollow microspheres by a hydrothermal process conducted at 120 °C for 24 h without using any surfactants or templates. The as-prepared CuO microspheres were used as the anode material in lithium ion batteries. The CuO hollow sphere electrodes exhibited a much better electrochemical performance compared to the CuO solid sphere or CuO/Cu_2O hollow sphere at a high current densityof 670mAg^{-1}. After 30 cycles, CuO hollow spheres delivered a reversible capacity of 457mAh g^{-1} with a high coulombic efficiency of 99% which is much higher than the theoretical capacity of graphite of 372mAh g^{-1}.

Comparatively, the reversible capacity of CuO solid spheres rapidly dropped from 373 to 193mAh g^{-1}, while CuO/Cu_2O hollow sphere only delivered a reversible capacity of 409mAh g^{-1}. These results were attributed by the authors to the synergetic effect of small diffusion lengths in building blocks of nanorods and proper void space that buffers the volume expansion.

Additionally, 3D hierarchical CuO hollow micro/nanostructures were realized by biomolecule-assisted green method using $CuSO_4$, $NH_3{\cdot}H_2O$, and tyrosine instead of any toxic and dangerous reagents [47]. It was found that the 3D hierarchical CuO hollow micro/nanostructurebased anode material for lithium-ion batteries can exhibit a high initial discharge capacity of 560 mAh/g and improved electrochemical performance toward Li uptake-release. The improved electrochemical performance was attributed by the authors to a three-dimensional current collector network and a chemical/mechanical robustness buffer of the as-formed novel CuO structure.

Furthermore, well-aligned arrays of CuO nanoplatelets and flower-like CuO/Ni nanocomposite films prepared by the hydrothermal route and electrodeposition method, respectively, were also employed as anode materials for lithium-ion batteries[22, 48].

The obtained CuO/Ni film exhibits higher initial coulombic efficiency (72.1%) than the pure CuO film (57.0%) and better capacity retention (96.3% of the 2nd cycle) than the pure CuO film (67.8% of the 2nd cycle) at the current density of 0.1mAcm^{-2}. While, the obtained well-aligned arrays of CuO nanoplatelets exhibit rather good electrochemical performances with a specific capacity of about 300 mAh/g after 50 cycles with a slow capacity fading rate and a relatively high Coulombic efficiency.

2.2. Sensors

2.2.1. Enzyme-Free Glucose Sensor

Currently, great attention is being paid to the development of enzyme-less electrochemical sensors with high sensitivity, fast response, and long-term stability. In particular, electrochemically active nanomaterials consisting of metals, metallic carbon nanotubes, transition metal oxides or conducting polymers have been extensively studied for non-enzymatic electrochemical sensors, because they have large surface-to-volume ratio and increased catalytic activities. Among these nanomaterials, CuO nanostructures are one of the promising candidates for the active electrode material of non-enzymatic electrochemical sensors [49].

CuO nanoflowers film, directly formed by chemical oxidation of copper foil under hydrothermal condition, was used as active electrode material of nonenzymic electrochemical sensors for H_2O_2 detection [49]. The authors noted that the sensitivity of the sensor with CuO nanoflowers electrode was 88.4μA/mMcm^2 with a linear response inthe range from 4.25×10^{-5} to 4×10^{-2} M.While the sensitivity of the sensor with CuO nanowires was 66.88 A/mMcm^2 (R = 0.9969) with the linear range between 6.4mM and 37.2 mM. The enhanced sensitivity of nanoflowers sensor compared to that of nanowires was attributed by the authors to the excellent electro-catalytic activity, large surface-to-volume ratio and efficient electron transport property of CuO nanoflowers.

In Ref. [50], CuO nanoflowers and 1D nanorods have been synthesized by the composite-hydroxide-mediated and the composite-molten-salt methods, respectively. Both nanostructures have been applied to modify the graphite substrates for nonenzymatic glucose detection. The electrochemical performance of G electrode, CuO nanoflowers/G electrode and CuO nanorods/G electrode were investigated by the electrochemical impedance spectrum and cyclic voltammograms (CV) curves. The authors noted that the order of electron transfer resistance (Ret) is as follows: G electrode> CuO nanoflowers/G electrode> CuO nanorod/G electrode, indicating that the electron transfer speed of the CuO nanorods/G electrode is faster than that of the CuO flowers/G electrode and G electrode. CV curves as given in the paper show that no redox peaks can be observed at the G electrode. While a gradual increase in current can be observed at around 0.6V(vs. Ag/AgCl) on the CuO flowers/G electrodeand CuO nanorods/G electrode. CuO nanostructures have greatly improved the performance of the electrode and increased the electrocatalytic ability towards glucose oxidation, which may be attributed to their large surface area, high surface energy, and enhanced electron transfer ability. Chronoamperometry technique was further used to ascertain the detection of glucose. The authors show that, although the CuO flowers/G electrode can produce a high sensitivity of 709.52μA/mMcm^2 with the initial injection of 0.5mM glucose and a low detection limit of 4μM, it cannot provide a large linear range due to its slow electron transfer and foulness of intermediate product. In contrast, the CuO nanorods/G electrode shows a steady sensitivity of 371.43μA/mMcm^2 within 10 s with successive injections of 0.5mM glucose, and a detection limit of 4μM to glucose, as well as a wide linear response from 4μM to 8mM.

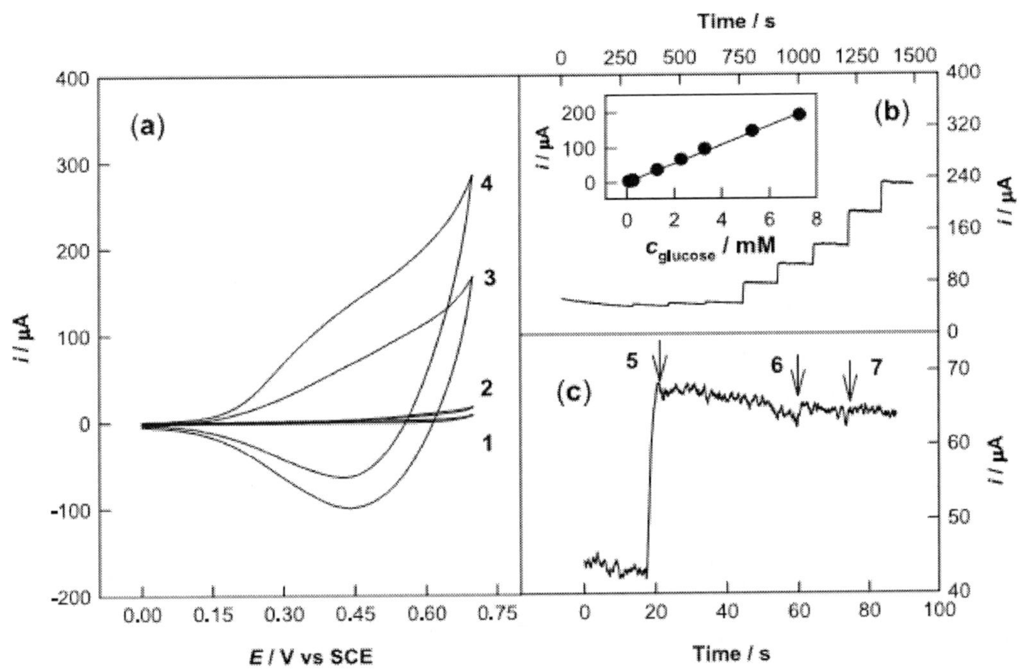

Figure 22. (a) CVs of a Nafion/commercial-CuO/GCE before (1) and after 2.0 mM glucose addition (2); CVs of aNafion/CuO-NLs/GCE before (3) and after 2.0 mM glucose addition (4). (b) Amperometriccurrent versus time for the detection of glucose at the Nafion/CuO-NLs/GCE. (c) Amperometric responses to the injection of 1.0 mM glucose (5), 0.10 mM AA (6) and 0.1 mM UA (7) at the Nafion/CuO-NLs/GCE [11].

The authors also noted that CuO nanostructure modified electrodes have good selectivity of glucose in the presence of dopamine and ascorbic acid.2D CuO nanoleaves (NL) prepared by a simple room temperature methodwere used in Nafion/GCE to construct an enzyme-free glucose sensor for detecting the electrocatalytic activity towards the oxidation of glucose in an alkaline solution [11]. As can be seen from Figure 22, the peak current of Nafion/CuO-NLs/GCE is about 15 times higher than that at Nafion/commercial-CuO/GCE. This reveals good electrocatalytic activity of the CuO-NL based electrode.

Well-defined, stable and fast amperometric response observed in Figure 22(b) indicates that Nafion/CuO-NLs/GCE sensor shows good sensitivity of detection of glucose. The tested results in Figure 22(c) show that no obvious current change occurs whenAA and UA are added to the system (points 6 and 7 on the curve), while the addition of glucose leads to an abrupt current change (point 5).These results suggest that the sensor containing CuO-NLs exhibits good selectivity to glucose and shows good anti-interferent ability.

CuO with three different nanostructures (wires, platelets, and spindles) synthesized by thermal conversion of $Cu(OH)_2$ were also modified on GC electrode to act as a sensor for application in electrochemical detecting of glucose [5]. The electrochemical impedance spectroscopies show that the order of Ret is: CuO nanospindles > CuO nanowires > CuO nanoplatelets modified electrodes.

This means that the electron transfer ability is: CuO nanospindles >CuO nanowires>CuO nanoplatelets. CV performances of different modified electrodes show that the electrochemical response of H_2O_2 on CuO nano-spindles is better than those on CuO nanowires and CuO nanoplatelets modified electrodes.

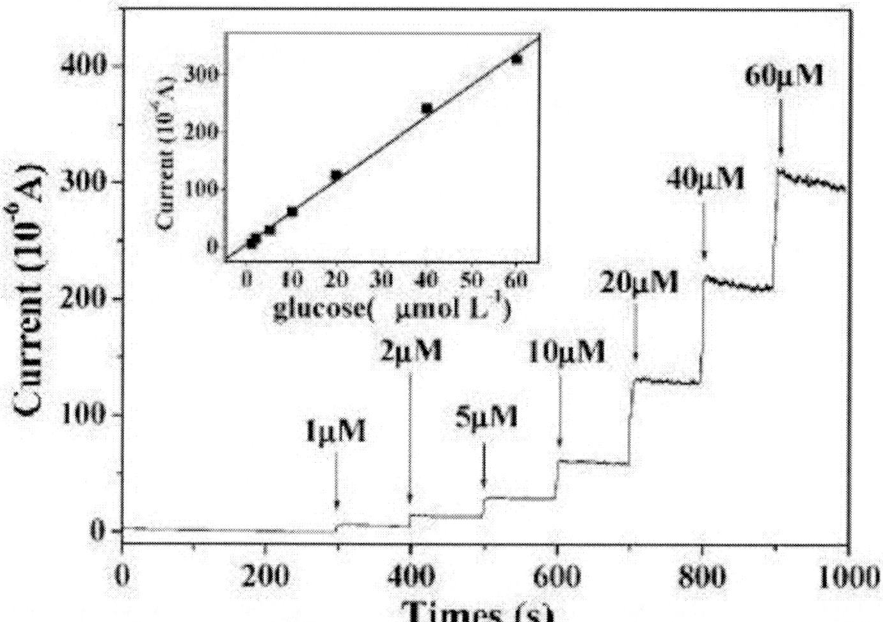

Figure 23.CA responses of CuO nanospindles/GOx/Nafion electrodesupon the addition of glucose solution. Inset shows a linear responsefor glucose concentrations between 1 μmol L^{-1} and 80 μmol L^{-1} [5].

These results suggest that the ability to accelerate the electron transfer between H_2O_2 and the GC electrode is: CuO nanospindles >CuO nanowires >CuO nanoplatelets. CA response curves (Figure 23) of CuO nanospindles/GOx/Nafion electrodes show a linear response for glucose concentrations between 1 and 80 μmol $L^{-1}$and display a sensitivity of 5.5675 × 10^{-6} A/μmolL^{-1}.The detection limit (S/N=3) is determined to be less than 1 μmol L^{-1}.

2.2.2. Field Emission and Humidity Sensors

Large-scale CuO honeycombs were synthesized via heating a $Cu(OH)_2$ precursor at 160 °C for 3 h under nitrogen protection. And then they were used as active electrode materials for application in field emission and humidity sensors [51]. Field emission (FE) measurements demonstrated that the CuO honeycombs have better FE properties with a turn-on field of ca 6–7Vμm^{-1} and a maximal current density of approximately 2.5mAcm^{-2}as shown in Figure 24a. While the turn-on fields of nanowires and flake-like nanostructures are ca 8Vμm^{-1} and ca 10Vμm^{-1}, and the maximal current densities are probably 0.8mAc$m^{-2}$and 0.2mAcm^{-2}, respectively. These results indicate that CuO honeycombs are favorable for reducing the field shielding effect and they exhibit good FE performance. The humidity sensitivity as shown in Figure 24(b) suggests that CuO honeycomb film exhibits good performance at room temperature due to the high surface-to-volume ratio.

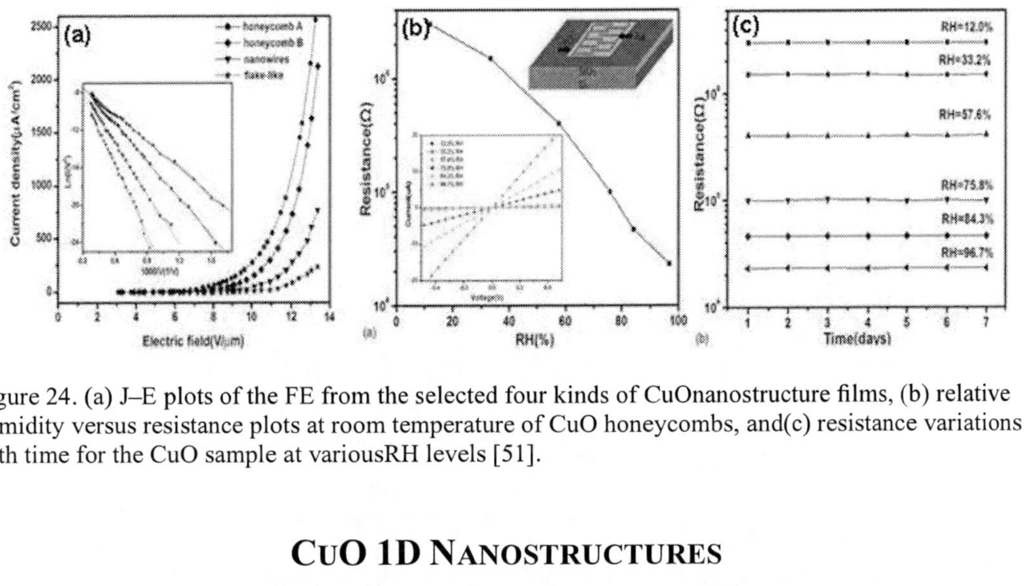

Figure 24. (a) J–E plots of the FE from the selected four kinds of CuOnanostructure films, (b) relative humidity versus resistance plots at room temperature of CuO honeycombs, and(c) resistance variations with time for the CuO sample at variousRH levels [51].

CuO 1D Nanostructures

One-dimensional (1D) CuO nanostructures including nanowires, nanorods, and nanobelts have been widely studied due to their unique properties and extensive applications. Various approaches of synthesizing 1D CuO nanostructures have been explored including thermal oxidation of copper in ambient air [52], thermal decomposition of CuC_2O_4 precursor [53], hydrothermal method [54-55], thermal oxidation of Cu_2S nanowires [56], self-catalytic growth mechanism [57], wet chemical route [58], and shape evolution via γ-irradiation [59].

Moreover, the applications of 1D CuO nanostructures have been explored in diverse areas such as sensing [60-63], field emission [64-65], solar cells [66-67] and nanoenergetic materials [68-69]. The synthesis methods, properties and applications of 1D CuO nanostructures will be presented in this section.

1. Synthesis and Characterization

1.1. Thermal Oxidation Method

Different methods have been employed to synthesize 1D CuO nanostructures. The mostly used way is to directly heat copper substrates in air and the reaction between copper and oxygen results in the growth of CuO nanowires.

Typically, the process of thermal oxidation just involves in heat treatment of pure copper substrates in either ambient air or oxygen atmosphere. The morphology of the grown CuO nanowires depends on the annealing temperature, growth time, and oxygen flow rate. In 2002, Jiang et al. [52] first reported a simple but effective method of synthesizing CuO nanowires by heating the copper grid and wire in ambient air for 4 hours at a temperature of 400–600 °C.

Large aspect ratio of CuO nanowires grew on the surface of grid or wire as shown in Figure 25. The diameter of the synthesized nanowires ranges from 30 to 100 nm while the length is up to 15 μm. Chen et al. [70] annealed the copper foil in air and reported that the density and the length of the nanowires will increase when prolonging the growth time at 400 °C as shown in Figure 26. Kumar et al. [71] synthesized single crystalline CuO nanowires by thermal annealing of a copper foil in an oxygen atmosphere and claimed that the oxygen flow rate and the annealing temperature both affect the morphology of the resulting nanowires, while the main effect of the annealing time is just on the aspect ratio. With the presence of oxygen flow during the thermal annealing, the necessary growth time for CuO nanowires has been reduced to no more than 1 hour.

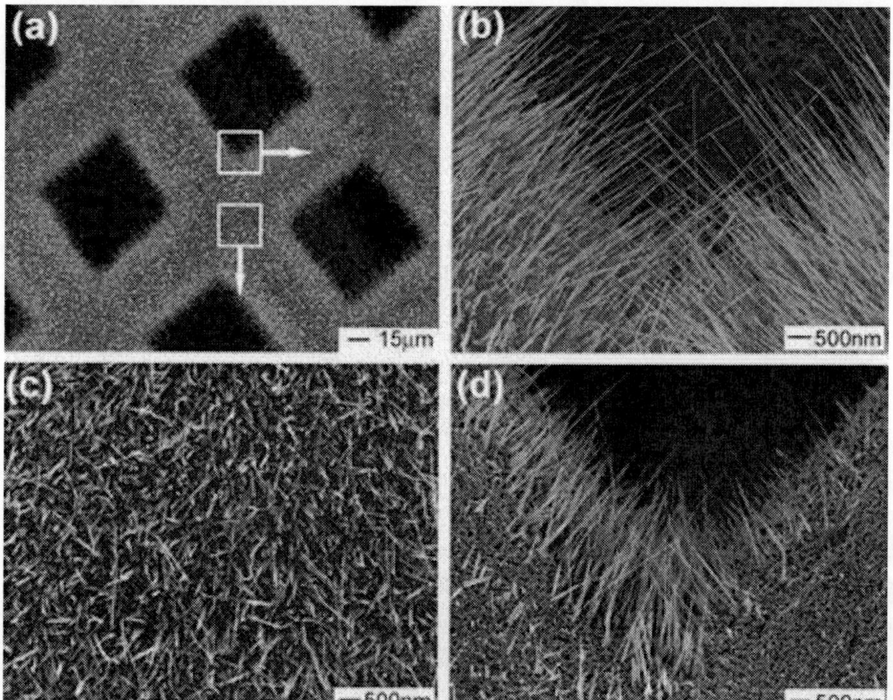

Figure 25. SEM images of the CuO nanowires synthesized by directly heating copper grids in air at 500°C for: (a-c) 4 h and (d) 2 h [52].

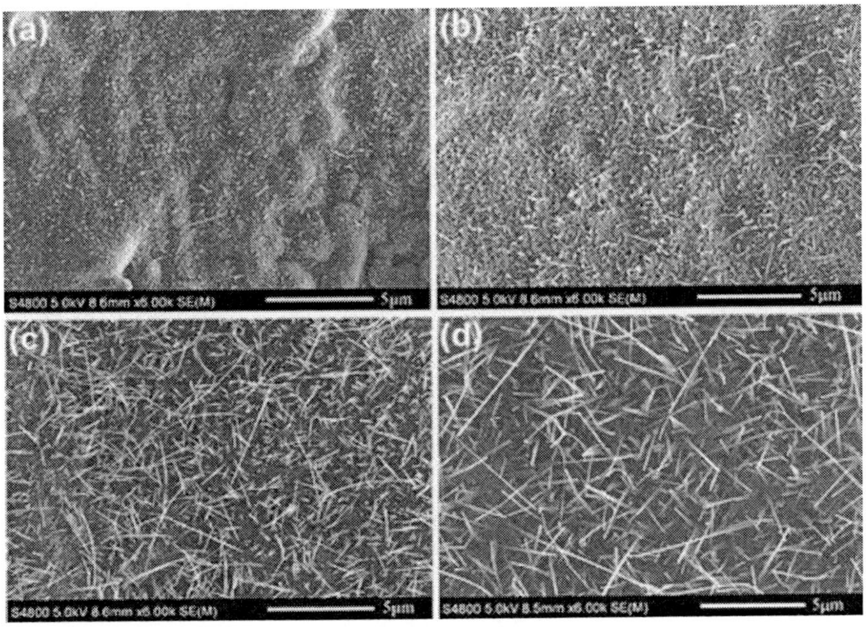

Figure 26 SEM images of copper foils thermally annealed at 400 °C for: (a) 0.5 h, (b) 2 h, (d) 4 h, and (d) 8 h, respectively [70].

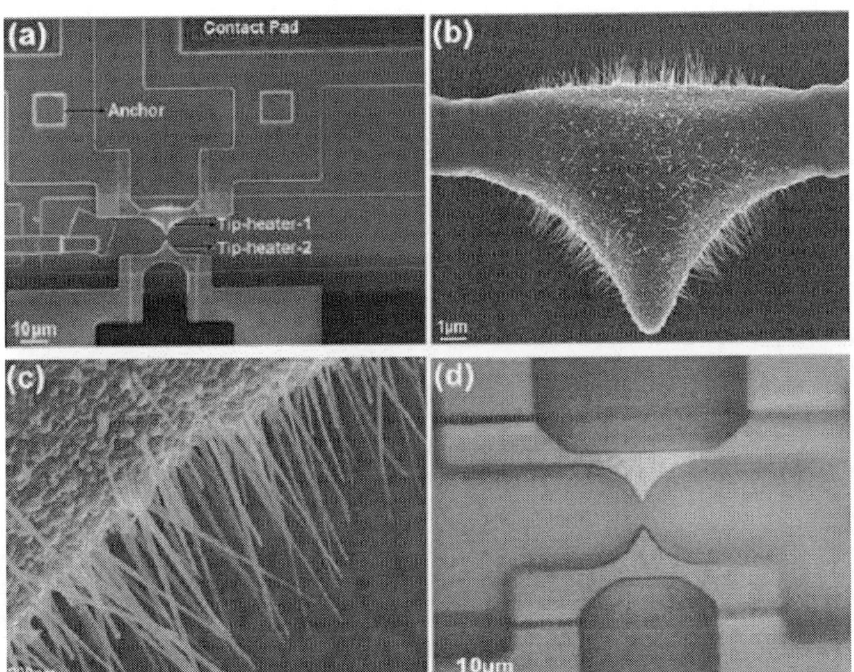

Figure 27. (a) SEM image of the suspended microheaters, (b) higher magnification SEM image of the tip in (a), (c) higher magnification SEM images of CuO nanowires in (b), and (d) optical image of one tip heater being heated [73].

Besides synthesizing CuO nanowires on copper substrates, Zhang et al. [72] reported large-scale and aligned CuO nanowires synthesized onto silicon that is the base material for semiconductor devices and microsystems. Furthermore, instead of heating the entire copper foils/films, Zhang et al. [73] presented a localized thermal heating method in ambient air to synthesize the CuO nanowires. Figure 27 shows that the CuO nanowires only grew on the surface of the heated area realized by localized joule heating. This method is CMOS-compatible and has the great potential of integrating CuO nanowires with microsystems to achieve functional devices.

Thermal oxidation is a simple, efficient, and low cost method of synthesizing CuO nanowires and it is suitable for batch fabrication and mass production. The achieved CuO nanowires are uniform, low impurity, vertically aligned and the morphology can be easily tailored by adjusting the synthesis parameters. Furthermore, the local synthesis of CuO nanowires achieved by local heating using a microheater is CMOS-compatible and has great potential to integrate the CuO nanowires with microsystems to realize functional devices.

1.2. Other Synthesis Methods

In addition to the synthesis of CuO nanowires by thermal oxidation of copper foils/grids/films, other methods have also been used in the literature. Hsieh et al. [57] reported the synthesis of well-ordered CuO nanowires by a self-catalytic growth mechanism.

The process employed polycarbonate (PC) nanoporous membrane as a template that was tightly mantled on the surface of a copper foil. And then a shot-time electro-deposition was performed to deposit copper nuclei on the copper foil surface in ambient temperature. After removing the PC membrane, the monocrystalline nanowires were grown at 500 °C under an oxygen gas flow.

The synthesized CuO nanowires are similar to those prepared by directly heating copper foils in air but the procedure is far more complex. Xu et al. [53] synthesized CuO nanorods by thermally decomposing the CuC_2O_4 precursor which was prepared by the chemical reaction between $Cu(CH_3COO)_2{\cdot}H_2O$ and $H_2C_2O_4{\cdot}2H_2O$ in the presence of nonyl phenyl ether (9)/(5)(NP-9/5) and NaCl. Figure 28 shows that the CuO nanorods are freestanding and interlaced.

The synthesized CuO nanorods are single crystalline, structurally uniform with the diameter of 30-100 nm and length of several microns.Dar et al. [55] reported a hydrothermal method of synthesizing CuO nanoneedles by the reaction between $Cu(NO_3)_2{\cdot}3H_2O$ and NaOH under continuous stirring followed by heating at 120-180 °C for 20-60 hours. Figure 29 shows the SEM images of the CuO nanoneedles which have very sharp tips and larger diameters at the bottom. Recently, Hai et al. [59] presented a novel approach of synthesizing CuO nanowires and Cu_2O crystals by γ-irradiation. The procedure starts with the preparation of $CuCl_2{\cdot}2H_2O$ solution added by NaOH and sodium dodecyl sulfate (SDS) in sequence. After stirring for 15 min, the sample is placed under γ-ray at a dose rate of 30 Gy/min for 72 hours. However, this approach is difficult to be popularized because of the potential danger of irradiation.

Anandan et al. [66] introduced a wet chemical method at room temperature to synthesize vertically aligned CuO nanorods on copper foils. The wet chemical method simply involves in immersing copper foils into the solution with selected pH containing ammonia and sodium hydroxide for a given reaction time (e.g. 96 hours). This approach has the advantages of simple procedure, uniform and aligned nanorods, and low-temperature.

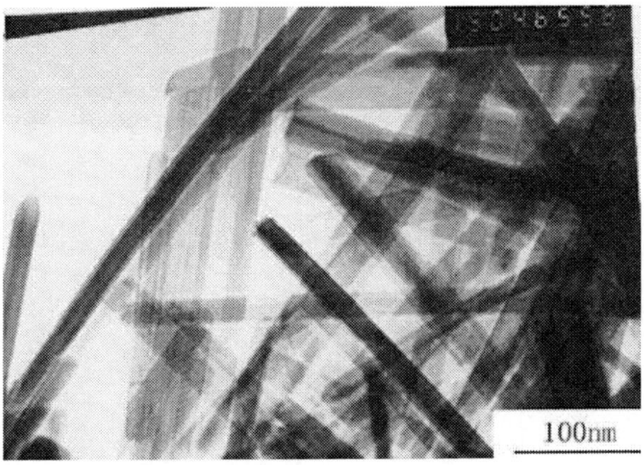

Figure 28. TEM image of the CuO nanorods [53].

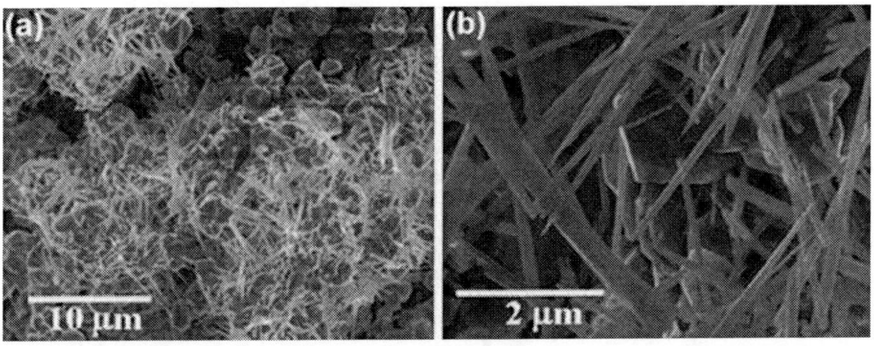

Figure 29. SEM imgaes of the grown CuO nanoneedles [55].

2. Applications

2.1. Application in Sensors

1D CuO nanostructures have found promising applications in various areas such as sensors, field emission, solar cells, and nanoenergetic materials due to their unique structures and properties. As a p-type semiconductor with narrow band gap (1.2 eV), the applications of CuO nanowires for gas sensing have been extensively studied. Kim et al. [60] fabricated CuO nanowire based gas sensors providing a novel and simple way to detect pollutant gases. The CuO nanowires were firstly synthesized on a copper foil by thermal oxidation as shown in Figs. 30(a, b). And then they were separated from the surface by ultrasonic treatment and immersed in a solution containing isopropyl alcohol. The droplet of the solution was dropped onto the alumina substrate and dried. The substrate had a microheater (beneath the CuO nanowires) and two Au electrodes as shown in Figs. 30(c, d). The resistance of CuO nanowires changes when CuO nanowires are exposed to CO and NO_2. Liao et al. further proved better performance of response to CO gas by fabricating single CuO nanowire based gas sensor in Ref. [61].Chen et al. [62] developed a gas sensor based on vertically aligned CuO nanowire arrays and investigated the H_2S sensing properties. This work proved that the sensor exhibited a good selectivity to H_2S compared to H_2, CO, and NH_3.

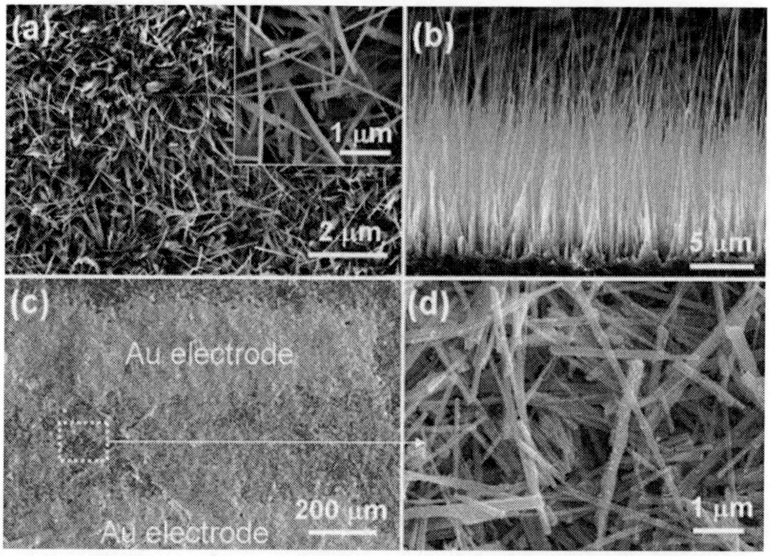

Figure 30. SEM images of: (a, b) top and side view of the CuO nanowires, (c) the sensor coated with CuO nanowires, and (d) a higher magnification image of the sensing surface [60].

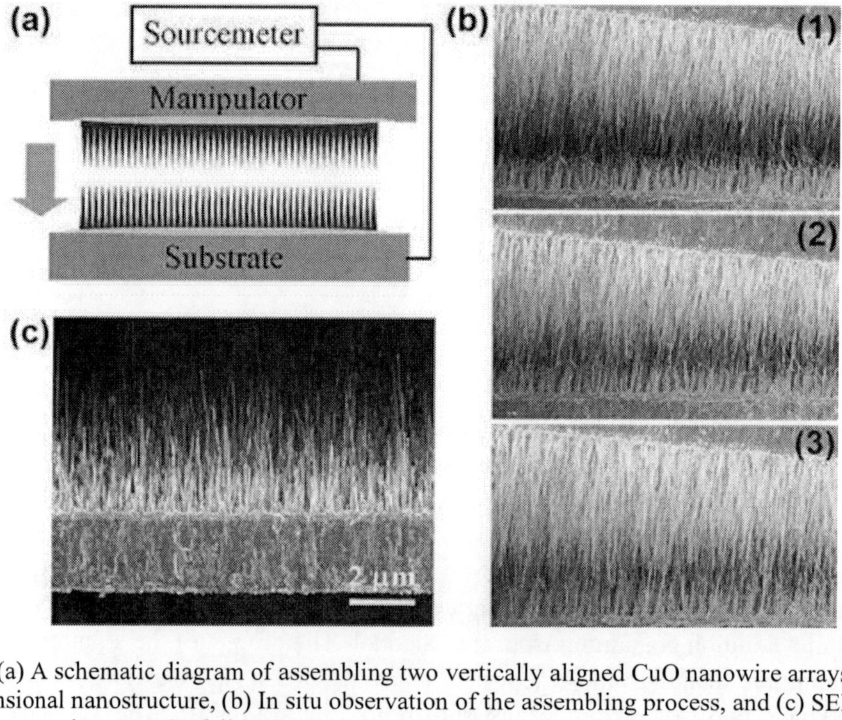

Figure 31. (a) A schematic diagram of assembling two vertically aligned CuO nanowire arrays into a three dimensional nanostructure, (b) In situ observation of the assembling process, and (c) SEM image of the CuO nanowires on a Cu foil [62].

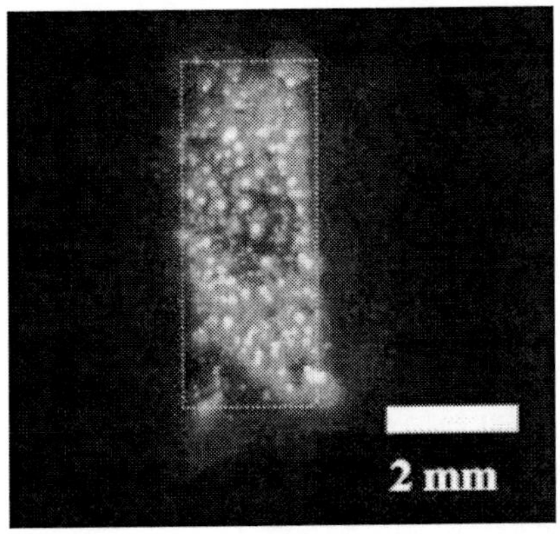

Figure 32. Flourescence field emission image of CuO nanowires under an applied voltage of 1000 V. The rectangular with dotted line represents the boundaries of the sample [64].

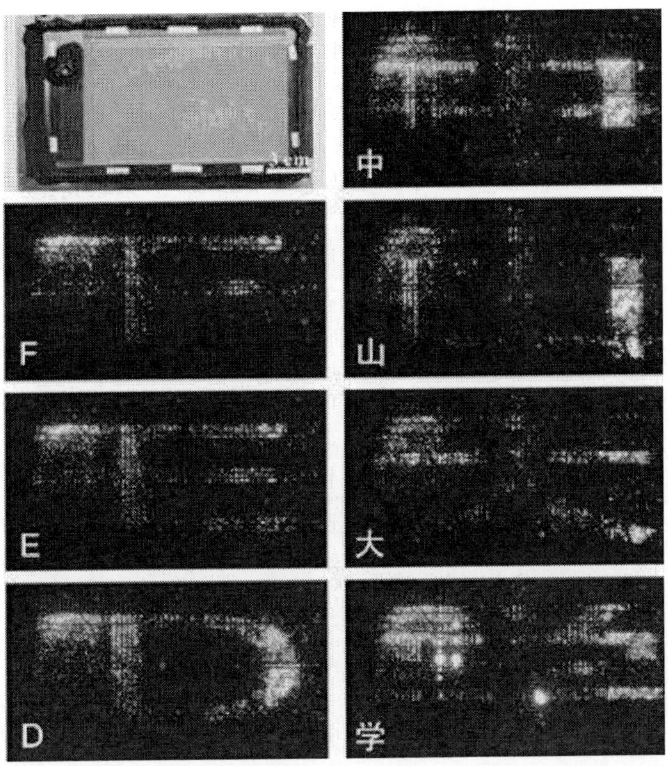

Figure 33. Pictures of the fabricated FED using CuO nanowires. The characters shown in the bottom left corner of each image indicate the display content [65].

The CuO nanowires arrays were firstly synthesized on copper foils as shown in Figure 31(c). And then two CuO nanowire arrays were assembled under the observation of SEM as can be seen in Figures31(a) and (b). Despite the applications in gas sensing, a non-enzymatic glucose sensor was reported by Zhuang et al. [63]. The fabricated glucose sensor presented a number of advantages, such as high sensitivity, fast response time, and good stability.

2.2. Applicationin Field Emission

The sharp tips of vertically aligned CuO nanowires have been found to effectively enhance local electric field, which makes them have promising applications in field emission (FE). FE properties of vertically aligned CuO nanowires have been studied by Zhu et al. [64] as shown in Figure 32. The fabricated FE emitters exhibited low turn-on field, high emission currentdensity, and uniform FE images.

Zhan et al. [65] developed the application for field emission display (FED) by fabricating gated CuO nanowire field emitter arrays on a glass substrate. The observed FED images are shown in Figure 33. However, more development is still needed on the FED made by CuO nanowires for a better image display with sharp boundaries.

2.3. Applicationin Solar Cells

Besides the innovative applications of CuO 1D nanostructures in microelectronic devices, potential uses in solar cells of CuO nanostructures have also been explored. Anandan et al. [66] presented a photovoltaic device based on vertically aligned CuO nanorods combining with TiO_2 nanoparticles.

The authors claimed that the photocurrent generation is caused by the high surface area and good hole mobility of the p-CuO nanorods. Wang et al. [67] developed the photovoltaic application by synthesizing ZnO-coated CuO nanowire arrays as shown in Figure 34. The optoelectronic properties and photovoltaic performance of the fabricated nanowire-based p-n heterojunction were investigated, which proved the possible use for durable and nontoxic photovoltaic components.

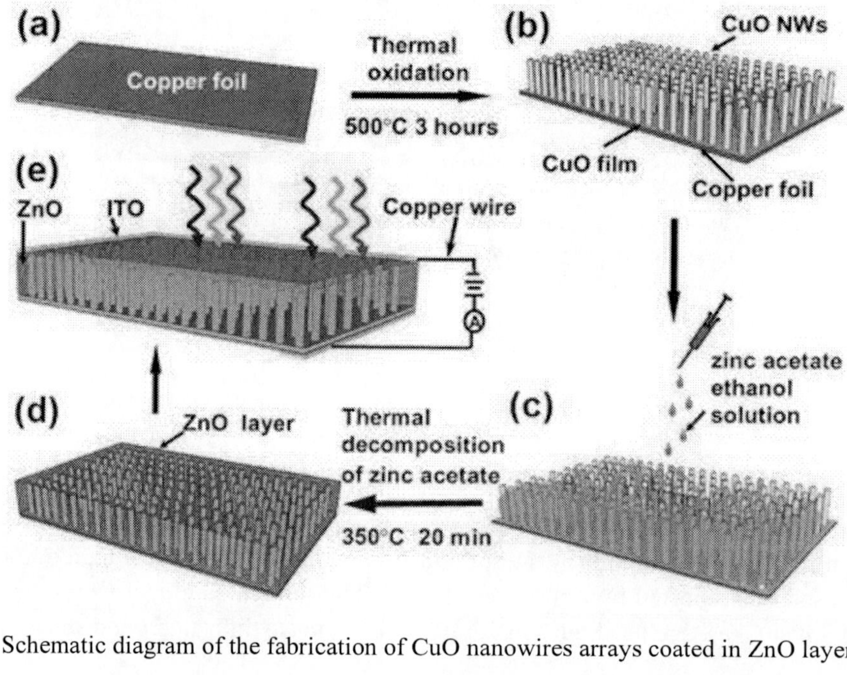

Figure 34. Schematic diagram of the fabrication of CuO nanowires arrays coated in ZnO layer [67].

2.4. Application in Nanoenergetic Materials

Energetic materials (EMs) including explosives, pyrotechnics, and propellants have found various applications in automobile airbag igniters, belt tensioners, mining, joining, soldering, and also in many defense-related areas. Nanoenergetic materials (nEMs) have shown improved performances in energy release rate and ignition properties compared to conventional EMs. This makes nEMs have promising applications in microactuation, microignition, micropropulsion, and micropower. Zhang et al. [68] synthesized nEMs based on CuO nanowires and nano-scale Al film on a silicon substrate. After thermal oxidation of a copper film, the nano-scale Al was deposited onto the grown CuO nanowires by thermal evaporation. Figures 35(a) and (b) show the CuO nanowires before and after Al integration, respectively. The large superficial area of the vertically aligned CuO nanowires results in more intimate contact between the two reactants of Al and CuO, thus significantly enhances the reactivity.

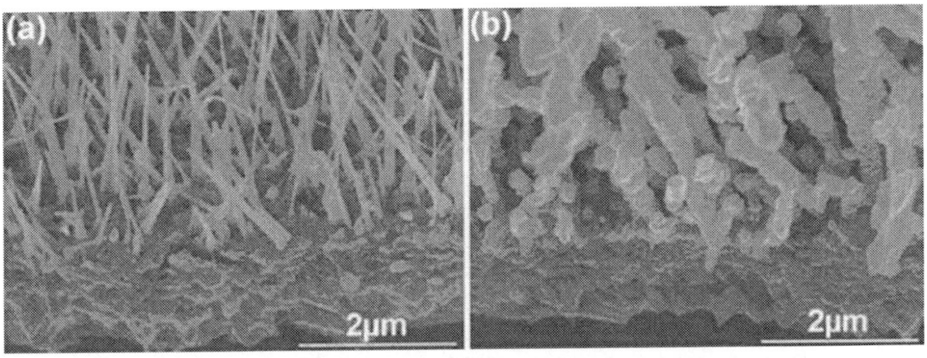

Figure 35. Cross-section view SEM images of the CuO nanowires (a) before and (b) after Al deposition [68].

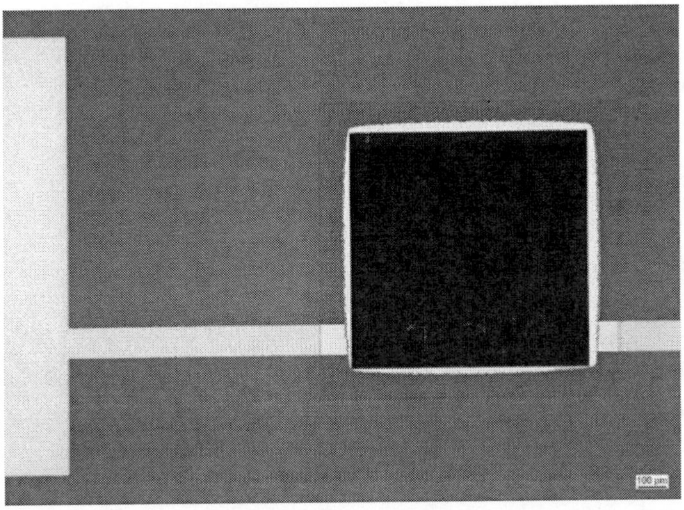

Figure 36. Optic image of the fabricated nano initiator.

Figure 37. Optic image of the combustion flame of the nano initiator.

By integrating Al/CuO-based nEMs with a microheater on a glass substrate, a nano initiator was reported by Zhang et al. [69]. Figure 36 shows the fabricated nano initiator. The thermite reaction between Al and CuO leads to high-temperature combustion in open-air when a current was input to the microheater as shown in Figure 37. This functional initiator has promising applications in electro-explosives devices at the micro and nanoscale.

CuO Nanoparticles

Nanoparticles are defined as particles with the size in the range of 1 to 100 nm at least in one of the three dimensions. Over the past years, CuO nanoparticles have been synthesized by various methods. CuO nanoparticles have found many applications in various fields such as catalysis, gas sensors, thermal conductivities, Li ion batteries, photoconductive devices, and magnetic storage media. The synthesis methods, properties and applications of CuO nanoparticles will be discussed in this section.

1. Synthesis and Characterization

Solution-based methodsare common and well-developed synthesis methods. They areoften used to fabricate off-substrate nanoparticles powder. One of the most important solution-based methods is chemical precipitation. The principle of this method is the chemical precipitation reaction between chemical compounds.For synthesizing CuO nanoparticles, cupric salt (normally nitrate or sulfate) and alkaline compounds (normally sodium hydroxide) are often used. A typical process of chemical precipitation is as follows [74]:300 mL 0.02 M $Cu(NO_3)_2$ solution was prepared by dissolving $Cu(NO_3)_2 \cdot 3H_2O$ in deionized water. The solution was put into a round-bottom flask equipped with a refluxing device. The $Cu(NO_3)_2$ solution was kept at an appropriate temperature($\leq 100^{\circ}C$) with vigorous stirring.

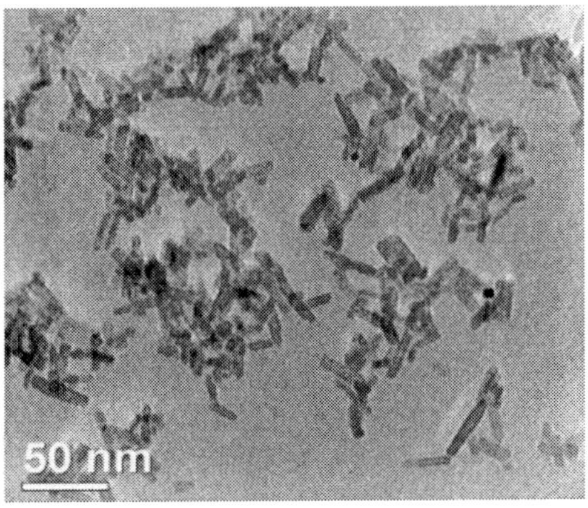

Figure 38. TEM image of CuO nanoparticles synthesized by chemical precipitation [74].

Then 0.50 g solid NaOH (platelets) was rapidly added into the solution, where a large amount of blue or black precipitates were simultaneously produced and maintained at the crystallization temperature for 10 min. Next, the precipitates were heated at 100°C for another 10 min. After the reactions were completed, the resulting products were centrifuged, washed with water and ethanol for several times, and dried in air at room temperature. Zhu et al.synthesized CuO nanoparticles by this route as shown in Figure 38 [74]. The average diameter of the particles is about 10nm.

It should be noticed that the proportion of reactive compounds and reactive temperature have great effects on the size and morphology of synthesized CuO nanoparticles. Usinga similar method, Zhou et al. [75] got three kinds of products with different sizes and morphologies: CuO nanoparticles, CuO nanobelts and CuO nanoplatelets (see Figure 39) by using three sets of recipes. They are (a) $Cu(NO_3)_2 \cdot 3H_2O$+NaOH, (b) $Cu(OAc)_2 \cdot H_2O$+ NaOH, and (c) $Cu(NO_3)_2 \cdot 3H_2O$ +Na_2CO_3.Shrestha et al. found that the physical parameters, concentration, temperature, and aging time greatly affected the size and morphology of the nanoparticles. They obtained different particles with a broad diameters range from 5 μm to 8 nm by controlling aforementioned parameters[54].

The synthesized CuO nanoparticles using ordinary chemical precipitation methods often have a common problem that the achieved nanoparticles are apt to agglomerate. To solve the problem, researchers have done a lot of work to improve the separation of the nanoparticles through applying for some external energy such as ultrasonic or high pressure. Sonochemical method is to apply ultrasonic during the synthesis procedure. Kumar et al. [76] synthesized well-separated 2~6 nm CuO nanoparticles by irradiating the reactive solution with a high-intensity ultrasonic horn under 1.5 atm of argon at room temperature for 3 h. However, this method employs expensive apparatus and excessive organic solvent and also needs severe conditions of reaction. In addition to the synthesis of dispersed CuO nanoparticles by the sonochemical method, Zhu et al. [77] developed a simple quick-precipitation procedure for the preparation of highly dispersed CuO nanoparticles with the size of about 6 nm in aqueous solution. The authors noted that the possible reasons of CuO nanoparticles apt to aggregate in the preparation process are the low nucleation and growth rates of CuO particles at mild reaction condition. By rapidly adding the NaOH solid to the mixture aqueous solution of copper acetate and glacial acetic acid at 100 °C, large amount of well-dispersed CuO nanoparticles were obtained.

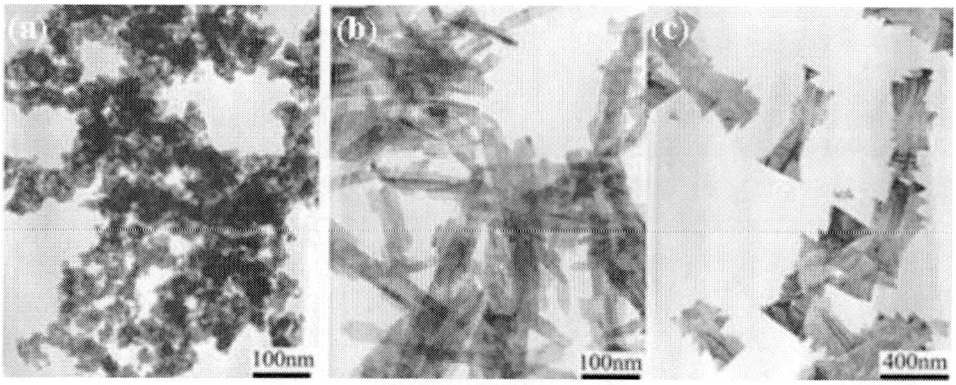

Figure 39. TEM images of (a)CuO nanoparticles,(b) CuO nanobelts, and (c) CuO nanoplatelets [75].

This result indicates that the higher temperature causes higher reaction rates, resulting in large amounts of nuclei to form in a short time, and the aggregation of crystals is inhibited. As a result, well-dispersed CuO nanoparticles with small size were achieved at relatively higher reaction temperature. On the other hand, Mahapatra et al. [78] introduced a wet chemical method to synthesize ultrafine dispersed CuO nanoparticles. The authors claimed that the size of as-prepared CuO nanoparticles can be controlled by simply changing the concentration of basic copper precursor. In addition, hot-solution decomposition and microemulsion process methods, which have been wildly applied to synthesize well-dispersed quantum dots, might show potentiality to prepare dispersed CuO nanomaterials with different shapes and sizes. Other synthesis technologies by introducing sonic waves or microwaves in the synthesis process can also be tried to get the well-dispersed CuO nanoparticles in the future [79].

Besides the chemical precipitation methods, there are many other solution-based methods to get CuO nanoparticles, such as thermal decomposition [80, 91], precipitation-pyrolysis method [81], and sol-gel process [82]. Each of them has its own advantages. Interesting microsphere cages composed of interlinked CuO nanoparticles have been fabricated viaa solution method by Qi et al. [91]as shown in Figure 40. The authors dissolved CuO into aqueous ammonia, diluted the solution by alcohol and then deposited onto a glass substrate bydip coating. After drying at 80°C and annealing at 500°C in air, CuO nanoparticle interlinked cages appeared. The CuO nanoparticle based cages have mesoporous hollowed structures and fine surface texture.

The CuO nanoparticles are easy to be synthesized by solution-based methods, and the productsare pure. After reduction by scorching CO/H_2, CuO nanoparticles can be transformedto Cu_2O [83]. In addition to the solution-based methods,chemical vapor deposition (CVD) is also an interesting method to get CuO nanoparticles. Barreca et al. [84] used $Cu(hfa)_2{\cdot}TMEDA$ complex as the precursor for the CVD synthesis of copper oxide nanosysterms. The authors synthesized CuO and Cu_2O+CuO nanoparticles on silicon substrates by regulation of reactive atmosphere and temperature.Both solution-based methods and CVD need long reactive time or multi-steps with complex equipment.

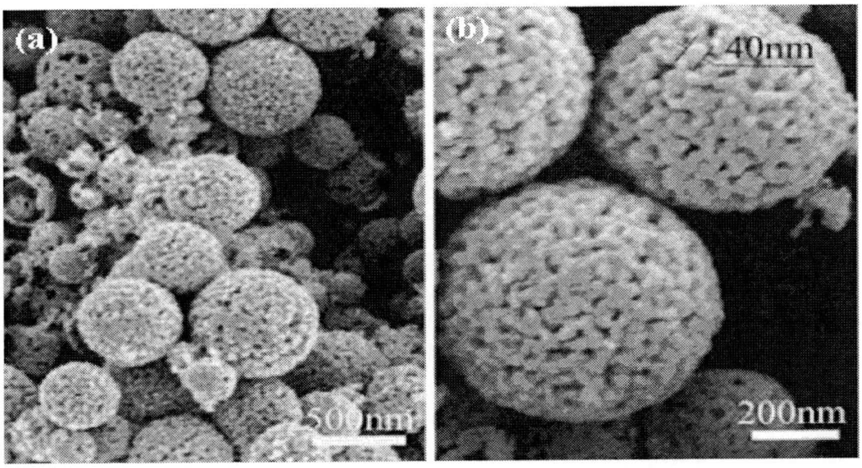

Figure 40. CuO nanoparticle interlinked cages [91].

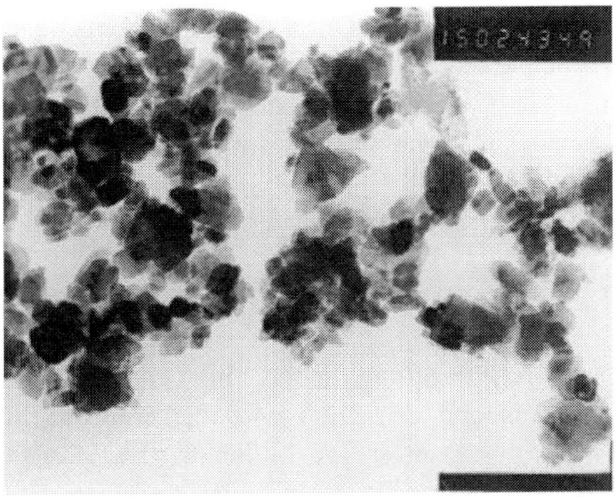

Figure 41. TEM image of the CuO nanoparticles by one step solid state method [85].

Asimple method based on solid state reaction was demonstrated by Xu et al. [85] to synthesize CuO nanoparticles. The authors prepared CuO nanoparticles by one step of grinding a proportional mixture of solid $CuCl_2{\cdot}2H_2O$ and NaOH for about 30min. As shown in Figure 41, the average size of the nanoparticles is 12 nm. However, this method has a disadvantage that it is difficult to get uniform nanoparticles. In general, compared to gas and solid phases based methods, solution-based synthesis methods are beneficial for getting uniform CuO nanoparticles with different morphologies for various applications, whereas the storage without aggregation is a problem that needs to be considered. CVD is good at controlling the ratio of Cu and O elements in the synthesized CuO nanoparticles, whichis useful for some applications. The solid state grinding method is a convenient and rapid route to get CuO nanoparticles, but it is not easy to achieve uniform nanoparticles so far.

2. APPLICATIONS

2.1. Application inCatalysis

CuO has been widely used in catalysis field. CuO nanoparticles typically have higher catalytic activity than bulk or micro CuO because thenanoparticles have large surface area and the Cu species are highly dispersed over the support. CO catalysis is one of the most important applications of CuO nanoparticles [75]. The authors showed that CuO nanoparticles with different morphologies had different catalysis properties. The (011) planes of CuO nanocrystal were found to have much higher catalysis activity of CO than other planes.

2.2. Application inThermal Conductivity Enhancement

Another important application of CuO nanoparticles is to enhance the thermal conductivity of organic fluids. Moghadassi et al. [86] investigated the effect of CuO nanoparticles on the thermal conductivities of paraffin and monoethylene glycol (MEG). The authors found that the effective thermal conductivity of the fluids increased with the increase in the volume fraction of nanoparticles in the suspension. The thermal conductivity of the fluids was further enhanced as more nanoparticles were added or the size of nanoparticles in the suspension was reduced.

2.3. Application inLi Ion Battery

Copper oxide electrode has been found to have high capacity density and good capacity retention for lithium insertion and extraction, thus the electrochemical capacity of CuO nanoparticles is a significant parameter for Li ion battery [80, 88]. Zhang et al. [80] synthesized well-dispersed CuO nanoparticles with the size from 10 to 100 nm by thermal decomposition of CuC_2O_4 precursor at 400°C. The electrochemical tests showed that the two times discharge capacity of the CuO nanoparticles arevery related to the size of the particles: thelargeraveragesizeofthe particles is, the smaller discharge capacity of the first time is.Seo et al. [88] discovered that the electrochemical performance of CuO nanoparticles could be further enhanced by the incorporation of multi-walled carbon nanotubes.

2.4. Application in Gas Sensors

CuO nanoparticles have also found applications in gas sensors. Zhang et al. [87] prepared nearly monodisperse Cu_2O and CuO nanospheres and then tested their sensitivity for alcohol vapor and gas oil at 210°C. The monodisperse Cu_2O and CuO nanospheres can form three-dimensional patterns by self-assembly and have potential usage in gas sensors because they have sufficient surface area and interspaces for gas absorption. Figure 44 shows a SEM image of the copper oxide nanoparticles with a photograph of the gas sensor.

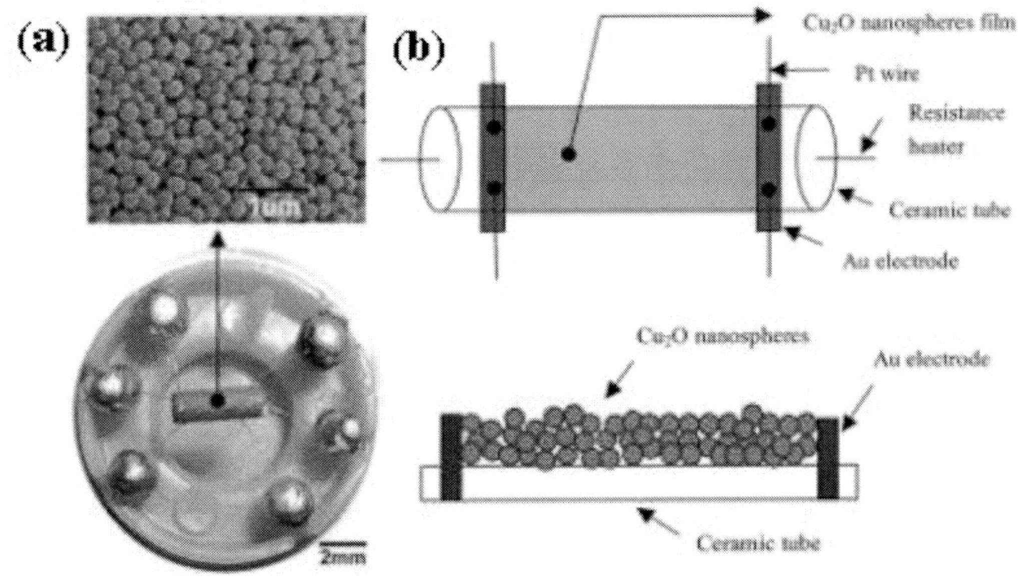

Figure 42. (a) Photograph of the gas sensor and the SEM image of thin films coated on it and (b) schematic diagram showing the structure of a typical Cu_2O nanosphere gas sensor by top view and sectional view [87].

The results in Ref. [87] indicated that the Cu_2O nanospheres based sensor offers a better and quick response/recovery time than CuO nanoparticles based sensor.However, the sensitivity testing for H_2S in [87] revealed the sensors are not sensitive enough for H_2S.

2.5. Other Applications

CuO nanoparticles have also been used in many other areas such as superconductors [89], field emission [90], magnetic storage media [81], and metallic bonding [92]. Rao et al. [82] tested the magnetic properties of the CuO nanoparticles indicating the presence of weak ferromagnetic interaction and the blocking behavior. Both the particle size and the morphology have been proved to have strong effects on the magnetic properties of CuO nanoparticles and the existence of uncompensated surface spins of the CuO nanoparticles causes its enhanced magnetic moments.

Lastly, it is worthy to mention that CuO nanoparticles have been proved to be highly toxic to human [93]. Compared with other metal oxide nanoparticles and carbon nanotubes, CuO nanoparticles are most potent on cytotoxicity, DNA damage and oxidative DNA lesions. If the body organs expose to CuO nanoparticles for long time, most of the cells will be killed and the DNA of the cells will be damaged or dissolved, which causes the inability of the body tissue.The small size of CuO nanoparticles leads to most reasons which cause body injury. Large surface areasof the CuO nanoparticles causehigh biological activity. And also the tinynanoparticles can easilyarrive to the deep area of the lung or penetrate many body barriers such as the skin and blood–brain barrier, and then translocate toother organs [94]. However,they are not very efficiently phagocytized by thealveolar macrophages, which is the main mechanism for particleclearance in human body. The deposition and aggregation of the

CuO nanoparticles cause lung irritation, chronic lung inflammations, exacerbation of asthma, and the formation of granulomas consisting of macrophage-like multinucleated cells. The toxicity of CuO nanoparticles is even higher than that of $CuCl_2$ solutions because cell membranes are good barriersfor most ions. Nevertheless, they cannot prevent small CuO nanoparticles.Therefore, when dealing with CuO nanoparticles, people should take necessary measures to protect them from direct exposure to the toxic dirty. Work or lab coat, rubber gloves, respirator and safety gogglesare the minimum appropriate level of protection. If potentialaerosol exposure threat exists, chemical fume hood or air-purifying respirator should be utilized for the processes. After the work, the hands should be thoroughly washed with soup and water.

CONCLUSION

CuO nanostructures have been widely studied and are receiving increasing interests due to their interesting properties and promising applications in various areas. This chapter presents a review on different CuO nanostructures including CuO 3D/2D nanostructures, CuO 1D nanostructures, and CuO nanoparticles. The synthesis methods of these CuO nanostructures are discussed in details. Their characterization and applications are also addressed.

CuO 3D/2D nanostructures will become important components for nanoscale devices and they have foundand will find more promising applications due to their large surface areas that can be exposed to gaseous environments and other unique properties. More work needs to be performed to further explore the applications of CuO 3D/2D nanostructures.

CuO 1D nanostructures have been paid special attention due to their unique properties and extensive applications. Among various synthesis methods of CuO 1D nanostructures, the most promising approach is the thermal oxidation of copper substrates or copper films deposited on other substrates in ambient air. This method is extremely simple and has the potential to be adapted for large scale synthesis of CuO 1D nanostructures with good crystal quality. However, the detailed growth mechanism of CuO 1D nanostructures using this method is still under debate, which needs further study. Moreover, the cracking problem existing in the synthesis process using the thermal oxidation method needs to be solved so as to achieve more CuO 1D nanostructures based functional devices, especially the devices that need to be used under librating, rubbing, and impactingenvironments.

Different methods have been employed to synthesize CuOnanoparticles. Compared to gas and solid phases based methods, solution-based synthesis methods are mostly used and they are good at obtaining uniform CuO nanoparticles with different morphologies for various applications. However, the storage without aggregation is a problem that needs to be considered. CVD is good at controlling the ratio of Cu and O elements in the synthesized CuO nanoparticles, whichis useful for some applications. The solid state grinding method is a convenient and rapid route to get CuO nanoparticles, but it is not easy to achieve uniform nanoparticles so far. Novel synthesis methods are desired for achieving well-separated CuOnanoparticles with promising applications.

REFERENCES

[1] P. Alivisatos, *Science*, 271, 933 (1996).
[2] S. J. Park, A. A. Lazarides, C. A. Mirkin, R. L. Letsinger, *Angew. Chem., Int. Ed.* 40, 2909 (2001).
[3] Y. Cui, C. M. Lieber, *Science*, 291, 851 (2001).
[4] G. Schmid, *Chem. Re.* 92, 1709 (1992).
[5] X. J. Zhang, G. F. Wang, X. W. Liu, J. J. Wu, M. Li, .J. Gu, H. Liu, B. Fang, *J. Phys. Chem. C.* 112, 16845 (2008).
[6] R. Jin, Y. C. Cao, E. Hao, G. S. M'etraux, G. C. Schatz, C. A. Mirkin, *Nature*, 425, 487 (2003).
[7] M. Bruchez, M. Moronne, P. Gin, S. Weiss, A. P. Alivisatos, *Science*, 281, 2013 (1998).
[8] A. Korgel, D. Fitzmaurice, *Adv. Mater.* 10, 661 (1998).
[9] L. K. Zheng, X. J. Liu, *Mater. Lett.* 61, 2222 (2007).
[10] W. Z. Jia, E. Reitz, P. Shimpi, Z. Rodrigue, P. X. Gao, Y. Lei, *Mater. Res. Bull.* 44, 1681 (2009).
[11] Y. Zhao, J. Z. Zhao, Y. L. Li, D. H. Ma, S. N. Hou, L. Z. Li, X. L. Hao, Z. C. Wang, *Nanotechnology*, 22, 115604 (2011).
[12] S. Y. Gao, S. X. Yang, J. Shu, S. X. Zhang, Z. D. Li, K. J. Jiang, *Phys. Chem. C.* 112, 19324 (2008).
[13] X. Y. Guan, L. P. Li, G. S. Li, Z. W. Fu, J. Zheng, T. J. Yan, *J. Alloys Compd.* 509, 3367 (2011).
[14] P. Singh, A. K. Ojha, O. N. Srivastava, *J. Phys. Chem. C.* 113, 3409 (2009).
[15] H. Deng, H. M. Hu, X. Q. Ge, C. L. Han, D. F. Zhao, G. Q. Shao, *Ultrason. Sonochem.* 18, 932 (2009).
[16] Z. Z. Zhuang, Q. Peng, Y. D. Li, *Chem. Soc. Rev.* 40, 5492 (2011).
[17] H. M. Xiao, S. Y. Fu, L. P. Zhu, Y. Q. Li, G. Yang, *Eur. J. Inorg. Chem.* 14, 1966 (2007).
[18] M. Q. Yang, J. H. He, *J. Colloid. Interface. Sci.* 355, 15 (2011).
[19] M. Q. Yang, J. H. He, X. C. Hu, C. X. Yan, Z. X. Chang, *Environ. Sci. Technol.* 45, 6088 (2011).
[20] Y. Liu, Y. Chu, Y. J. Zhuo, M. Y. Li, L. L. Li, L. H. Dong, *Cryst. Growth Des.* 7, 469 (2007).
[21] J. Y. Li, B. Xi, J. Pan, Y. Qian, *Ad. Mater. Res.* 79, 1059 (2009).
[22] Q. Liu, H. J. Liu, Y. Y. Liang, Z. Xu, G. Yin, *Mater. Res. Bull.* 41, 697 (2006).
[23] G. Y. Zou, H. Li, D. W. Zhang, K. Xiong, C. Dong, Y. T. Qian, *J. Phys. Chem. B.* 110, 1632 (2006).
[24] J. Wang, W. D. Zhang, *Electrochim. Acta.* 56, 7510 (2011).
[25] M. Zhang, X. Xu, M. Zhang, *Mater. Lett.* 62, 385 (2008).
[26] Y. Mu, J. Yang, S. Han, H. Hou, Y. Fan, *Mater. Lett.* 64, 1287 (2010).
[27] Z. H. Yang, J. Xu, W. X. Zhang, A. P. Liu, S. P. Tang, *J. Solid State Chem.* 180, 1390 (2007).
[28] J. W. Zhu, H. P. Bi, Y. P. Wang, X. Wang, X. J. Yang, L. D. Lu, *Mater. Chem. Phys.* 109, 34 (2008).

[29] M. Vaseem, A. Umar, S. H. Kim, Y. B. Hahn, *J. Phys. Chem. C*. 112, 5729 (2008).

[30] L. P. Xu, S. K. Sithambaram, Y. S. Zhang, C. H. Chen, L. Jin, R. Joesten, S. L. Suib, *Chem. Mater*. 21, 1253 (2009).

[31] H. Xu, W. Wang, W. Zhu, L. Zhou, M. Ruan, *Cryst. Growth Des*. 7, 2720 (2007).

[32] J. Y. Li, S. L. Xiong, B. J. Xi, X. G. Li, Y. T. Qian, *Cryst. Growth Des*. 9, 4108 (2009).

[33] W. Z. Jia, E. Reitz, H. Sun, B. K. Li, H. Zhang, Y. J. Lei, *J. Appl. Phys*. 105, 064917 (2009).

[34] W. Z. Jia, E. Reitz, H. Sun, H. Zhang, Y. Lei, *Mater.Lett*. 63, 519 (2009).

[35] J. S. Xu, D. F. Xue, *J. Phys Chem B*. 109, 17157 (2005).

[36] L. Z. Zhang, J. C. Yu, A. W. Xu, Q. Li, K. W. Kwong, S. H. Yu, *J. Cryst Growth*. 266, 545 (2004).

[37] P. Volanti, M. O. Orlandi, J. Andres, E. Longo, *CrystEngComm*. 12, 1696 (2010).

[38] D. Keyson, D. P. Volanti, L. S. Cavalcante, A. Z. Simoes, J. A. Varela, E. Longo, *Mater. Res. Bull*. 43, 771 (2008).

[39] C. H. Deng, H. M. Hu, W. L. Zhu, C. H. Han, G. Q. Shao, *Mater. Lett*. 65, 575 (2011).

[40] J. X. Xia, H. M. Li, Z. J. Luo, H. Shi, K. Wang, H. M. Shu, Y. S. Yang, *J. Phys. Chem. Solids*. 70, 1461 (2009).

[41] Xu, X. D.; Zhang, M.; Feng, J.; Zhang, M. L.*Mater. Lett*.62, 2787 (2008).

[42] Z. H. Liang, Y. J. Zhu, *Chem. Lett*. 33, 1314 (2004).

[43] O. V. Kharissova, B. I. Kharisov, *Ind. Eng. Chem. Res*. 49, 11142 (2010).

[44] Y. H. Ni, H. Li, L. Jin, J. M. Hong, *Cryst. Growth Des*. 9, 3868 (2009).

[45] J. Y. Xiang, J. P. Tu, L. Zhang, Y. Zhou, X. L. Wang, S. J. Shi, *J. Power Sources*. 195, 313 (2010).

[46] X. F. Guan, L. P. Li, Q. S. Li, Z. W. Fu, J. Zheng, T. J. Yan, *J. Alloys Compd*. 509, 3367 (2011).

[47] S. Y. Gao, S. X.Yang, J. Shu, S. X. Zhang, Z. D. Li, K. Jiang, *J. Phys. Chem. C*. 112, 19324 (2008).

[48] J. Y. Xiang, J. P. Tu, Y. F.Yuan, X. L.Wang, X. H. Huang, Z. Y. Zeng, *Electrochim. Acta*. 54, 1160 (2009).

[49] M. J. Song, S. W. Hwang, D. M. Whang, *Talanta*. 80, 1648 (2010).

[50] X. Wang, C. G. Hu, H. Liu, G. J. Du, X.S. He, Y. Xi, *Sens. Actuators, B*. 144, 220 (2010).

[51] J. W. Xu, K. Yu, J. Wu, D. J. Shang, L. J. Li, Y. Xu, Z. Q. Zhu, *J. Phys. D: Appl. Phys*. 42, 075417 (2009).

[52] X. Jiang, T. Herricks, and Y. Xia, *Nano Lett*. 2, 1333 (2002).

[53] C. K. Xu, Y. K. Liu, G. D. Xu, and G. H. Wang, *Mater. Res. Bull*. 37, 2365 (2002).

[54] K. M. Shrestha, C. M. Sorensen, and K. J. Klabunde, *J. Phys. Chem. C* 114, 14368 (2010).

[55] M. A. Dar, Y. S. Kim, W. B. Kim, J. M. Sohn, and H. S. Shin, *Appl. Surf. Sci*. 254, 7477 (2008).

[56] S. Wang, Q. Huang, X. Wen, X. Li, and S. Yang, *Phys. Chem. Chem. Phys*. 4, 3425 (2002).

[57] C. T. Hsieh, J. M. Chen, H. H. Lin, and H. C. Shih, *Appl. Phys. Lett*. 82, 3316 (2003).

[58] D. P. Singh, A. K. Ojha, and O. N. Srivastava, *J. Phys. Chem. C* 113, 3409 (2009).

[59] Z. Hai, C. Zhu, J. Huang, H. Liu, and J. Chen, *Inorg. Chem*. 49, 7217 (2010).

[60] Y. S. Kim, I. S. Hwang, S. J. Kim, C. Y. Lee, and J. H. Lee, *Sens. Actuators B* 135, 298 (2008).
[61] L. Liao, Z. Zhang, B. Yan, Z. Zheng, Q. L. Bao, T. Wu, C. M. Li, Z. X. Shen, J. X. Zhang, H. Gong, J. C. Li, and T. Yu, *Nanotechnology* 20, 085203 (2009).
[62] J. Chen, K. Wang, L. Hartman, and W. Zhou, *J. Phys. Chem. C* 112, 16017 (2008).
[63] Z. Zhuang, X. Su, H. Yuan, Q. Sun, D. Xiao, and M. F. Choi, *Analyst* 133, 126 (2008).
[64] Y. W. Zhu, T. Yu, F. C. Cheong, X. J. Xu, C. T. Lim, V. B. C. Tan, J. T. L. Thong, and C. H. Sow, *Nanotechnology* 16, 88 (2005).
[65] R. Z. Zhan, J. Chen, S. Z. Deng, and N. S. Xu, *J. Vac. Sci. Technol. B* 28, 558 (2010).
[66] S. Anandan, X. Wen, and S. Yang, *Mater. Chem. Phys.* 93, 35 (2005).
[67] P. Wang, X. Zhao, and B. Li, *Opt. Express* 19, 11271 (2011).
[68] K. Zhang, C. Rossi, C. Tenailleau, P. Alphonse, and G. Rodriguez,*Appl. Phys. Lett.* 91, 113117 (2007).
[69] K. Zhang, C. Rossi, M. Petrantoni, and N. Mauran,*J. Microelectromech. S.* 17, 832 (2008).
[70] J. T. Chen, F. Zhang, J. Wang, G. A. Zhang, B. B. Miao, X. Y. Fan, D. Yan and P. X. Yan, *J. Alloys Compd.* 454, 268 (2008).
[71] Kumar, A. K. Srivastava, P. Tiwari and R. V. Nandedkar, *J. Phys.: Condens. Matter* 16, 8531 (2004).
[72] K. Zhang, C. Rossi, C. Tenailleau, P. Alphonse, and J. Ching, *Nanotechnology* 18, 275607 (2007).
[73] K. Zhang, Y. Yang, E. Y. B. Pun and R. Shen,*Nanotechnology* 21, 235602 (2010).
[74] J. Zhu, H. Bi, Y. Wang, X. Wang, X. Yang, L. Lu. *Mater. Chem. Phys.* 109, 34 (2008).
[75] K. Zhou, R. Wang, B. Xu, Y. Li, Synthesis. *Nanotechnology*. 17, 3939 (2006).
[76] R. V. Kumar, Y. Diamant, A. Gedanken. *Chem. Mater.* 12, 2301(2000).
[77] J. W. Zhu, D. Li, H. Q. Chen, X. J. Yang, L. D. Lu, X, Wang. *Mater. Lett.* 58,3324 (2004).
[78] O. Mahapatra, M. Bhagat. C. Gopalakrishnan, K. D. Arunachalam. *J. Exp. Nanosci.*3, 185 (2008).
[79] D. Bera, L. Qian, T. K. Tseng, P. H. Holloway. *Materials.* 3, 2260 (2010)
[80] X. Zhang, D. Zhang, X. Ni, J. Song, H. Zheng. *J. Nanopart Res.* 10, 839 (2008).
[81] S. Rehman, A. Mumtaz, S.K. Hasanain. *J. Nanopart Resvol.* 13, 2497 (2011).
[82] N. Rao, Y. D. Yao, J. W. Chen. *J. Appl. Phys.* 105, 093901 (2009).
[83] J. Pike, S. Chan, F. Zhang, X. Wang, J. Hanson. Appl. *Catal.A-Gen.* 303, 273 (2006).
[84] D. Barreca, A. Gasparotto, C. Maccato, E. Tondello, O. I. Lebedev, G. V. Tendeloo. *Cryst. Growth. Des.* 9, 5, 2470 (2009).
[85] J. F. Xu, W. Ji, Z. X. Shen, S. H. Tang, X. R. Ye, D. Z. Jia, X. Q. Xin. J. *Solid. State. Chem.* 147, 516 (1999).
[86] R.. Moghadassi, S. M. Hosseini, D. E. Henneke. Ind. Eng. *Chem. Res.* 49, 1900 (2010).
[87] J. Zhang, J. Liu, Q. Peng, X. Wang, Y. Li. *Chem. Mater.*, 18, 867 (2006).
[88] S. D. Seo, Y. H. Jin, S.H. Lee, H. W. Shim, D. W. Kim. Nanoscale. *Res. Lett.*, 6,397 (2011).
[89] A.H. MacDonald. *Nature.*, 414, 409 (2001).
[90] J. Chen, S.Z. Deng, N.S. Xu, W.X. Zhang, X.G. Wen, S.H. Yang. Appl. *Phys. Lett.* 83, 746 (2003).
[91] J. Q. Qi, H. Y. Tian, L. T. Li, H. L. W. Chan. *Nanoscale Res Lett.* 2, 107 (2007).

[92] Y. Kobayashi, T. Maeda, K. Watanabe, K. Ihara, Y. Yasuda, T. Morita. DOI 10.1007/s11051-011-0522-9.

[93] L. Karlsson, P. Cronholm, J. Gustafsson, L. Moller. *Chem. Res. Toxicol.* 21, 1726 (2008).

[94] Z. Yang, Z. W. Liu, R. P. Allaker, P. Reip, J. Oxford, Z. Ahmad and G. Ren, *J. R. Soc. Interface* 7, 411(2010).

In: Nanostructures
Editor: Yu Dong

ISBN: 978-1-62618-081-9

Chapter 3

RADIATION METHODS OF NANOSTRUCTURES PRODUCTION

Gennady Gerasimov*
Institute of Mechanics, Moscow State University,
Moscow, Russia

ABSTRACT

This chapter gives a brief review of recent progress in the application of radiation techniques for synthesis and modification of nanostructured materials. The review includes the examination of available experimental data as applied to the radiation-induced formation of ion track membranes as nanoporous filters and templates for the synthesis of metallic nanostructures, preparation of radiation-cured polymeric nanocomposites and nanogels, and production of nanopowders with different substances. The considerable attention is drawn to the methods of radiation lithography (fabrication of 3D nanostructures, elements of nano- and microelectronics and optical and communication systems) and the radiation treatment of surfaces (etching and reinforcing of materials, cleaning of surfaces before deposition, thin films deposition, etc.), which have been subjected to widespread utilization in the present industrial applications. This chapter also reviews the main directions in radiation-induced modification of carbon nanostructures such as welding of crossing carbon nanotubes, polymerization of fullerenes layers and nanotubes bundles, formation of new nanostructures, and so on. The special consideration is particularly devoted to safety problems associated with radiation production of nanomaterials.

INTRODUCTION

Nanotechnology is one of rapidly developed areas in science and engineering, which arose from cross-disciplinary area of physics, electronics, chemistry, biology and material science [1]. It deals with materials and structures, which are controlled at the nanometer scale

* Corresponding author's email address: gerasimov@imec.msu.ru

(less than 100 nm). The current interest in such materials and structures comes from their unique electronic, optical, magnetic, and mechanical properties associated with quantum effects in the matter. These properties are the fundamentals for the practical use of nanostructures as new effective catalysts, sensors, displays, systems of energy storage and conversion, active and passive electronic components, biomedical materials and devices, etc. [2].

Recent advances in the creation of new carbon nanomaterials and methods of their functionalization solve the problem of miniaturization of electronic and optical devices that can enhance new research areas such as nanophotonics, nanoelectronics, or terahertz devices [3]. The use of radiation technique plays a significant role in the investigation and fabrication of nanostructured systems with high resolution since the radiation beams can be focused into less than a few nanometers and scanned with quite high speeds [4]. The matter irradiation by *X*- and γ-rays, electron and focused ion beams is essential in nanolithography, 3D fabrication, production of nanopores and nanowires, cross-linking of nanopolymers and nanocomposites, etc. [5]. The radiation modification of graphene may play the most important role in the appearance of industrial microchips on the basis of graphene transistors [6]. This chapter gives a brief review of the recent progress in the application of radiation techniques for nanostructures production.

Since the number of publications on this subject is too great, the review quotes the most typical work on each of the considered methods. The development of any innovative technology must be accompanied by the estimation of its influence on human health and safety. The chapter also analyses latest data on the investigation of induced radioactivity from industrial radiation processing as well as the potential problems of carcinogenic hazard of nanomaterials.

Ion-Track Technology

The modification of various materials using heavy ions beams is one of the examples for the industrial application of ionizing radiation [7]. Heavy ions produce latent tracks with the radiation damage of high-density matter along their path in the material. The diameter of such ions is in range of 1-10 nm depending on the ion type and its kinetic energy. The ion tracks are experimentally found in the limited class of materials, particularly in polymers, insulators, some metal alloys and semiconductors [8-10].

Ion-Track Membranes

The latent tracks may be used directly for creating conductive nanowires in insulators due to the possibility of inserting any conducting material into these tracks [11]. These tracks can also be implemented to produce track-etch membranes [12] after etching procedure for the purpose of the separation of toxic metal ions, biomolecules, and biological cells [13] or as templates for the synthesis of metallic and organic nanostructures [14]. Furthermore, the grafting functionalization of ion tracks can lead to the formation of highly proton conductive electrolyte membranes for fuel cell applications [15].

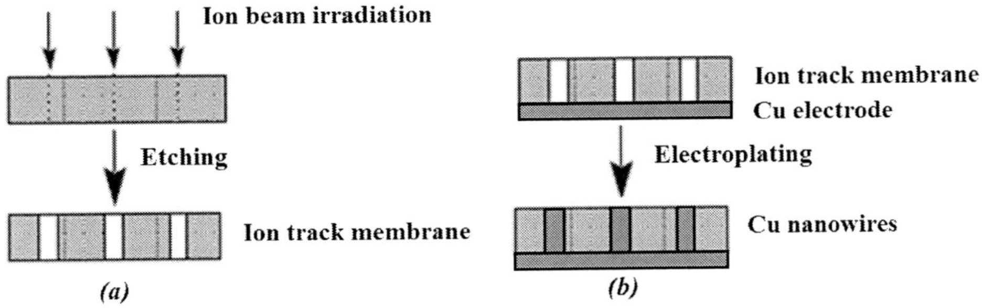

Figure 1. The production of ion-track membranes *(a)* and electroplating of cooper nanowires within the pores of polymer membranes *(b)*.

The widely used polymer materials for the production of ion track membranes are polyethylene terephthalate (PET) and polycarbonate (PC) thin films. The commercially available polymer membranes contain randomly distributed pores with uniform diameter from 0.02 to 5.0 μm, length of 10 – 50 μm and pore density from a single pore up to 10^9 cm^{-2} [4]. There are some other examples of t focused ion beams application for the nanostructures production. In particular, amorphous diamond-like carbon films contain high-energy heavy ions like U or Au with energy ~ 1 GeV. Their irritation leads to a transformation of the material from insulating diamond-like carbon to conducting graphite-like carbon [16]. The ion tracks form thin straight channels (nanowires) with a diameter of approximately 8 nm. The conductivity of such nanowires becomes up to eight orders of the magnitude larger than that for the surrounding material.

NANOWIRE STRUCTURES

The hollow track space of membranes can be easily filled practically with atoms of any kind. This can be achieved by using either the electrochemical deposition from the electrolyte solution, or the chemical polymerization from solution of the desired monomer and initiator, or the capillary action of the pores. The capillary action enables the solution of the dissolved material entering into the pores, which is followed by slow evaporation of the solvent [12]. The electrochemical deposition is carried out through the pores on the metallic layer that is coated on a back side of the template and used as a cathode for the electroplating. The micro- and nanomaterials produced in this way take the form of wires and tubules (see Figure 1b). After finishing the deposition process, the polymer membrane is dissolved in the corresponding solvent (e.g. in alkali solution at temperature 70 – 80 °C).

Nanowire structures formed on the massive metallic substrate give considerable (by tens and hundreds times) increase of the object surface activity, especially increase of the radiation, absorption, and desorption abilities. These nanostructured systems are capable of absorbing certain wave lengths of the electromagnetic radiation and may have potential applications in some devices such as microwave filters for shielding microwave ovens and mobile phones [4]. Unique optical, electronic, and magnetic properties of such structures may result in the appearance of new generation of electronic and optoelectronic devices [17]. Another example of the nanowire structures application is the production of high-performance components of heat exchange systems [18].

The ion-track membrane in this technology is wrapped around a cooper heat exchange tube that serves as a cathode in the galvanic cell. The subsequent deposition process fills the pores with cooper. The tube with a nanostructured surface provides a very high heat transfer coefficient between tube and boiling water after membrane dissolving.

Radiation Processing of Polymeric Materials

The radiation technique has already become one of the standard tools for the production of nanostructured polymer-based materials such as polymeric nanocomposites and nanogels [19, 20]. The radiation processing of polymeric materials includes radiation-induced polymerization (graft polymerization and curing), radiation cross-linking, and the degradation of polymeric chains [21].

Polymeric Nanocomposites

The recent advance in the plastic industry is associated with the creation of new polymeric nanocomposite materials [22]. The properties of polymeric nanocomposites differ advantageously from the usual composite materials. The small dimension of filler particles and, accordingly, large surface of the phase separation give the final product characteristics, which considerably exceed traditional ones at a low filler concentration. The advantages of polymeric nanocomposites include enhanced stiffness and strength [23], thermal and radiation stability [24] as well as improved catalytic and electro-conductive properties [25].

The wide assorted synthetic resins (polypropylene, poly(methyl methacrylate), polyepoxide, polystyrole, polycarbonate, etc.) are used as polymeric matrix for the nanocomposites production [22]. The most well-known fillers of polymeric matrix are alumosilicates (nanoclays and, in particular, montmorillonite) [26], carbon nanotubes [27], and metallic nanoparticles [28]. The formation of polymeric nanocomposites may be achieved by the process of filling free space in disperse phase with melted or dissolved polymer, or with monomer followed by its *in-situ* polymerization that has chemical or radiation influence on the formed composite structure. The scheme of polymeric nanocomposite formation under radiation is shown in Figure 2.

The radiation methods of nanocomposite synthesis attract the increasing attention at present. In addition to *in-situ* polymerization of monomers [29], they are also used for reinforcement of the available polymeric matrix owing to the radiation cross-linking between polymeric chains of the matrix [30, 31] and between the polymeric matrix and the filler particles [32]. In these conditions, the radiation, as an initiating process stage, influences monomers and polymers either directly with formation of free valences or through the active particles formed under solvent radiolysis [21]. Apart from conventional polymeric nanocomposites, the ionizing radiation is able to receive regulated nanocomposite structures based upon heterogeneous polymeric materials (hydrophilic and hydrophobic polymers) [33].

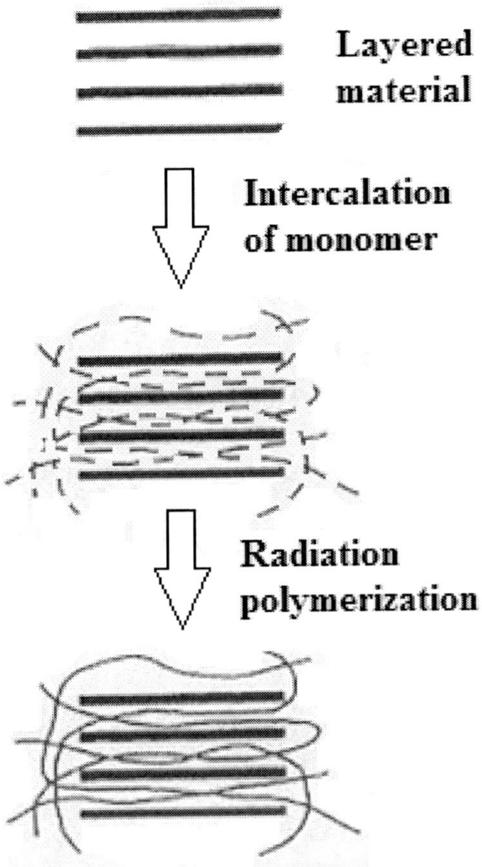

Figure 2. The schematic view of radiation induced formation of polymeric nanocomposite.

The radiation methods are widely used for the synthesis of new nanocomposite materials. In particular, the irradiation of graphite surfaces with simultaneous Fe supply promotes the formation of various types of carbon nanocomposites, consistingof well-aligned Fe-C fibers [34]. The electron-beam treatment of the aqueous solution containing Ag^{+} and Cu_2^{+} ions also leads to the fabrication of Ag/Cu_2O nanocomposite [35].

POLYMERIC NANOGELS

The polymeric gel represents a two-component system consisting of a stable 3D network of cross-linked polymer chains, and molecules of the solvent filling the pores in this network [36]. Nanoparticles of the polymeric gel are called nanogels. The other approach defines the polymeric nanogel as a macromolecule, in which a separate fragments of the polymeric chains are connected by inter- and intramolecular cross-linking [37]. The cross-linking reactions are shown in Figure 3, leading to the formation of polymeric nanogel. The chemical methods of nanogels synthesis presuppose the addition of sensitizers to the chemical system, which at thermal decomposition form active intermediates that initiate chain reactions of polymerization and intramolecular recombination.

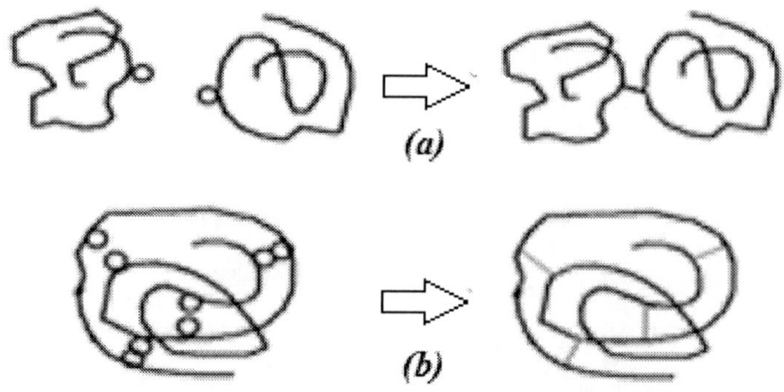

Figure 3. The scheme of the polymeric nanogel formation by intermolecular *(a)* and intramolecular *(b)* recombination of the polymeric radicals.

But these methods have a number of shortcomings. For instance, a pollution of the final product with fragments of the sensitizer is not permitted for some applications (biomaterials, optics, and electronics) [38]. When the process initiation is put into effect by the ionizing radiation, the radiation polymerization shows the main advantage of the process realization in the clear polymer (monomer)/solvent system. Such system is free from various sensitizing and stabilizing admixtures, which is especially important at the production of high-clear biomedical products [39].

The method shows the advantage both at synthesis of the polymeric nanogel by the intramolecular recombination of the polymeric radicals in separate macromolecules (see, for example, [40, 41]) and at the polymerization of various monomers (see, for example, [42-44]). The critical review of radiation synthesis of hydrogels, when the main initiators of polymerization are hydroxyl radicals formed at water radiolysis, is given in [45]. The unique properties of nanogels are the fundamental of their commercial applications. The most intensive investigations for the purpose of nanogels practical use are linked up with biomedicine (e.g. soft contact lenses, controlled drug delivery, wound dressing, immunodiagnostics, genetic engineering, etc.), cosmetics, printing and photographic technology, surface coating industry, and so on [36, 46].

Radiation Synthesis of Nanoparticles

The production and investigation of highly dispersed powders of various materials is an important branch of the nanoindustry. This is associated with the appearance of new perspective possibilities of their application in many fields of science and technology, especially for the production of new effective catalysts, the construction of elements of microelectronic and optical devices, and the synthesis of new materials. The methods of nanoparticles synthesis in gas and solid phases with an application of high-energy influence on the raw material are related to the physical techniques (see, for example, [47]); whereas the synthesis of nanoparticles in aqueous solutions is related to the chemical methods (see, for example, [48]). In particular, the methods, which are based on the reduction of metal ions in solutions with high-molecular compounds and surfactants as stabilizers, are predominant among the chemical ones [49].

ELECTRON-BEAM EVAPORATION

The method of the electron-beam evaporation of various artificial and natural materials in different atmospheres, followed by their vapor cooling and catching of formed nanoparticles, was investigated by many authors at various process conditions (see, for example, [50-53]). The method is one of the condensation physical methods to produce ultra-thin refractory powders. As opposed to laser and plasma heat sources, the use of electron beam offers much higher efficiency for the ablation in that the heating process takes place due to the volumetric penetration of electrons into the raw material. A new high-efficiency method of nanopowder production based on industrial electron beam accelerator was carried out in [47]. A high power density of the accelerator (up to 5000 kW/cm^2) is able to create such high temperatures that are sufficient for evaporation of any refractory material. Nanopowder production is a continuous process with the efficiency up to tens of kilograms per hour. The investigated materials include oxides (SiO_2, MgO, Al_2O_3, TiO_2, Cu_2O, WO_3, etc.), metals (W, Ta, Mo, Co, Al, Ni, Ag, Cu, etc.), as well as semiconductors, nitrides, fullerenes, carbon nanotubes and so on.

RADIATION INDUCED REDUCTION OF METAL IONS

The radiolysis of aqueous solution for various metal salts underlies the radiation-chemical reduction of metal ions M^+ with the formation of metallic nanoparticles. The solution irradiation results in the formation of active species, among which the hydrated electron e^-_{aq} has a high reduction potential, and hydroxyl radical OH possesses a high oxidizing ability. To obtain a favorable condition for the reduction of metal ions, an organic scavenger (for example, alcohol RCH_2OH) needs to be added in the salt solution, which is the effective acceptor of hydroxyl radicals [54]. The interaction of alcohol molecule with hydroxyl radical provides its transformation into organic radical RCHOH with the reduction properties [55]. A set of intermediate stages with the formation of unstable metallic clusters foregoes to the synthesis of metallic nanoparticles. The cluster coagulation leads to the formation of larger clusters with the appearance of quasi-metallic particles (germs of the metallic phase). The main distinctive features are the existence of the physical division surface as well as the ability for the reduction of adsorbing metal ions and, consequently, for the growth of these germs. The variation of process conditions (e.g. level of oversaturation or supercooling, presence of admixtures and so on) gives the formations of both amorphous and crystal nanoparticles.

Appropriate amounts of surfactants like polyvinyl alcohol are added in the solution for the process stabilization, and this also influences the crystallization process [49]. The radiation process of metallic nanoparticle production in the solution is sufficiently productive. Its use gives a possibility to considerably increase the range of metals that may be received in the nanometer scale. The variety of nanopowders of pure metals and alloys [56-58] as well as bimetallic structures of core-shell type [59, 60] has been synthesized up to present. The main peculiarity for the chemical properties of metallic nanoparticles is their high reactivity, conditioned by their heightened inclination to ionic and atomic exchange, adsorption on various surfaces, and formation of the surface bonds with other particles, etc. This can make new materials have unique properties.

Radiation Lithography

Lithography is a part of the technology for the production of silicon integrated circuits. It involves the fabrication of specific silicon regions with given electrical characteristics. The methods of photolithography are usually used for this purpose. Nevertheless, a tendency to miniaturization of electronic technique puts itself on the forefront of the problem to produce semiconducting structures on submicron and nanometer levels. This problem can be solved only with use of the radiation lithographic methods (electron-beam, ion-beam, and X-ray lithography) [61].

Electron-Beam Lithography

The base of electron-beam lithographic process is a sensitive scanning to irradiation surface (resist) in a patterned fashion by focused electron beam [62]. This process has an extremely high-resolution capability due to more short wave length of electrons in comparison with a light [63]. The resist is layered on the surface of oxidized layer of silicon substrate. There are positive resists received on the base of poly(methyl methacrylate) and its derivatives, and negative resists received on the base of chloromethylated polystyrene and poly(glycidyl methacrylate), which are used in practice [64].

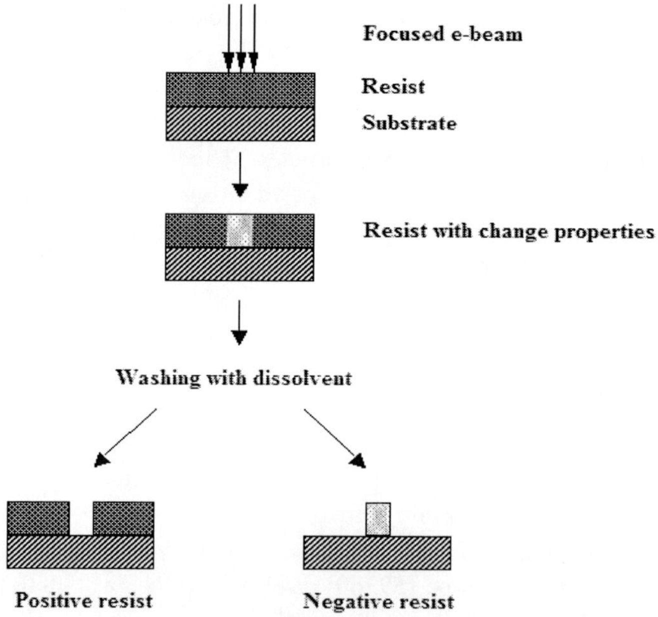

Figure 4. The scheme of the electron-beam lithographic process.

In the case of positive resists, the electron irradiation breaks the resist of polymer structure that increases a solubility of the irradiated part of surface. The irradiation of negative resists leads to strengthening of the exposed surface due to cross-linking of polymer chains. More soluble material of the resist is then removed by its washing with corresponding dissolvent. The pattern received on the resist is transferred on the surface of oxidized layer of the silicon substrate by way of wet or dry etching. The general scheme of the process is shown in Figure 4.

The possible application of the electron-beam technique in micro- and nanoelectronics are widespread (see, for example, [65-67]). The electron-beam lithography is used for the fabrication of photolithographic masks [68], small batches of special devices such as optical waveguides [69], micromechanical switches and cantilevers for the atomic force microscopy [70], as well as in the scientific practice. The commercial electron-beam lithographic systems are very expensive. For research applications, there is a common practice to convert an electron microscope into the lithography system, which can further reduce the installation cost more than the order [71]. Such altered systems have a linewidth on the order of 20 nm; whereas the specialized systems have the linewidth on the order of 10 nm or smaller [63].

Ion-Beam Lithography

The ion-beam lithographic systems offer a higher resolution than the electron-beam counterparts due to a weaker angle scattering of ions in the beam [72]. These systems usually use the light ions (protons and He^+ ions) for the exposition of polymer resists. The application of higher ions can help to alloy the substrate surface or to create thin layers of new chemical compounds on it.

The irradiation of crystalline semiconductor materials with high energy ions produces defects of the interstitial-vacancy (Frenkel pair) type along the ions paths, which changes the solubility of the damage regions. This enables to exclude the resist application since the material itself behaves as inorganic resist [73]. The losses of the ion beam energy in the resist are approximately two orders higher than those of electron beam irradiation. This considerably decreases the irradiation dose required for exposition. The methods of the ion-beam lithography are used for the fabrication of integrated circuits, lithographic nanoimprint templates and 3D nanostructures [74-76].

X-Ray Lithography

This type of the lithographic process is characterized by insignificant radiation damage of formed structures owing to the small energy used in X-ray irradiation, as well as by higher depth of sharpness and weak influence of the substrate material and its topography on the resolution capability [77].

The X-ray lithography has a high productivity because of a possible simultaneous processing in large areas. The installations on its base are cheaper and have a simpler construction than the electron-beam ones. The modern methods of X-ray lithography are used to generate the elements of micro- and nanoelectronics [78], optical and communication systems [79].

Radiation Treatment of Surfaces

There are a great number of technological processes elaborated at present for the radiation treatment of solid surfaces for the purpose of superficial nanostructure production. They include in the first place (a) the reactive ion-beam etching of materials, (b) the cleaning and activation of the surfaces before deposition, (c) the polishing and reinforcement of the surfaces, and (d) the thin films deposition directly from the ion beams by the target sputtering under the action of high energy charged particles (electrons and ions).

Ion-Beam Etching

The methods of chemical and physical etching are widely used for the fabrication of nanometer hollows with a complex profile on the flat surfaces of the friction-stable materials. The ion-beam etching is based on the physical sputtering of the surface material by energetic ions due to the kinetic interaction between ions and the atoms or molecules of the target surface [80]. This method is widely used for the lithographic production of silicate integrated circuits (the design is transferred from the resist on the surface of the oxidized silicon layer) [81], for the elimination of the structural defects of various surfaces [82, 83], and so on. To eliminate some imperfections associated with a slow etching rate and an increase of the surface roughness, various modifications of the method are used, namely, the reactive ion-beam etching (RIBE), the ion-beam assisted chemical etching (IBAE), and their combination [84].

Ion-Beam Polishing

Some problems of nanoelectronics and microwave optics can be tackled by means of sufficient superficial smoothness of a number of materials. The surface relief of the given roughness can be achieved with use of the combined chemical and mechanical methods as well as methods of ion-beam polishing [85]. The effective method of ion-beam polishing used foroptical materials up to subnanometer level is described in [86]. The special features of the method consist of (a) ion-beam assisted deposition of a nanometer-thick layer into cavities of the initial relief, which utilizes oxygen-ion sputtering of a target with a composition identical to that of the processed object; (b) sputtering of resulting surface structure by a normally incident low-energy oxygen ion beam to a depth of approximately two deposited material layers ; and (c) reiteration of the deposition-sputtering cycles until the necessary final state of the surface is attained. The nanoscaled polishing of hard materials and, in particular, of diamond is carried out using a gas-cluster ion beams, as well as the combination of chemical and mechanical polishing, ion-beam sputtering, and plasma etching techniques [87-89].

Ion-Beam Reinforcement

The improvement of material wear resistance is a critical problem encountered in the industry. A variety of physical and chemical methods are used for increasing the surface durability of constructional metals and alloys [90]. Among them, the ion-beam reinforcement attracts an increasing attention to specialists. The fundamental of the process lies in the implantation of alloy elements and their compounds into processing surface on submicron depth, which can change the material crystalline structure with the formation of amorphous state [91]. This provides a positive impact on the surface resistance. On the other hand, the irradiation of amorphous films made of metal oxides such as Al_2O_3 leads to the crystallization of amorphous structure [92]. It must be pointed out that the implantation of various ions into treated surface has different influence on its durability characteristics. For example, the irradiation of Ti surface with O^+ ions results in the double increase of its hardness; whereas Ca^+ implantation does not show such effect that can be explained by the formation of amorphous CaO and $CaTiO_3$ layers [93]. The low-energy ion irradiation of instrumental steel surface leads to the formation of complex multilayer structure consisting of amorphous, micro- and nanocrystalline metal layers. Such structure changes physical and mechanical properties of irradiated materials [94].

Surface Coating

The radiation methods are widely used to produce qualitative thin films on the surface of various solid materials. Its basic process is heating of the initial specimen (target) by high energy beam of charged particles (electrons or ions), followed by the specimen material evaporation and its condensation on the coating surface in a form of thin film. In the method of electron-beam ablation for the target material, the treatment of coating surface with ion beam is usually used for improving the film quality (see, for example, [95-99]). In this case, the ion beam provides the surface cleaning and assists in the surface coating (ion-beam assisted deposition). The process is carried out with use of ions of inert gases, which modify the surface structure and composition and make the produced film more durable and stable. The surface coating can be performed immediately from the beam by ions implantation in the substrate surface [100].

The advantage of this method is due to the possibility to operate with ions energy within the wide limits. This permits the change of properties and structure of the formed film as well as a possible control of the incidence angle in the ion beam. The latter is quite important at the film formation on walls of deep channels. The radiation methods are also used for the production of thin polymer films on the surface of various substrates [101]. The irradiation in this case plays a stage-initiating role that induces the process of monomer polymerization from the gas phase on the substrate surface (i.e. grafting polymerization). The radiation grafting is at present well established and accepted as an industrial method for the polymer coating. This kind of coating can achieve coat thickness at a submicron level with elasticity and corrosion stability [102]. Moreover, it is used for the modification of porous polymeric membranes [103], silicate surfaces [104], polymeric fibers [105], and polymeric electrolyte membranes in fuel cells [106].

RADIATION MODIFICATION OF CARBON NANOSTRUCTURES

The use of radiation technique also plays a significant role in the fabrication of new carbon nanomaterials. Due to their unique properties, carbon nanostructures such as carbon nanotubes (CNTs), fullerenes, and graphene sheets have a great potential for a new technological applications. These applications include electronic devices [107, 108], CNTs-based polymers [109, 110], composite materials [111, 112], hydrogen storage systems [113], biosensors [114], devices with highly focused ion beams [115], and so on.

CARBON NANOTUBES AND FULLERENES

The electron beam irradiation of crossing single-walled CNTs at high temperatures induces structural defects, leading to the formation of stable junctions of various geometries involving seven- and eight-membered carbon rings at the surface between the tubes [116]. The welding and controllable preparation of some special nanostructures such as nanospindles and nanodumbbells at electron beam irradiation of multiwalled CNTs were reported in [117]. The structural transformation of fluorinated single-walled CNTs bundles into multiwalled-like nanotubes was observed in [118]. This transformation can be explained by the local strains induced by fluorination. A welding of CNTs is also induced by the ion beam irradiation [119, 120]. The ion beam irradiation of multiwalled CNTs can be considered as an effective way for the formation of nanowire junctions [121]. The polymerization of C_{60} fullerene films has lately drawn the great attention because this new form of carbon material exhibits physical and chemical properties of both diamond and graphite [122]. In the case of radiation-induced polymerization, the unifying influence of energetic particles is accompanied by the destruction of fullerene-cage structure, which makes the polymerization phenomenon more complicated [123]. The electron beam irradiation of C_{60} intercalated graphite leads to its conversion in diamond without the application of high pressure and temperature [124]. The thermal activation and electron beam irradiation of hexanitro (60) fullerene film deposited on Au substrate produce a new polymeric material with unique characteristics [125].

GRAPHENE AND ITS DERIVATIVES

The majority of previous publications on the investigation of carbon nanostructures are devoted to graphene and its derivatives. Graphene is an isolated monolayer of carbon atoms packed into a dense honeycomb crystal structure. It can be considered as an ideal realization of two-dimensional material [126, 127]. Recent advances in micromechanical extraction and fabrication techniques for graphite structures [128] enable to investigate such exotic system experimentally. As shown in measurements, individual graphene sheets have extraordinary electronic transport properties, making this material a good candidate for future chips instead of silicon [129]. The hydrogenised form of graphene, graphane, takes place in contrast to electrically conducting graphene with promising semiconductor properties [130]. The controllable hydrogenation of graphene using the electron-beam lithography with high spatial resolution and reversible dehydrogenation are good characteristics for the applications in patterning of the graphene sheets into semiconducting or insulating nanodomains [131, 132].

Safety Problems of Radiation Nanotechnology

Many of the most important health and safety concerns of radiation nanotechnology are due to a lack of knowledge to health effects and of levels of occupational and other types of exposure to nanomaterials during their radiation production and use. The main safety problems that arise at the radiation processing of nanomaterials are associated with a risk assessment of induced radioactivity in the matter and exposure to nanoparticles and their release from products.

Induced Radioactivity

The induced radioactivity in the matter is created generally by neutron activation when free neutrons are captured by nuclei. The industrial irradiation facilities used in production of nanomaterials operate with electromagnetic radiation, electron and focused ion beams. The probability of γ-quanta and charged particles to induce nuclear reactions is much less than that of neutron activation because these particles interact basically with electronic structure of atom [133]. Three mechanisms are considered significantly for inducing radioactivity in processed materials under the action of γ-quanta and charged particles: photoneutron production, neutroncapture, and isomeric excitation [134]. Above threshold energy, the fast particle interaction with certain isotopes can result in the ejection of a neutron from the nucleus of the target atom. The production of this photoneutron can induce radioactivity in two ways: the target atom with one neutron removed may be a radioactive species, and the ejected neutron can be absorbed by another nucleus to form a radionuclide. The induced radioactivity from isomeric excitation is determined by the ejection of photons instead of neutrons.

The investigation of safety problems due to induced radioactivity under radiation processing of materials is associated mainly with food sterilization and medical applications [135, 136]. According to the recent IAEA report, the increase in radiation background dose from consumption of food, which was irradiated with 10 MeV electrons or with X-rays produced by electron beams with energy below 5 MeV, is insignificant [137]. The analysis of radioactivity from the processing of meat with 10 MeV electrons shows that the induced activity is 4 orders of magnitude less than the natural background activity found in meat [138]. The evaluation of potential induced radioactivity in medical products, as a function of the electron energy in electron beam sterilization process, confirms that the amount of induced activity at the electron energy 12 MeV can be considered insignificant [139]. The induced radioactivity generated from the metal probe, which is bombarded by proton beam, can be ignored when the beam energy is below 5 MeV [140]. The radiation methods of nanostructure production are operated as a rule with radiation sources, which do not originate the induced radioactivity from the processed material.

Additionally, the radiation processing of polymeric materials is carried out with the application of electron beams in the energy range between 0.15 keV and 10 MeV and with sources of radio-nuclides Co-60 and Cs-137 emitting γ-rays 1.17/1.33 and 0.662 MeV, respectively [21]. The energy of heavy ions that create ion tracks in thin polymer films is about 1 MeV/nucleon [7]. The radiation treatment of surfaces is carried out with the help of focused ion beams with energies below 50 keV [81]. The production of nanopowders in different substances, based on the technology of solid evaporation in the industrially employed electron-beam facility [47], operates with 1.4 MeV electrons.

RISK EFFECTS OF ENGINEERED NANOMATERIALS

The unique properties of nanomaterials are essential for their wide application in a number of industrial sectors such as micro-electronics, medicine, paper, textile, energy, and cosmetics. However, very little is known concerning risks, which are associated with impacts of these materials on human health [141]. The main route of engineered nanomaterials translocation into the body is inhalation process. As shown in recent investigations, different types of inhaled nanoparticles including radioactive ones and carbon nanotubes can reach systemic circulation in the body. Such nanoparticles can be distributed via this route in different body organs, provoking various negative effects (pulmonary inflammation, genotoxicity, carcinogenic effects, etc.) [142].

The inhalation of fine TiO_2 particles as well as single and multi-walled CNTs increases the number of inflammatory cells in the broncho-alveolar fluid and the levels of inflammatory mediators in this fluid [143]. The genotoxicity of nanoparticles was examined in several papers. However, the use of different types of particles, various cell systems, and variable assay conditions complicates the comparison of the existing studies [144]. As an example, the existing genotoxicity data on TiO_2 particles and CNTs appear to be positive [145]. The carcinogenic effect of nanoparticles including CNTs is suggested to be due to the local generation of reactive oxygen and nitrogen species in association with emerging inflammation [146]. An assessment of the carcinogenic potential of nanoparticles and its relevance for humans are currently fraught with uncertainty [147].

CONCLUSION

The brief review of the recent progress in the application of radiation techniques for nanostructures production shows that this direction of scientific investigations is of the fundamental importance due to their unique electronic, mechanical and chemical properties. It makes them possibly as attractive candidates for a variety of uses ranging from molecular electronics to composite materials. The application of *X*- and γ-rays, electron and focused ion beams is essential in the ion-track technology, nanolithography, production of nanopores and nanowires, cross-linking of nanopolymers and nanocomposites, etc. The radiation modification of graphene may play the most important role in the advent of the industrial microchips based on graphene transistors.

ACKNOWLEDGMENTS

This work was performed in the framework of the IAEA Technical Cooperation Project "Supporting Radiation Synthesis and the Characterization of Nanomaterials for Health Care, Environmental Protection and Clean Energy Applications", Project No. RER/8/014.

REFERENCES

[1] *Springer Handbook of Nanotechnology,* Ed. B. Bhushan, Springer, Berlin (2010).
[2] *Progress in Nanotechnology,* Ed. A. T. Larimar, Nova Sci. Publ., New York (2008).
[3] M. Dragoman, D. Dragoman, *Prog. Quant. Electron.* 33, 165 (2009).
[4] G. Chmielewski, D. K. Chmielewska, J. Michalik, and M. H. Sampa, *Nucl. Instr. Meth.* B265, 339 (2007).
[5] *Radiation Chemistry: from Basics to Applications in Material and Life Science,* Eds. M. Spotheim-Maurizot, M. Mostafavi, T. Douki, J. Belloni, EDP Sci., Paris (2008).
[6] S. Barnard and I. K. Snook, *Carbon,* 48, 981 (2010).
[7] Waheed, D. Forsyth, A. Watts, et al. *Radiat. Measur.* 44, 1109 (2009).
[8] D. K. Avasthi, Y. K. Mishra, F. Singh, and J. P. Stoquert, *Nucl. Instr. Meth.* B268, 3027 (2010).
[9] F. Staufenbiel, G. Schiwietz, K. Czerski, et al. *Nucl. Instr. Meth.* B230, 426 (2005).
[10] F. Bergamini, M. Bianconi, S. Cristiani, et al. *Nucl. Instr. Meth.* B266, 2475 (2008).
[11] D. Fink, A. Chandra, W. R. Fahrner, et al. *Vacuum,* 82, 900 (2008).
[12] S. K. Chakarvarti, *Radiat. Measur.* 44, 1085 (2009).
[13] H. Hanot and E. Ferain, *Nucl. Instr. Meth.* B267, 1019 (2009).
[14] L. Dauginet-De Pra, E. Ferain, R. Legras, et al. *Nucl. Instr. Meth.* B196, 81 (2002).
[15] M.-C. Clochard, T. Berthelot, C. Baudin, et al. *J. Power Sources.* 195, 223 (2010).
[16] J. Krauser, J.-H. Zollondz, A. Weidinger, and C. Trautmann, *J. Appl. Phys.* 94, 1959 (2003).
[17] K. Hoppe, W. R. Fahrner, D. Fink, et al. *Nucl. Instr. Meth.* B266, 1642 (2008).
[18] Schultz, G. N. Akapiev, V. V. Shirkova, et al. *Nucl. Instr. Meth.* B236, 254 (2005).
[19] J.-H. Choi, C.-H. Jung, D.-K. Kim, et al. *Radiat. Phys. Chem.* 78, 517 (2009).
[20] P. Ulanski and J. M. Rosiak, *Nucl. Instr. Meth.* B151, 356 (1999).
[21] G. Chmielewski, M. Haji-Saeid, and S. Ahmed, *Nucl. Instr. Meth.* B236, 44 (2005).
[22] E. T. Thostenson, C. Li, T.-W. Chou, *Compos. Sci. Technol.* 65, 491 (2005).
[23] R. J. Hemply, G. W. Crabtree, and M. V. Buchanan, *Phys. Today.* 62(11), 32 (2009).
[24] E. Najafi and K. Shin, *Colloid. Surf.* 257-258, 333 (2005).
[25] A. Rozenberg and R. Tenne, *Prog. Polym. Sci.* 33, 40 (2008).
[26] S. Pavlidou and C. D. Papaspyrides, *Prog. Polym. Sci.* 33, 1119 (2008).
[27] Y. Hu, O. A. Shenderova, Z. Hu, et al. *Rep. Prog. Phys.* 69, 1847 (2006).
[28] B. Zenin, V. B. Rogacheva, V. I. Feldman, et al. *Adv. Colloid Interface Sci.* 158, 84 (2010).
[29] V. Thakur, A. Leuteritz, U. Gohs, et al. *Appl. Clay Sci.* 49, 200 (2010).
[30] H.-J. Gläsel, F. Bauer, E. Hartmann, et al. *Nucl. Instr. Meth.* B208, 303 (2003).

[31] J. Sharif, K. Z. M. Dahlan, and W. M. Z. W. Yunus, *Radiat. Phys. Chem.* 76, 1698 (2007).
[32] E. Planes, L. Chazeau, G. Vigier, et al. *Compos. Sci. Technol.* 70, 1530 (2010).
[33] E. J. Robinette and G. P. Palmese, *Nucl. Instr. Meth.* B236, 216 (2005).
[34] Z. Wang, M. Z. M. Yusop, T. Hihara, et al. *Appl. Surf. Sci.* 256, 6371 (2010).
[35] X. Lin, R. Zhou, J. Zhang, and S. Fei, *Appl. Surf. Sci.* 256, 889 (20090.
[36] J. M. Rosiak, P. Ulanski, and S. Kadłubowski, IAEA TECDOC-1438, IAEA, Vienna (2005), 99.
[37] J.-C. An, *J. Ind. Eng. Chem.* 16, 657 (2010).
[38] F. J. Xu, K. G. Neoh, and E. T. Kang, *Prog. Polym. Sci.* 34, 719 (2009).
[39] J. M. Rosiak, I. Janik, and S. Kadłubowski, *Nucl. Instr. Meth.* B208, 325 (2003).
[40] K. F. Arndt, T. S. Schmidt, and R. Reichelt, *Polymer.* 42, 6785 (2001).
[41] P. Ulanski, S. Kadłubowski, and J. M. Rosiak, *Radiat. Phys. Chem.* 63, 533 (2002).
[42] V. Vijayabaskar, S. Bhattacharya, and V. K. Tikku, V. K. et al. *Radiat. Phys. Chem.* 71, 1045 (2004).
[43] Henke, S. Kadłubowski, and P. Ulanski, *Nucl. Instr. Meth.* B236, 391 (2005).
[44] S. Mitra, S. Chattopadhyay, S. Sabharwal, et al. *Radiat. Phys. Chem.* 79, 289 (2010).
[45] K. Makuuchi, *Radiat. Phys. Chem.* 79, 267 (2010).
[46] T. Xu, N. Zhang, H. Nichols, et al. *Mater. Sci. Eng.* C27, 579 (2007).
[47] S. P. Bardakhanov, A. I. Korchagin, N. K. Kuksanov, et al. *Mater. Sci. Eng.* B132, 204 (2006).
[48] S. Singh, M. C. Rath, A. K. Singh, et al. *Radiat. Phys. Chem.* 80, 736 (2011).
[49] F. Zhou, R. Zhou, X. Hao, et al. *Radiat. Phys. Chem.* 77, 169 (2008).
[50] L. Castaldi, K. Giannakopoulos, A. Travlos, et al. *J. Magn. Magn. Mater.* 290, 544 (2005).
[51] J. Li, Y. N. Kim, E. J. Lee, et al. *Nucl. Instr. Meth.* B251, 425 (2006).
[52] J. H. Sohn, L. Q. Pham, H. S. Kang, et al. *Radiat. Phys. Chem.* 79, 1149 (2010).
[53] T.-F. Hsieh, C.-C. Chuang, Y.-C. Chou, et al. *Mater. Design.* 31, 1684 (2010).
[54] Meisel, IAEA TECDOC-1438, IAEA, Vienna (2005), 125.
[55] M. Kumar, L. Varshney, and S. Francis, *Radiat. Phys. Chem.* 73, 21 (2005).
[56] K. Naghavi, E. Saion, K. Resaee, et al. *Radiat. Phys. Chem.* 79, 1203 (2010).
[57] H. Remita, A. Etcheberry, and J. Belloni, *J. Phys. Chem.* B107, 31 (2003).
[58] Q. Yang, K. Tang, F. Wang, et al. *Mater. Lett.* 57, 3508 (2003).
[59] C.-C. Kim, C. Wang, Y.-C. Yang, et al. *Mater. Chem. Phys.* 100, 292 (2006).
[60] K. Dick, T. Dhanasekaran, Z. Zhang, et al. *J. Am. Chem. Soc.* 124, 2312 (2002).
[61] *Handbook of Microlithography, Micromachining and Microfabrication: Microlithography,* Ed. P. Rai-Choudhury, SPIE Press, Bellingham (1997).
[62] T. R. Groves, D. Pickard, B. Rafferty, et al. *Microelectron. Eng.* 61-62, 285 (2002).
[63] Vieu, F. Carcenac, A. Pepin, et al. *Appl. Surf. Sci.* 164, 111 (2000).
[64] L. Scalia, IAEA TECDOC-1438, IAEA, Vienna (2005), 213.
[65] T. Borzenko, C. Gould, G. Schmidt, et al. *Microelectron. Eng.* 75, 210 (2004).
[66] W. Weber, G. Ilicali, J. Kretz, et al. *Microelectron. Eng.* 78-79, 206 (2005).
[67] Y. Chen, Z. Lu, X. Wang, et al. *Microelectron. Eng.* 84, 1499 (2007).
[68] Y. Sohda, H. Ohta, F. Murai, et al. *Microelectron. Eng.* 67-68, 78 (2003).
[69] W. Zhang, A. Pott, D. M. Bagnall, et al. *Thin Solid Films.* 515, 3714 (2007).
[70] G. Rius, J. Llobet, J. Arcamone, et al. *Microelectron. Eng.* 86, 1046 (2009).

[71] K. Mølhave, D. Nørgaard, and P. Bøggild, *Ultramicroscopy.* 102, 215 (2005).
[72] K. Arshak, M. Mihov, A. Arshak, et al. *Microelectronic,* 2, 459 (2004).
[73] J. Dienelt, K. Zimmer, and H. Neumann, *Surf. Coat. Technol.* 174-175, 157 (2003).
[74] J. Taniguchi, K. Koga, Y. Koga, et al. *Microelectron. Eng.* 83, 940 (2006).
[75] R. Kometani, S. Warisawa, and S. Ishihara, *Microelectron. Eng.* 87, 1044 (2010).
[76] S. Matsui, *Nucl. Instr. Meth.* B257, 758 (2007).
[77] P. Meyer, J. Schulz, and V. Saile, *Micromanufacturing Engineering and Technology,* Ed. Y. Qin, Amsterdam, Elsevier (2010), 202.
[78] Romanato, M. Tormen, L. Businaro, et al. *Microelectron. Eng.* 73-74, 870 (2004).
[79] T. Mappes, S. Achenbach, and J. Mohr, *Microelectron. Eng.* 84, 1235 (2007).
[80] N. Savvides, *J. Appl. Phys.* 97, 053517 (2005).
[81] T. Sadoh, H. Eguchi, A. Kenjo, et al. *Nucl. Instr. Meth.* B206, 478 (2003).
[82] S. Xu, W. Zheng, X. Yuan, et al. *Nucl. Instr. Meth.* B266, 3370 (2008).
[83] Perentos, A. Mitchell, A. Holland, *Appl. Surf. Sci.* 252, 1006 (2005).
[84] Y. Kawabata, J. Taniguchi, I. Miyamoto, *Diamond Relat. Mater.* 13, 93 (2004).
[85] V. S. Smentkowski, *Prog. Surf. Sci.* 64, 1 (2000).
[86] I. Stognij, N. N. Novitskii, and O. M. Stukalov, *Tech. Phys. Lett.* 28, 17 (2002).
[87] Bourelle, A. Suzuki, A. Sato, *Nucl. Instr. Meth.* B241, 622 (2005).
[88] Z. Insepov, A. Hassanein, J. Norem, et al. *Nucl. Instr. Meth.* B261, 664 (2007).
[89] B. Koslowski, S. Strobel, and P. Ziemann, *Diamond Relat. Mater.* 9, 1159 (2000).
[90] V. V, Uglov, V. M. Anishchik, V. V. Astashinski, et al. *Surf. Coat Technol.* 180-181, 108 (2004).
[91] E. Weiser, M. Peikert, C. Wenzel, et al. *Thin Solid Films.* 410, 121 (2002).
[92] N. N. Cherenda, V. V. Uglov, G. V. Litvinovich, et al. *Nucl. Instr. Meth.* B211, 219 (2003).
[93] M. Ikeyama, S. Nakao, H. Morikawa, et al. *Surf. Coat Technol.* 128-129, 400 (2000).
[94] V. Tereshko, V. V. Abidzina, I. E. Elkin, et al. *Surf. Coat Technol.* 201, 8552 (2007).
[95] Y. Gotoh, K. Kagamimori, H. Tsuji, et al. *Surf. Coat Technol.* 158-159, 729 (2002).
[96] Mitsuo, T. Mori, Y. Setsuhara, et al. *Nucl. Instr. Meth.* B206, 366 (2003).
[97] P. Mosaner, M. Bonelli, and A. Miotello, *Surf. Coat Technol.* 180-181, 41 (2004).
[98] W. Ensinger and M. Kiuchi, *Surf. Coat Technol.* 203, 2763 (2009).
[99] Leng, Z. Yu, Y. Li, et al. *Appl. Surf. Sci.* 256, 5832 (2010).
[100] V. Byeli, M. A. Belotserkovski, and V. A. Kukareko, *Wear,* 267, 1757 (2009).
[101] P. Roose, I. Fallais, C. Vandermiers, et al. *Prog. Org. Coat.* 64, 163 (2009).
[102] E.-S. A. Hegazy, H. A. AbdEl-Rehim, H. Kamal, et al. *Nucl. Instr. Meth.* B185, 235 (2001).
[103] Z. Y. Xi, Y. Y. Xu, L. P. Zhu, et al. *J. Membr. Sci.* 339, 33 (2009).
[104] Yu. Shmykova, S. V. Mjakin, I. V. Vasiljeva, et al. *Appl. Surf. Sci.* 255, 6391 (2009).
[105] Vahdat, H. Bahrami, N. Ansari, et al. *Radiat. Phys. Chem.* 76, 787 (2007).
[106] Chen, D. Li, H. Koshikawa, et al. *J. Membr. Sci.* 362, 448 (2010).
[107] P. Anantram, and F. Leonard, *Rep. Prog. Phys.* 69, 507 (2006).
[108] Y.-W. Son, M. L. Cohen, and S. G. Louie, *Nature,* 444, 347 (2006).
[109] Wang, Z.-X. Guo, S. Fu, et al. *Prog. Polym. Sci.* 29, 1079 (2004).
[110] S. V. Ahir, Y. Y. Huang, and E. M. Terentjev, *Polymer,* 49, 3841 (2008).
[111] Y. Hu, O. A. Shenderova, Z. Hu, et al. *Rep. Prog. Phys.* 69, 1847 (2006).
[112] T. Ramanathan, A. A. Abdala, S. Stankovich, et al. *Nature Nanotech.* 3, 327 (2008).

[113] Hirscher and M. Becher, *J. Nanosci. Nanotech. 3*, 3 (2003).
[114] Shim, N. W. S. Kam, Y. Chen, et al. *Nanolett.* 2, 285 (2002).
[115] V. M. Biryukov, S. Bellucci, and V. Guidi, *Nucl. Instr. Meth.* B231, 70 (2005).
[116] M. Terrones, F. Banhart, N. Grobert, et al. *Phys. Rev. Lett.* 89, 075505 (2002).
[117] Q. Yang, S. Bai, G. Wang, and J. Bai, *Mater. Lett.* 60, 2433 (2006).
[118] K. H. An, K. A. Park, J. G. Heo, et al. *J. Am. Chem. Soc.* 125, 3057 (2003).
[119] R. Adhikari, H. Bakhru, P. M. Ajayan, et al. *Nucl. Instr. Meth.* B265, 347 (2007).
[120] Kim, E. Kim, J. Lee, et al. *Thin Solid Films* 516, 3474 (2008).
[121] Z. Wang, L. Yu, W. Zhang, et al. *Phys. Lett.* A324, 321 (2004).
[122] J. Onoe, T. Nakayama, M. Aono, et al. *J. Phys. Chem. Solids* 65, 343 (2004).
[123] V. Krasheninnikov, K. Nordlund, J. Keinonen, et al. *Nucl. Instr. Meth.* B202, 224 (2003).
[124] V. Gupta, P. Scharff, and N. Miura, *Mater. Lett.* 59, 3259 (2005).
[125] R.-S. Zhai, A. Das, C.-K. Hsu, et al. *Carbon,* 42, 395 (2004).
[126] K. Geim, *Science,* 324, 1530 (2009).
[127] Soldano, A. Mahmood, and E. Dujardin, *Carbon,* 48, 2127 (2010).
[128] K. S. Novoselov, A. K. Geim, S. V. Morosov, et al. *Science,* 306, 666 (2004).
[129] R. Van Noorden, *Nature,* 442, 228 (2006).
[130] C. Elias, R. R. Nair, T. M. G. Mohiuddin, et al. *Science,* 323, 610 (2009).
[131] S. Ryu, M. Y. Han, J. Maultzsch, et al. *Nano Lett.* 8, 4597 (2008).
[132] J. D. Jones, K. K. Mahajan, W. H. Williams, et al. *Carbon,* 48, 2335 (2010).
[133] Banhart, *Rep. Prog. Phys.* 62, 1181 (1999).
[134] M. A. Smith, *Radiat. Phys. Chem.* 77, 1079 (2008).
[135] J. S. Findlay, T. V. Parsons, and M. R. Sene, *Radiat. Phys. Chem.* 42, 417 (1993).
[136] Cesana, E. Mauro, and M. Silari, *Nucl. Instr. Meth.* B268, 2272 (2010).
[137] *Natural and Induced Radioactivity in Food*, IAEA TECDOC-1287, IAEA, Vienna (2002).
[138] M. A. Lone, *Nucl. Instr. Meth.* A299, 656 (1990).
[139] M. A. Smith, *Radiat. Phys. Chem.* 81, 57 (2012).
[140] W. Tao, Z. Kaizhi, L. Qing, et al. *Nucl. Instr. Meth.* B269, 1512 (2011).
[141] K. Savolainen, H. Alenius, H. Norppa, et al. *Toxicology*, 269, 92 (2010).
[142] K. Aschberger, C. Micheletti, B. Sokull-Klüttgen, and F. M. Christensen, *Environ. Int.* 37, 1143 (2011).
[143] A. Shvedova, E. Kissin, A. R. Murray, et al. *Am. J. Physiol. Lung Cell Mol. Physiol.* 295, 552 (2008).
[144] C. M. Falck, H. K. Lindberg, S. Suhonen, et al. *Hum. Exp. Toxicol.* 28, 339 (2009).
[145] Y. Sakamoto, D. Nakae, N. Fukumori, et al. *J. Toxicol. Sci.* 34, 65 (2009).
[146] Takagi, A. Hirose, T. Nishimura, et al. *J. Toxicol. Sci.* 33, 105 (2008).
[147] Becker, F. Herzberg, A. Schulte, and M. Kolossa-Gehring, *Int. J. Hygiene Environ. Health*, 214, 231 (2011).

In: Nanostructures
Editor: Yu Dong

ISBN: 978-1-62618-081-9

Chapter 4

HIERARCHICALLY ORDERED COLLOIDAL CRYSTALS: FABRICATION, STRUCTURES, AND FUNCTIONS

Qin Li** and Ehsan Eftekhari

Environmental Engineering and Queensland Micro- and Nanotechnology Centre,
Griffith University, Australia

ABSTRACT

Marrying nanochemistry with optics has opened a new gateway to design and fabricate novel nanostructures that can interact with light in unprecedented manners. Self-assembled colloidal photonic crystal is one of the emerged nanostructured photonic materials, which possess well-defined, highly ordered structure and periodicities of light wavelength. Various 2D and 3D ordered structures have been fabricated through self-assembly methods by employing latex colloidal particles, or functionalized latex particles with more sophisticated structural features such as core-shell and anisotropy. The unique ordered structure and light-interactive properties have afforded them versatile functionalities in light manipulation, catalysis, sensing, and light harvesting. In this book chapter, we review recent research progress on hierarchical colloidal photonic crystals: the preparation methods, structural properties and applications. By introducing a hierarchy of interconnected structures at different length scales, the surface-to-volume ratio can be dramatically increased and the materials properties tailored accordingly. Moreover, the band-gap features of the photonic crystals may also be adjusted by the structure hierarchy.

INTRODUCTION

Colloidal crystals are a kind of periodically structured lattices assembled from monodisperse colloidal particles [1]. They appear as analogous entities to atomic crystals but

* Email address: qin.li@griffith.edu.au

with constituents in the size range of several nanometers to several micrometers. The prerequisite to their formation is the monodispersity of colloidal particle size, which is enabled by the increased control of particle size and shape in synthesis. The formation of these ordered structures in general is facilitated by intermolecular and surface forces as opposed to covalent forces or ionic bonds in atomic crystals.

In this chapter, we focus our attention on the colloidal crystals made of spheres of several hundred nanometers. Because this periodicity range matches with light wavelength, these colloidal crystals can diffract light in UV, visible, and near-infrared regions, and prevent light from being transmitted in a range of frequencies known as the photonic bandgap (PBG) [2, 3]. Subsequently, they are also referred to as colloidal photonic crystals (CPC). Photonic crystals have been first theoretically proposed by Yablonovich[2] andJohn [3] and gradually experimentally proven and demonstrated as a fascinating material for guiding, confining and manipulating light at low energy losses,which leads to their strong potential in optoelectronics, all-optic devices, and sensing [4-6].Physicists first employed microfabrication methods such as drilling to obtain 2D structures with periodic contrast in dielectric constants, which are generally capital cost intensive and laborious. When chemists made the link between photonic crystals and self-assembled colloidal crystals, it has raised significant hope in developing self-assembly into an effective parallel production method for 2D and 3D photonic crystals [7-9].

Self-assembled CPCs often refer to the inverse porous structures; the monodisperse particle packings serve as sacrificial template to allow infiltration and solidification of a second material that is of high reflective index or possesses other desired properties. Ordered porous networks through colloidal templates alone is a topical subject in its own right [10]. However, self-assembly also implies minimum free energy, which means most structures obtained from self-assembly are of face centered cubic (*fcc*) or hexagonal close-packed (*hcp*) lattice symmetry.These kinds of crystal structures require the refractive index of the backbone material to be greater than 2.8 in order to have complete photonic band gap [11]. Overcoming the *fcc* structure limit and forming diamond lattice has been one of the key challenges in colloidal self-assembly [12]. The other very active area in CPC research has been the exploration of ways to impart various functionalities to the material by utilizing the structural regularities. Hierarchically ordered colloidal crystal (HOCC) represents one of the promising routes serving for both objectives.

HOCCs are self-organized structures combining features of several hierarchical ordering characteristic length scales. Hierarchically ordered porous structures are widely present in nature, e.g., diatoms, bones and lungs. The open, interconnected pores of different sizes are orderly arranged. They offer efficient transport of fluids for facilitating molecular-level biophysicochemical processes, typically using macropores to enhance the mass transport and meso- and micropores to enlarge the surface area for reactions. Yang et al. [13] have first demonstrated this concept by combining copolymer templating (~10 nm), latex sphere templating (~100 nm) and micromolding (~μm), and fabricated hierarchically ordered oxides. Recent years have seen significant progress of this concept and the categorical shift of colloidal crystallization from thermodynamics-driven to well-controlled assembly through hierarchical organization [14]. HOCCs also offer great potential for the design of new materials and devices for various applications due to the interconnected cavities at different length scales, increased specific surface area and higher density of reactive sites. Given the importance and strong potential of this field, this chapter intends to provide a brief account of the recent progress on HOCCs, fabrication strategies, their properties and potential applications.

FABRICATIONS AND APPLICATIONS OF HOCCS

Building on extensively studied fabrication methods for colloidal crystals [10, 15, 16], a major challenge in the synthesisof HOCCsis on how to introduce other feature hierarchical length scales, particularly on the nanoscale.

The general fabrication strategy for functional colloidal crystals is through colloidal templating[10]: first, self-assemble colloidal crystal from monodisperse silica or latex submicron spheres;second, it is infiltrated with the precursor of a targeted material, followed by polymerization or solidification;subsequent removal of the sphere-assembly template results in an inverse structure with ordered and interconnected macropores of hundreds nanometers in size.

The meso- and micro-size features are mostly introduced by surfactants, amphiphilic block copolymers, multiple-size colloidal particles, anisotropic colloid particles, ionic liquids, and/or pre-functionalized constituent particles.

1. Block Copolymer and Surfactant Templating

Similar to surfactants, amphiphilic block copolymers consist of at least two segments with different chemical natures. While the size of surfactants or surfactant micelles is more comparable to micropores (<2 nm) different to surfactants, amphiphilic copolymers can cover a length scale from several nanometers to a couple of tens nanometers, therefore, they are more suitable for being used as mesopore (2 – 50 nm) templates. Since surfactant templating is relatively mature, here we place more emphasis on block copolymer templating. Because of the difference in interface activity between the two kinds of polymer segments with their surroundings, block copolymers tend to self-organize into ordered structures such as lamellar, cylindrical and vesicular to reduce the interfacial energy between the immiscible blocks and to maximize the chain conformation entropy. Their polymeric nature affords versatility and diversity in both chemistry and geometry to this class of materials [17].

Therefore, block copolymers have been broadly explored in 2 dimensional (2D) and 3 dimensional (3D) nanopatterning. Yang et al. [13] investigated for the first time the hierarchical templating combination of top-down and bottom-up fabrication methods. They produced 3D HOCCs by concurrently or sequentially combining micromolding, colloidal self-assembly with polystyrene (PS) spheres and cooperative assembly of hydrolyzed inorganic species and amphiphilic block copolymers. In a typical sequential process, latex particles were first self-assembled in a pattern imposed by a silicon mold. Then, a drop of sol-gel-block copolymer precursor solution was introduced into the patterned colloidal crystal, followed by gelation. Subsequent calcination removed both latex particle and block copolymer templates.

Figure 1 shows the electron microscopic images of the resultant silica structure at different length scales.

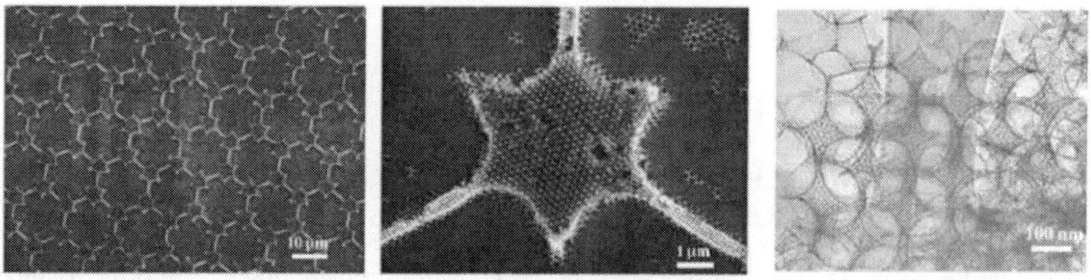

Figure 1.SEM images (left and middle) of hierarchically ordered mesoporous silica displaying organization over three discrete characteristic dimensions. TEM images of the same samples (right image), showing that the framework of the macroporous skeleton is made up of ordered cubic mesoporous silica [13].

In this process, while the silicon mold provided the macro-scale patterning and latex particles self-assembled into ordered structure on hundreds nanometer scale, block copolymer, here Pluronic F127 ($EO_{106}PO_{70}EO_{106}$), also self-organized with regularity in the interstitial voids among the latex particles, affording mesopores to the final silica film.

Silica-based hierarchically ordered macro-/meso-, macro-/micro- or macro-/meso-/microporous structures have since been fabricated through colloidal templating with sol-gel in combination with surfactants, or liquid crystals, or ionic liquid, or block copolymers [18-20]. Among these efforts, Villaescusa et al. have shown that when a good control is applied over the filling fraction of the meso- or microporous guest material by employing the right infiltration method, the optical functionality conferred by the colloidal particle regular packing can be well preserved for the HOCCs[20]. In addition, Chmelka and coworkers have also demonstrated that the catalytic selectivity of meso-macroporous bimodal aluminosilicameterials is higher than microporous zeolite beta and mesoporous SBA-15, owing to the improved diffusive resistance to mass transfer [21].

These successes in silica-based porous materials have naturally led to strong interests in synthesising hierarchically ordered porous carbon materials, because of the usefulness of porous carbon in separation, catalysts and catalyst supports, electrodes, sensors, super capacitors, batteries, gas storage , and many other applications [22- 25]. Stein and coworkers [26] prepared hierarchically ordered biomodal porous carbon with macropores of 200 – 400 nm and mesopores of 2.8 nm by a multistep templating method combining colloidal templating and nanocasting. As Figure 2 illustrates, their procedure involves: 1. Preparation of 3D ordered macro-/mesoporous silica replica by infiltrating a sol of nonionic oligomer surfactant (Brij 56, $C_{16}H_{33}EO_{10}$) and tetramethylorthosilicate into a PMMA colloidal crystal and calcinations; 2. Infiltration of the macro-/mesoporous silica replica with phenol resin precursor as a carbon source; 3. Pyrolysis of the composite sample and removal of silica scaffolds by hydrofluoric acid (HF). It should be appreciated that where the surfactants were applied, various blockcopolymers can substitute depending on the requirement of the mesopore size.

By introducing mesopores into the skeleton structure, the surface area has reached >1200 m^2/g in contrast to 20 m^2/g for a macropore only structure. However, such a multiple templating procedure is highly laborious. Zhao and coworkers have developed an organic–organic self-assembly between block copolymers and phenolic resin precursors as a direct templating approach to ordered mesoporous carbons with various pore structures [27, 28].

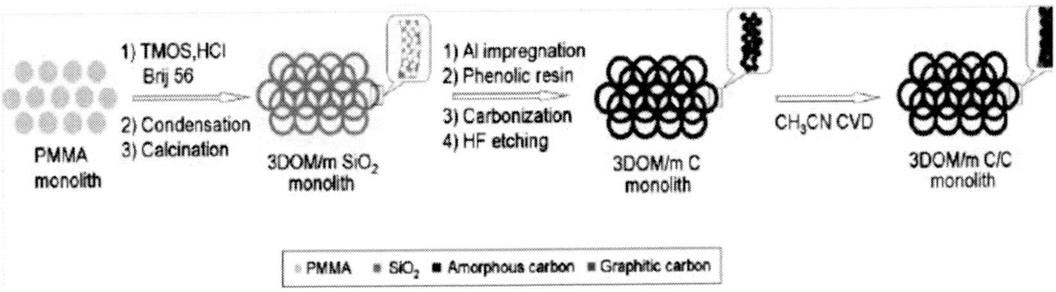

Figure 2.Diagram of the Synthesis of hierarchically ordered carbon monolith [26].

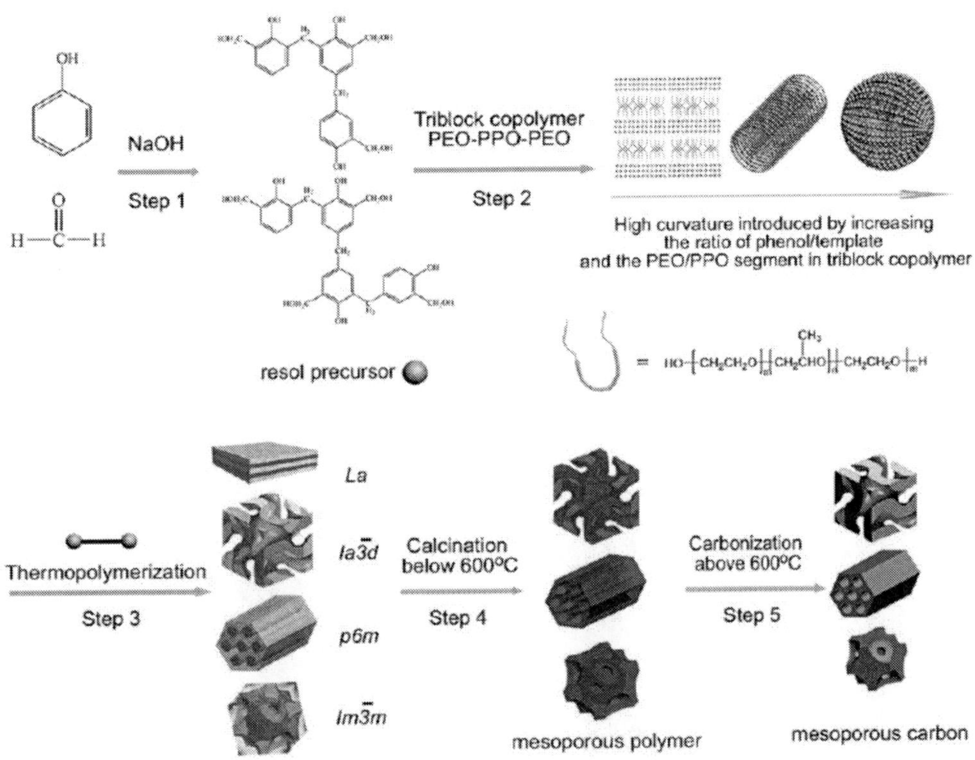

Figure 3.Preparation Schematic of organic-organic self-assembly to mesoporous carbons.Reproduced with permission from Reference [28].

This technique typically involves the following procedures as illustrated in Figure 3: a. polymerizing phenol and formaldehyde in alkaline condition to form the low-molecular-weight phenolic resol; b. mixing low-molecular-weight resol with PEO-PPO-PEO triblock copolymer in an ethanolic solvent and thermally polymerizing the resol; c. evaporating the solvent, which induces self-assembly of block copolymer to ordered structure, hence ordered assembly of resol nanoparticles.

The copolymer template can be removed by calcination at 350–450 °C and carbonization of resol precursor takes place when calcination over 600 °C. Studies have shown that the formed mesostructures are highly stable and can be retained at a temperature as high as 1400 °C under nitrogen atmosphere [29].

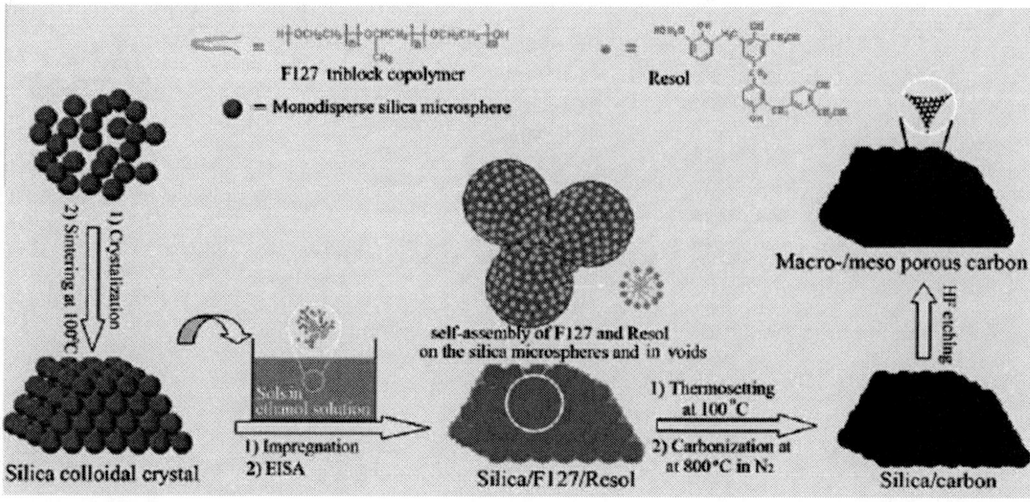

Figure 4. Synthesis scheme used for the preparation of the ordered macro-/mesoporous carbons using monodisperse silica colloidal crystals and triblock copolymer F127 as dual templates and phenolic resols as sol precursors [27].

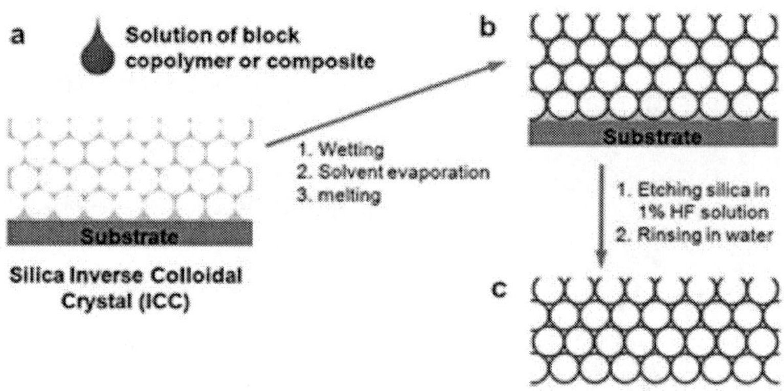

Figure 5.Schematic representation of the solution wetting method for fabricating copolymer surface-patterned HOCCs [37].

This organic-organic mesopore formation method has been neatly combined with colloidal templating and a facile synthesis method for producing hierarchically ordered macro-/mesoporous carbon materials was developed [29]. As shown in Figure 4, a mixture of F127 triblock copolymer and resol is infiltrated into a preformed silica colloidal crystal by a sol-gel process. Subsequent sol evaporation induces ordered vesicle formation on silica sphere surface as well as in the voids among silica spheres. After carbonization and etching off silica template, a carbon-based HOCC is obtained. Owing to the copolymer template size, Deng et al. [27] obtained ~11 nm mesopores within the matrix of hundreds nanometer macropores, which makes the materials more compatible when dealing with large molecules such as biomolecules.

It needs to be stressed that copolymers alone can create post-mesoscale structures in the macropores of colloid crystals and their inverse structures. These pores provide strong spatial confinement to copolymer self-organization, which may break the symmetry of a structure, thus allow the copolymers to demonstrate new behavior [30-32]. Ozin, Manners and coworkers [33] have systematically studied the various mesoscale structures that lamella- and cylinder- forming diblock copolymers in silica colloidal and inverse colloidal crystals. 3D confined geometries such as octapods and tetrapods, core-multishell, golf ball; belt hexagonal packed spheres were observed when the ratio of 3D interspacing distance over the equilibrium periodicity of the diblock copolymer microphase separation (D/L_0) varied, where D stands for interspacing distance and L_0 denotes the copolymer characteristic equilibrium period.

The previous examples have shown block copolymer templating can afford a mesoporous structure in the backbone of the ordered marcoporous framework. Questions also arise as to the possibility of introducing mesostructures on the inner walls of the inverse colloidal crystal. The motivations are two-folds: *a.* from fundamental research point of view, it is highly interesting to understand the self-assembly of block copolymers on surface in confined geometry as opposed to the case of filling the entire confined space; *b.* from application point of view, the mesostructures on the walls of macropores are most accessible sites for reactants, hence, by introducing mesostructures only on the macropore walls it is highly probable to generate materials with similar functionalities but of higher mechanical strength.

As a versatile surface patterning template, block copolymer can self-assemble into well-defined nanostructures such as hexagonally ordered cylinders or stripe-like lamella, which are highly interesting for 2D ordered patterning of nanoparticles [34] and proteins [35] with several to tens nanometer inter-distance as well as synthesis of nanofibrils[36]. By wetting a pre-formed silica inverse colloidal crystal with a PS-*b*-PMMA solution as illustrated in Figure 6, Fu et al. [37] have obtained hierarchically ordered 3D arrays of block copolymer hollow spheres with nanostructured shells and nanopatterns on inner walls. We have produced unique copolymer-coated silica inverse colloidal crystals and their daughter hollow copolymer sphere arrays (by etching away silica with HF solution), owing to the fact that solution wetting method generated a thin layer of the block copolymer solution on the silica walls of inverse colloidal crystal templates and thus created thin block copolymer films on the internal surface of templates after solvent evaporation. By systematically varying the symmetry of the diblock copolymers PS-*b*-PMMA, Fu et al. [37] have observed intriguing copolymer patterns as shown in Figure 6 compared with their behavior under un-confined conditions. Together with the data on varying D_3/L_0, this study showed that the interplay among copolymer microphase separation, geometrical confinement imposed by template macropores and the additional free surface between polymer and air adds more complexity to copolymer self-organization, in turn provides additional routes for forming hierarchically ordered functional materials.

2. Multiple-Size Particle Templating

Co-deposition of colloidal particles of different diameters also presents an effective and facile fabrication approach to HOCCs. Jonas and coworkers [38] reported for the first time the preparation of hierarchical ordered structure by vertical lifting co-deposition of PS particles (465 nm), PMMA particles (84 nm), and silica nanoparticles (6 nm), which resulted in macropores of two distinct sizes (460 and 80 nm) with interconnecting holes of 140 and 25 nm, as shown in Figure 7.

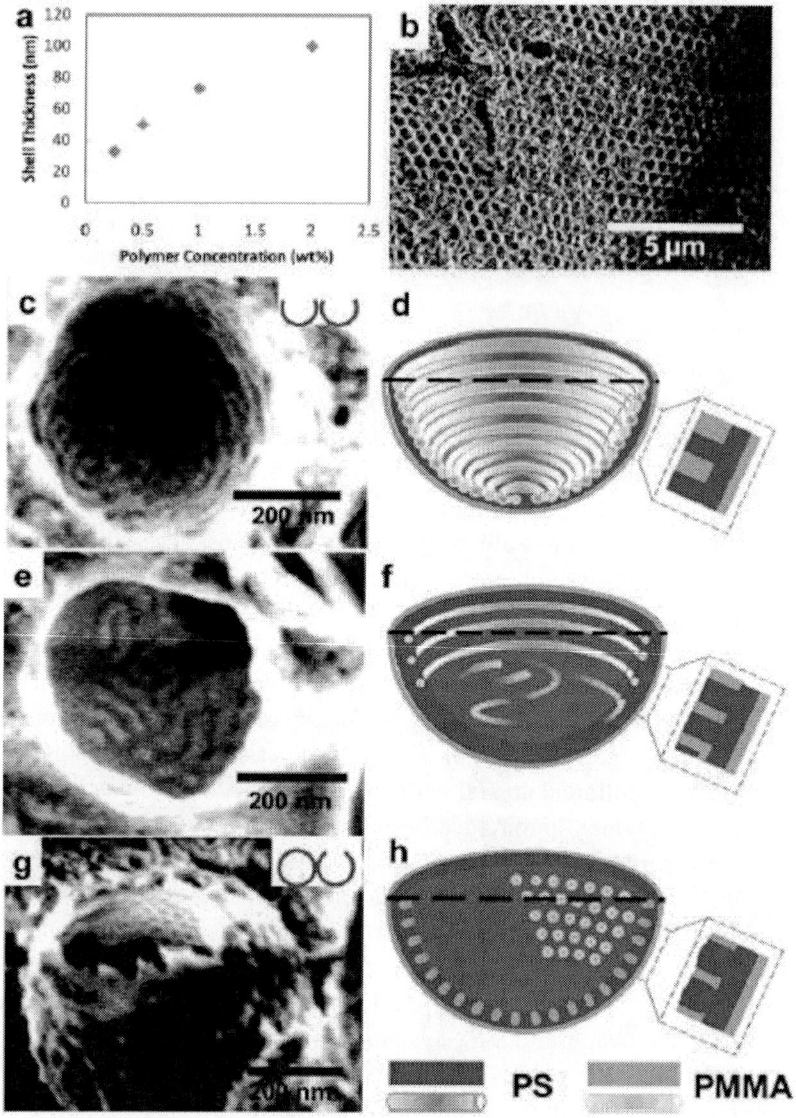

Figure 6. (a) Concentration dependence of shell thickness and the SEM images of, after removal of silica templates, (b) the 3D arrays of hollow PS-*b*-PMMA spheres and theannealed hollow spheres of (c) S_{326}-$PMMA_{301}$, (e) S_{450}-MMA_{182}, and (g) S_{1037}-MMA_{322}. The PS- rich domains appear bright and the PMMA-rich domains appear dark in the SEMimages. The schemes in (d), (f), and (h) represent the schematic cross-section view of the nanostructured shells observed in (c), (e), and (g) [37].

The entire procedure only requires two steps, *a*. co-crystallization of template and matrix particles, and *b*. template removal, which is simple and elegant. As shown by our vis-NIR spectroscopy characterization (Figure 7), the colloidal crystals formed by multiple-size templating method possess distinct pseudo stop bands as mono-colloidal crystals.

Li, Zhao and co-workers [39]have further improved the co-crystallisation technique. We found that the multiple-size co-crystallization can take place by horizontal deposition, therefore, the technique can readily prepare centimetre-size binary and ternary colloidal crystals without requirement of any sophisticated equipment, as illustrated in Figure 8.

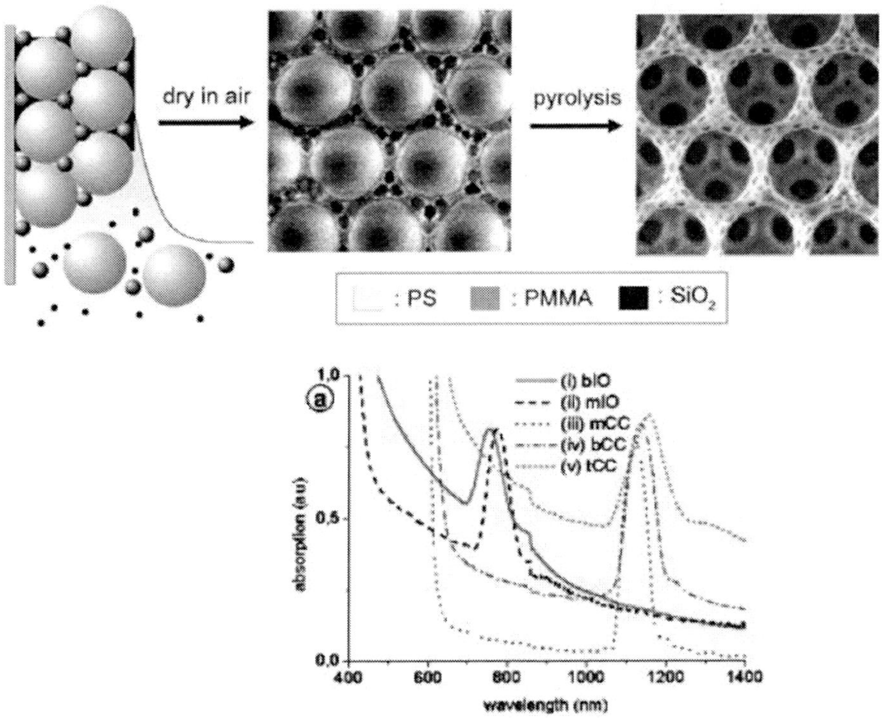

Figure 7. Schematic and SEM image of multiple-size particle templating of hierarchically ordered binary colloidal particles and vis-NIR spectra [38].

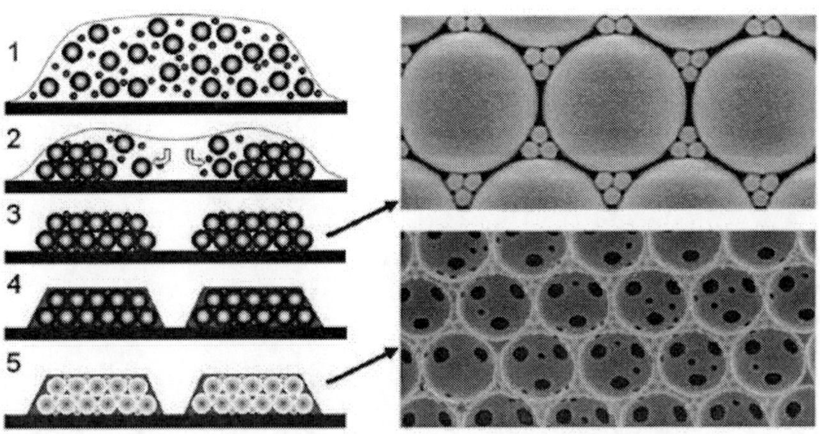

Figure 8.Horizontal deposition method for fabrications of HOCCs [39].

The uniform distribution of the small particles in the voids of large particle packing has been attributed to the combined ‘halo effect’ in colloidal suspension [40] and void constriction effect [39]. The operational window of utilizing the void constriction effect in forming HOCC was found to be around 0.154 <diameter ratio< 0.225.

HOCCs can also be prepared by contact printing with pintool plotter [41], which makes it highly promising for protein array application. Kingshott and co-workers [42] have employed the binary colloidal crystal packing for highly ordered mixed protein patterns by first coating the two sizes colloidal particles with different proteins before mixing and co-crystallisation. This work has provided a new concept on utilising the hierarchically ordered binary system.

On the utilisation of binary inverse colloidal crystals, Cai et al. [43] have demonstrated the advantage of the hierarchically ordered structure. TiO_2 inverse colloidal crystals have been prepared by multisize-particle (1 μm and 170 nm polystyrene spheres) cocrystallisation. In this study, TiO_2 was introduced by infiltrating the pre-formed binary colloidal crystals with a TiO_2 precursor, which was prepared by adding titanium isopropoxide to a mixture of trifluoroacetic acid and of hydrochloric acid under vigorously stirring. Photo-active anatase TiO_2 inverse structure was formed upon calcination at 500°C as shown in Figure 9. By varying the small to large particle number ratio, the surface areas of HOCCs were altered. Tests on benzoic acid degradation under UV irradiation (Figure 10) showed that the TiO_2 HOCCs have exhibited enhanced photo-activity, mostly attributed to their increased surface area [43].

In the application field of photocatalysis, so far the slow photon effect in photonic crystals has been underutilised in HOCC. It has been observed first by Imhof et al. [44] that at the frequency edges of the stop bands of photonic crystals, photons propagate with strongly reduced group velocity, hence, they are called slow photons. Ozin and coworkers [45]have shown that slow photons in photonic crystals can significantly increase the interaction of light with TiO_2. By reducing the group velocity of light at energies near the edge of the photonic stop-band, a higher probability of absorption can be achieved and the photodegradation efficiency of TiO_2 doubled [45]. Therefore, there is still plenty of room to further improve HOCCs for the applications of photocatalysis and solar cells.

Multisize-particle co-crystallisation strategy has also been employed in preparation of hierarchically ordered porous carbon. Wu and co-workers [46] co-crystallised colloidal particles of two sizes (large polystyrene spheres and small silica particles) in sucrose solution as shown in Figure 11.

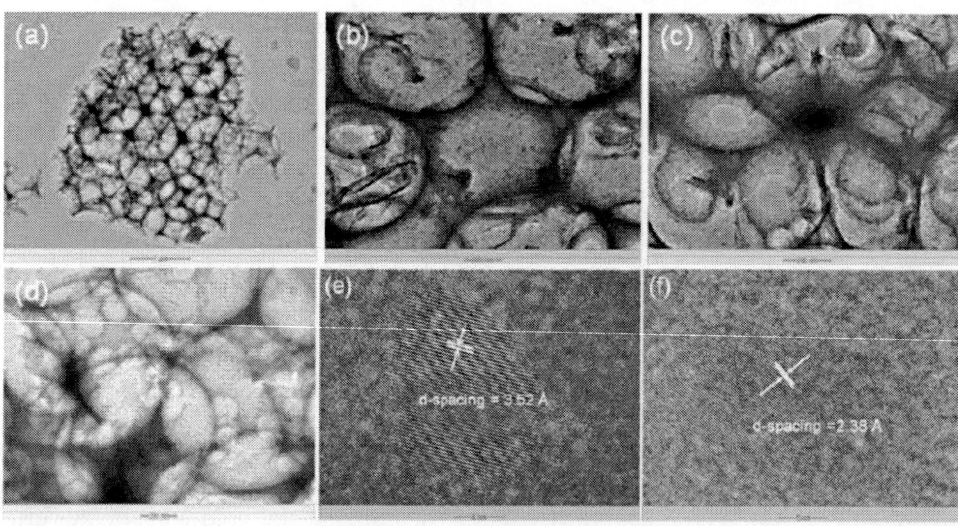

Figure 9. FETEM images of TiO_2 binary inverse opals with different porous structures. The number ratios of small to large spheres are (a-d) 2, 2, 4, and 8. (e, f) High-resolution TEM images showing the TiO_2 crystal lattice [43].

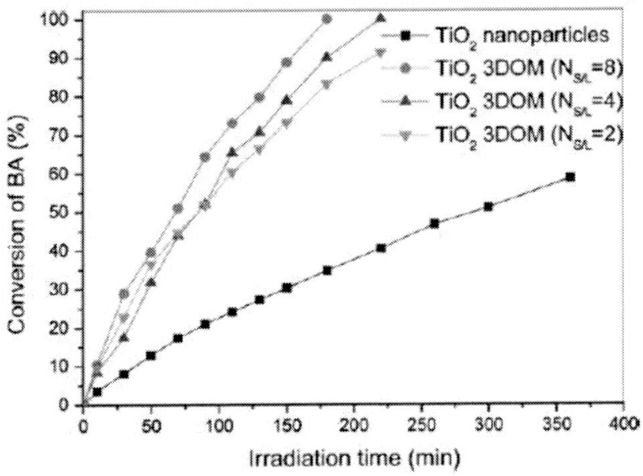

Figure 10. Photodegradation of Benzoid Acid using TiO_2 HOCCs and TiO_2 nanoparticles [43].

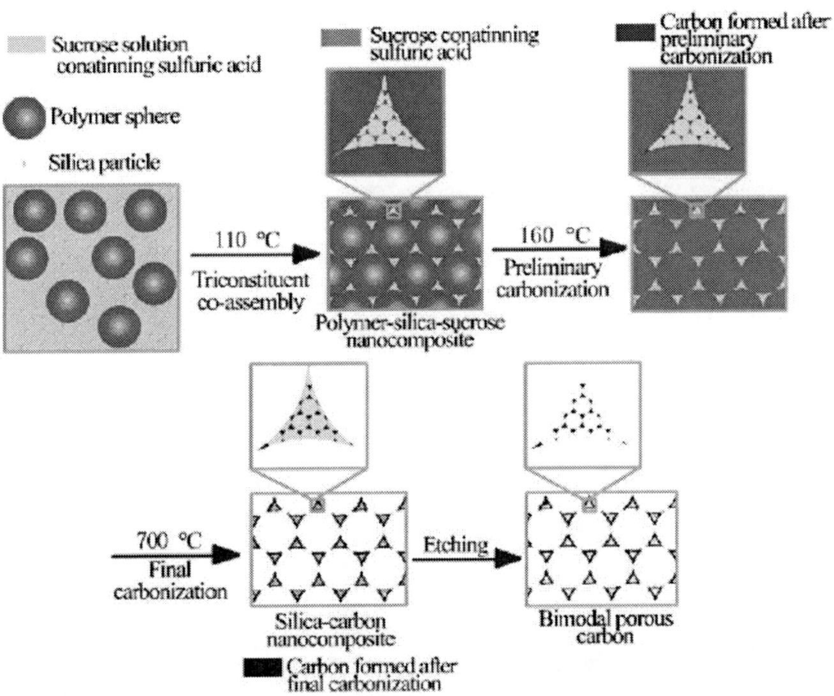

Figure 11. Schematic illustration for the preparation of hierarchically ordered porous carbon [46].

The sucrose in the interstitial voids was then turned into partially graphitized frameworks, which were shown as very good support for Pt-Ru alloy catalyst in direct methanol fuel cells. Their work has further confirmed the simplicity and versatility of this method for preparation of useful HOCC functional materials.

It is worth mentioning that in a theoretical study conducted by Wan et al. [47], binary inverse colloidal crystals show wider photonic band gaps compared to their mono-inverse colloidal crystal counterparts. The increased width is dependent on the distribution density of the small voids resulting from the smaller particle templates.

3. Directed Assembly from Pre-Functionalised Colloids

The third promising HOCC fabrication route is through directed assembly from pre-functionalised colloids or colloidal clusters.

The aforementioned two classes of methods have focused on thermodynamics-controlled self-assembly, primarily mediated by van der Waals force, electric double layer repulsions as well as capillary forces [48]. The new exploratory approach is to devise the colloidal particles into true atom analogues, which are capable of mimicking molecular self-assembly through covalent and non-covalent interactions to build designer materials for specific applications. As illustrated by Wang and co-workers in Figure 12 [14], colloidal particles can be first either decorated with nematic materials or functionalised to mimick molecular structures before self-organising through directed assembly into cluster structures, which then can be further assembled into super lattice structures. The final superlattice structure is determined by the functionality of the colloidal particle and the cluster spatial arrangement.

Nelson [49] has theoretically postulated that spherical colloidal particles coated with liquid crystals or other materials can exhibit nematic degrees of freedom, which may form composite materials that exhibit point defects with sp and sp^3 valences. Zhang et al. [50] have decorated PS spheres with gold dots analogous to sp valence. A leading work produced by Pine and co-workers [51] has shown that by employing emulsions colloidal particles can be assembled into clusters due to the confinement of dispersant droplets, which illuminates a new pathway to building designer materials from hierarchically organized colloids.

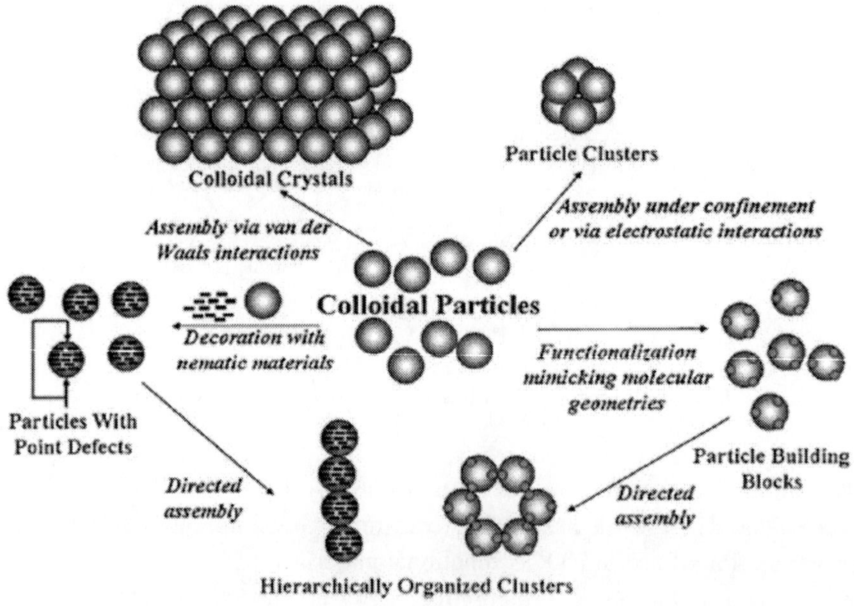

Figure 12. Schematic diagram illustrating various strategies for the generation of hierarchically organized colloidal particle structures [14].

CONCLUSIONS AND OUTLOOK

Colloidal crystals and hierarchically ordered colloidal crystals have captured significant interests from research community, owing to their beauty, simplicity and usefulness. Besides being a good analogue system to atoms [52], colloids have shown fascinating self-organization behavior as summarized in this chapter. HOCCs will remain as an active area in the coming years with research focuses shifting more towards directed self-assembly of pre-functionalized colloids with designer materials in mind. The tunabilities in photonic, electronic and magnetic properties [53-56] still require further explorations to exhibit their full potential.

REFERENCES

[1] Y.N. Xia, B. Gates, Y.D. Yin, Y. Lu, *Advanced Materials*, 12, 693-713 (2000).
[2] E. Yablonovitch, *Physical Review Letters*, 58, 2059-2062 (1987).
[3] S. John, *Physical Review Letters*, 58, 2486-2489 (1987).
[4] D.J. Norris, Y.A. Vlasov, *Advanced Materials*, 13, 371-376 (2001).
[5] C. Lopez, *Advanced Materials*, 15, 1679-1704 (2003).
[6] K. Busch, G. von Freymann, S. Linden, S.F. Mingaleev, L. Tkeshelashvili, M. Wegener, *Physics Reports-Review Section of Physics Letters*, 444, 101-202 (2007).
[7] P.L. Flaugh, S.E. Odonnell, S.A. Asher, *Applied Spectroscopy*, 38, 847-850 (1984).
[8] A.A. Zakhidov, R.H. Baughman, Z. Iqbal, C.X. Cui, I. Khayrullin, S.O. Dantas, I. Marti, V.G. Ralchenko, *Science*, 282, 897-901 (1998).
[9] J. Wijnhoven, W.L. Vos, *Science*, 281, 802-804 (1998).
[10] Q. Li, M. Retsch, J. Wang, W. Knoll, U. Jonas, *Topics in Current Chemistry*, 287, 135-180 (2009).
[11] K. Busch, S. John, *Physical Review E*, 58, 3896-3908 (1998).
[12] A.-P. Hynninen, J.H.J. Thijssen, E.C.M. Vermolen, M. Dijkstra, A. Van Blaaderen, *Nature Materials*, 6, 202-205 (2007).
[13] P.D. Yang, T. Deng, D.Y. Zhao, P.Y. Feng, D. Pine, B.F. Chmelka, G.M. Whitesides, G.D. Stucky, *Science*, 282, 2244-2246 (1998).
[14] E.W. Edwards, D. Wang, H. Mohwald, *Macromolecular Chemistry and Physics*, 208, 439-445 (2007).
[15] O.D. Velev, S. Gupta, *Advanced Materials*, 21, 1897-1905 (2009).
[16] O.D. Velev, A.M. Lenhoff, *Current Opinion in Colloid and Interface Science*, 5, 56-63 (2000).
[17] S. Forster, M. Antonietti, *Advanced Materials*, 10, 195-+ (1998).
[18] C.G. Oh, Y.Y. Baek, S.K. Ihm, *Advanced Materials*, 17, 270-+ (2005).
[19] O. Sel, D.B. Kuang, M. Thommes, B. Smarsly, *Langmuir*, 22, 2311-2322 (2006).
[20] L.A. Villaescusa, A. Mihi, I. Rodriguez, A.E. Garcia-Bennett, H. Miguez, *Journal of Physical Chemistry B*, 109, 19643-19649 (2005).
[21] J.J. Chiu, D.J. Pine, S.T. Bishop, B.F. Chmelka, *Journal of Catalysis*, 221, 400-412 (2004).
[22] R. Ryoo, S.H. Joo, M. Kruk, M. Jaroniec, *Advanced Materials*, 13, 677-681 (2001).

[23] H.F. Yang, D.Y. Zhao, *Journal of Materials Chemistry*, 15, 1217-1231 (2005).
[24] T. Kyotani, *Carbon*, 38, 269-286 (2000).
[25] R. Ryoo, S.H. Joo, S. Jun, *Journal of Physical Chemistry B*, 103, 7743-7746 (1999).
[26] Z. Wang, F. Li, N.S. Ergang, A. Stein, *Chemistry of Materials*, 18, 5543-5553 (2006).
[27] Y. Deng, T. Yu, Y. Wan, Y. Shi, Y. Meng, D. Gu, L. Zhang, Y. Huang, C. Liu, X. Wu, D. Zhao, *Journal of the American Chemical Society*, 129, 1690-1697 (2007).
[28] Y. Meng, D. Gu, F. Zhang, Y. Shi, L. Cheng, D. Feng, Z. Wu, Z. Chen, Y. Wan, A. Stein, D. Zhao, *Chemistry of Materials*, 18, 4447-4464 (2006).
[29] Y. Deng, C. Liu, T. Yu, F. Liu, F. Zhang, Y. Wan, L. Zhang, C. Wang, B. Tu, P.A. Webley, H. Wang, D. Zhao, *Chemistry of Materials*, 19, 3271-3277 (2007).
[30] K. Shin, H.Q. Xiang, S.I. Moon, T. Kim, T.J. McCarthy, T.P. Russell, *Science*, 306, 76-76 (2004).
[31] J.Y. Cheng, C.A. Ross, H.I. Smith, E.L. Thomas, *Advanced Materials*, 18, 2505-2521 (2006).
[32] Y.Y. Wu, G.S. Cheng, K. Katsov, S.W. Sides, J.F. Wang, J. Tang, G.H. Fredrickson, M. Moskovits, G.D. Stucky, *Nature Materials*, 3, 816-822 (2004).
[33] D.A. Rider, J.I.L. Chen, J.-C. Eloi, A.C. Arsenault, T.P. Russell, G.A. Ozin, I. Manners, *Macromolecules*, 41, 2250-2259 (2008).
[34] S. Horiuchi, M.I. Sarwar, Y. Nakao, *Advanced Materials*, 12, 1507-1511 (2000).
[35] K.H.A. Lau, J. Bang, D.H. Kim, W. Knoll, *Advanced Functional Materials*, 18, 3148-3157 (2008).
[36] X. Li, S.J. Tian, Y. Ping, D.H. Kim, W. Knoll, *Langmuir*, 21, 9393-9397 (2005).
[37] J. Fu, J. Wang, Q. Li, D.H. Kim, W. Knoll, *Langmuir*, 26, 12336-12341 (2010).
[38] J. Wang, Q. Li, W. Knoll, U. Jonas, *Journal of the American Chemical Society*, 128, 15606-15607 (2006).
[39] L. Wang, Y. Wan, Y. Li, Z. Cai, H.-L. Li, X.S. Zhao, Q. Li, *Langmuir*, 25, 6753-6759 (2009).
[40] V. Tohver, J.E. Smay, A. Braem, P.V. Braun, J.A. Lewis, *Proceedings of the National Academy of Sciences of the United States of America*, 98, 8950-8954 (2001).
[41] K. Burkert, T. Neumann, J. Wang, U. Jonas, W. Knoll, H. Ottleben, *Langmuir*, 23, 3478-3484 (2007).
[42] G. Singh, S. Pillai, A. Arpanaei, P. Kingshott, *Advanced Materials*, 23, 1519-1523 (2011).
[43] Z. Cai, J. Teng, Z. Xiong, Y. Li, Q. Li, X. Lu, X.S. Zhao, *Langmuir*, 27, 5157-5164 (2011).
[44] A. Imhof, W.L. Vos, R. Sprik, A. Lagendijk, *Physical Review Letters*, 83, 2942-2945 (1999).
[45] J.I.L. Chen, G. von Freymann, S.Y. Choi, V. Kitaev, G.A. Ozin, *Advanced Materials*, 18, 1915-+ (2006).
[46] S. Zhang, L. Chen, S. Zhou, D. Zhao, L. Wu, *Chemistry of Materials*, 22, 3433-3440 (2010).
[47] Y. Wan, Z. Cai, L. Xia, L. Wang, Y. Li, Q. Li, X.S. Zhao, *Materials Letters*, 63, 2078-2081 (2009).
[48] Q. Li, U. Jonas, X.S. Zhao, M. Kappl, *Asia-Pacific Journal of Chemical Engineering*, 3, 255-268 (2008).
[49] D.R. Nelson, *Nano Letters*, 2, 1125-1129 (2002).

[50] G. Zhang, D.Y. Wang, H. Mohwald, *Angewandte Chemie-International Edition*, 44, 7767-7770 (2005).
[51] V.N. Manoharan, M.T. Elsesser, D.J. Pine, *Science*, 301, 483-487 (2003).
[52] W. Poon, *Science*, 304, 830-831 (2004).
[53] T. Ruhl, P. Spahn, C. Hermann, C. Jamois, O. Hess, *Advanced Functional Materials*, 16, 885-890 (2006).
[54] J.H. Holtz, S.A. Asher, *Nature*, 389, 829-832 (1997).
[55] J. Ge, Y. Hu, Y. Yin, *Angewandte Chemie-International Edition*, 46, 7428-7431 (2007).
[56] J. Ge, Q. Zhang, T. Zhang, Y. Yin, *Angewandte Chemie-International Edition*, 47, 8924-8928 (2008).

NANOPARTICLES/NANOMATERIALS

In: Nanostructures
Editor: Yu Dong

ISBN: 978-1-62618-081-9

Chapter 5

ORGANOMETALLIC-METALLIC-CYCLOTRIPHOSPHAZENE MIXTURES: SOLID STATE METHOD FOR METALLIC NANOPARTICLE GROWTH

***C. Diaz*[1,*], *M. L. Valenzuela*[2] *and C. O'Dwyer*[3,4,5*]**

[1]Departamento de Química, Facultad de Ciencias,
Universidad de Chile, Santiago, Chile
[2]Universidad Andrés Bello, Departamento de Ciencias Química,
Facultad de Ciencias Exactas, Santiago, Chile
[3] Applied Nanoscience Group, Department of Chemistry,
University College Cork, Cork, Ireland
[4] Micro and Surface Science Institute,
University of Limerick, Limerick, Ireland
[5]Micro & Nanoelectronics Centre,
Tyndall National Institute, Dyke Parade, Cork, Ireland

ABSTRACT

We review a recent general solid state method to obtain metallic, metal oxide and phosphate nanoparticles and crystals by pyrolysis at 800 °C using organometallic derivatives of cyclo and polyphosphazene precursors containing diverse organometallic fragments linked to polymeric or oligomeric phosphazenes. When the preparation of the molecular precursor is not possible or results in low yield, an alternative method using solid state mixtures of the type MLn/$N_3P_3[O_2C_{12}H_8]_n$, where MLn can be a single metallic salt, and a coordination compound or an organometallic, is possible. For $AuCl(PPh_3)/[NP(O_2C_{12}H_8)]_n$ mixtures, single crystal cubic Au nanoparticles form, whose morphology, crystal shape, size and distribution strongly depends on deposition quantity and the mixture molar ratio. Nanoparticles as small as 3.5 nm are observed if the mixture is prepared in a crucible and varied geometries of microcrystals found when the mixture

* Corresponding author's email address: cdiaz@uchile.cl.

was deposited on Si or SiO_2 wafers, including single-crystal gold fullerene structures. Extension to Ag, Pd and Re-containing precursor mixtures such as $Ag(PPh_3)(CF_3SO_3)$/ $[NP(O_2C_{12}H_8)]_3$, $PdCl_2/N_3P_3[O_2C_{12}H_8]_3$, and $KReO_4/N_3P_3[O_2C_{12}H_8]_3$ allows microcrystal formation during pyrolysis. A thermally induced phase demixing mechanism describes the evolution of the crystal growth, aided microphase separation of the polymer mixture. This microphase demixing is shown to be an overarching mechanism involved in the nano to micro scale growth of crystals. A probable mechanism of the atomic and molecular-level chemistry is also proposed based on decomposition of the macromolecular polymeric, trimer and oligomeric precursors for the initial stages.

INTRODUCTION

Nanoparticles have attracted extensive research interest, mainly due to their size-dependent properties and great potential for many applications [1-3]. There are two main ways for the preparation of nanomaterials: those in solution and those in solid state. The greatest development has been in solution, while solid state approaches are scarce. The main solid state methods to prepare and synthesize nanoparticles have been recently discussed, [4, 5] and in stark contrast with the hundreds of solution-based reports, as little as four routes to grow Au nanoparticles completely in the solid state are known at the time of writing. A similar situation holds for Ag nanoparticles where only two solid-sate methods have been reported. The aim of developing solid state methods to prepare nanoparticles stems mainly from their possible application in solid state materials and powder-oriented applications, from thin film metal deposition to noble metal nanoparticle-carbon catalysts, oxide growth, photonic and dielectric materials, to new materials for Li-ion rechargeable batteries.

The ability to rationally prepare metallic and metal oxide nanoparticles stems from the exploring methods for alternative nanoscale metal deposition in solid state nanoelectronics and nanotechnology [6-8] and the benefit of being able to deposit both metals and dielectric or semiconducting oxides, both from the same base route.

Issues including limitations on good mechanical and thermal stability of nanoscale metals have been found to be related to certain deposition methods for these metals [8]. A new solid state method to obtain metallic nanostructured have reported recently for materials of the type: M, M_xO_y and $M_xP_yO_z$ for a range of transition, valve and noble metals, using organometallic derivatives of poly and cyclotriphosphazenes as molecular precursors, summarized in Figure 1. [9-26]

Briefly, the method consists of the pyrolysis of the molecular or macromolecular precursors at 800 °C under air. The resulting product depends strongly on the nature of the metal. Thus for metals such as Cr [9, 10], Ru [17] and Mo [21] where the metal oxide is easily formed, the respective nanostructured metal oxides are observed. On the other hand, in other cases such as with Mn [11], Sn [22, 23] and Fe [24], only the respective nanostructured pyrophosphate was observed. With metals and metalloids such as Ti [15], Si [18, 22, 23] and W [12], both phases were observed, *i.e.* the metal oxide and the pyrophosphate. With noble metals for which the formation of oxides or pyrophosphates is thermodynamically unfavored, the reduction with CO produced during molecular decomposition in the pyrolysis process, allowed the formation of nanostructured M° [19, 20, 25].

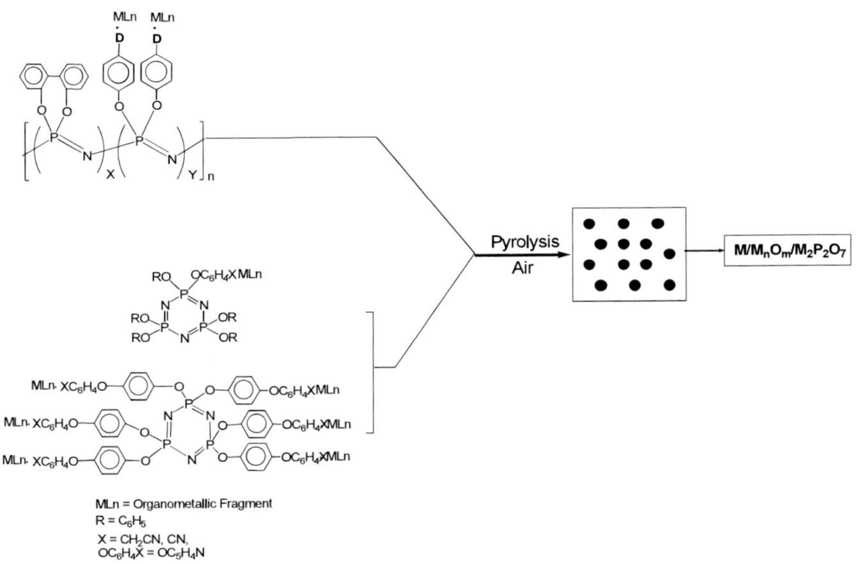

Figure 1. Schematic representation of the solid state method for nanostructured material formation using oligomeric and polymeric precursors [9-26].

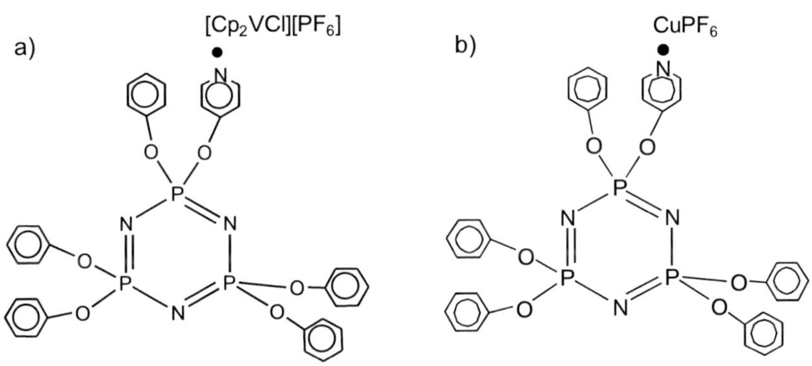

Figure 2. Representative formulas of $[N_3P_3(OC_6H_5)_5OC_5H_4N{\bullet}Cp_2VCl][PF_6]$ and $N_3P_3[OC_6H_5]_5[OC_5H_4N{\bullet}Cu][PF_6]$.

While the process has now been well established for a range of metals, giving a smorgasbord of oxide and phosphate based nanostructured materials all by a similar route, some metals can be difficult to prepare in such form and some are often unsuccessful. Good examples are the sensitively multivalent vanadium and copper derivatives of cyclotriphosphazene, as shown in Figure 2, and only $[N_3P_3(OC_6H_5)_5OC_5H_4N{\bullet}Cp_2VCl][PF_6]$, $N_3P_3[OC_6H_5]_5[OC_5H_4\ N{\bullet}Cu][PF_6]$ and $N_3P_3[OC_6H_4CH_2CN{\bullet}CuCl]_6[PF_6]_6$ have been prepared, but in relatively poor yields compared to metals and metalloids described earlier.

Furthermore by using the polymer $[\{NP(O_2C_{12}H_8)\}_{0.7}\{NP(OC_5H_4N\text{-}4\{\ Au(C_6F_5)_2\}_{0.3}]_n$ as a precursor, a 3-D network of porous Au can be obtained instead the expected Au nanoparticles [20]. This and the previously mentioned V and Cu cases, could be solved using adequate $MLn/N_3P_3[O_2C_{12}H_8]_n$ mixtures. In fact, by pyrolyzing the $AuCl(PPh_3)/[NP(O_2C_{12}H_8)]_n$ mixtures, nanoparticles as small as 3.5 nm were obtained [26].

This chapter reviews the latest progress using these new methods with several important noble metals: Au, Ag, Pd and Re.

THE SOLID STATE PRECURSOR MIXTURES (SSPM) METHOD

The method is highly sensitive to the preparation procedure. Two main procedures can be used: pyrolyzing the material in a crucible or pyrolyzing a deposited sample on Si or SiO_2 wafers, as summarized in Figure 3.

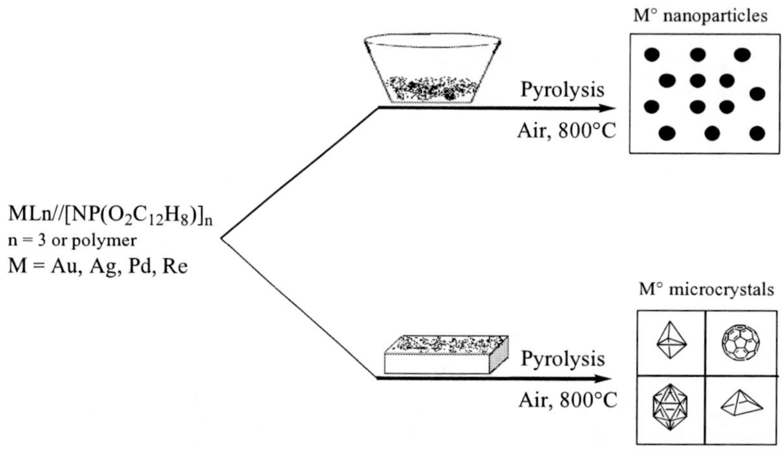

Figure 3. Schematic representation of the SSPM method.

As detailed elsewhere [9-26], specific differences in overall organometallic and cyclotriphosphazene chemistry allows a range of systems. The following describes details of this new approach for Au, Ag, Pd and Re.

MIXTURES OF $AuCl(PPh_3)$ AND $[NP(O_2C_{12}H_8)]_N$

Some of these results have been recently reported [26] and here we briefly describe the method itself and summarise the unique observation of a variety of gold morphologies possible through a single solid state method. We used mixtures comprising $AuCl(PPh_3)/[NP(O_2C_{12}H_8)]_3$. The results indicate that both the morphology and the size particles of the products depend strongly on the preparation procedure. The mixtures are made in dichloromethane as solvent and separation of the solid product is performed using the following methods:

a) *Film*: A CH_2Cl_2 solution of the respective mixtures of $AuCl(PPh_3)$ and the trimer or polymer is allowed to evaporate slowly to cast a film that is dried at room temperature.

b) *Powder (I)*: The respective molar mixture of $AuCl(PPh_3)$ and the cyclotriphosphazene or polyphosphazene is dissolved in dichloromethane stirred for 24 hours and evaporated to dryness in a vacuum at 70 °C.
c) *Powder (II)*: The respective molar mixture of $AuCl(PPh_3)$ and the respective trimer or polymer is dissolved in THF, stirred for 24 hours and evaporated to dryness in a vacuum at room temperature.

The resultant solids from procedures a) to c) are then placed in a crucible and pyrolyzed at 800 °C under air. In all cases the products were identified to be Au by their XRD patterns, which show the (111), (200), (220), and (311) crystal reflections that can be indexed as fcc. The $I_{(220)}/I_{(111)}$ ratio was somewhat higher that observed for Au bulk (0.32) indicating an preferred (110) termination in resulting materials. A representative XRD pattern is shown in Figure 4a. The red bars are the patterns of standard cubic gold (JCPDS 004-0784).

Consistently, the diffraction electron patterns exhibit polycrystalline rings with diameters in ***k***-space corresponding to the (111), (200), (220) and (311) planes. While crystallographically, all products can be classified as single crystal or polycrystalline cubic Au, the principal difference in all Au products using this method are the resulting particle morphologies; variants including fused granular networks, low density metallic foams and nanoparticles can be formed. A summary of the main morphological variants and particle size (where relevant) are summarized in Table 1.

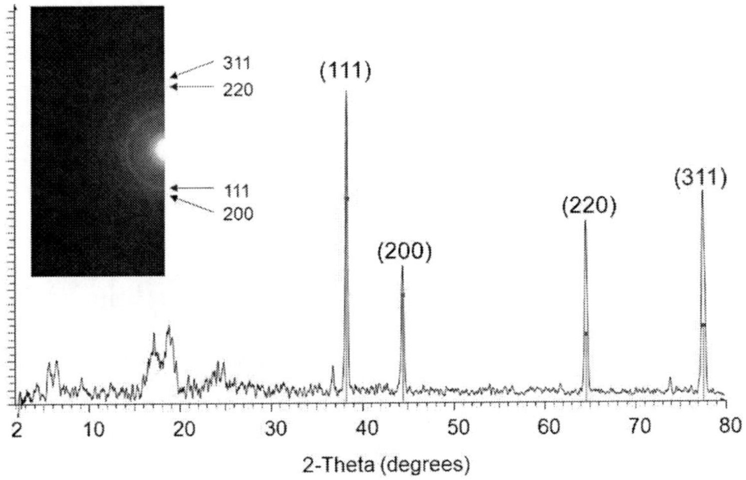

Figure 4. (a) X-ray diffraction pattern of the pyrolytic residue from 1:1 $AuCl(PPh_3)/[NP(O_2C_{12}H_8)]_3$ mixture. The inset shows the corresponding electron diffraction pattern.

Table 1. Summary of the main properties of the as-prepared Au nanoparticles from precursor mixtures

Mixture[d]	Polymer/Au	Trimer/Au	Morphology	Particle size (nm)[e]
1a	1//1[a]			3.5
2a	3//1 [a]		Dense grains	
1b	1//1[b]			75

2b	3//1 [b]		Dense grains	
1c	1//1[c]		Porous	
2c	3//1 [c]		Foam	
3c	5//1 [c]		Foam	
1'c		1//1 [c]	Foam	4.7
2'c		3//1 [c]	Fused grains	
3'c		5//1 [c]	Fused grains	6.5
4'c		10//1[c]	3-D Porous network	

[a] Film; [b] Powder (I); [c] Powder (II); [d] Mixture Polymer/Au (1,2,3…n) or Trimer/Au (1',2' 3'…n' etc.) prepared according to method a), b), or c).[e] Measured by TEM using a histogram analysis.

As summarized in Table 1, the smallest individual entities are nanoparticles obtained using a 1:1 phosphazene/$AuCl(PPh_3)$ ratio using preparation methods a) to c). In several cases metal foams [27-33] are obtained.

Previous solid state methods for preparing Au nanoparticles involved heating by the Brust method, which gives nanoparticles with diameters in the range 3.4-9.7 nm [34-36] and in some cases, nanoparticles trapped in solid glass materials are possible by irradiating a solid mixture containing the silicate glasses and Au_2O_3 [37]. Using this approximation, gold nanoparticles in the range 6-8 nm was obtained. Also, glasses containing Au nanoparticles with composition $15CaO{\bullet}5P_2O_5{\bullet}80SiO_2$ have been reported [38]. Au nanoparticles and microparticles were obtained, tuned by the calcination temperature: 1.5-2.5 nm at 150 °C, ~30 nm at 600 °C and between 30-200 nm in diameter when formed at 900 °C. More recently [39], in a similar approximation the glass $80SiO_2{\bullet}20Na_2O{\bullet}10CaO$ was doped with $AuCl_3{\bullet}2H_2O$ and irradiated using synchrotron radiation. After calcination at 1450 °C, the resulting Au nanoparticles ~3.2 nm in size were generated inside the glass material.

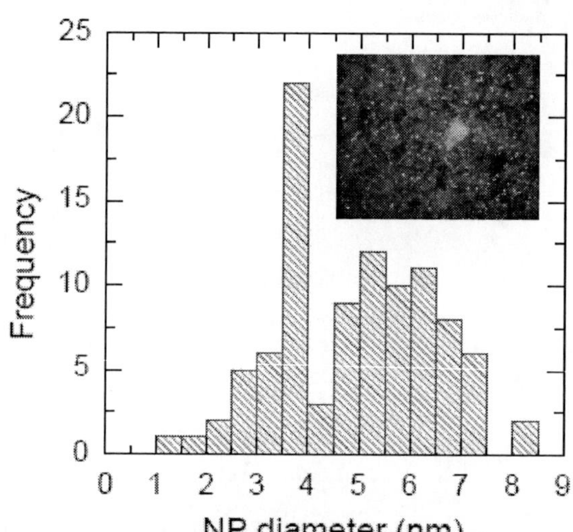

Figure 5. The diameter distribution profile confirms a bimodal Gaussian spread in diameters with 95% of NPs having diameters in the range 3.7-4.0 nm. *Inset*: Dark field TEM image of gold NPs formed in the demixing polymer mixture. In dark field, the brighter labyrinthine regions constitute the polymer and bright points are the Au NPs.

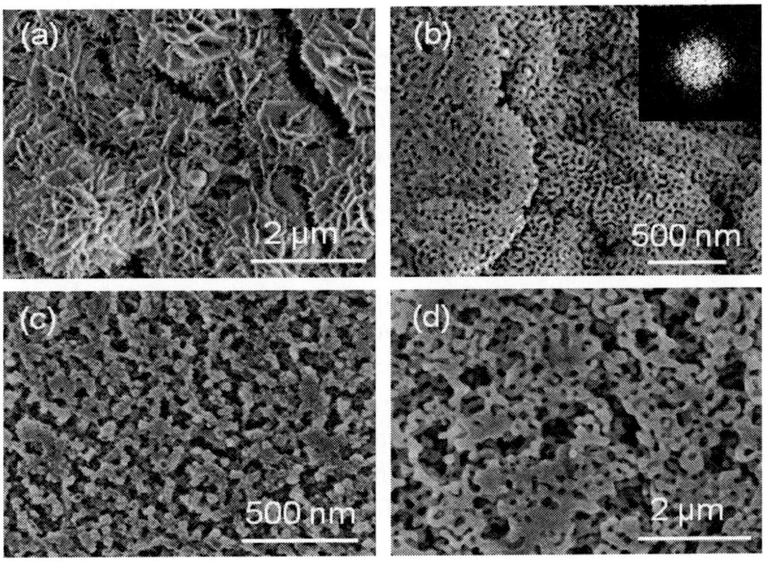

Figure 6. (a-d) Various density high surface area foam-like polycrystalline gold structures formed after pyrolysis of respective precursor mixtures with a high Au content. Inset: FFT of the porous gold showing a characteristic wavelength indicative of spinodal decomposition during the demixing of the two polymeric phases, one of which contains gold.

For Au, using this SSPM method, we have shown that the variation of the phosphazene/Au ratios allows for the formation of either metallic foams of high density or discrete Au NPs, whose mean size and morphology are controlled by the structure of the precursor constituents and the relative ratio in the mixtures. With the $[NP(O_2C_{12}H_8)]_n$/AuCl(PPh_3) mixtures, Au NPs of 3.5 nm diameter were obtained from films formed through pyrolysis of the 1:1 mixture ratio, while with the analogous 1:1 mixture of $N_3P_3(O_2C_{12}H_8)_3$/ AuCl(PPh_3) resulted in Au NPs with a mean diameter of 4.7 nm from a powder deposit.

An example of Au NP size distribution and physical location throughout the decomposed polymeric matrix using HAADF TEM is shown in Figure 5. Comparison with previous studies shows that the formation of metallic nanostructures does not require the covalent linking of the organometallic fragment to the polymeric or trimeric phosphazenes, although the presence of the cyclic or polymeric phosphazene as solid state template is necessary. The method allows the tuning of the resultant Au product; the Au is phase pure, single crystal and can be formed as small NPs or as low density, high surface area foams completely in the solid state. Interestingly, using the SSPM method, in some deposits where the density of Au is kept highest, the formation of gold metal foams was observed and attributed to spinodal decomposition during demixing of the Au-containing phase and the polymer phase, resulting in a 3D spinodal network of polycrystalline high surface area gold. Metal foams are a relatively new class of materials with unique combinations of properties such as high stiffness, low density, gas permeability and thermal conductivity [27-33]. As such these materials promise to enable new technologies in areas as diverse as catalysis, fuel cells, hydrogen storage and thermal and acoustical insulation. Preparative methods for making metal foams to date have somewhat limited in scope, the most common typically that of de-alloying of the bimetallic $Au_{0.22}Ag_{0.58}$ [27,28]. The here presented method could be a new useful way to these type of materials. Figure 6 shows the various foam-like morphologies possible at high Au loading of the organometallic-polymer mixture. In each porous structures shown in Figure 6, the entire porous material is pure monocrystalline gold.

More interesting results were obtained pyrolyzing the Au mixtures on Si and SiO_2 wafers [20] and observing the products by SEM. The methods for anchoring pyrolyzed products have been reported previously [20].

Briefly, a solution of the mixture was dropped on Si or SiO_2 wafers and pyrolyzed at 800 °C under air. SEM or AFM measurements were performed on the crystals formed on the wafers. When the phosphazene/$AuCl(PPh_3)$ mixtures were deposited on Si or SiO_2 wafers, Au microcrystals were observed. Previous attempts to obtain structured gold NPs by pyrolysis of polyphosphazenes containing AuCl covalently bonded to the polymer backbone by coordination to pendant -O-C_6H_4-PPh_2 substituents [25] were unfruitful; only diffuse and randomly shaped and sized Au particles were found.

We performed analogous pyrolytic experiments with physical mixtures of $AuCl(PPh)_3$ and the cyclic or polymeric phosphazenes $[N_3P_3(O_2C_{12}H_8)_3]$ or $[NP(O_2C_{12}H_8)_2]_n$ ($O_2C_{12}H_8$ = 2,2′-dioxy-1,1′-biphenyl) (Figure 7) that have λ^5-dioxydiaryl groups in the repeating units (poly-spirophosphazenes) [40]. The growth steps of the tetrahedral gold microcrystals from triangular platelets to full tetrahedra are shown in Figure 7(a,b), where we observe coarsening of Au NPs formed in the intial stages of decomposition, to the pinhole voids formed during nucleation dewetting of the polymer mixture. The gold comes from one phase, the other being polymeric and demixing occurs under thermal excitation. Compositional analysis using EDX and XPS confirms that the features within pionhold voids are pure gold and the resulting polyhedral crsytals that are formed (Figure 7) are also pure cubic gold. Higher resolution analysis of individual crystals (Figures 7b-d) shows the variety of crystal shapes possible through coarsening of NPs formed by cleavage of metal centers through a decomposing polymer without needing surfactants [41]. The remarkable observation is the subsequent growth of crystals (uniquely from these pinholes), subsequent Ostwald ripening of nanoparticles by migration of the NPs to each pinhole void formed during demixing, and the eventual growth of the crystals.

A variety of individual crystal shapes are found; icosahedra, tetrahedra (which also develop into octahedra), heptahedra, cuboctahedra, decahedra and truncated icosahedra (fullerene-shaped) crystals of similar size are all formed through simple pyrolysis.

The fullerene-like crystals shown in Figure 7d [42,44] are, to the best of our knowledge, the first to be observed for Au on any length scale. The tetrahedra evolve symmetrically, where the edge length is first defined by forming a platelet which evolves to three-fold rotational symmetry from asymmetric hexagonal platelets where the vertices are terminated by a {111} facet; subsequent growth upfills the structure until a geometrically perfect tetrahedron is formed.

The method relieves the constraint on solution based methods that use surfactants to control particle size and shape and the approach and can be directly applied to many other transition and noble metals capable of coordination to the organometallic derivative. This bottom-up single-step approach for molecular level metal-polymer building blocks provides the often sought solid state route to polyhedral metallic crystal growth without the need for ligands or surfactants to influence shape.

Figure 7. (a) Structure of mixtures using in conjunction with Au-containing organometallics. (b) AFM and SEM analysis of surface pinholes features formed during microphase demixing of the mixture and subsequent coarsening and ripening of Au NPs for form individual crystals. (c, d) A range of exotic polyhedral single crystal gold crystals such as a decahedron, tetrahedron, truncated icosahedra (fullerene) and EDX line mapping analysis confirming single phase gold composition.

As recently discussed anisotropy in micro and nanocrystals can led to an unprecedented diversity in new exotic crystalline shapes with interesting and unexpected properties [45, 46]. These findings outline the possibility of controlling morphology, shape and size of gold crystals obtained by solid state reactions.

The marriage of polymer dewetting, demixing and spinodal decomposition to metal-coordinated macromolecular polymers augers well for a variety of important applications. Further insight into this issue would be gained by understanding the influence of the mixing parameters and structure of the precursors so controlled microphase demixing and dewetting processes, which are largely used in the understanding of thin liquid polymer film. Methods to stabilize unwanted dewetting, would allow bespoke metallic crystals in a fully solid state process and also have the potential for surface patterning inbuilt into the technique. Optimization of thickness and molecular weight will eventually allow more controlled instability formation in the polymer, which will lend itself to metal crystals and structures not possible through solution based routes.

Silver, Palladium and Rhenium Crystal Growth: Solid State Pyrolysis of $AgPPh_3[CF_3SO_3]$ /$[NP(O_2C_{12}H_8)]_N$ and $PdCl_2/[NP(O_2C_{12}H_8)]_N$ Mixtures

Similar to the Au-phosphazene mixtures, the pyrolysis of $AgPPh_3[CF_3SO_3]/[NP(O_2C_{12}H_8)]_3$ and $PdCl_2/[NP(O_2C_{12}H_8)_3$ mixtures at 800 °C under air can result in Ag and Pd nanoparticles and microcrystals identified mainly by powder XRD and HRTEM with electron diffraction. Specifically, the mechanism of the macromolecular precursor deposition under heating is similar, with cleavage of metal centers and their subsequent agglomeration to single crystal nanoparticles. Demixing of the metal- and non-metal-containing phases within the mixture allows spatial patterning due to nucleation dewetting processes. In all cases for all metals thus far studied, carbonization and graphitization of the resulting carbon residue prevent full dewetting of the layer which gives a high density of crystals on the surface. Figure 8 summarizes the stages of decomposition, nanoparticles formation and resulting high surface area porous Ag and Pd foams formed from each precursor at high metal loading.

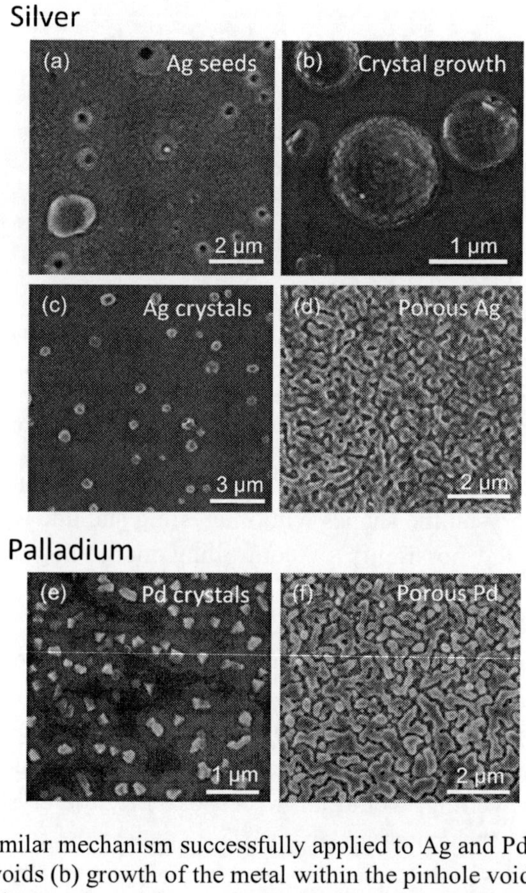

Figure 8. Evidence for a similar mechanism successfully applied to Ag and Pd. (a) NP and seed formation within pinhole voids (b) growth of the metal within the pinhole voids (c) faceted crystals that grow from pinholes (d) and porous metal foams. (e) nucleation dewetting formation of sub-micron Pd crystals and (f) corresponding porous Pd.

It is interesting to note that although several solution-based methods also exist for prepare Pd nanoparticles have been reported [47-57], relatively few solid state routes are known [57]. Usefully, thermolysis of $[Pd(S_{12}C_{12}H_{22})_2]_6$ gives Pd nanoparticles but mixed with PdO and PdS [58]. Using our method, both microcrystals and nanoparticles can be formed in addition to porous foams, depending on the concentration of Pd, which affects not only the chemical decomposition but also the demixing of the two polymeric-type phases during heating; low concentration results in NPs, high concentration results in porous metals.

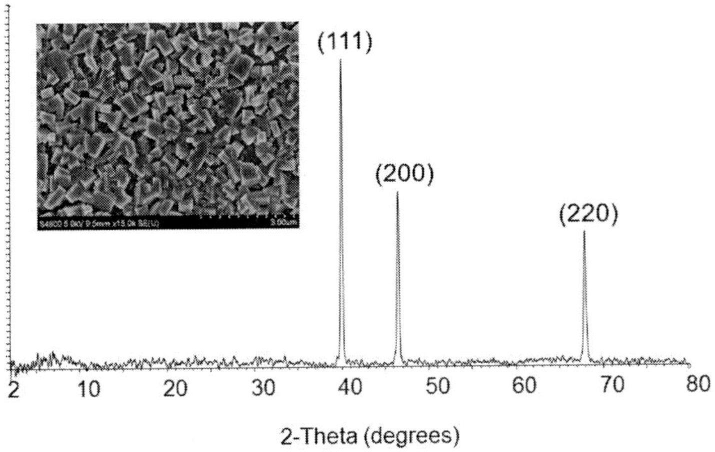

Figure 9. X-ray diffraction pattern for $PdCl_2/N_3P_3[O_2C_{12}H_8]_3$ precursor after pyrolysis confirming single crystal Pd. The inset is a SEM image of the cubic-like Pd microcrystals formed.

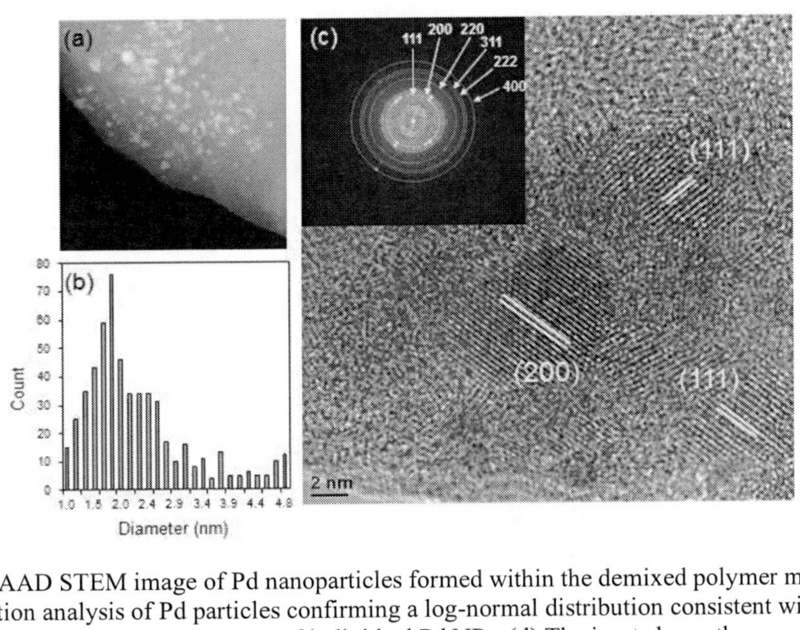

Figure 10. HAAD STEM image of Pd nanoparticles formed within the demixed polymer matrix. (b) Size distribution analysis of Pd particles confirming a log-normal distribution consistent with Ostwald ripening of the NPs. (c) HRTEM image of individual Pd NPs. (d) The inset shows the corresponding selected area electron diffraction pattern confirming the cubic Fm3m space group.

SEM and XRD show in in Figure 9 confirms this. Additionally, HRTEM analysis of the Pd nanoparticle and microcrystals confirms that they form as single crystals, much like the situation for Au and Ag, and be a similar overall chemical and physical decomposition mechanisms, and without unwanted oxide and sulfide phase formation. Figure 10 shows the variation in nanoparticle diameters for Pd synthesized using the SSPM method. With low Pd content and high mixture or C content, small nanoparticles with diameters as small as ~2.5 nm (averaged from ~800 NPs) are found with a log-normal size distribution.

THE CASE OF RHENIUM CRYSTAL GROWTH: PYROLYSIS OF $K[ReO_4]/[NP(O_2C_{12}H_8)]_3$

The practical importance of Rhenium arises from its multiple uses as an element of superalloys typically employed in rotating elements of aircraft engines, as barrier layers and can be used in the enhancement of interfacial properties of microelectronic conductors.

Additionally, Re compounds have useful catalytic activity in several hydrocarbon transformations and selective reduction of NO_x. It also has catalytic applications for hydrodesulfurization, hydrodenitrogenation and highly selective oxidation of methanol. However few reports on nanostructured Re have appeared [59-68].

In a review by Méndez [59], only two reports on Re films [60, 61] are mentioned and were limited to spectroscopic identification. This comprises the first report that claims the formation of nanostructured Re. Later, Hassel and coworkers reported the preparation of Re nanowires from a selected dissolution of a Ni/Al/Re alloy [62].

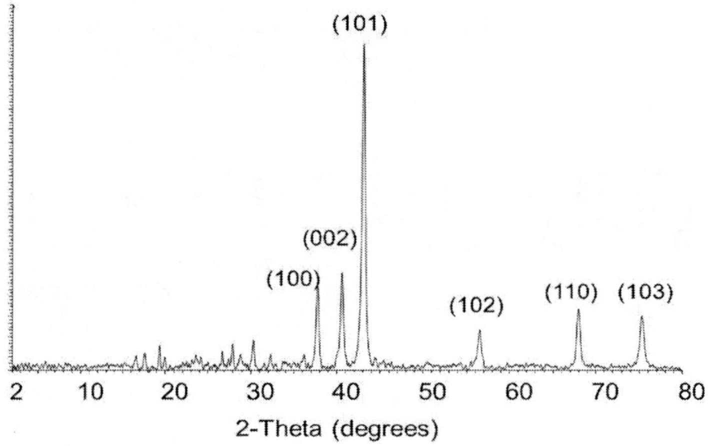

Figure 11. X-ray diffraction pattern from the pyrolytic product from 1:5 $K[ReO_4]$ /$[NP(O_2C_{12}H_8)]_3$ mixture.

However, the characterization was performed only by SEM and EDAX techniques. Subsequently Philippe [63] studied the hardness properties of Re nanowires prepared by this latter method. From deposition of films of NH_4ReO_4, films of Re were identified by SEM and XPS [64]. Nanoparticles of 3 nm in size have also been observed by TEM and characterized by powder XRD. More recently, nanoparticles of Re were prepared from $ReCl(CO)_3(N,N)$

[65] and $Re_2(CO)_{10}$ [66] using ionic liquids as stabilizing agents. Some catalytic applications on ammonia synthesis [67-68] and glycerol conversion have been reported [68] have been reported.

Using the SSPM method, pyrolysis of the $K[ReO_4]/[NP(O_2C_{12}H_8)]_3$ mixtures satisfyingly produced pure Re as single crystal microcrystals and nanoparticles.

Again, incubation of NPs that coarsen to pinholes formed within the carbon residue allow the ripening growth of individual structure on surfaces, thus forming the first successfully solid state approach to size control individual single crystals of Re. X-ray as indicated by the XRD pattern see Figure 11. The planes corresponding to (100), (002), (101), (102), (110), and (103) reflections from hexagonal Re are clearly observed.

Figures 12a and b show that Re microcrystals with quasi-spherical morphologies and rough, layered outer surfaces similar to a nautilus structure formed by growth from within a carbon pinhole on the surface. Each particle grows from within each of these pinholes. For a given Re quantity and mixture ratio, the shape and yield of the Re particles can be varied. For instance, in Figure 12c using half the quantity of Re with respect to the overall *i.e.* a 1:5 mixture ratio of $K[ReO_4]/[NP(O_2C_{12}H_8)]_3$, deposited on SiO_2, faceted polyhedral microcrystals are found, as opposed to circular crystals. In Figure 12a, the spherical microcrystals are found to grow from pinhole voids following the typical route for this solid state method. For lower Re quantities, such as with the 1:10 $K[ReO_4]/[NP(O_2C_{12}H_8)]_3$ mixture for example, the resulting crystals are nucleated on the surface as shown in Figure 12b, and subsequent growth allow the definition of polyhedral crystals of Re.

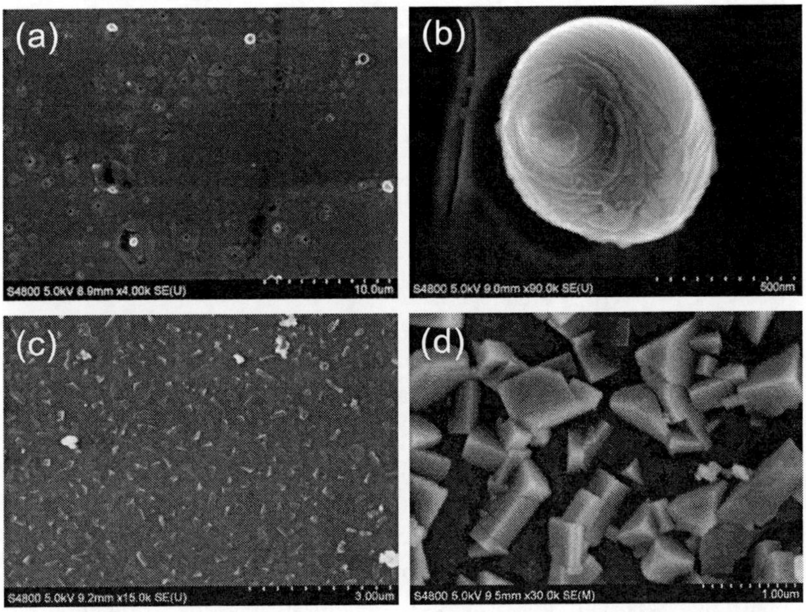

Figure 12. SEM images of Re crystal growth on surfaces using the SSPM method for 1:5 $K[ReO_4]/[NP(O_2C_{12}H_8)]_3$ precursor mixtures. Single crystal Re is formed (a,b) quasi-spherical micron sized particles from pinhole voids in the demixed and decomposed polymeric mixture. (c,d) polyhedral Re crystals formed after pyrolysis of the mixture with half the quantity (1:10 ratio) of Re.

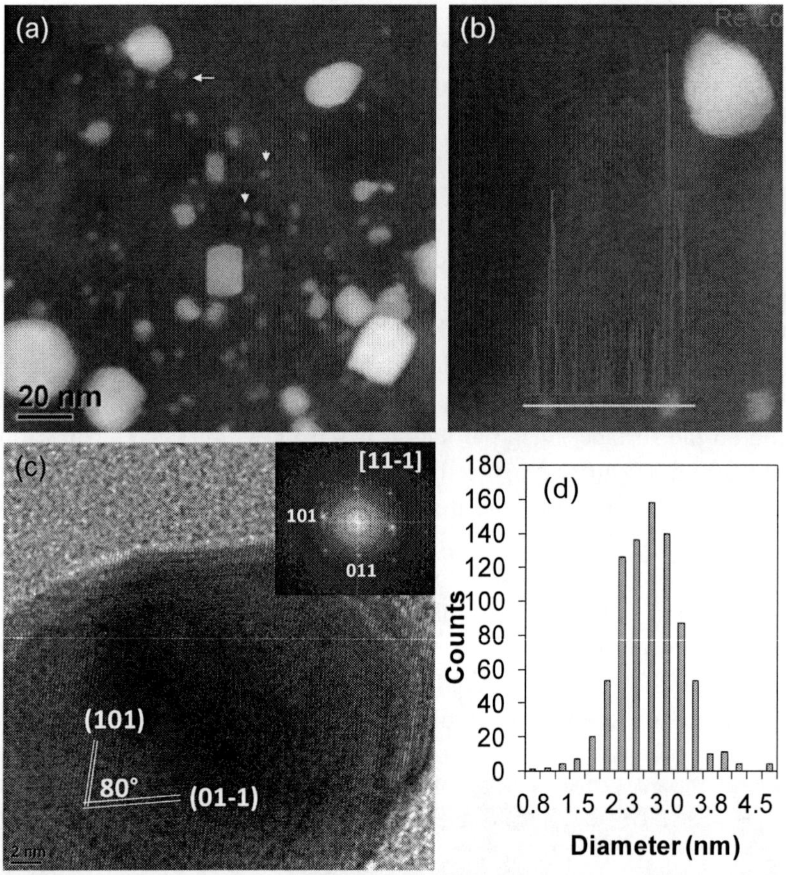

Figure 13. (a) Dark field STEM image of a range of nanoparticles of Re. (b) Nanoscale EDX of the Re L confirming that crystals are composed of Re. (c) Lattice resolution HRTEM image of a single Re nanoparticle confirming single crystal nature and (d) size distribution of ~700 nanoparticles from TEM measurements.

Additionally, the size control extends down to the nanoscale where high quality nanoparticles and nanocrystals of pure Re are possible by this method. Dark-field scanning TEM was used to highlight high atomic number materials such as noble metals by their propensity for high angle electron scattering, and also to avoid carbon degradation or in-situ formation of particles under the beam. Dark field STEM shown in Figure 13a indicates nanoparticle formation with a range of sizes from pyrolysis of a 1:5 $K[ReO_4]/[NP(O_2C_{12}H_8)]_3$ precursor mixture. Analysis of >700 particles confirms that the mean diameter is 2.9 nm (when the larger particles >15 nm are ignored), see Figure 13d. HRTEM EDX shown in Figure 13b also confirms the identification of pure Re.

The nanoparticles were all confirmed to be single crystal. Typically, when growing metallic nanoparticles from solution, thermodynamically unfavoured shapes are very common and control over the subsequent growth depends on the liquid-borne seed being single crystal. All NPs formed with the metals reported here are single-crystal, and Re is no different. The lattice resolution HRTEM image in Figure 13c and corresponding FFT pattern are indexed to Re in the hexagonal unit cell with space group $P6_3/mmc$ and cell parameters a = 0.276 nm and c = 0.445 nm.

Insights into the Formation Mechanism

Although the formation mechanism of Au nanoparticles has been widely studied and recently revised [69-73], the solid state mechanism for the formation of Au nanoparticles in solid state [35-40] has not been significantly investigated or clarified [5]. The most probable mechanism of formation of Au nanostructures involve the energetically easy loss of Cl to give a PPh_3Au^+ species which is known to interact the π system of the aryl groups of the bispyro both of the polymer as well as the trimer [74-77]. These in turn afford a stable solid matrix which is not lost, as a volatile, in the first stage of the annealing.

Subsequently oxidation of the organic matter gives rise to holes where the metallic gold begins to agglomerate, nucleate and grow. It is at this stage, the mixture contains single crystal NPs of gold. In parallel, thermal decomposition of the mixture in the form of phase separation also occurs, where the instability is driven by nucleation dewetting and the formation of pinhole voids within the solidifying mixture. The result in the growth of crystal from within these incubation sights by Ostwald ripening of the migrating Au nanoparticles.

The pinholes act as a nucleation point and are pinned to the surface and thus all nanoparticles for Au and Ag, for example, migrate towards these holes. This occurs for cases in which the metal content is sufficiently low; higher metal content results in a spinodal decomposition of the demixed polymer mixture to give porous metal foam structures. Further evidence for Au-content influence on nanoscale decomposition is found in Figure 14a, where for polymer mixture with little or no gold content, characteristic Voronoi tessellation patterns consistent with classical polymeric dewetting is found. The reason for the observation of unfinished tessellation or beading pattern is due to the solidification and carbonization of the organic matter, which eventual graphitizes.

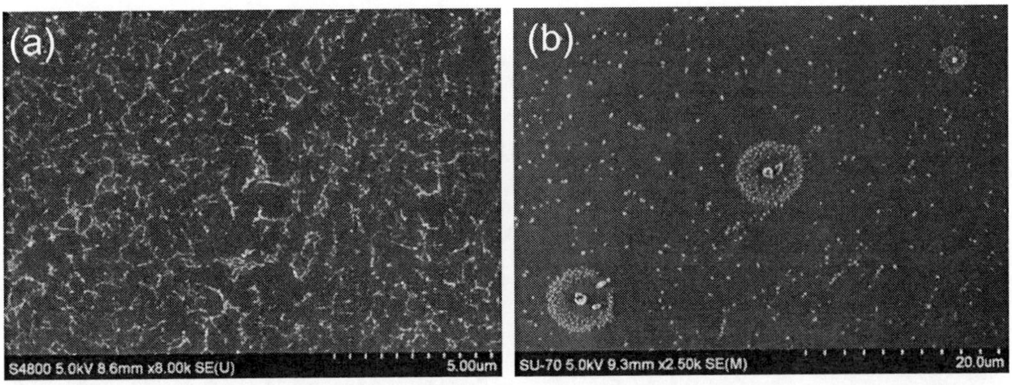

Figure 14. (a) Voronoi-like dewetting of a polymer film without metal content. Unlike liquid polymer films, complete tesselation pattern formation and dewetting cannot occur since a constant temperature ramp results in carbonization and solidification of the polymer. (b) SEM image of NP coalescence around a dust particle defect. This confirms a polymeric dewetting mechanism where in this case nucleation dewetting (up to the point of polymer solidification via carbonization and subsequent graphitization) causes NPs to characteristically form around the defects.

While specific decomposition chemistry occurs in all cases to result in the initial nanoparticles of Au, detailed below, the overall pattern is driven by physical dewetting and most of the characteristics are that of nucleation dewetting occurring during the demixing of two bicontinuous phases. Introducing surface defects has a specific influence on such systems, where dirt acts as a nucleation point for mobile surface-bound (NPs at the surface-air interface) and polymer-bound nanoparticles. Figure 14b shows this precisely, where nucleation of surface bound NPs surrounds surface defects and mimics the profile of the defect, confirming NP coarsening to surface pinned defects through a competition between particle and surface interaction potential mediated by the viscosity, mobility, mixture thickness that ultimately tune the Hamaker constants between each material. If the defect is a pinhole formed during demixing, as seen in Figures 7 and 8, then the NPs aggregated and ripen to form single crystals of the respective metal coordinated to the organometallic fragment.

At the molecular level, the case of the $N_3P_3(O_2C_{12}H_8)_3/AuCl(PPh_3)$ and $[NP(O_2C_{12}H_8)]_n/AuCl(PPh_3)$ mixtures shows that a strong Au-arene interaction between the $Au(PPh_3)^+$ cationic centers and biphenyl rings could allow a regular ordering of the Au centers along with the $NP(O_2C_{12}H_8)$ groups of the chains of the polymer or around the cyclic molecules of the trimer, as depicted in Figure 7b. Figure 15 summarizes a plausible general structure of the molecular level decomposition and cyclomatrix formation and the nanoscale demixing of phases, for the case of Au. Au(I)-arene interactions have been recently discussed by Laguna *et al.* [74-77]. Intermolecular gold-arenes have been reported in the trinuclear complex Au_3(p-tolN=COEt)$_3$ [75]. Also metal-arene interactions have been observed in the one dimensional array supramolecular structures of the gold(I) complex [Au(CN){2,6-$NC_5H_3(C_6H_4)_2$-C,C'N}] [76]. Intramolecular metal-arene interactions have been also observed in some complexes. Recently strong M-arene interactions (M = Au, Ag, Cu) have been reported in metal-dialkylbiarylphosphane complexes [78]. It is likely that a similar mechanism operates for all metals reviewed here so as to form the nanoparticles which influence the ultimate demixing and dewetting of the polymer film.

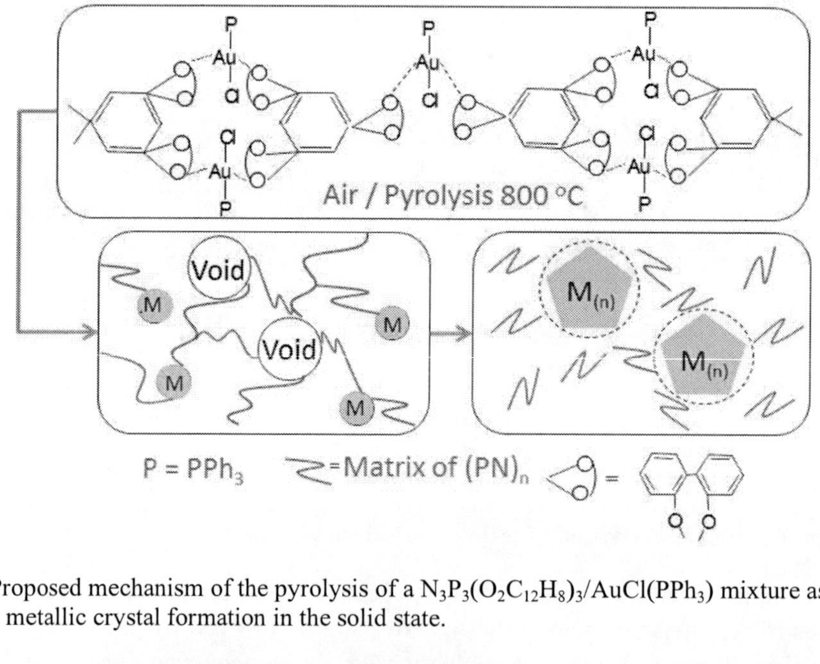

Figure 15. Proposed mechanism of the pyrolysis of a $N_3P_3(O_2C_{12}H_8)_3/AuCl(PPh_3)$ mixture as a precursor to metallic crystal formation in the solid state.

The results presented here show that the formation of metallic nanostructures does not require the covalent linking of the organometallic fragment to the polymeric or trimeric phosphazenes structures. Thus the size of the Au nanoparticles can be controlled by selecting the appropriate mixture for the precursor, *e.g.* $[NP(O_2C_{12}H_8)]_n/AuCl(PPh_3)$ or $[N_3P_3(O_2C_{12}H_8)_3]/ AuCl(PPh_3)$ and an adequate preparation procedure.

It is interesting to note that using this method, Au nanoparticles as small as 3.5 nm can be obtained, with a mean size of ~5 nm. As is known for gold, at ultrasmall NP sizes (typically <2 nm) quantum size effects start to be manifested. The main advantage of the preparation of solid Au nanoparticles is that the nanoparticles do not requires stabilizing molecules. In this method the presence of the $N_3P_3(O_2C_{12}H_8)_3$ or $[NP(O_2C_{12}H_8)]_n$ is crucial because solid state pyrolysis of $AuCl(PPh_3)$ results in bulk Au agglomerates.

Phase separation is expected in asymmetric polymer-particle mixtures [78-80] for large particle-to-monomer size ratios. As in polymer mixtures, increasing the molecular mass should increase the critical point [81] and result in phase separation. Further, an entropically driven surface phase transition has been observed [82], and predicted [34,84] for low-Mw polymer-particle thin films, resulting in the expulsion and layering of NPs along the solid substrate. Early stages of spinodal decomposition are generally described by Cahn-Hilliard theory [85] yielding a dominant wavelength, which then grows via a number of coarsening mechanisms [86].

We rationalize these observations in terms of the entropic-driven phase separation of $[NP(O_2C_{12}H_8)]_n/AuCl(PPh_3)$ or $[N_3P_3(O_2C_{12}H_8)_3]/AuCl(PPh_3)$ mixtures, followed by Au association, observed in the bulk above a threshold temperature [87] into a ripening phase that is surface pinned, enveloped by a $[NP(O_2C_{12}H_8)]_n$ or $[N_3P_3(O_2C_{12}H_8)_3]$ rich phase that converts to carbon during pyrolysis, much like in thin film polymer mixtures with asymmetric surface attraction [88,89]. We interpret this behavior as an interplay between polymer-particle miscibility and particle-surface attraction via entropic pushing, occurring in a polymer matrix with known viscosity and film thickness. Low-Mw thin composite films have been shown to form a diffuse layer of NPs at the substrate interface after spin coating which failed to be redispersed by subsequent thermal annealing. The formation of micron-sized irregular NP clusters at the substrate interface, instead of a uniform surface monolayer, has also been recently reported on relatively thick spun-cast films of PS(600k) with 5% C_{60} [90]. In the SSMP method, the situation is markedly improved where this bicontinuous phase separation is driven by dewetting where NPs form due to the chemical decomposition and crystals eventually grow due to the physical decomposition. While all other know systems separate to form aggregates, this method allows single crystal growth via ripening, rather than simply disordered coalescence. Nanoparticle diffusion in thin films is strongly anisotropic (considerably faster in the plane of the film direction) and the viscosity of high-Mw composite films correspondingly higher, tuning the interplay between surface attraction and phase separation.

CONCLUSION

Metallic foams and nanoparticles can now be easily prepared from mixtures of polymeric or cyclic phosphazenes and organometallic complexes containing noble metals. For Au, it has been shown that the variation of the phosphazene/Au ratios allows for the formation of either metallic foams of high density or discrete Au polyhedral crystals and also nanoparticles, whose mean size and morphology are controlled through the structure of the precursor constituents and the relative ratio in the mixtures. With the $[NP(O_2C_{12}H_8)]_n/AuCl(PPh_3)$ mixtures, Au nanoparticles of 3.5 nm diameter are obtained from films formed by pyrolysis of the 1:1 mixture ratio, while with the analogous 1:1 mixture of $N_3P_3(O_2C_{12}H_8)_3/AuCl(PPh_3)$ resulted in Au nanoparticles with a mean diameter of 4.7 nm from a powder deposit. For a given Re quantity and mixture ratio, the shape and yield of the Re particles can be varied. Using a 1:5 mixture ratio of $K[ReO_4]/[NP(O_2C_{12}H_8)]_3$, deposited on SiO_2, faceted polyhedral microcrystals are found, as opposed to circular crystals. For lower Re quantities, such as with the 1:10 $K[ReO_4]/[NP(O_2C_{12}H_8)]_3$ mixture, the resulting crystals are nucleated on the surface, and subsequent growth allow the definition of polyhedral crystals of Re. Nanoparticle formation with a range of sizes was possible from pyrolysis of a 1:5 $K[ReO_4]/[NP(O_2C_{12}H_8)]_3$ precursor mixture. Analysis of >700 particles confirms that the mean diameter is 2.9 nm. Extension to Ag and Pd also resulted in micro and nanoparticle crystals using similar polymeric precursors, including high surface area porous Ag and Pd, nanoparticles and microcrystals. Additionally, a comparison with previous studies shows that the formation of metallic nanostructures does not require the covalent linking of the organometallic fragment to the polymeric or trimeric phosphazenes, although the presence of the cyclic or polymeric phosphazene as solid state template is necessary to incubate seeds and crystal growth.

This method for preparing Au, Ag, Pd, and Re noble metal nanostructured materials completely in the solid state is potentially useful in the fabrication of solid-sate device contacts and interconnects in the field of electronics, as nanoparticle embedded materials for chemical sensing/detectors and catalysis, and also as surface plasmon-mediated sensors or support materials. Controlling surface bound NP concentrations and position through dewetting allow the possibility for single-step SERS substrates exploiting plasmon 'hot-spots' from NP pairs or clusters in close proximity. Importantly, the method allows the deposition of metallic materials in the complete absence of water or liquids. Phase demixing also has potential for precursor tuning (Mw, viscosity, dielectric constant, metal concentration etc.) for patterning purposes similar to block-copolymer patterning via phase separation or demixing, with the added ability to form perfect polyhedral crystals in the solid state in parallel.

ACKNOWLEDGEMENTS

The authors thank Fondecyt (project 1085011) for financial support. We also thank Professor G.A. Carriedo from the University of Oviedo, Spain, for valuable collaboration in the work cited in this chapter.

REFERENCES

[1] C. N. R. Rao, A. Muller, A. K. Cheetham, (Eds). *The Chemistry of Nanomaterials*, Wiley-VCH, Weinheim, Germany (2004).

[2] S. Edelstein, R. C. Cammarata, (Eds). *Nanomaterials: Synthesis, Properties and Applications*, J.W. Arrowsmith Ltd, Bristol, UK (1998).
[3] K. J. Klabunde, (Ed). *Nanoscale Materials in Chemistry,* Wiley and Sons, New York (2001).
[4] C. Díaz, M. L. Valenzuela, Chapter 16, *Encyclopedia of Nanoscience and Nanotechnology*, H. S. Nalwa, (Ed). American Scientific Publishers (2011).
[5] P. E. Chow, Chapter 14, *Gold Nanoparticles: Properties, Characterization and Fabrication*, Nova Science Publishers (2010).
[6] G. B. Khmutov V. V. Kislov, M. N. Antipirina, R. V. Gainutdinov, S. P. Gubin, A. N. Sergeev-Cherenkov, E. S. Soldatov, D. B. Suyatin, A. L. Tolstikhina, A. S. Trifonov and T. V. Yurova, *Microelectron. Eng.* 69, 373 (2003).
[7] E. C. Walters, K. Ng, M. P. Zach, R. M. Penner and F. Favier, *Microelecton. Eng.* 61, 62 (2002).
[8] G. Walters and I. Parkin, *J. Mater. Chem.* 19, 574 (2009).
[9] C. Díaz and M. L. Valenzuela, *J. Chil. Chem. Soc.* 50, 417 (2005).
[10] C. Díaz, P. Castillo and M. L. Valenzuela, *J. Cluster Sci.* 16, 515 (2005).
[11] C. Díaz and M. L. Valenzuela, *J. Inorg. Organomet. Polym. Mater.* 16, 123 (2006).
[12] C. Díaz and M. L. Valenzuela, *Macromolecules* 39, 103 (2006).
[13] C. Díaz and M. L. Valenzuela, *J. Inorg. Organomet. Polym. Mater.* 16, 216 (2006).
[14] C. Díaz and M. L. Valenzuela, *J. Inorg. Organomet. Polym. Mater.* 6, 419 (2006).
[15] C. Díaz, M. L. Valenzuela, L. Zuñiga, C. O'Dwyer, *J. Inorg. Organometallic Polymer*, 19, 507 (2009).
[16] C. Diaz, M. L. Valenzuela, D. Bravo, V. Lavayen and C. O'Dwyer, *Inorg. Chem.* 47, 11561 (2008).
[17] C. Díaz, M. L. Valenzuela, E. Spodine, Y. Moreno and O.Peña, *J. Clust. Sci.* 18, 831 (2007).
[18] C. Díaz, M. L. Valenzuela and S. Ushak, *J. Clust. Sci.* 19, 471 (2008).
[19] J. Jimenez, A. Laguna, M. Benouazzane, J. A. Sanz, C. Díaz, M. L. Valenzuela, P. G. Jones, *Chem. Eur. J.* 15, 13509 (2009).
[20] C. Diaz, M. L. Valenzuela, A. Laguna, V. Lavayen, J. Jimenez, L. Power and C. O'Dwyer, *Langmuir* 26, 10223 (2010).
[21] C. Diaz, V. Lavayen, and C. O'Dwyer, *J. Solid State Chem.* 183, 1595 (2010).
[22] G. A. Carriedo, M. L. Valenzuela, C. Diaz and S. Ushak, *Eur. Polym. J.* 44, 686 (2008).
[23] C. Díaz, M. L. Valenzuela, S. Ushak, V. Lavayen and C. O'Dwyer, *J. Nanosci. Nanotechnol.* 9, 1825 (2009).
[24] C. Díaz, M. L. Valenzuela and N. Yutronic, *J. Inorg. Organomet. Polym. Mater.* 17, 577 (2007).
[25] C. Díaz, M. L. Valenzuela, G. A. Carriedo, F. J. Garcia Alonso, A. Presa, *Polym. Bull.* 57, 913 (2006).
[26] C. Diaz, M. L. Valenzuela, G. A. Carriedo, L. Zuñiga and C. O'Dwyer, *J. Inorg. Organomet. Polym. Mater.*22, 447 (2011).
[27] M. Hodge, J. R. Hayes, J. A. Caro, J. Biener, A. Nhamza, *Adv. Eng. Mater.* 8, 853 (2006).
[28] C. Tappan, M. H. Huyuh, M. A. Hiskey, D. E. Chavez, E. P. Luther, I. T. Mang, S. F. Son, *J. Am. Chem. Soc.* 128, 6589 (2006).

[29] Banhart, *Adv. Eng. Mater.* 8, 781 (2006).
[30] H. Ehang, A. I. Cooper, *J. Mater. Chem.* 15, 2157 (2005).
[31] Biener, G. W. Nyce, A. M. Hodge, M. M. Biener, A. V. Hamza, S. A. Maier, *Adv. Mater.* 20, 211 (2008).
[32] Y. Ding, J. Enlabacher, *J. Am. Chem. Soc.* 125, 772 (2003).
[33] F. Khan, S. Mann, *J. Phys. Chem. C* 113, 19871 (2009).
[34] T. Teranishi, S. Hasegawa, T. Shimizu, M. Miyake, *Adv. Mater.* 13, 1699 (2001).
[35] T. Shimizu, T. Teranishi, S. Hasegawa, M. Miyake, *J. Phys. Chem. B.* 107, 2719 (2003).
[36] M. Miyake, W. Zheng, F. L. Leibowitz, N. K. Ly, Ch. Zhong, *Langmuir* 16, 490 (2000).
[37] Oiu, X. Jiang, C. Zhu, M. Shirai, J. Si, N. Jiang, K. Hirao, *Angew. Chem. Int. Ed.* 43, 2230 (2004).
[38] G. Lusvardi, G. Malavasi, V. Aina , L. Bertinetti, G. Cerrato, G. Magnacca, C. Monterra and L. Menabue, *Langmuir* 26, 10303 (2010).
[39] Eichelbaum, K. Rademann, R. Muller, M. Radtke, H. Riesemeier, W. Gorner *Angew. Chem. Int. Ed.* 44, 7905 (2005).
[40] G. A. Carriedo, L. Fernández-Catuxo, F. J. García Alonso, P. Gómez-Elipe, P. A. González, *Macromolecules* 29, 5320 (1996).
[41] Sánchez-Iglesias, I. Pastoriza-Santos, J. Pérez-Juste, B. Rodríguez-González, F. J. García de Abajo and L. M. Liz-Marzán, *Adv. Mater.* 18, 2529 (2006).
[42] R. Levi, M. Bar-Sadan, A. Albu-Yaron, R. Popovitz-Biro, L. Houben, Ch. Shahar, A. Enyashin, G. Seifert, Y. Prior, R. Tenne, *J. Am. Chem. Soc.* 132, 11214 (2010).
[43] R. Tenne, G. Seifert, *Ann. Rev. Mater. Res.* 39, 387 (2009).
[44] R. Tenne, *Chem. Eur. J.* 8, 5297 (2002).
[45] S. C. Glotzer, M. J. Solomon, *Nat. Mater.* 6, 557 (2007).
[46] Halder, P. Kundu, B. Viswanath, N. Ravinshankar, *J. Mater. Chem.* 20, 4763 (2010).
[47] J. Turkevich, G. Kim, *Science* 169, 873 (1970).
[48] J.G. de Vries, *J.C.S. Dalton Trans* 421 (2006).
[49] T. Teranishi and M. Miyake, *Chem. Mater.* 10, 594 (1998).
[50] P. John Thomas, G. U. Kulkardin, and C. N. R. Rao, *J. Phys. Chem.* 104, 8138 (2000).
[51] Y. Sun, L. Zhang, H. Zhou, Y. Zhu, E. Sutter, Y. Ji, M. H Rofailovich, J.C Sokolov, *Chem. Mater.* 19, 2065 (2007).
[52] M. T. Reetz, W. Heilbirg, *J. Am. Chem. Soc.* 116, 7401 (1994).
[53] E. Ramirez, S. Jonsat, K. Philippol, P. Lecante, M. Gomez, A. M Masden Bulto, B. Chaudret; *J. Organomet. Chem.* 589, 4601 (2004).
[54] N. Anil Dhor and A Godenken, *Chem. Mater.* 8, 445 (1998).
[55] G. Cardenas Trivino, K. J. Klabunde and E. Broca Dale, *Langmuir* 3, 986 (1987).
[56] Y. Yu, Y. Zhao, T. Huang and H. Liu, *Pure Appl. Chem.* 81, 2377 (2009).
[57] H. Chen, G.Wei, A. Ispas, S. G. Hickey and A. Eychmuller, *J. Phys. Chem.* C 114, 21976 (2010).
[58] D. Jose, B. R Jagida, *J. Solid State Chem.* 183, 2059 (2010).
[59] E. Mendez, M. F. Cerdà, A. M. Castro Luna, C. F. Zinola, C. Kremer, M. E. Martins, *J. Colloid Interface Sci.* 263, 119 (2003).
[60] R. Schrebler, P. Cury, M. Orellana, H. Gòmez, R. Còrdova, E.A. Dalchiele, *Electrochim. Acta* 46, 4309 (2001).

[61] J. Zerbino, A. M. Castro Luna, C. F. Zinola, M. E. Martins, *React. Kinet. Chem.* 521, 763 (2002).
[62] W. Hassel, B. Bello, Rodriguez, S. Milenkovic, A. Schneider, *Electrochim. Acta* 51, 795 (2005).
[63] L. Philippe, Z. Wang, I. Peyrot, A.W. Hassel, J. Michler, *Acta Materialia* 57, 4032 (2009).
[64] O. A. Nikonova, K. Jansson, V. G. Kessier, M. Sundberg, A. I. Baranov, A. V. Shevelkov, D. V. Drobot, G. A. Seisenbaeva, *Inorg. Chem.* 47, 1295 (2008).
[65] R. R. Dykeman, N. Yan, R. Scopelliti, P. J. Dyson, *Inorg. Chem.* 50, 717 (2011).
[66] Vollmer, E. Redel, K. Abu-Shandi, R. Thomann, H. Manyar, C. Hardacre, C. Janiak, *Chem. Eur. J.* 16, 3849 (2010).
[67] F. Hayashi, M. Iwamoto, *Micro. Meso. Mater.* (2011). DOI: 10.10161/j.micromeso.2011.03.035.
[68] E. L. Kunkes, D. A. Simonetti, J. A. Dumesic, W. D. Pyrz, L. E. Murillo, J. G. Chen, D. J. Buttrey, *J. Catalysis* 260, 164 (2008).
[69] J. E. Millstone, S. J. Hurst, G. S. Metraux, J. I. Cutler, Ch. A. Mirkin, *Small* 5, 646 (2005).
[70] L. Malfatti, D. Morongiu, S. Costacurta, P. Falcaro, H. Amenitsch, B. Marmiro, J. Li, G. Crenci, M. F. Casula, P. Innocenzi, *Chem. Mater.* 22, 2132 (2010).
[71] J. Polte, T. T. Ahner, F. Delissen, S. Sokolov, F. Emmerling, A. F. Thunemann, R. Kraehnert, *J. Am. Chem.Soc.* 132, 129 (2010).
[72] X. Ji, X. Song, J. Li, Y. Bai, W. Yang, X. Peng, J. Am.Chem.Soc. 129, 13939-13948 (2007).
[73] K. Pong, H. I. Elim, J. X. Chong, W. Ji, B. L. Trout, J. Y. Lee *J.Phys.Chem. C* 111, 628 (2007).
[74] Laguna, *Modern Supramolecular Gold Chemistry,* Chapter 5, p. 318, Wiley-VCH, Wenheim (2008).
[75] R. A. Rawashdeh-Omary, M. A. Omary, J. P. Fackler, R. Galassi, B. R. Pietroni, A. Burini, *J. Am. Chem. Soc.* 123, 9689 (2001).
[76] A. Mohamed, R. A. Rawashdeh-Omary, M. A. Omary, J. P. Fackler, *Dalton Trans.* 2597 (2005).
[77] P. Perez-Galan, N. Delpont, H. Guerrero-Gomez, F. Maseras, A. M. Echavarren, *Chem. Eur. J.* 16, 5324 (2010).
[78] P. Paricaud, S. Varga, and G. Jackson, *J. Chem. Phys.* 118, 8525 (2003).
[79] G. J. Fleer and R. Tuinier, *Adv. Colloid Interface Sci.* 143, 1 (2008).
[80] J. B. Hooper, K. S. Schweizer, T.G. Desai, R. Koshy, and P. Keblinski, *J. Chem. Phys.* 121, 6986 (2004).
[81] J. Dudowicz, J. F. Douglas, and K. F. Freed, *J. Phys. Chem. B* 113, 3920 (2009).
[82] M. E. Mackay, Y. Hong and M. Jeong, S. Hong and T. P. Russell, C. J. Hawker, R. Vestberg, J. F. Douglas, *Langmuir* 18, 1877 (2002).
[83] S. McGarrity, A. L. Frischknecht, L. J. D. Frink, and M. E. Mackay, *Phys. Rev. Lett.* 99, 238302 (2007).
[84] S. McGarrity, A. L. Frischknecht, and M. E. Mackay, *J. Chem. Phys.* 128, 154904 (2008).
[85] J. W. Cahn and J. E. Hilliard, *J. Chem. Phys.* 31, 688 (1959).

[86] J. D. Gunton, M. San-Miguel, and P. S. Sahni, *Phase Transitions and Critical Phenomena*, Vol. 8, p. 267, C. Domb and J. L. Lebowitz (Eds). Academic, New York (1983).
[87] H. C. Wong, A. Sanz, J. F. Douglas, and J. T. Cabral, *J. Mol. Liq*. 153, 79 (2010).
[88] L. Sung, A. Karim, J. F. Douglas, and C. C. Han, *Phys. Rev. Lett.* 76, 4368 (1996).
[89] B.D. Ermi, A. Karim, and J. F. Douglas, *J. Polym. Sci. B* 36, 191 (1998).
[90] J. T. Han, G.-W. Lee, S. Kim, H.-J. Lee, J. F. Douglas, and A. Karim, *Nanotechnology* 20, 105705 (2009).

In: Nanostructures
Editor: Yu Dong
ISBN: 978-1-62618-081-9

Chapter 6

EFFECT OF ANNEALING ON PHYSICAL CHARACTERISTICS OF TiO_2 NANOTUBES BY ELECTROCHEMICAL ANODIZATION

Wen-Yang Chang[1], Te-Hua Fang[2,*] and Zhe-Wei Chiu[2]

[1]Department of Mechanical and Computer-Aided Engineering, National Formosa University, Yunlin, Taiwan

[2]Department of Mechanical Engineering, National Kaohsiung University of Applied Sciences, Kaohsiung, Taiwan

ABSTRACT

The physical characteristics of array TiO_2 nanotubes based on electrochemical anodization growth are investigated. The characteristics include the TiO_2 growth mechanism, contact angle, nanoindentation load, and photoluminescence and XRD spectra.

Field emission scanning electron microscopy, a nanoindenter, a contact angle analyzer, X-ray diffraction, and photoluminescence (PL) spectroscopy, were used to identify the structures and the characteristics of the TiO_2 nanotubes. Results show that the morphologies of the TiO_2 nanotubes exhibit nearly regular vertical alignment when the anodization time is over 90 minutes.

The nanotubes tapered off to a point and had ripples in their side walls. TiO_2 nanotube surfaces possess a hydrophilic phenomenon and the contact angle decreased with increasing electrochemical anodization time. A hysteresis loop during loading and unloading was observed, indicating that the TiO_2 nanotube surface has elastic energy dissipation.

The Young's modulus decreased with increasing annealing temperature before 400°C, but increased after annealing at 500°C. The intensities of PL increased with increasing annealing temperature.

[*] Corresponding author's email address: fang.tehua@msa.hinet.net.

1. Introduction

Within the last decades, titanium dioxide (TiO_2) nanotubes have attracted a lot of attention for catalysis [1], photoelectrolysis [2], biological coating [3], self-cleaning [4], gas sensing [5], and absorption [6] applications due to their strong oxidizing power, electrochemical stability, non-toxicity, and excellent optical and electronic properties [7]. TiO_2 nanotubes are usually synthesized using the sol-gel technique [8], chemical vapor deposition [9], templating [10], or electrochemical anodization [11], etc. Electrochemical anodization produces highly ordered TiO_2 nanotube arrays that have remarkable properties [12]. The growth of TiO_2 nanotubes have been well studied in the past [13-15], but there is still a general lack of understanding of the physical characteristics. Investigating the physical characteristics of TiO_2 nanotubes for sensing applications is important because durability, reliability, and deformation are problems for fabrication and sensing applications. In addition, according to our present study, adhesive forces and mechanical characteristics are critical factors for producing TiO_2 sensors during fabrication.

In this study, we investigated the mechanical characteristics, the contact angle, and the photoluminescence (PL) and X-ray diffraction (XRD) spectra of TiO_2 nanotubes. The mechanical properties were measured using a nanoindentation tip to obtain the hardness and Young's modulus characteristics of TiO_2 nanotubes at various annealed temperatures. The contact angle was investigated at various anodization times and annealing temperatures, while the un-annealed TiO_2 nanotubes and those annealed at various temperatures were investigated using PL and XRD spectra. In addition, the TiO_2 nanotube growth mechanisms of electrochemical anodization were also investigated. The effects of anodization time, anodization voltage, and annealing temperature are determined. A better understanding of the physical characteristics and growth mechanisms of electrochemical anodization can provide useful information to design better TiO_2 applications for photodetectors or sensors.

2. Experiments

2.1. Sample Production

TiO_2 nanotubes based on electrochemical anodization were produced using a two-electrode system with a DC voltage, cathode for graphite rod and anode for titanium foil, at room temperature, as shown in Figure 1. The system included an electrochemical cell, a holder, titanium foil, a graphite rod, and a power supply, which was used as the voltage source/current measuring device during the electrochemical tests. Titanium foil fixed on a copper plate, which was placed in a holder, was used as the anode and the graphite rod was used as the cathode. A distance of 2 cm was set between the cathode and the anode. The diameter and thickness of the titanium foil (99.7 % purity) were 3 cm and 0.1 mm, respectively.

The electrochemical reaction was driven by an external applied voltage in the electrolyte. The electrolyte was a mixture of glycerol (400 mL with 99% purity), ammonium fluoride (NH_4F) (1.25 g with 96% purity), and deionized water (100 mL). NH_4F was firstly mixed with the glycerol using a magnetic stirring bar for 24 hours.

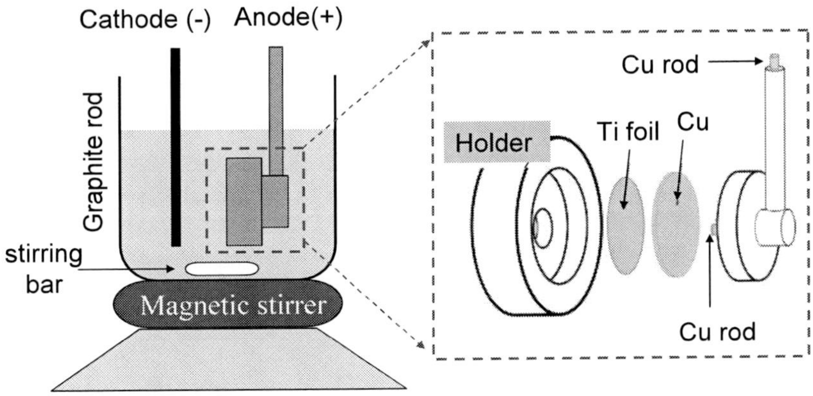

Figure 1. Electrochemical anodization system for TiO_2 nanotubes.

Deionized water was then slowly added and stirred into the solution of NH_4F and glycerol for 5 hours. Before the growth process, the titanium foil was twice ultrasonically cleaned in deionized water at a frequency of 30 kHz and a power of 50 W for 10 minutes, acetone, and then ethanol to remove contaminants.

2.2. TiO_2 Growth Mechanism

The TiO_2 formation mainly consists of chemical dissolution and electrochemical oxidation. The chemical dissolution process controls the formation structures and the geometric size of the TiO_2 nanotubes. The anodization voltage, anodization time, and acid concentration are critical parameters for electrochemical oxidation. In the growth mechanism of TiO_2 nanotubes, the ammonium fluoride in the solution firstly dissociates into ammonia and hydrofluoric acid. The titanium foil at the anode then dissolves with the electrolysis of water. Ti^{4+} ions (see Eq. 1.) ejected from the anode migrate through the oxide under the application of an electric field and accumulate at the entrances of the nanopores.

The electrolysis of water releases O_2 due to the interaction of Ti metal with O^{2-} or OH^- ions [16]. The F^-, O^{2-}, and OH^- species move toward the anode and the Ti^{4+} ions move toward the electrolyte. Then, $Ti(OH)_4$ precipitate (see Eq. 2.) forms via the instantaneous hydrolysis reaction at the entrances of the nanopores. These anions migrate through the oxide layer to the Ti metal/oxide interface and react with the Ti metal. Bowl-like pores initially form on the Ti foil surface due to the pitting corrosion of F^- ions. At the same time, the TiO_2 dissolution in the presence of fluoride also occurs corresponding to the chemical etching, and hydrogen fluoride is consumed. The chemical reaction is as follows:

$$Ti \rightarrow Ti^{+4} + 4e^- \tag{1}$$

$$Ti^{+4} + 4H_2O \rightarrow Ti(OH)_{4(aq)} + 4H^+ \tag{2}$$

$$Ti(OH)_{4(aq)} \rightarrow TiO_2 + 2H_2O \tag{3}$$

The TiO_2 nanotubes are described as the formation of the anodic layer under an applied electric field (see Eq. 4) and as the chemical field assisted dissolution of the forming oxide (see Eq. 5). Fluoride ions dissolve the TiO_2 forming a complex compound. The TiO_2 at the bottom of the bowl-like pores combines with F^- ions and then dissolves into the electrolyte in the form of soluble TiF_6^{2-}. The oxidation rate is limited by the thickness of the barrier layer at the Ti metal/oxide interface and is kept active by the chemical dissolution process at the oxide/electrolyte interface.

$$Ti + 2H_2O \rightarrow TiO_2 + 2H_2 \quad (4)$$

$$TiO_2 + 4H^+ + 6F^- \rightarrow TiF_6^{2-} + 2H_2O \quad (5)$$

2.3. Nanoindentation Tests

The nanoindentation experiment was conducted using a nanoindenter system (MTS XP, US) with a Berkovich diamond tip of radius 20 nm. For mechanical characteristics, the TiO_2 nanotube specimens were put on the platform and then penetrated using a needle that was pressed into the sample surface. The nanoindenter, with a displacement resolution of 0.01 nm, applied a constant force on the specimen's surface to evaluate Young's modulus and hardness of the TiO_2 nanotubes. Young's modulus is a measure of the material stiffness. Hardness, load divided by the area, is a measure of a material's resistance to surface penetration. During the nanoindentation experiment, the indenter approached a strain rate of 0.05 s^{-1} to a pre-defined maximum force and used load-unload displacement. The maximum penetrating load was 3 mN at the indentation velocity of 10 nm/s and frequency target of 45 Hz.

2.4. Contact Angles

In order to understand how the interfacial properties affect the introduction of water molecules on the TiO_2 nanotube, it is important to perform a careful study of the liquid-solid interface interactions. Here, the contact angle has been used as a measure of wetting between a liquid and a solid surface. The surface wettability of TiO_2 nanotubes was investigated using a contact angle analyzer at room temperature. DI water droplets with a total volume of 0.4 μL were made for the subsequent contact measurement. The contact angle of Young's equation was obtained by estimating the tangent to the drop at its intersection with the surface, as shown in Figure 2. The equilibrium contact angle θ, which results from a balance of interfacial forces acting on the contact line, can be calculated by Young's equation:

$$\gamma_{SV} = \gamma_{SL} + \gamma_{LV} \times \cos(\theta) \quad (6)$$

where γ_{LV} is the surface tension of the liquid-vapor energy which has a constant value of 74 mJ/m^2, γ_{SV} is the surface tension of the solid-vapor in air, and γ_{SL} is the interfacial surface tension of the solid-liquid.

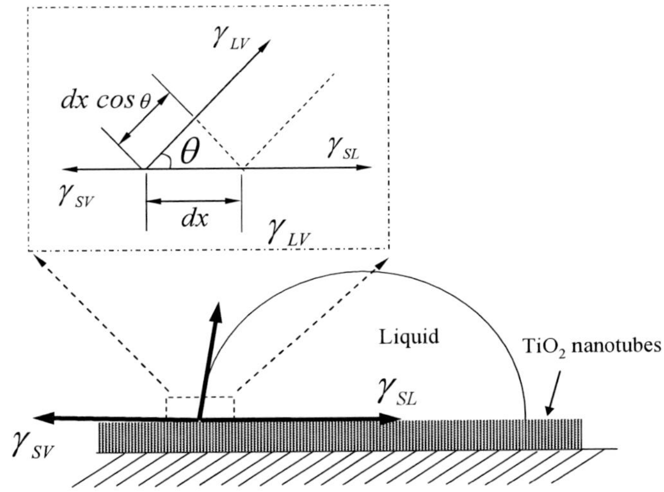

Figure 2. Contact angle measurement on TiO_2 nanotube surfaces based on Young's equation.

2.5. PL and XRD Characteristics

Photoluminescence spectra are closely related to the TiO_2 nanotube surface stoichiometry and surface states. The photoreaction performance of TiO_2 nanotubes was investigated using a spectrometer (Hitachi F-4500). The performance was measured in the wavelength range of 350~600 nm in air at room temperature. The TiO_2 nanotubes were excited using monochromic light with a resolution of 1.5 nm from a 150 W xenon lamp at room temperature. A 350-nm filter was used to obtain the real emission of the TiO_2 nanotubes.

The characteristics analyses of the TiO_2 nanotubes with and without annealing, respectively, were conducted using X-ray diffraction (XRD, MO3XHF22, Mac Science Co.). The patterns were run with nickel-filtered copper radiation with a wavelength of 1.54056 A at 40 *kV* and 30 *mA*. The measured 2è range was between 20° and 80°and the scanning rate was 0.02°/s. The content percentage of the phase constitution was calculated using the integrated intensities of the anatase (A), titanium (Ti), and rutile (R) peaks [17]. The weight fractions of the A and R phases of the TiO_2 nanotubes subjected to annealing were estimated from the relative intensities of the strongest peaks.

3. RESULTS AND DISCUSSION

3.1. TiO_2 Nanotubes Growth

TiO_2 nanotubes were grown with the various electrochemical anodization times of 10, 30, 50, and 90 minutes at a constant anodization voltage of 30 V. After growing, the surface morphologies of the TiO_2 nanotubes were observed using scanning electron microscopy (SEM), as shown in Figure 3. Initially, there are bowl-like pores that form on the Ti foil surface due to the pitting corrosion of F^- ions under an anodization growth time of 30 minutes, as shown in Figs. 3(a) and 3(b).

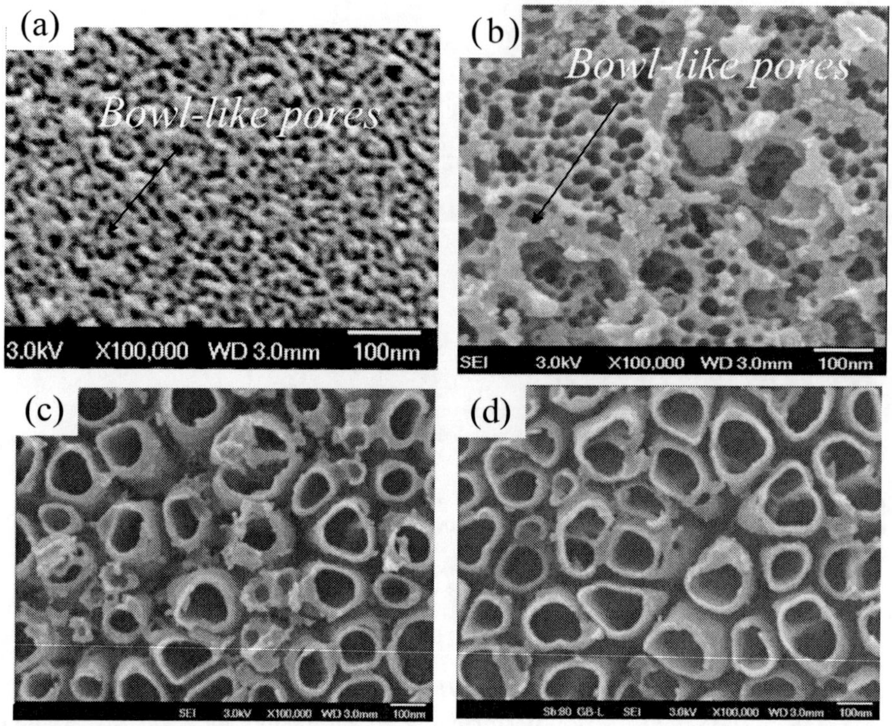

Figure 3. SEM images of TiO_2 nanotubes grown for the electrochemical anodization times of (a) 10, (b) 30, (c) 50, and (d) 90 minutes at an anodization voltage of 30 V.

The anodization times limited the thickness of the barrier layer for growing TiO_2 nanotubes. There have almost same nanopores size at an anodization time of 10 minutes. For an anodization time of 30 minutes, the sizes of nanopores at the top of nanotubes were non-homogeneous and appear to be poorly organized.

However, the morphologies in Figs. 3(c) and 3(d) show a nearly regular vertical alignment of the TiO_2 nanotubes. The nanotubes tapered off to a point and had ripples in their side walls [18], resulting in a non-homogeneous tube wall thickness. The ripples may be the result of oxygen evolution due to the electrolysis of water at the anode, which only occurs when sufficient water is present in the electrolyte and the barrier layer oxide at the metal/oxide interface is relatively thin, allowing electrons to pass through the oxide [19]. In addition, the TiO_2 nanotube adherence between the porous TiO_2 film and its substrate was relatively weak. The nanopores completely dissolved and the tops of the nanotubes can be clearly observed after an anodization time of 90 minutes. The TiO_2 nanotubes also tapered off to a point and had ripple rings in their side walls for TiO_2 nanotubes grown with various electrochemical anodization voltages. Experimental results show that TiO_2 nanotube characteristics at different anodization voltages were similar, thus only the anodization times are shown.

The average diameter and length of the TiO_2 nanotubes increased with increasing anodization time, as shown in Figure 4. The average diameter and length were calculated using 20 adjacent nanotubes. The average diameters at the anodization times of 10, 30, 50, and 90 minutes are 10.5, 52.6, 105.2, and 121.7 nm, respectively. For an electrochemical anodization time of over 60 minutes, the increase in nanotube diameter became slow.

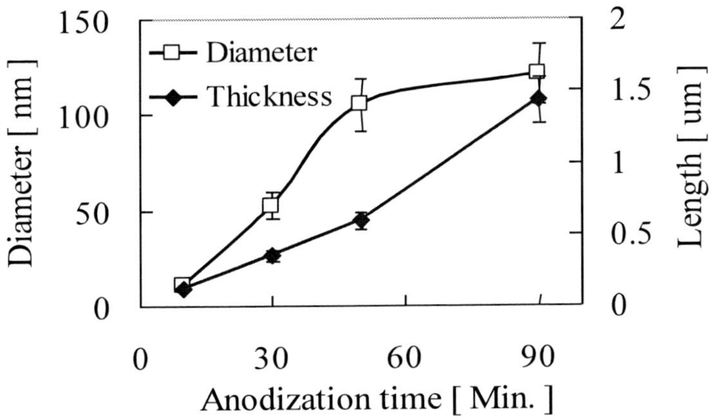

Figure 4. Diameter and length of TiO_2 nanotubes grown for the electrochemical anodization times of (a) 10, (b) 30, (c) 50, and (d) 90 minutes at an anodization voltage of 30 V.

It can be predicted that the stable anodic current density would quickly decrease with increasing electrochemical anodization time [20]. The drop in current density is attributable to the oxidation of the Ti foil which formed the TiO_2 nanotubes. The average lengths at the anodization times of 10, 30, 50, and 90 minutes are 0.125, 0.35, 0.592, and 1.431 μm, respectively. The average length of TiO_2 nanotubes increased almost linearly with increasing electrochemical anodization time due to the slow diffusion of F^- ions. The results show that the electrolytic concentration controlled the reaction equilibrium of electrochemical oxide formation, indicating that the length and diameter of TiO_2 nanotubes is relative to the chemical dissolution rate [18].

3.2. Contact Angles

For wettability analysis of TiO_2 nanotubes, the liquid of deionized water was dropped on the TiO_2 nanotube surface to measure the contact angle, which is defined as the angle between two of the interfaces at the three-phase line of contact. The contact angle directly provides information on the interaction for identifying the influence of roughness on the TiO_2 nanotube surface.

The contact angle was measured by estimating the tangent to the drop at its intersection with the surface, as shown in the insets of Figure 4. The contact angles of TiO_2 nanotube surfaces with the various anodization times of 10, 30, 50, and 90 minutes at a constant anodization voltage of 30V are 32.2°, 30.9°, 26.1°, and 18.9°, respectively, as shown in Figure 5. TiO_2 nanotube surfaces obviously possess a hydrophilic phenomenon because the contact angles are less than 90°. In addition, the contact angles of TiO_2 nanotube surfaces are dependent of the anodization times. The contact angle of TiO_2 surfaces decreased with increasing electrochemical anodization time.

Figure 6 shows contact angles of un-annealed TiO_2 nanotubes and those annealed at temperatures of 300, 400, and 500°C. The un-annealed TiO_2 nanotubes were grown at an electrochemical anodization time of 90 minutes at a constant anodization voltage of 30 V. The contact angles of un-annealed TiO_2 nanotubes and those annealed at temperatures of 300, 400, and 500°C are 18.9°, 16.4°, 5.8°, and 5.5°, respectively.

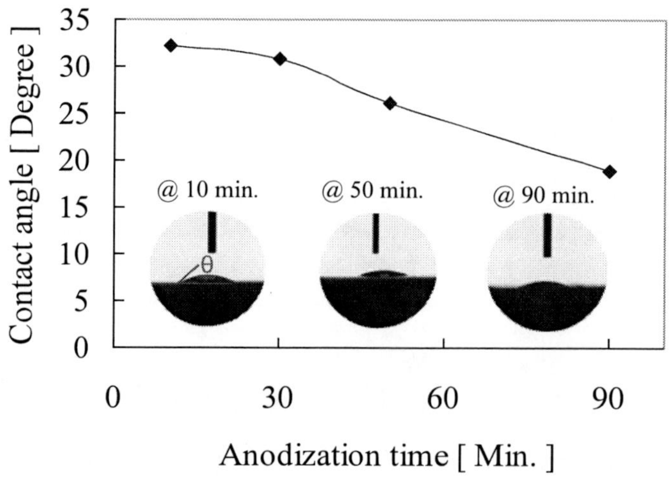

Figure 5. Contact angles of TiO_2 nanotubes grown for the electrochemical anodization times of 10, 30, 50, and 90 minutes at an anodization voltage of 30 V.

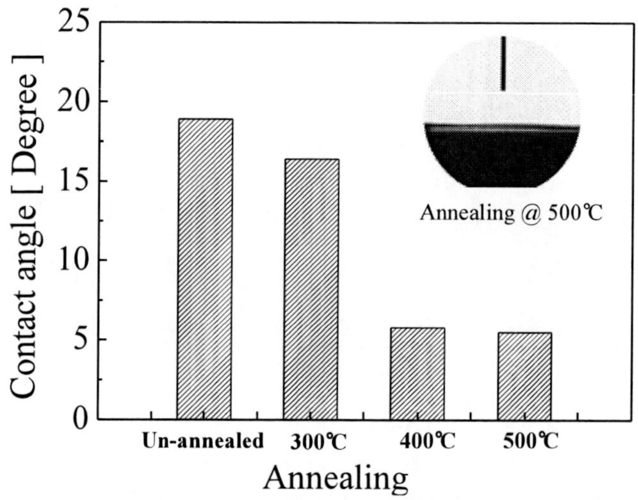

Figure 6. Contact angles of un-annealed TiO_2 nanotubes and those annealed at temperatures of 300, 400, and 500°C.

TiO_2 nanotube surfaces also display hydrophilic characteristics after annealing. In addition, TiO_2 nanotube surfaces are similar to super hydrophilic characteristics when the annealing temperature is over 400°C. The results agree with previous studies [21].

3.3. Nanoindentation Tests

The nanoindentation tests measured the mechanical characteristics of the TiO_2 nanotubes, including hardness and Young's modulus characteristics. The maximum load is 3 mN during the nanoindentation test, while the TiO_2 nanotubes were grown at an anodization time of 90 minutes at a constant anodization voltage of 30 V. The curves of the load-unload forces versus the nanoindentation depths of un-annealed TiO_2 nanotubes and those annealed at temperatures of 300, 400, and 500°C are shown in Figure 7. A hysteresis loop during loading and unloading was observed, indicating that the TiO_2 nanotube surface has elastic energy dissipation. The elastic energy dissipation is attributed to higher stiffness, elastic response, frictional energy, and compressive plastic deformation [22]. The area of the hysteresis loop increased with an increasing annealing temperature before 400°C, due to bigger grain size and density of TiO_2 [21].

During nanoindentation tests, hardness, *H*, can be measured by the maximum applied load, P_{max}, divided by the contact area, $A(h)$ [23]. A constant of 24.56, $A(h) = 24.56h^2$, was used since it is assumed that a Berkovich indenter has a perfect tip [24]. Figure 8(a) shows hardness values of un-annealed TiO_2 nanotubes and those annealed at temperatures of 300, 400, and 500°C. The hardness values decreased with increasing annealing temperature before 400°C. This phenomenon is in good agreement with the energy dissipation trends of the load-unload curve. Un-annealed TiO_2 nanotubes have an amorphous structure, indicating that the amorphous TiO_2 dominates the hardness property. TiO_2 transformed into the anatase crystal structure, causing hardness to decrease, after annealing and the anatase decreased with an increasing annealed temperature before 400°C. However, the intensity of the rutile polycrystalline increased with increasing annealed temperature after 500°C. The annealing treatment of TiO_2 would induce crystal phase transition from amorphous structure to anatase structure or rutile structure.

Young's modulus was performed at maximum loads of 1, 2, and 3 mN using the nanoindentation tests and can determined using the following formula [25]:

$$E = \frac{k\sqrt{\pi}}{2\sqrt{24.56h}} \times \frac{dP_{\max}}{dh_{\max}} \tag{7}$$

where dP_{max}/dh_{max} is the contact stiffness, which is determined from the maximum load and maximum depth at the initial unloading slope. *k* is the shape constant of a Berkovich indenter that depends on the geometry of the indenter, which is 1.034 for the Berkovich tip [26].

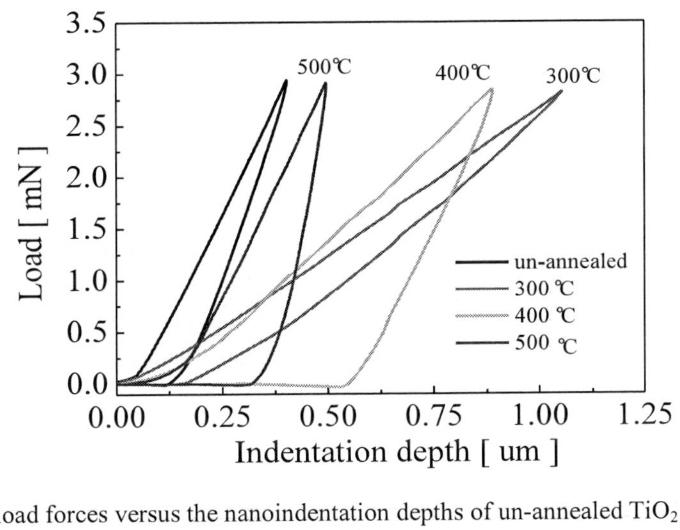

Figure 7. Load-unload forces versus the nanoindentation depths of un-annealed TiO_2 nanotubes and those annealed at temperatures of 300, 400, and 500°C.

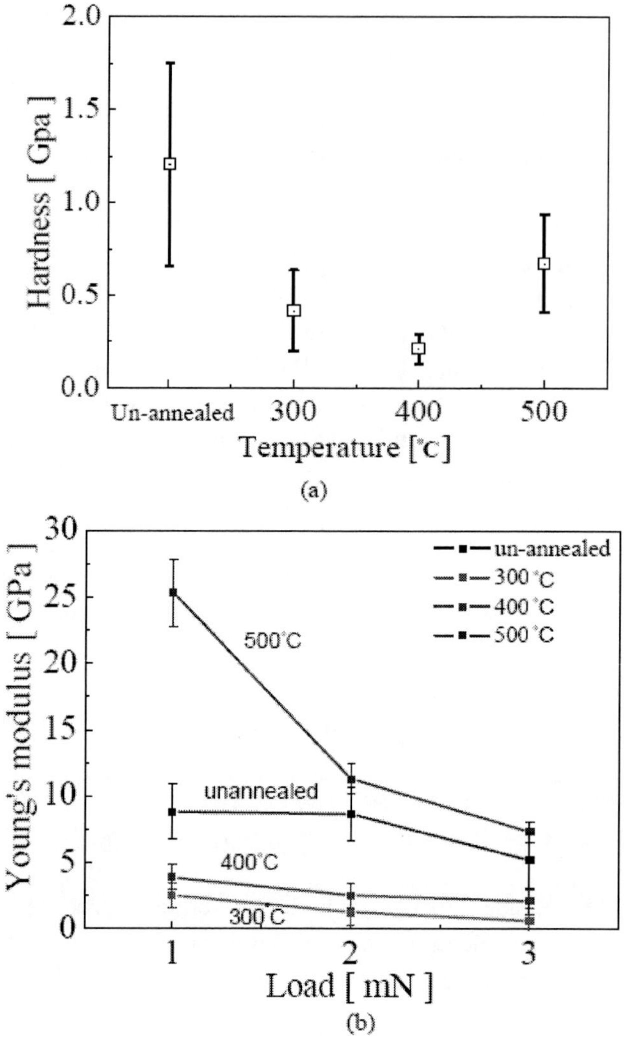

Figure 8. Mechanical characteristics of un-annealed TiO_2 nanotubes and those annealed at temperatures of 300, 400, and 500°C, (a) Hardness and (b) Young's modulus.

Figure 8(b) shows Young's modulus values of un-annealed TiO_2 nanotubes and those annealed at temperatures of 300, 400, and 500°C. The Young's modulus decreased with increasing nanoindentation force. For lower nanoindentation loads on TiO_2 nanotubes, there is a higher Young's modulus because the surface effect and elastic recovery energy are greater [27]. The Young's modulus decreased with increasing annealing temperature before 400°C, but Young's modulus increased after an annealing temperature at 500°C. This phenomenon is in good agreement with the hardness trend.

3.4. PL and XRD Spectra

PL spectra are closely related to the TiO_2 nanotube surface stoichiometry and surface states. The effects of the crystalline phases on the PL spectra and surface properties for various annealing temperature were obtained. Figure 9 shows PL spectra of un-annealed TiO_2 nanotubes and those annealed at temperatures of 300, 400, and 500°C. The TiO_2 nanotubes were also grown at an anodization time of 90 minutes at a constant anodization voltage of 30 V. Results show that the curves have similar trends. Only one peak of the luminescence band appeared, which was in the range of UV light (200~400 nm). In addition, the peak of the wavelength gradually shifted to the left with an increasing annealed temperature. The intensities of PL increased with increasing annealed temperature. Therefore, the photoluminescence spectra of anatase and titania can be attributed the surface states, self-trapped excitons, and oxygen vacancies [28-30].

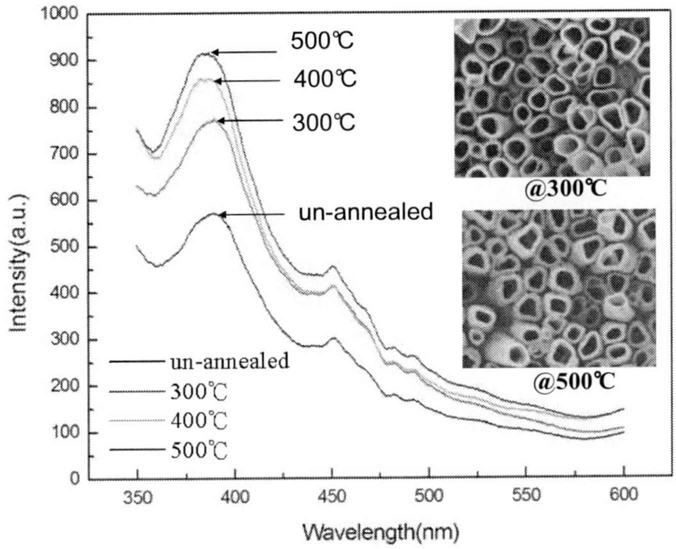

Figure 9. Photoluminescence spectra of un-annealed TiO_2 nanotubes and those annealed at temperatures of 300, 400, and 500°C.

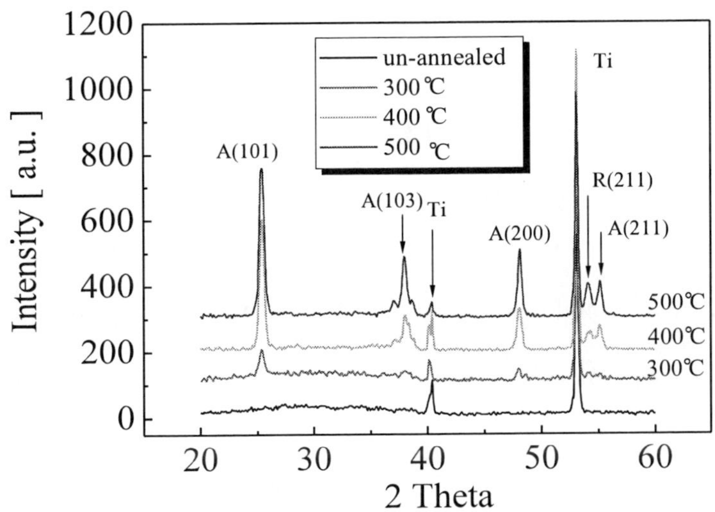

Figure 10. XRD spectra of un-annealed TiO_2 nanotubes and those annealed at temperatures of 300, 400, and 500°C.

Figure 10 shows the XRD patterns for un-annealed samples and samples annealed at temperatures of 300, 400, and 500°C. The un-annealed sample had an amorphous TiO_2 structure since only Ti peaks were found. The amorphous TiO_2 transformed into the anatase crystal structure at an annealing temperature of 400°C, which is indicated by the diffraction peak (101) at 2è = 25.3°. The peak amplitudes of the anatase decreased with an increasing annealed temperature. However, the intensity of broadened peaks of the rutile polycrystalline increased with an increasing annealing temperature. These results show that the amorphous TiO_2 transformation into the crystalline anatase or rutile phase depends on the annealing temperature. When the temperature was increased, an increasing number of anatase and rutile peaks were obtained.

CONCLUSION

The physical characteristics and growth mechanisms of electrochemical anodization of TiO_2 nanotubes were investigated, including the effects of the growth mechanism, the contact angle, the nanoindentation load, and PL and XRD spectra. TiO_2 nanotubes after annealing over 400°C display super hydrophilic characteristics. The hardness array of TiO_2 nanotubes decreased with increasing annealed temperature before TiO_2 nanotubes transformed into the rutile polycrystalline. The average diameter and length of the TiO_2 nanotubes increased with increasing anodization time. The peak amplitudes of the anatase decreased with an increasing annealed temperature. The investigation of the physical characteristics plays a fundamental role for TiO_2 nanotube applications in detectors or sensors.

ACKNOWLEDGEMENTS

This work was partially supported by the National Science Council of Taiwan, under Grant No. NSC 100-2221-E-151 -018 -MY3.

REFERENCES

[1] X. Quan, S. Yang, X. Ruan, H. Zhao, *Environ. Sci. Technol.* 39, 3770 (2005).

[2] K. Shankar, G. K. Mor, H. E. Prakasam, S. Yoriya, M. Paulose, O. K. Varghese, C. K. Grimes, *Nanotechnology* 18, 065707 (2007).

[3] S. H. Oh, R. R. Finones, C. Daraio, L. H. Chen, S. H. Jin, *Biomaterials*. 26, 4938 (2005).

[4] R. Wang, K. Hashimoto, A. Fujishima, M. Chikuni, E. Kojima, A. Kitamura, *Nature* 388, 431 (1997).

[5] Y. Y. Song, P. Schmuki, *Electrochem. Commun.* 12, 579 (2010).

[6] J. C. Yu, J. Lin, D. Lo, S. K. Lam, *Langmuir*. 16, 7304 (2000).

[7] J. Shi, J. Chen, Z. Feng, T. Chen, Y. Lian, Z. Wang, C. Li, *J. Phys. Chem. C*. 111, 693 (2007).

[8] C. Anderson, A. J. Bard, *J. Phys. Chem.* 99, 9882 (1995).

[9] V. G. Bessergenev, I. V. Khmelinskii, R. J. F. Pereira, V. V. Krisuk, A. E. Turgambaeva, I. K. Igumenov, *Vacuum*. 64, 275 (2002).
[10] J. Choi, R. B. Wehrspohn, J. Lee, U. Gosele, *Electrochim. Acta*. 49, 2645 (2004).
[11] R. Beranek, H. Hildebrand, P. Schmuki, *Electrochem. Solid-State Lett.*, 6, B12 (2003).
[12] S. Sreekantan, Z. Lockman, R. Hazan, M. Tasbihi, L. K. Tong, A.R. Mohamed, *J. Alloy. Compounds* 485, 478 (2009).
[13] S. C. Han, J. M. Doh, J. K. Yoon, G. H. Kim, J. Y. Byun, S. H. Han, K. T. Hong, S. I. Kwun, *Met. Mater. Int.* 15, 493 (2009).
[14] S. Q. Li, G. M. Zhang, D. Z. Guo, L. G. Yu, W. Zhang, *J. Phys. Chem. C*. 113, 12759 (2009).
[15] S. H. Kang, H. S. Kim, J. Y. Kim, Y. E. Sung, *Nanotechnology* 20, 355307 (2009).
[16] G. K. Mor, O. K. Varghese, M. Paulose, K. Shankar, C. A. Grimes, *Sol. Energy Mater. Sol. Cells* 90, 2011 (2006).
[17] N. Todorova, T. Giannakopoulou, T. Vaimakis, C. Trapalis, *Mater. Sci. Eng. B*. 152, 50 (2008).
[18] G. Kontos, A. I. Kontos, D. S. Tsoukleris, V. Likodimos, J. Kunze, P. Schmuki, P. Falaras, *Nanotechnology*, 20, 045603 (2009).
[19] M. Paulose, H. E. Prakasam,O. K. Varghese, L. Peng, K. C. Popat, G. K. Mor, T. A. Desai, C. A. Grimes, *J. Phys. Chem. C*. 111, 14992 (2007).
[20] D. Wang, Y. Liu, B. Yu, F. Zhou, W. Liu, *Chem. Mater.* 21, 1198 (2009).
[21] Q. Ye, P. Y. Liu, Z. F. Tang, L. Zhai, *Vacuum* 81, 627 (2007).
[22] C. L. Liu, T. H. Fang, J. F. Lin, *Mater. Sci. Eng. A* 452, 135 (2007).
[23] W. Y. Chang, T. H. Fang, Y. C. Lin, *Appl. Phys. A* 92, 693 (2008).
[24] J. S. Chen, J. G. Duh, *Surf. Coat. Technol.* 139, 6 (2001).
[25] W. Y. Chang, T. H. Fang, Y. C. Lin, *J. Polym. Sci. B Polym. Phys.* 46, 949 (2008).
[26] W. Y. Chang, T. H. Fang, Y. C. Lin, *Applied Physics A* 92, 693 (2008).
[27] T. H. Fang, W. Y. Chang, J. J. Huang, *Acta Materialia* 57, 3341 (2009).
[28] H. Yuwono, Y. Zhang, J. Wang, X. H. Zhang, H. M. Fan, W. Ji, *Chem. Mater.* 18, 5876 (2006).
[29] L. V. Saraf, S. I. Patil, S. B. Ogale, S. R. Sainkar,S. T. Kshisarger, *Int. J. Mod. Phys.* 12, 2635 (1998).
[30] L. Forss, M. Schubnell, *Appl. Phys B*. 56, 363 (1993).

In: Nanostructures
Editor: Yu Dong

ISBN: 978-1-62618-081-9

Chapter 7

FILTRATION, SEPARATION AND GAS-PHASE PROCESSING OF NANOPARTICLES AND NANOMATERIALS

Benjamin J. Mullins*

Fluid Dynamics Research Group, Curtin Health Innovation Research Institute, Curtin University, Australia

ABSTRACT

Separation, size classification and collection are vital process operations for the production and preparation of high quality nanoparticles and nanomaterials. Additionally, the removal of undesired nanoparticles from gases and liquids is vital for the protection of humans and the environment. Such materials present opportunities and challenges, due to their small size and therefore extremely high diffusivity in fluids. Recent research has comprehensively studied the filtration and separation of nanoparticles down to sizes as low as 1 nm. The influence of morphology, surface charge state and temperature on filtration, have been thoroughly examined. Equations to predict capture or separation efficiencies were developed and validated, and have been presented below. Furthermore, the processing, size classification, agglomeration, de-agglomeration, charging and sintering of nanoparticles in the gas phase has been comprehensively studied. Such research is also presented in this chapter. Finally, current research into the development of nanofibre filter media has also been reviewed.

INTRODUCTION

Filtration is the single most efficient method for the removal of solid or liquid aerosols from gas streams. Current filtration technology is capable of complete or near complete removal of particulates, although the capture efficiency of filters is characteristically reduced in the 100-500 nm size region. However, compared to alternative technologies, such as

---

* Email: B.Mullins@curtin.edu.au.

impactors, cyclones and electrostatic separators, their efficiency is typically greater, even in such size ranges.

Nanofiltration, or the capture of nanometre-sizes particles, has developed into a separate branch of filtration, coincident with the development of nanotechnology. Nanofiltration refers to the filtration of particles <300 nm in size. Indeed, nanofiltration (particularly sub 20 nm) has received significant attention in the last few years, prompted largely (in the Author's case) by a conference paper by Prof. Albert Podgórski [1], who unfortunately passed away recently. This chapter is therefore dedicated to Prof. Podgórski.

Filters function almost entirely through physical/mechanical mechanisms [2]. Therefore they are suited to the collection of any type of nanoparticles or nanomaterials, be they metal or metal oxides, carbonaceous particles or nanotubes, polymers, silicates, or ceramics, primary (spherical or cubic) particles or fibres/tubes, They can also collect liquid "nanoparticles" produced by atomisation processes [3] or electrosprays [4-5]. However, in this case the filters will serve to coalesce liquid particles to form droplets, on the order of hundreds of microns or larger [6-7]. Nanoparticle size, aspect ratio, density (or agglomerate bulk) density, and to a lesser extent charge state, are the only factors which influence filtration performance.

Particles in the 'nano' size range are predominantly collected in filters by diffusional or Brownian forces [2]. However, it is generally accepted that as particles decrease in size they will - at some point - begin to behave more like (fluid) molecules and less like particles, with the resulting effect of a decrease in adhesion probability when coming into contact with surfaces such as a filter fibre. This effect has been termed "thermal rebound" or "diffusional bounce" [8-9] since, unlike particles, gas molecules do not generally adhere to each other, nor are they captured by filters.

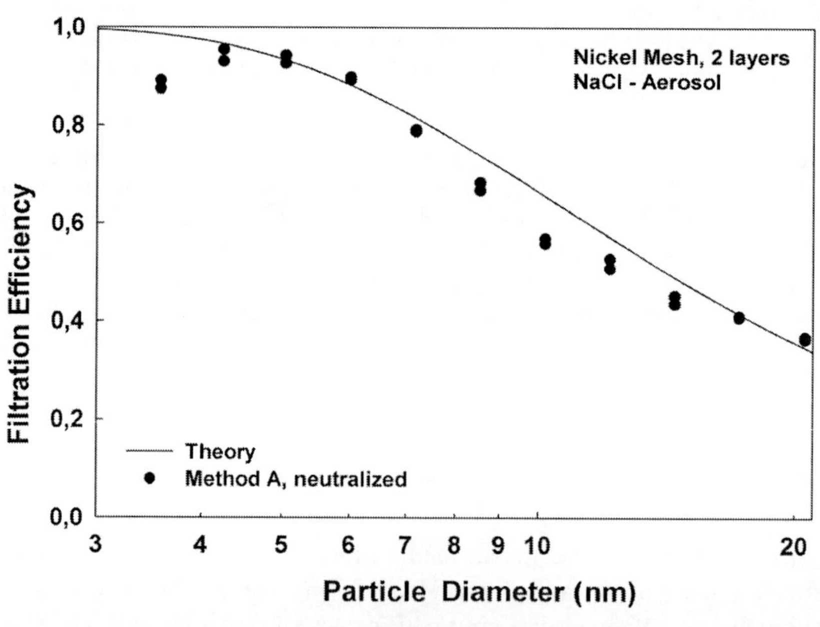

Figure x.1. Experimental artefact which many authors have mistaken for "thermal rebound" of nanoparticles. Filtration efficiency measurements using more accurate equipment will result in data which does not deviate systematically from the theoretical single fibre efficiency curve (see Eqn. (9) below). After [10].

However, there exists considerable conjecture over the point at which such rebound effects begin to occur, as it is closely associated with particle adhesion energy and other material properties. Studying the onset and magnitude of this diffusional or thermal bounce effect is very important when attempting to design a highly efficient filter for nano-sized particles.

Wang and Kasper [9] produced a seminal theoretical work, which predicted that below 10 nm, thermal rebound would occur, resulting in a decrease in filter capture efficiency. However, at the time measurements in such a size range were difficult using current aerosol technology. A number of studies, e.g. [11-13] were then conducted which attempted to measure thermal rebound, with conflicting results – some detected none, others detected thermal rebound below ~5 nm. Podgorsky and Balazy [1], then published a paper which purported to show measurement of thermal rebound commencing at 30 nm, with supporting (empirically fitted) theory. This paper prompted a reinvigoration of thermal rebound research, coupled with recent advances in nano- and aerosol technology. Heim et al. [10] produced a comprehensive study on nanoparticle filtration between 4.5 and 20 nm, which explained a number of sources of error in previous measurements (see Figure x.1), and conclusively showed no thermal rebound down to 2.5 nm. Heim et al. [14] re-analysed a previous study which had supposedly detected thermal rebound, proving the analysis of the data had been in error and the results actually showed no thermal rebound, down to 1.3 nm. Several studies followed, [15-23] which conducted experiments using a range of particle types, particle sizes, filter and mesh types and (elevated) temperatures. These studies invariably did not detect thermal rebound, when robust methods were used. A comprehensive treatment of the previous 25 years of thermal rebound research and new theory, has recently been published [24] which suggests that any thermal rebound could only occur in the particle size range <1 nm, (likely <<1 nm), even at elevated temperatures. Such a size range is below the realms of nanoparticles and nanotechnology, as it now approaches single molecules and atoms. Heim et al. [15] have also now measured the filtration of particles down to ~1 nm using molecular standards, and detected no thermal rebound. We can therefore conclude, after more than a quarter of a decade of research, that conventional filtration theory is sufficient for the collection of all relevant nanoparticles on clean filters, under all filtration conditions. Nevertheless, more research is required on the long-term loading of filters with nanoparticles.

Types of Filters

There are a wide range of filter types that can be used for the removal of contaminants from gas or liquid streams. These can be separated into two broad categories: surface filters and depth filters [2]. Figure x.2 shows a schematic of both types of filter. As the name suggests, surface filters collect contaminants on the upstream face of the filter, while depth filters collect fibres throughout the filter. The principal types of filter are: membrane filters, granular filters, foam filters and fibrous filters. Membrane filters [25] are simply filters with suitably small pore sizes, which operate by sieving the particles from the air. Many nano-fibre filters can also be considered membrane filters, however these will be discussed below. Whilst conventional membrane filters are common in water treatment and aerosol sampling equipment, they are seldom used for control of aerosols, as they clog readily and are difficult or impossible to regenerate. The processes involved in this type of filtration have been extensively studied [2]. Granular filters or packed bed filters consist of packed beds of isometric particles which work on the principle of depth filtration. Foam filters are most easily categorised as non-fibrous depth filters, they are constructed from porous foam and appear as a matrix of bubbles within a constituent material. The interesting feature of this type of filter is that the packing density (relative proportion of void space to constituent material) of the filter does not vary much between classes, however the mean bubble size can increase by a factor of 10 [26]. This type of filter, although commonly used for automotive and other similar aerosol cleaning applications, principally due to low cost and ease of handling, is seldom used industrially, principally due to low tensile strength, poor performance and high maintenance requirements.

Fibrous filters, the type to be discussed in most detail here, are usually made from fine textile fibres (fibre diameter <100 μm). The fibres are usually manufactured from materials such as polyester, polypropylene, Teflon, glass fibre, polyethylene terepthallate (PET), metals and specially developed patented fabrics such as Nomex® (DuPont). Asbestos was formerly used as a filter material in high temperature applications, however this has been discontinued for obvious health reasons. All fibrous filters generally operate as depth filters for the initial part of their life, however as filtration continues and the mass of captured particles increases, the filtration regime changes to surface filtration, and further influent particles are captured by the 'cake' of collected particles on the surface of the filter, rather than the fibres themselves. This switch from depth to surface filtration is accompanied by an increase in filtration efficiency, however at the cost of increased pressure drop across the filter.

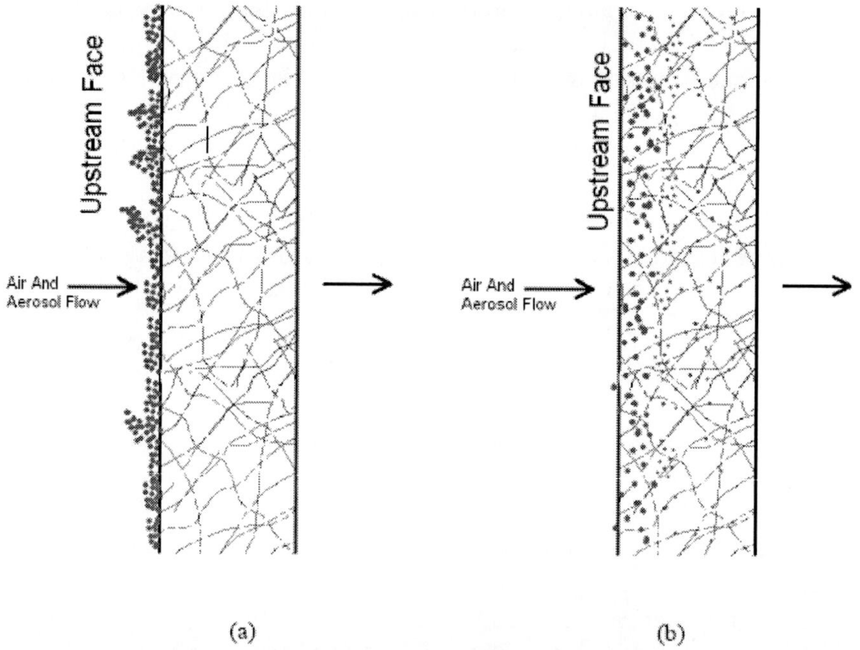

Figure x.2. Representation of Aerosol Particle Collection By Fibrous Filters. (a) surface filtration, (b) depth filtration using a homogeneous fibrous filter (uniform fibre size and packing density from front-rear faces).

Figure x.2 shows a schematic of the particle collection in two fibrous filters (one a depth filter and the other a surface filter). Fibrous filters designed to act as surface filters will typically possess a higher packing density (α – also termed solidity) on the order of 10-30%, whereas depth filters will typically have α<5% [26].

Microscopically, most fibre filters are random arrangements of fibres that have no particular orientation, and with the fibres having varying lengths, and usually a distribution of fibre thicknesses. To increase their strength, most commercially available fibrous filters are either woven, felted, sintered, or contain a (polymer) binder [26].

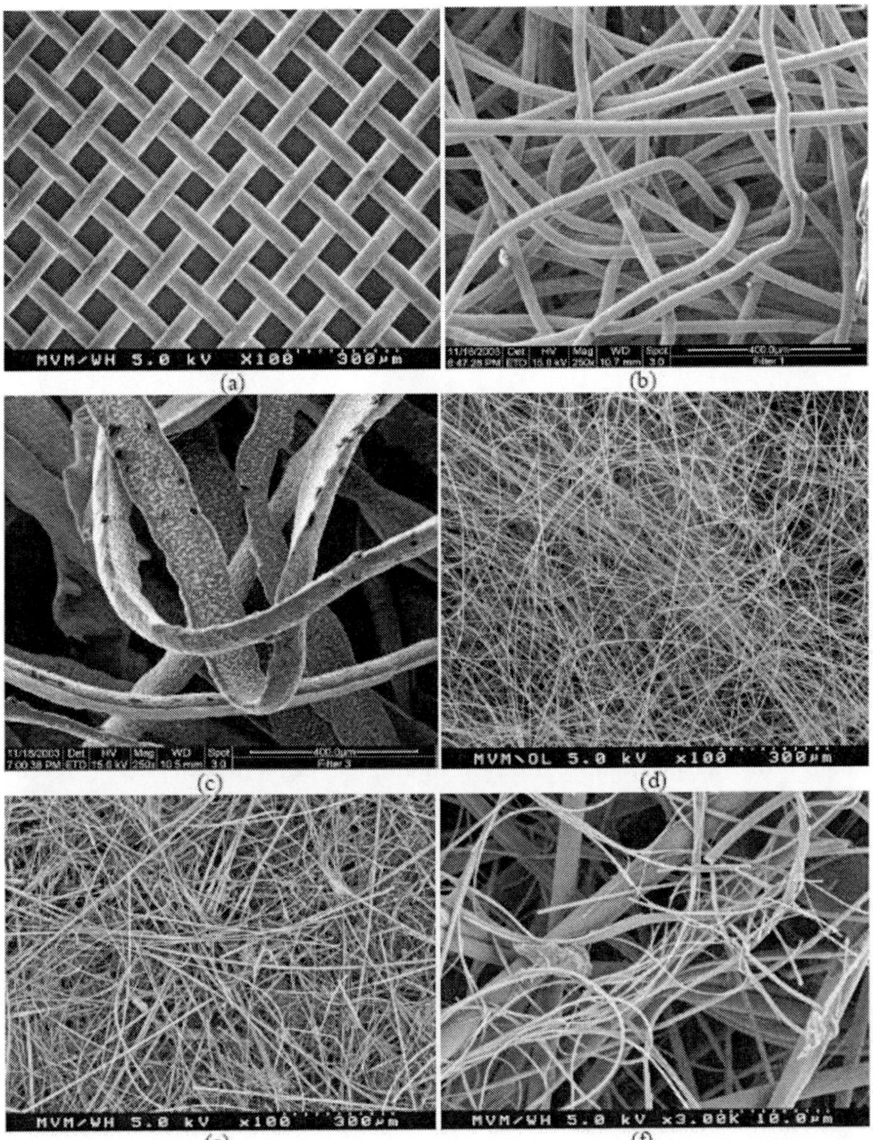

Figure x.3. Spectrum of Fibrous Filters used for Nanofiltration. (a) Mesh 'filter' (nickel); (b) Typical (polyester) felted depth filter; (c) Stainless steel depth filter with crescent-shaped fibres; (d) Fine stainless steel fibre filter; (e) Bimodal glass micro-fibre media; (f) Enlargement of (e) showing micro/nano fibres and polymer binder.

Relatively recent advances in filter technology include the development of ceramic filters and nanofibre filters [27], which are capable of extremely high filtration efficiencies (99.9995% and higher) and, together with certain classes of fibrous filters, are suitable for use as high efficiency particulate air (HEPA - 99.995%) and ultra low penetration air (ULPA - 99.9995%). Another key advantage of ceramic filters is their ability to withstand extremely high temperature applications, and as such can be used for processes where asbestos fibres were previously utilised. Another option for such high temperature applications is the use of filters constructed of stainless steel wires or fibres [28-29].

Figure x.3 below shows scanning electron micrographs of the range of filter structures, from highly ordered meshes such as those used in diffusion batteries and industrial sieves, to (near) randomly ordered filters such as needle felts.

All of the above are depth filters, excluding (a) which is a single layer so cannot be classified. Filter (a) would commonly be utilised in diffusion batteries or industrial (liquid) sieving, Filters (b, e, f) are be used for a range of (low temperature) filtration applications, whereas filters (c) and (d) are intended for high temperature and moderately high temperature applications respectively.

Manufacture and Mechanical Properties of Filters

The commercial manufacture and testing of fibres and filters is a highly specialised field, conducted mainly by paper and textile manufactuers, as similar methods are used for producing both. Most filter production operations start with staple fibres, which are then matted together by either wet or dry processes. Dry processes include needle felting (Figure x.3(b)), weaving (Figure x.3(a)), compressed air blowing (Figure x.3(d)) and electrospinning [30-33] (noting that electrospinning simultaneously manufactures the fibres from solution). Needle felting is a particular variation of the traditional (wet) felting process as the latter is ineffective with synthetic fibres [26]. Needling mats the fibres together, to form a (seemingly) random arrangement of fibres into a substance with sufficient tensile strength (caused by intertwined fibres) for its intended application.Wet processes include traditional (wet) felting, and the filtration of a suspension of fibres in liquid, leaving a matted filter (Figure x.3(e-f)). Raynor and Leith [34] describe a laboratory scale version of the latter method, which utilises as dispersant to ensure the fibres do not agglomerate, and which can be performed with or without binder to increase structural integrity of the filter.

Filters must of course possess sufficient mechanical strength to enable them to be fit for the task to which they are applied, in addition to resistance to the range of chemicals to which they may be exposed. However these requirements vary greatly depending on application. In general, felted polymers e.g. (Figure x.3(b)) will have a stress resistance on the order of 1-10 kN, using standard ball-stress testing methods. Cellulose or glass-fibre filter media will have a much lower stress resistance, making them unsuited to many applications (such as regenerable/cleanable filters). Metal media typically can resist the highest stresses, and along with polymer media tends to have good fatigue resistance. Fibre diameter is usually a strong determinant of stress resistance though, as both are usually directly correlated. There are a range of standards and requirements for filters, which are largely dependant on the application. The filters which are generally subjected to the highest stresses and highest extremes of temperature are generally surface filters for industrial gas cleaning. Such filters are expected to have long servive lives, and are cleaned repeatedly throughout their life. The cleaning is normally conducted using a "pulse" of compressed air, which obviously imparts a rapid stress on the filter. Competing standards exist for the testing of such filters, including VDI 3926, ASTM D6830-02, JIS Z 8909-1 and GB 12625 [35].

Filtration Theory

The fundamental equations for flow through a filter are the Navier-Stokes equations. The Navier-Stokes equations are a fundamental series of nonlinear partial differential equations which relate the forces acting on a fluid to the rate of change of momentum. They are of the form,

$$\frac{Du_i}{Dt} = F_i - \frac{1}{\rho}\frac{\partial p}{\partial x_i} + \frac{\partial}{\partial x_j}\left[2\nu\left(e_{ij} - \frac{1}{3}e_{ij}\delta_{ij}\right)\right], \tag{1}$$

where $i, j = 1, 2, 3$ for three-dimensional flow, δ_{ij} is the Kronecker delta tensor, F_i is the external force and $\delta p/\delta x_i$ are the pressure gradient components. Also,

$$\frac{D}{Dt} = \frac{\partial}{\partial t} + u_i\nabla, \tag{2}$$

so the acceleration of a fluid element may be written as Du_i/Dt where $u_i(x_i, t)$. The viscous stress tensor is of the form,

$$e_{ij} = \frac{1}{2}\left(\frac{\partial u_i}{\partial x_j} + \frac{\partial u_j}{\partial x_i}\right) \quad \text{and} \quad e_{ij} = \nabla u_i \tag{3}$$

Two dimensional steady state flow can be described by the stream function, ψ, which can be defined by its x and y component vectors,

$$U_x = \frac{\partial \psi}{\partial y}; \qquad U_y = -\frac{\partial \psi}{\partial x}, \tag{4}$$

in Cartesian co-ordinates. Furthermore, for any flow that can be described in terms of (5), the air velocity obeys the continuity equation,

$$\nabla \cdot \underline{U} = \frac{\partial U_x}{\partial x} + \frac{\partial U_y}{\partial y} + \left(\frac{\partial U_r}{\partial r} + \frac{U_r}{r} + \frac{1}{r}\frac{\partial U_\theta}{\partial \theta}\right) = 0. \tag{5}$$

The equations were first termed the Navier-Stokes equations by Lamb, who attributed the equations to Navier, Poisson, Saint-Venant and Stokes, who dropped the molecular action component [36-39].

Due to the complexity of flow through filters, most work has used two dimensional representations of filters consisting of cylindrical fibres, either uniformly or randomly arranged, and occasionally with differing fibre diameter.

Most air filters operate within the laminar regime, and are usually considered to be in the Stokes flow regime (Reynolds number (Re) close to zero so inertia can be neglected) [26]. The Reynolds number for the flow through a filter is represented by equation, which is defined as a ratio of inertial and viscous forces of flow,

$$\mathrm{Re} = \frac{2a\rho_g U}{\eta}, \tag{6}$$

where a is the fibre radius (2 fibre diameters is considered to be the greatest flow obstruction in a filter), U is the flow velocity, ρ_g is the gas viscosity and η is the dynamic viscosity of the gas.

The drag force F_D on a particle is described by the following equation,

$$F_D = 3\pi\mu V d_p = C_D \frac{\pi}{8}\rho_g, \qquad \text{for Re} < 1 \tag{7}$$

where V is the fluid velocity, d_p is the particle diameter and the C_D is the drag coefficient and μ is the viscosity,

$$C_D = \frac{24\mu}{\rho_g V d_p} = \frac{24}{\mathrm{Re}}. \tag{8}$$

Single Fibre Efficiency Theory

The most fundamental model for filtration efficiency calculations is the single fibre efficiency model. The model is so termed because it views the void space inside the filter as so great compared to the fibre diameter, that the only interactions that need be considered are those between the particle and a single fibre. The single-fibre efficiency (SFE) theory is a semi-empirical model which has been proven accurate for predicting initial (clean / new filter) collection efficiency in well ordered fibrous filters, at low airflow velocities.

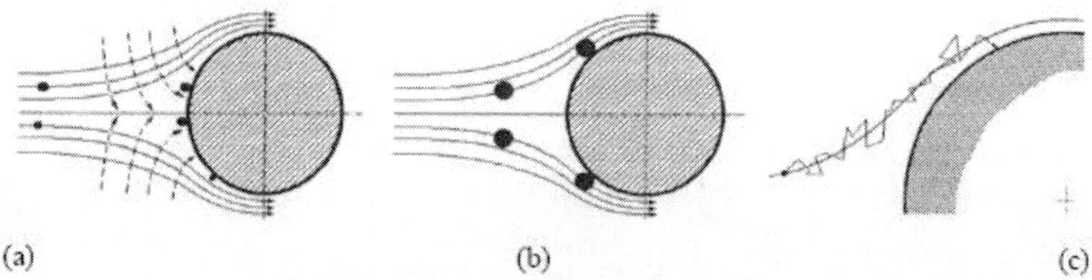

Figure x.4. Predominant Single-Fibre Capture Mechanisms – Inertia (a), Interception (b) and Diffusion (c).

Figure x.4 shows the predominant mechanical single fibre collection mechanisms. Non-mechanical mechanisms will be dealt with below. For particles in the range 1-100 nm, diffusion is by far the dominant capture mechanism.

The single-fibre efficiency E_Σ is the sum of each mechanism (interception, inertial impaction and diffusion), and represents a dimensionless particle deposition rate on a unit length of fibre, thus describing the efficiency, with which a fibre removes particles from an aerosol stream. The efficiency η for the whole filter can be calculated using η_Σ, the packing density (α), the fibre diameter (d_f), and the filter thickness (t).

$$\eta = 1 - \exp\left(\frac{-4 * \alpha * \eta_\Sigma * t}{\pi * d_f}\right) \tag{9}$$

The single-fibre efficiency η_Σ can be calculated using equation (10). Since only interception, inertial impaction and diffusion are important here, the equation for η_Σ can be written as,

$$\eta_\Sigma = 1 - (1 - \eta_R) * (1 - \eta_t) * (1 - \eta_D)\,. \tag{10}$$

For interception it is assumed that the particles follow the velocity field i.e. the streamline of the gas perfectly. If this streamline comes within a particle radius of the surface of a fibre, the particle is collected.

The single-fibre efficiency for interception η_R [2] can be expressed of the form,

$$\eta_R = \frac{(1 - \alpha) * R^2}{Ku * (1 + R)}, \tag{11}$$

where R is a dimensionless parameter, given as the ratio of particle diameter (d_p) to d_f,

$$R = \frac{d_p}{d_f}, \tag{12}$$

and Ku is the Kuwabara hydrodynamic factor,

$$Ku = -\frac{\ln \alpha}{2} - \frac{3}{4} + \alpha - \frac{\alpha^2}{4}\,. \tag{13}$$

Inertial impaction is when a particle is unable to follow the streamlines of the gas as it passes a flow obstruction (fibre) and therefore strikes the fibre. The classical equation for the single-fibre collection by inertial impaction η_t is,

$$\eta_t = \frac{Stk * J}{2 * Ku^2} \tag{14}$$

The above equation shows that this collection mechanism is governed by the Stokes number Stk,

$$Stk = \frac{\rho_p * d_p^{\,2} * C_c * U_0}{18 * \eta * d_f}. \tag{15}$$

Hinds [2] gives the Cunningham correction factor C_c as,

$$C_c = 1 + \frac{2.52 * \lambda}{d}. \tag{16}$$

The parameter J is,

$$J = (29.6 - 28 * \alpha^{0.62}) * R^2 - 27.5 * R^{2.8}. \text{ If R} < 0.4 \tag{17}$$

Naturally impaction is the dominant collection mechanism for large particles. Nevertheless interception is also significant for large particles, and also for larger nanoparticles (80 nm +) in some flow regimes.
The single-fibre collection due to diffusion E_D is given by the following equation,

$$\eta_D = 2.7 * Pe^{\frac{-2}{3}}. \tag{18}$$

The Peclet number Pe is defined as,

$$Pe = \frac{d_f * U_0}{D}, \tag{19}$$

where the diffusion coefficient D is,

$$D = \frac{k_b * T * C_c}{3 * \pi * \eta * d_p}. \tag{20}$$

However, a separate theory for screen-type diffusion batteries (e.g. structures such as Figure x.2(a)) has been derived from the single-fibre efficiency theory [40], which gives the penetration (P) of a series of n screens as,

$$P = \exp\left[-nS\eta\right], \tag{21}$$

where S is the screen parameter, defined as,

$$S = \frac{4\alpha h}{\pi(1-\alpha)d_f}, \tag{22}$$

where h and d_f are the screen thickness, and wire/fibre diameter, respectively. The diffusional collection (η_D) efficiency is then calculated as per equation (18). It should be noted that Hinds [2] gives the constant 2.0 in equation (18), rather than the 2.7 given by Cheng and Yeh [40], which was originally derived by Kirsch and Fuchs [41] for fan shaped arrays of filter fibres, and later confirmed analytically by Cheng et al. [42]. In the Author's experience 2.0 gives much more accurate results when simulating nanoparticle collection on filters.

From the above equations it can be seen that η_D increases as particle diameter and thus Peclet number decrease. η_D is the only collection mechanism which increases while d_p decreases. Therefore, for nanoparticle filtration, η_D is typically orders of magnitude more significant than other collection mechanisms.

In some cases, other filtration mechanisms may be significant. These include, the interaction of diffusion and interception (η_{DR}), gravitational settling (η_G), electrostatic attraction or repulsion (η_Q). The equations for these terms will not be presented here as they are rarely significant for nanoparticle filtration [2]. For the filtration of colloidal dispersions of nanoparticles in liquids, the electrokinetic potential (zeta potential) may also be of significance. As demonstrated by the equations above, the fluid viscosity, particle and fibre diameter, filter solidity and particle density are the factors which govern the (mechanical) collection efficiency of a filter for nanoparticles. Figure x.5 shows the collection efficiency of an identical filter for 1-100 nm particles in both air and water at 20°C. It should be noted that all of the above assumes that the filter is homogenous. It has been found that in real filters, inhomogeneity is a factor which commonly influences filtration efficiency. Work in this area is ongoing [43-44], and whilst filter manufacturers attempt to make their filters as homogeneous as possible, it remains an issue.

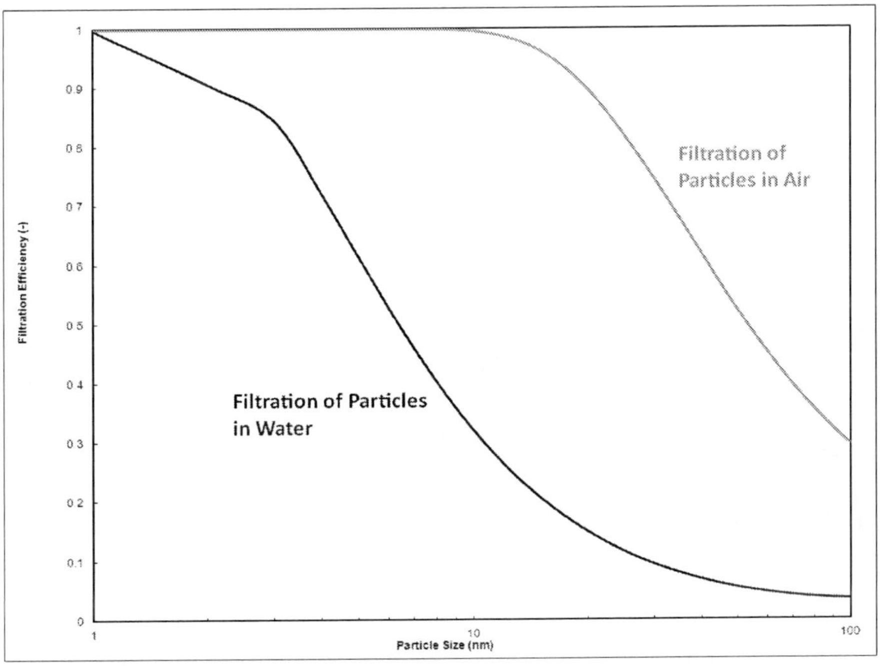

Figure x.5. Collection of nanoparticles via mechanical means on an identical filter in air and water (d_f=1000 nm, α= 0.005, h=0.5 mm U_o= 5 cms^{-1}).

Fibre Loading and the Influence of Fibre Orientation

Figure x.6(a) shows the typical dendrite structures which particles form when they collect on filter fibres. These structures, and the particle cake which forms on surface filters, has been shown to approach a maximum packing density/solidity of 30% [28, 45]. It has been found that for filters collecting solid particles, capture efficiency increases rapidly as the filter loads, however will eventually reach a constant (maximum) value.

Figure x.6(b) shows the two extremes of fibre loading, deposition of particles on the upstream side of each fibre in the inertial case, and uniform deposition in the diffusional case. The influence of particle bounce on the shape of the inertial deposit has been neglected in this case, and will be discussed below [28, 45]. Nanofiltration would almost invariably conform to the diffusional case, as diffusion is the dominant capture mechanism for nanoparticles, unless they are filtered at very high velocities.

We can conclude from the above, that as nanoparticles would tend to coat fibres uniformly, the performance of nanoparticle filters is less susceptible to variations in fibre orientation and fibre shape than filters collecting larger particles. Several authors have examined the orientation of fibres in fibrous filters [43-44, 46], finding that even supposedly random media such as needle-felts possess a preferred fibre orientation. Studies have also examined the performance of filters using non-cylindrical fibres [47-48] (such as Figure x.3(c)) or bimodal fibre distributions [49-50] (such as Figure x.3(e-f)), developing useful equations for these specific filter types. However, in the nano-regime, such filters tend not to differ significantly from traditional filter types (monomodal cylindrical fibres) with respect to performance.

It can therefore be seen that in most nanoparticle filtration applications, fibre shape/uniformity and fibre orientation is less important than filter homogeneity. This can otherwise be termed as the average (equivalent) pore size of filters, which can be measured directly for thin media or indirectly for thicker media [51].

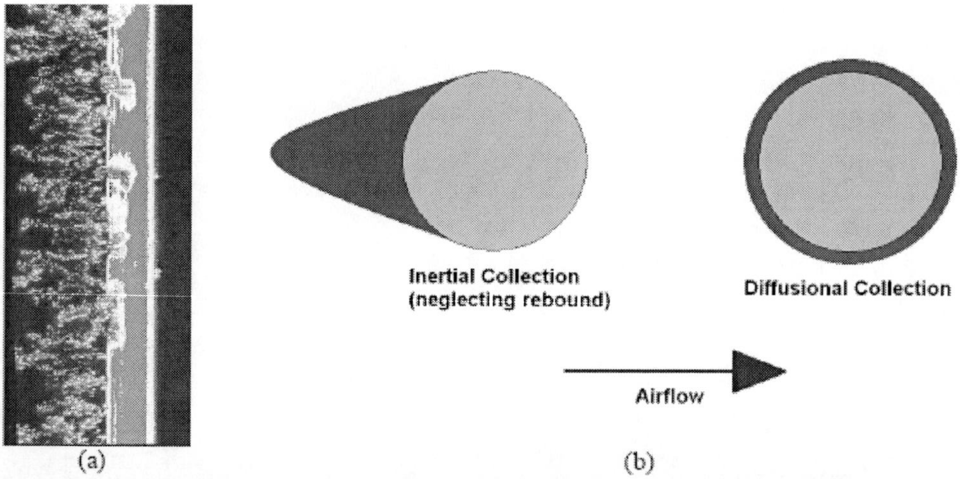

Figure x.6. Collection of Particles on Single Fibres. (a) Deposition of particles on a (~20 micron) filter fibre under the inertial filtration regime, noting the typical dendritic structures which are formed; (b) A cross section through filter fibres and particle deposits, for purely inertial deposition (left) and for diffusional deposition (right).

Meshes such as x.3(a) are evidently ideally homogeneous (and also completely uniform). However, for filters collecting liquid micro/nano-particles, orientation of the fibres becomes critical, to ensure drainage of the coalesced liquid [52].

It was found that the optimum fibre angle θ (with respect to a plane in the airflow direction) was,

$$\theta = \tan^{-1}\left(\frac{F_d}{F_g}\right) + \frac{3\pi}{2},$$

where F_d represents the drag force on the coalesced liquid, and is of the same form as (7) and F_g gravitational force. Both forces are of course dependant on the mean droplet size which the particles coalesce to form.

Pressure Drop

Figure x.7 below shows the typical (experimental) pressure drop (ΔP) evolution for a surface filter. It will be noted however, that the filter initially undergoes a period of depth filtration, characterised by a very low rate of increase in ΔP over time. The filter eventually transitions to surface filtration, where ΔP increases much more rapidly, however linearly. Depth filters experience a much longer period of depth filtration before transitioning to surface filtration.

Depth filters will usually be replaced at or before reaching the transition point. Therefore in industry, differential pressure sensors are utilised to determine when depth filters should be replaced and also when surface filters should be regenerated.

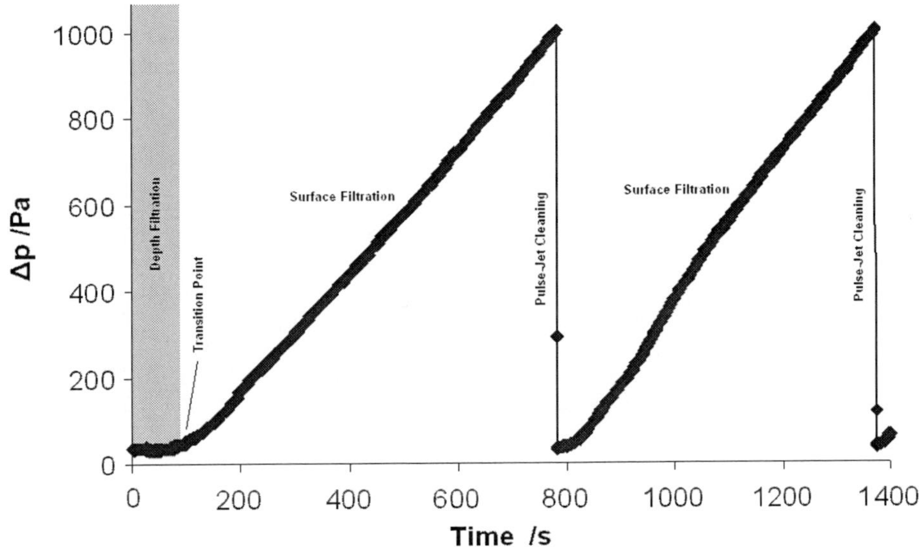

Figure x.7. Filter Pressure Drop Evolution Over Time. This filter is a polyester surface filter regenerated (cleaned) when the pressure drop (ΔP) reaches 1000 Pa.

The pressure drop in a clean filter (operating in depth filtration mode) can be calculated as,

$$\Delta P = \frac{4\mu_g cUH}{X^2\left(c - \frac{3}{4} - \frac{1}{2}\ln c - \frac{c^2}{4}\right)}, \tag{23}$$

where X is the dimensionless ratio of the particle diameter to the fibre diameter ($X = d_p/d_f$)

The application of Darcy's law is the standard method of describing single phase fluid flow in disordered porous media. Darcy's law is of the form:

$$V = -\frac{k}{\mu}\frac{\Delta P}{L}, \tag{24}$$

where: k = global permeability, V = fluid velocity, L = length of the medium (filter) in the flow direction, ΔP = pressure drop across the system. It is stated [53] that Darcy's equation should only be used to describe viscous flow conditions, as other factors tend to dominate the flow at higher values of Re. It is further stated that experimental studies have found that the transition from the linear (Darcy) regime to the nonlinear does not happen at a defined value of Re, however is very gradual [54]. It is argued that inertial (convective) forces should be incorporated into calculations in the upper end of the laminar regime. Andrade et al. [53] have produced numerical approximations for the 2D Navier-Stokes equations and continuity equations for low porosity structures in order to determine and quantify the deviation from Darcy's law. They have shown that the nonlinear deviation at high velocities can be explained in terms of inertial contribution to laminar fluid flow, meaning that the system can be successfully described/modelled without considering turbulence. It is stated that nonlinear changes to the streamlines orthogonal to the flow direction can be described by the Forchheimer equation:

$$-\frac{\Delta P}{L} = \alpha_f \mu V + \beta\rho V^2, \tag{25}$$

where: α_f is the reciprocal permeability of the porous material and β is termed an "inertial parameter". The Forchheimer equation can be re-arranged into the form:

$$f = \frac{1}{\mathrm{Re'}} + 1, \tag{26}$$

where:

$$f \equiv -\frac{\Delta P}{L\beta\rho V^2}, \tag{27}$$

and

$$\text{Re}' \equiv \frac{\beta \rho V}{\alpha \mu}. \tag{28}$$

The authors [53] term this a "friction factor Reynolds number" which is stated to be universal and has been tested under a wide range of conditions. It is reported that the point of departure from linear (Darcy's law) to nonlinear flow is in the range $10^{-2} < \text{Re}' < 10^{-1}$. Thus the authors state that real flows can be successfully modelled under laminar conditions, using the nonlinearity corrections, regardless of whether real turbulence has been detected or not. However, it has been determined that the Forchheimer equation generally overestimates the computed values of the friction factor in the region $6.2 \times 10^{-2} < \text{Re}' < 1.8$, which limits the applicability of the modelling methodology.

Particle Bounce or Reentrainment

Currently, most filtration models assume complete capture of all particles contacting a fibre, however this is not necessarily the case. It is possible for particles to be re-entrained after capture, or strike the fibre but immediately bounce off again. However, as discussed in the introduction, the concept of thermal rebound or “diffusional bounce” has been thoroughly refuted for particles > 1 nm in size. This section deals with particle rebound at high particle inertia, which would only apply to nanoparticles travelling at very high velocities. The end of this section will, however deal with recent research on the influence of nanoparticle shape on capture probability.

When a particle strikes a fibre, there is a complex system of plastic and elastic deformation as the fibre absorbs the force imparted by the particle. The principles which govern the mechanics of particle bounce are quite complex. Many of the previous attempts to quantify and describe particle bounce have focused on aerosols contacting flat plates rather than fibres. Gillespie and Rideal [55] examined the impact and bounce of liquid (di-octylphthallate – DOP) particles on glass slides and attempted to quantify bounce by examining the traces left on the impact slide and droplets which fell onto another slide. Other experiments have been conducted using more advanced techniques such as laser Doppler velocimetry to measure the bounce of polystyrene latex (PSL) particles on polished quartz and stainless steel surfaces [56-58]. Techniques involving impaction of aerosols on flat plates to compare bounce properties have also been developed [59-60]. Other work has examined the effect of particle deformation on adhesion to a flat surface [61-62]. If particles are able to deform to absorb the impact force they will be less likely to bounce.

While such work is crucial for understanding of the general principles involved, and very useful for impactor design/research where the particles are collected on flat surfaces, it is not as useful in the fibrous filtration field. Since many of the above authors have emphasised that the impaction surface has a significant effect on particle bounce, then impact on fibres (which may be able to deform to absorb impact) will be different from that on a solid surface. Also, the flow field around a filter fibre will be completely different to the flow field across a flat plate.

There are classically two approaches to describe particle bounce. The first defines a critical velocity V_c, above which bounce will occur, of the form [26, 63],

$$V_c = \frac{\beta}{d_p} = \frac{1}{d_p}\frac{(1-e_{pl}^2)^{1/2}}{e_{pl}^2}\frac{A}{\pi x^2 (6 p_{pl}\rho)^{1/2}}, \tag{29}$$

where β is a constant for a particular impaction surface, d_p is the particle diameter, e_{pl} is the coefficient of restitution (for plastic deformation only), A is the Hamaker constant, p_{pl} is the microscopic yield pressure, ρ is the particle density, and x is the separation distance of the particle and surface. Hamaker constants are given in the literature for a limited number of elements and compounds, including PSL (6.37 x 10^{-13} ergs) [62]. However values for liquid aerosols such as DEHS are not available.

The other method involves the kinetic energy (KE_b) required for bounce to occur when a particle (d_p) collides with a surface [64],

$$KE_b = \frac{d_p A(1-e^2)}{2xe^2}, \tag{30}$$

where e is the coefficient of restitution (plastic and elastic deformation), which is equal to the ratio of the rebound velocity to the approach velocity. The value of e is reported to range from 0.73 to 0.81, although these values were typically derived using hard impaction surfaces such as glass and metal [65]. A and e depend only on the material of the particle and the surface. It is reported that these constants must be determined experimentally, as it is very difficult to determine them theoretically.

Some experimental results for irregularly shaped fly ash particles (mean diameter 140 nm) give the probability of bounce as [66],

$$P_b = 1 - 0.000224\ (KE)^{-0.233}, \tag{31}$$

where KE is expressed in Joules. These values, however will not be representative, since it is reported that spherical particles are much more likely to bounce than irregular particles [26]. One of the main factors governing adhesion of particles to fibres is the particle surface area, thus flakes are the most likely to adhere, hence least likely to bounce [67].

For liquid particles, the most significant force is the surface tension acting on the liquid droplet. As particle size diminishes, the surface tension force increases to the point where the particle is so rigid its behaviour is identical to that of solid particles. Liquid particles are better able to absorb energy than solid particles, principally through deformation [26]. It is reported that liquid particles are far more likely to break up on impact than solid particles. This is principally through shear when the particle impact is not normal to the fibre [55].

The continuing work of researchers such as [68] shows that there is still work to accomplish in describing simple aerosol/plate collision processes, which have yet to be adapted to complex processes inside air filters. To this end there has been very little work which has examined the practical effects of particle bounce in real filter systems, principally due to the complex particle and fibre interactions involved, making theoretical calculations difficult. Maus and Umhauer [69] have, however examined the fractional filtration efficiencies of biological aerosols and compared these to non-biological aerosols with equivalent aerodynamic diameter, such as PSL and limestone dust. Kasper et al. [28, 45] have recently published a comprehensive study measuring particle collection on single stainless steel fibres, and have been able to quantify rebound in terms of the shape of the dendrite deposits on the fibres.

There has been some research into the effect that coating filter fibres with a liquid has on particle adhesion/bounce. Walkenhorst [70] examined the effect of coating model wire filters with vegetable oils, mineral oils and Vaseline, and reported that the former two substances increased filtration efficiency (particle adhesion) while the latter did not. Mullins [71] found that a water coating on the filter fibres could negate particle bounce.

Recent research [72-74] has found that capture efficiencies differ between nanoparticles of different morphology. The work compared the collection efficiency of cubic, spherical and elongated nanoparticulate structures. It was concluded that the shape of the nanoparticle influences adhesion forces, and whether the particle rolls or tumbles on the fibre on adhesion. Whilst the work is highly interesting and relevant, it has not conclusively proven that other factors such as the location of elementary charges on the different particle morphologies, or the differences in Differential Mobility Analyser (DMA) classification of such particles (to condition the particles prior to challenging the filter) do not permit a more complete explanation of the results.

Filtration of Liquid Nano-"Particles"

It is evident from the previous sections that some aerosols will be present as liquid droplets, while others will be solid. This is the case even for nanoparticles – at least with respect to environmental exposures – however nanotechnological applications rarely utilize liquid nano-aerosols, with the exception of electrosprays or atomisers as mentioned previously.

Since liquids are able to coalesce on filter fibres, the physics and dynamics of "mist" or "coalescing" filters are completely different to that of filters collecting solid particles. This field will therefore not be considered in detail here, and readers are instead referred to other works [7, 34, 51-52, 75-86]. Of particular note, is a work [83] which utilized filters coated by tee-tree oil, to simultaneously capture and disinfect nano-sized virons, as well as larger bacterial cells/spores and fungal spores. This approach appears more effective than other methods which combine silver ions or UV light with filters to disinfect biological or pathogenic aerosols.

Nanofibres - Filtration and Use As Filter Fibres

It should be evident from the previous sections that filtration efficiency is inversely proportional to fibre diameter. Therefore, the use of nanofibres in filters has become increasingly common, as there is an obvious advantage to nanofibres, as shown in Figure x.8, which shows the filtration efficiency curves for (otherwise identical) filters, however one is composed of 1000 nm fibres, and the second of 100 nm fibres (all other parameters kept the same as Figure x.5).

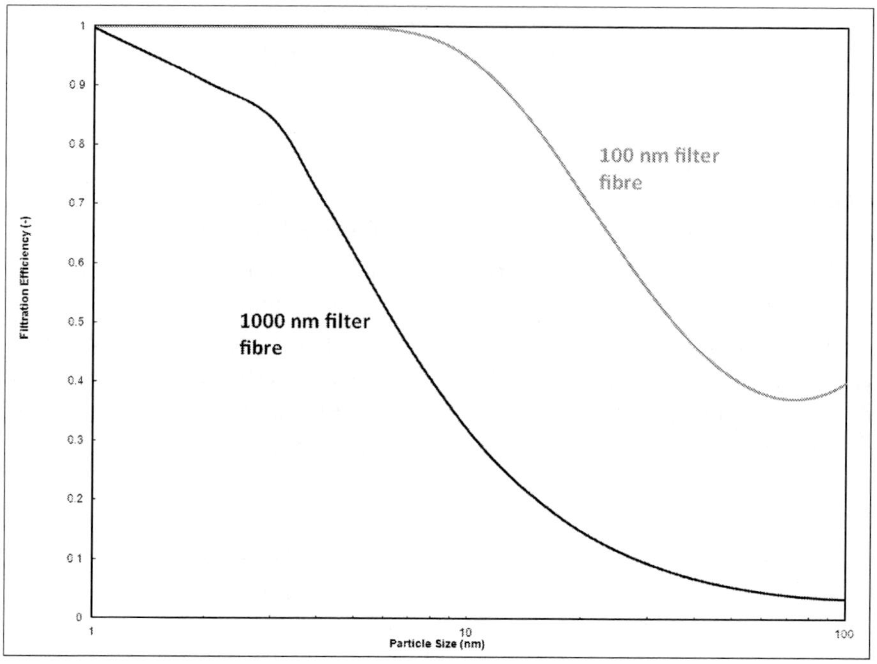

Figure x.8. The Influence of Nanofibres on Filter Capture Efficiency (for a water filter).

We can see from the above, that particle capture efficiency increases dramatically as fibre diameter decreases. However, constructing filters entirely from nanofibres is not so simple, and does not allow us to retain such properties as ideally as one would wish. Decreasing fibre diameter (d_f), increases packing density (α) of the overall filter, as the fibres pack together more closely. This results in unacceptably high pressure drop. Filters constructed entirely of nanofibres are also prohibitively expensive. Both of these features have meant that filter media manufacturers generally use micro-fibres as a substrate, and add nano-fibres either throughout the media, or (more commonly) as a surface layer on the upstream face, in order to achieve the desired filtration efficiency [25, 87-91].

However, the development of nanofibres for filtration media, as well as technical nano-"fibres" such as carbon nanotubes, has created concerns that a "new asbestos" may be created, though there is little evidence of this to date [92]. In general, fibres are an interesting aerodynamic and hydrodynamic case, as they generally behave like particles of the same cross sectional diameter. This means that fibres 100–500 nm diameter would be of greater concern then fibres <50 nm diameter, as the latter are much more readily captured by filters.

PARTICLE AND FILTER CHARGE

Particle (and filter) charge-state are other factors which can influence collection efficiency in filters, however this depends on a number of factors. Figure x.9 shows the result for a metal filter (mesh), located in a grounded metal chamber, collecting nanoparticles with a single elementary charge, vs uncharged particles (or those with a Boltzmann equilibrium charge distribution). It can be seen that a difference only appears below 8 nm, and even then remains of marginal significance. Particles with a higher elementary charge level would of course be likely to produce a greater deviation, however unless charged on purpose, nanoparticles of that size would be unlikely to retain a high number of elementary charges.

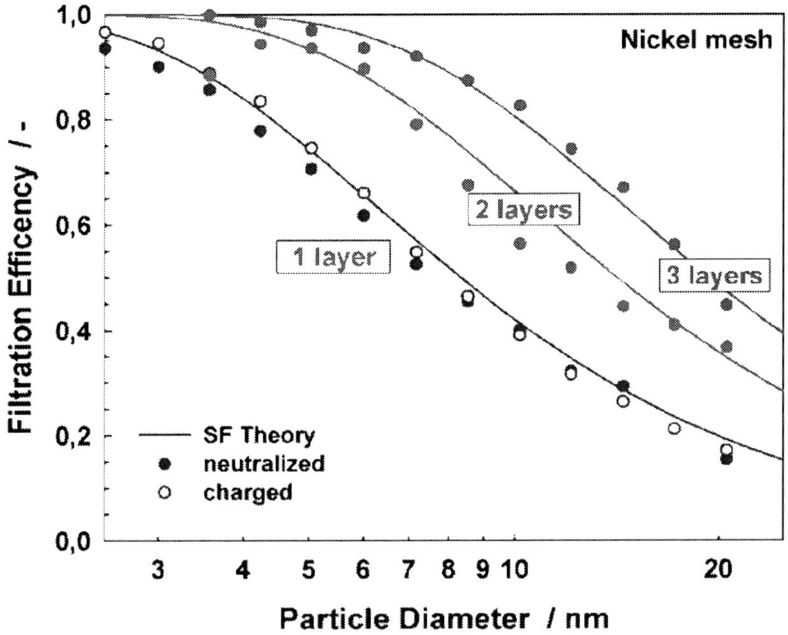

Figure x.9. Nanoparticle capture efficiency for layered conductive (metal) mesh, using uncharged and particles with a single elementary charge per particle. After [10].

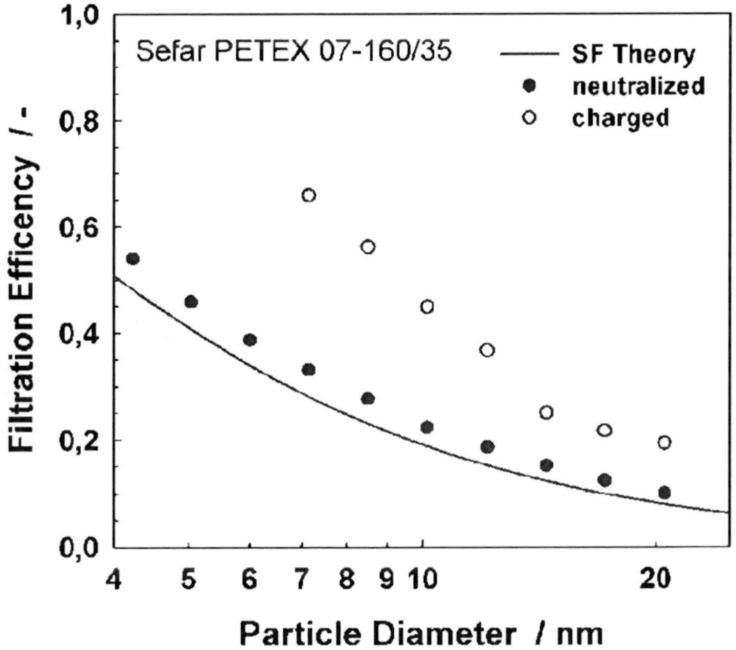

Figure x.10. Nanoparticle capture efficiency for polypropylene mesh, using uncharged and particles with two elementary charges per particle. After [10].

Figure x.10 shows a polypropylene mesh, which was not charged deliberately, in the same grounded metal housing. The difference here is more dramatic, due to the polymer filter. Such effects have been widely utilised in filters for air conditioning (HVAC) systems, since in theory the efficiency of a filter can be "electrostatically enhanced" to provide a greater efficiency for a given pressure drop. It has however been shown, that this enhanced collection efficiency is very short lived, and once the fibres of the filter have been loaded with particles to any extent, the efficiency drops down to the same level as for the neutral particle and uncharged filter case [93]. Therefore, charging of filters (and or particles), should only be considered a viable means of enhancing collection efficiency for short-term collection.

Gas Phase Processing and Classification of Nanoparticles

Aside from filters, there are a range of other devices and methods which are useful for gas phase nanoparticle and nanomaterials processing and handling. In general, gas-phase processing is advantageous because all operations can be conducted on-line, without undesired agglomeration and without contamination – as the particle purity is only limited by the purity of the carrier gas(es). The disadvantage of gas phase processing is that particle concentrations (and therefore production rates) are typically limited if agglomeration and or generation of larger than desired particles is to be avoided. Common gas phase nanoparticle process operations include evaporation, condensation, and sintering [94-96]. Diffusion batteries (constructed of carefully selected banks of mesh similar to that in Figure x.3(a)) and low pressure impactors [97-101], are commonly used for size segregated collection of nanoparticles. The fragmentation of loosely agglomerated particles and sintered agglomerates has also been successfully conducted in the gas phase [102-103].

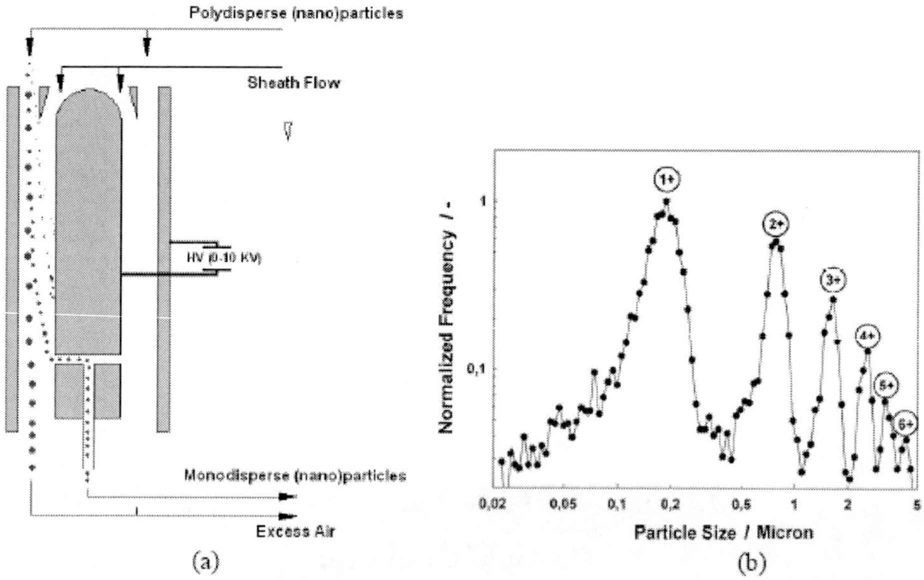

Figure x.11. (a) Schematic of a DMA; (b) A DMA classified distribution measured using an Optical Particle Counter (OPC), showing multiply charged particle peaks (1+ to 6+ elementary charges) which have the same electrical mobility as the singly charged particles.

One of the most important nanoparticle processing devices is the Differential Mobility Analyser (DMA) [101, 104-120]. The DMA is widely used to classify nanoparticles into monodisperse size classes. The device classified particles using a combination of electrostatic and hydrodynamic forces, termed "electrical mobility" classification. A high voltage DC field (0-10 kV) is applied between two plates or concentric cylinders, through which an aerosol is passing. Under ideal conditions a DMA will produce a highly monodisperse (single size) particle with either a single positive or negative elementary charge, from a polydisperse distribution. Depending on the voltage applied, smaller or more highly charged particles than desired will be collected on the walls (as they possess higher mobility than the other particles in the stream). However, if the influent distribution is too broad and contains multiply charged particles, then the these may be selected if they possess the same electrical mobility as the desired particles [121]. When combined with a suitable particle counter, the combined device can scan particle size distributions and produce a full size spectrum in near-real-time [122-129].

CONCLUSION

It is evident from current research that when accurate sampling methods are used, no thermal rebound can be detected in fibrous filters down to <1 nm. When filters or meshes can be neutralized or grounded, only slight particle-charge induced effects on the collection efficiency can be observed. However for plastic mesh (which cannot be reliably grounded or neutralised for long periods), a significant difference in the capture efficiency can be observed with charged particles compared to uncharged, resulting in an increased capture efficiency.

Further work to examine the filtration of particles ~1 nm and smaller would be academically interesting, however not useful for any practical applications in nanoscience or environmental protection. However, the author proposes that the optimal method to examine diffusional collection efficiency in this region would be to use electrosprayed monodisperse molecular standards [129] as the test aerosol, classified using a high resolution nano-DMA [105], and measured using a turbulent mixing type Condensation Particle Counter (CPC) [123] or accurate Faraday Cup Electrometer (FCE) to perform measurements.

Based on the above, it is clear that conventional diffusional collection (Single Fibre Efficiency) theory can be applied to predict the capture efficiency of new/clean filters for nanoparticles, when the filter media is sufficiently homogeneous and Re <1, all important properties are known. Prediction of the long-term performance (efficiency and pressure drop) of filters collecting nanoparticles, is an issue which requires further research. Work to date on the loading characteristics of filters has typically focused on particles >>100 nm. Likewise, the performance (both short and long term) of nanofibre filters, and optimization of nanofibre filter structures and production methods, requires significantly more research. In particular, optimisation of the incorporation of nanofibres into filters and the adhesion of nanofibres to larger (substrate) fibres.

REFERENCES

[1] A. Podgorsky and A. Balazy. *EAC 2004 - Post Deadline Abstracts*. 2004. Budapest.
[2] W.C. Hinds, *Aerosol Technology*. 1999, New York: John Wiley and Sons.
[3] A. Hirabayashi, *Int. J. Mass Spectrom.* 175, 241 (1998).
[4] J. Fernández de la Mora, et al., *J. Aerosol Sci.* 21, S673 (1990).
[5] S. Ude and J.F. de la Mora, *J. Aerosol Sci.* 36, 1224 (2005).
[6] T. Frising, et al., *Chem. Eng. Sci.* 60, 2751 (2005).
[7] B.J. Mullins and G. Kasper, *Chem. Eng. Sci.* 61, 6223 (2006).
[8] H.-C. Wang, *Aerosol Sci. Technol.* 18, 180 (1993).
[9] H.-C. Wang and G. Kasper, *J. Aerosol Sci.* 22, 31(1991).
[10] M. Heim, et al., *Aerosol Sci. Technol.* 39, 782 (2005).
[11] H. Ichitsubo, et al., *Aerosol Sci. Technol.* 24, 119 (1996).
[12] H.-C. Wang, *Aerosol Sci. Technol.* 24, 129 (1996).
[13] H.C. Wang, *Aerosol Sci. Technol.*18, 180 (1993).
[14] M. Heim, et al., *Aerosol Sci. Technol.* 40, 144 (2006).
[15] M. Heim, et al., *J. Aerosol Sci.* 41, 207 (2010).
[16] S.-H. Huang, et al., *J. Aerosol Sci.* 38, 719 (2007).
[17] V.A. Kirsh, *Colloid Journal* 73, 478 (2011).
[18] V.A. Kirsh and A.A. Kirsh, *Colloid Journal* 72, 491 (2010).
[19] A. Podgorski, *J. of Nanopart. Res.* 11, 197 (2009).
[20] A. Podgorski, et al., *Inzynieria Chemiczna I Procesowa* 28(3), 615 (2007).
[21] A. Podgorski, et al., *Aerosol Sci. Technol.* 45, 215 (2011).
[22] W.G. Shin, et al., *J. Aerosol Sci.* 39, 488 (2008).
[23] S. Yamada, et al., *Aerosol and Air Qual. Res.* 11, 155 (2011).
[24] G. Mouret, et al., *Sep. Pur. Technol.* 78, 125 (2011).
[25] R.S. Barhate and S. Ramakrishna, *J. of Membrane Sci.* 296, 1 (2007).
[26] R.C. Brown, *Air Filtration, an Integrated Approach to the Theory and Applications of Fibrous Filters*. 1993, Oxford, UK.: Pergamon.
[27] S.-J. Park, et al., *J. Eng. Fibers Fabrics* 5, 50 (2010).
[28] G. Kasper, et al., *J. Aerosol Sci.* 41,1167 (2010).
[29] Z.-M. Li, et al., *Aerosol Sci. Technol.* 45, 1199 (2011).
[30] S.-J. Park, et al., *J. of Sol-Gel Sci. and Technol.* 54, 188 (2010).
[31] R. Teye-Mensah, et al., *J. Phys-Condensed Mat.* 16, 7557 (2004).
[32] O.O. Dosunmu, et al., *Nanotechnology* 17, 1123 (2006).
[33] P. Katta, et al., *Nano Letters* 4, 2215 (2004).
[34] P. Raynor and D. Leith, *J. Aerosol Sci.* 31, 19 (2000).
[35] J. Binnig, et al., *Powder Technol.* 211, 275 (2011).
[36] M.M. Zdravkovich, *Flow Around Circular Cylinders - Vol 1: Fundamentals*. Vol. 1. 1997, Oxford: Oxford University Press.
[37] G.K. Batchelor, *An Introduction to Fluid Dynamics*. 1967, Cambridge: Cambridge University Press.
[38] C.N. Davies, *Proceedings of the Physical Society of London* 57, 259 (1945).
[39] V.A. Gushchin, et al., *J. .Wind Eng. Ind. Aerodyn* 90, 341(2002).
[40] Y.S. Cheng and H.C. Yeh, *J. Aerosol Sci.* 11, 313 (1980).

[41] A.A. Kirsch and N.A. Fuchs, *Annals Occ. Hygiene* 11, 299 (1968).
[42] Y.S. Cheng, et al., *Aerosol Sci. Technol.* 12, 286 (1990).
[43] M.J. Lehmann, et al., *Chemie Ingenieur Technik* 75, 1283 (2003).
[44] M.J. Lehmann, et al., *Mag. Res. Imaging* 23, 341(2005)
[45] G. Kasper, et al., *J. Aerosol Sci.* 40, 993 (2009).
[46] J. Hoferer, et al., *Chem. Eng. Technol.* 29, 816 (2006).
[47] P.C. Raynor, *Aerosol Sci. Technol.* 36, 1118 (2002).
[48] P.C. Raynor, *Aerosol Sci. Technol.* 42, 357 (2008).
[49] S. Fotovati, et al., *J. Aerosol Sci.* 41, 295 (2010).
[50] H.V. Tafreshi, et al., *Chem. Eng. Sci.* 64, 1154 (2009).
[51] B.J. Mullins, et al., *Chem. Eng. Sci.* 62, 6191 (2007).
[52] B.J. Mullins, et al., *J. Colloid Int. Sci.* 269, 449 (2004).
[53] J. Andrade, et al., *Physical Review Letters* 82, 5249 (1999).
[54] F.A.L. Dulien, *Porous Media: Fluid Trans. Pore Struct.* 1992, San Diego: Academic Press.
[55] T. Gillespie and E. Rideal, *J. Colloid Sci.* 10, 281 (1955).
[56] B. Dahneke, *J. Colloid Int. Sci.* 45, 584 (1973)
[57] B. Dahneke, *J. Colloid Int. Sci.* 51, 58 (1975)
[58] X. Li, et al., *J. Aerosol Sci.* 30, 439 (1999).
[59] H.-C. Wang and J. Walter, *Aerosol Sci. Technol.* 7, 285 (1987).
[60] M. Xu and K. Willeke, *Aerosol Sci. Technol.* 18, 129 (1993).
[61] D.S. Rimai, et al., *J. of Adhe.Sci. and Technol.* 8, 1333 (1994).
[62] C.-J. Tsai, et al., *Aerosol Sci. Technol.* 15, 293 (1991).
[63] Y.S. Cheng and Y.C. Yeh, *Env. Sci. Technol.* 13, 1392 (1979).
[64] B. Dahneke, *J. Colloid Int. Sci.* 37, 342 (1971).
[65] S. Wall, et al., *Aerosol Sci. Technol.* 12, 926 (1990).
[66] M.J. Ellenbecker, et al., *J. Air Poll. Cont. Assoc.* 30, 1224 (1980).
[67] M.E. Mullins, et al., *Aerosol Sci. Technol.* 17, 105 (1992)
[68] B. Dahneke, *Aerosol Sci. Technol.*23, 25 (1995).
[69] R. Maus and H. Umhauer, *J. Aerosol Sci.* 28, 401 (1997)
[70] W. Walkenhorst, *Staub Reinhaltung der Luft* 34, 153 (1974).
[71] B.J. Mullins, et al., *Aerosol Sci. Technol.* 37, 587 (2003).
[72] L. Boskovic, et al., *J. Aerosol Sci.* 39, 635 (2008).
[73] L. Boskovic, et al., *J. Aerosol Sci.* 38, 1220 (2007).
[74] L. Boskovic, et al., *Clean-Soil Air Water* 37, 843 (2009).
[75] R. Mead-Hunter, et al., *Langmuir* 27, 227 (2011).
[76] B.J. Mullins, et al., *J. Colloid Int. Sci.* 279, 213 (2004).
[77] B.J. Mullins, et al., *Math. Comp. Simul.* 81, 1257 (2011).
[78] B.J. Mullins, et al., *J. Colloid Int. Sci.* 300, 704 (2006).
[79] B.J. Mullins, et al., *J. Colloid Int. Sci.* 284, 245 (2005)
[80] B.J. Mullins, et al., Modsim 2003. Vol 4: General Systems, 2003. 1829-1834.
[81] B.J. Mullins, et al., 18th World Imacs Congress and Modsim09 International Congress on Modelling and Simulation: 2009. 1746-1752.
[82] B.J. Mullins, et al., *J. Colloid Int. Sci* 312, 333 (2007).
[83] O.V. Pyankov, et al., *Clean-Soil Air Wat.* 36, 609 (2008).
[84] S. Payet, et al., *J. Aerosol Sci.* 23, 723 (1992).

[85] C. Shin and G.G. Chase, *J. Colloid Int. Sci.* 272, 186 (2004).
[86] D. Thomas, et al., *Chem. Eng. Sci.* 56, 3549 (2001).
[87] H. An, et al., *J. of Membrane Sci.* 283, 84 (2006).
[88] J. Fang, et al., *Chin. Sci. Bull.* 53, 2265 (2008).
[89] C. Shin, et al., *Aiche. J.* 51, 3109 (2005).
[90] C. Ye, et al., *Desal.* 270, 188 (2011).
[91] G. Zhang, et al., *Mat. Sci. Eng. B-Solid State Mat. Adv. Technol.* 116, 353 (2005).
[91] K. Savolainen, et al., *Safety Sci.* 48, 957 (2010).
[92] P.C. Raynor and S.J. Chae, *J. Occ. Env. Hygiene* 1, 463 (2004).
[93] A. Binder, et al., *J. Phys. Chem. C* 114, 7816 (2010).
[94] M. Seipenbusch, et al., *Helvetica Chimica Acta* 84, 3686 (2001).
[95] V. Goertz, et al., *Aerosol Sci. Technol.* 45, 1287 (2011).
[96] B. Asgharian, et al., *Aerosol Sci. Technol* 16, 96 (1992).
[97] S.A. Grinshpun, et al., *J. Env. Monit.* 9, 855 (2007).
[98] S.A. Grinshpun, et al., *J. Aerosol Sci* 36, 575 (2005)
[99] I.G. Loscertales, *J. Aerosol Sci.* 31, 923 (2000).
[100] G.P. Reischl, *J. Aerosol Sci.* 13, 216 (1982).
[101] S. Froeschke, et al., *J. Aerosol Sci.* 34, 275 (2003).
[102] M. Seipenbusch, et al., *Part. Part. Sys. Char.* 24, 193 (2007).
[103] J. Fernández de la Mora, *J. Aerosol Sci.* 33, 411 (2002).
[104] J. Fernández de la Mora, et al., *Method and Apparatus to Increase the Resolution and Widen the Range of Differential Mobility Analyzers (DMA's)*. 2004: United States.
[105] R.C. Flagan, *Aerosol Sci. Technol.* 30, 556 (1999).
[106] D.E. Hagen and D.J. Alofs, *Aerosol Sci. Technol.* 2, 465 (1983).
[107] D.E. Hagen and D.J. Alofs, *Aerosol Sci. Technol* 2, 230 (1983).
[108] D. Hummes, et al., *Part. Part. Sys. Char.* 13, 327 (1996)
[109] J. Jiang, et al., *Aerosol Sci. Technol.* 45, 480 (2011).
[110] W.P. Kelly and P.H. McMurry, *Aerosol Sci. Technol.* 17,199 (1992).
[111] Y. Kousaka, et al., *J. Chem. Eng. Japan* 19, 401 (1986).
[112] Y. Kuga, et al., *J. of Nanoparticle Res.* 3, 175 (2001).
[113] I.G. Loscertales, *J. Aerosol Sci.* 29, 1117 (1998).
[114] G.W. Mulholland, et al., *Aerosol Sci. Technol.* 31, 39 (1999).
[115] Y. Otani, et al., *J. Chem. Eng. Japan* 30, 1065 (1997).
[116] A. Reineking and J. Porstendorfer, *Aerosol Sci. Technol.* 5, 483 (1986).
[117] G.P. Reischl, *J. Aerosol Sci.* 22, 297 (1991).
[118] G.P. Reischl, *Aerosol Sci. Technol.* 14, 5 (1991).
[119] G.P. Reischl, et al., *Aerosol Sci. Technol.* 27, 651 (1997).
[120] M. Heim, et al., *J. Aerosol Sci.* 39, 1019 (2008).
[121] H. Bartz, et al., *J. Aerosol Sci.* 16, 443 (1985)
[122] M. Gamero-Castano and J. Fernández de la Mora, *J. Aerosol Sci.* 31, 757 (2000).
[123] B. Giechaskiel and A. Bergmann, *J. Aerosol Sci.* 42, 195 (2011).
[124] J. Kesten, et al., *Aerosol Sci. Technol.* 15, 107 (1991).
[125] T. Niida, et al., *J. Aerosol Sci.* 19, 1417 (1988).
[126] M.T. Saros, et al., *Aerosol Sci. Technol.* 25, 200 (1988).
[127] G.J. Sem, *Atmos. Res.* 62, 267 (2002).
[128] J. Wang, et al., *Aerosol Sci. Technol.* 36, 678 (2002).

In: Nanostructures
Editor: Yu Dong

ISBN: 978-1-62618-081-9

Chapter 8

MECHANICAL PROPERTIES OF CONTINUOUS NANOFIBERS: CHARACTERIZATION AND MECHANICS

Xiang-Fa Wu**, Zhengping Zhou, Arifur Rahman and Amol Bedarkar

Department of Mechanical Engineering, North Dakota State University, Fargo, North Dakota, USA

ABSTRACT

Mechanical properties are fundamental to the potential structural and functional applications of continuous nanofibers produced by electrospinning. Due to their unique continuity, high specific surface area, tailorable material properties and microstructures, low cost in manufacturing and excellent scalability, electrospun nanofibers have found rapidly expanding applications in structural, biomedical and bioengineering, environmental, and renewal energy sectors, among others. This chapter focuses on the production, mechanical characterization and mechanics modeling of these continuous nanofibers.

A brief introduction to electrospinning and continuous nanofibers as well as scalable nanofiber manufacturing is given. Several cutting-edge experimental techniques for mechanical characterization of continuous nanofibers are reviewed. The unique mechanical responses of electrospun nanofibers are discussed including the size effect in mechanical characterization, failure mechanisms, surface instability, etc. In addition, several recently developed continuum mechanics models are presented to correlate the experimental observations of electrospun nanofibers such as the size effect in mechanical properties, wave propagation, surface rippling, adhesion and contact, surface wetting, hydroelastic response, etc. These mechanics models are expected to place the manufacturing, characterization and applications of continuous electrospun nanofibers on a rational foundation.

---

* Corresponding author's email address: Xiangfa.Wu@ndsu.edu.

INTRODUCTION

Nanomaterials or nanostructured materials are defined as materials or structures with at least one dimension in the size of nanoscale (<100 nm). The morphology of nanomaterials can be zero-dimensional (0D) such as nanoparticles, nanoclusters and quantum dots, one-dimensional (1D) such as carbon nanotubes (CNTs), silicon nanowires (SNWs), magnetic nanorods and continuous polymer nanofibers, two-dimensional (2D) such as biological membranes and graphene, and complicated three-dimensional (3D) nanostructures. Among these, each nanomaterial corresponds to one or several specifically developed nanofabrication techniques. For example, chemical vapor deposition (CVD) can be utilized for synthesis of CNTs and nanowires of silicon, metal and metal oxides; electric nanolithography can be used to produce well patterned nanostructures. Due to the unique size effect in their chemical and physical properties, nanomaterials are expected to provide promising properties that their counterpart bulk materials do not carry, including improved stiffness, strength, thermal and electrical conductivities, chemical activity, etc.

Among various nanomaterials, continuous nanofibers produced by electrospinning [1-7] represent a relatively new type of nanostructured materials. To date, hundreds of natural and synthetic polymers as well as polymer-derived metals, metal oxides, carbon, silicon, and ceramics have been successfully electrospun into nanofibers and nanowires with the diameters in the range of 1-2,000 nm [6, 8-13]. The broad spectrum of electrospinnable materials and relatively low fabrication cost of the electrospinning technique have attracted fast growing research and industrial interests in this field in the recent years. As evidenced, a rapidly increasing number of materials scientists, engineers and chemists in academic communities and industries have joined the intensive research and commercial development of electrospun nanofibers for various applications. These applications include filter media, nanocomposites, wound dressing, biomedicine and tissue engineering, sensors and transducers, renewable energy harvesting, conversion, and storage devices, etc. [7-17]. As a result, the number of publications and article citations relevant to electrospinning and electrospun nanofibers increases very fast since the first paper on modern electrospinning published by Doshi and Reneker in 1995 [1].Moreover, recent technological breakthroughs in the bottleneck of scalable mass production of continuous nanofibers such as use of the needleless electrospinning technique (e.g., the Nanospider™ of Elmarco Inc., Czech) have paved the path for industry-level applications of a large quantity of continuous electrospun nanofibers.

Yet, compared to the fast development of nanofiber fabrication and applications, the fundamental understanding of the mechanical properties of electrospun nanofibers and relevant mechanics models are still highly desired since the mechanical properties of electrospun nanofibers are fundamental to fulfilling their structural and functional applications. Therefore, in this chapter, we are going to discuss the production, mechanical characterization and mechanics modeling of continuous nanofibers. To do so, a brief introduction will be given on electrospinning technique, scalable nanofiber manufacturing, and simplified process modeling. Recently developed techniques for mechanical characterization of continuous nanofibers will be reviewed. Several unique mechanical responses of electrospun nanofibers will be addressed such as the size effect observed in the mechanical measurements, failure mechanisms, surface instability, etc. Moreover, to understand the unique mechanical behavior of electrospun nanofibers such as the size effect in axial quasi-static mechanical behavior, wave propagation, surface rippling, adhesion and contact, surface wetting, hydroelastic response, etc., several continuum mechanics models developed recently by the authors and others will be presented. These phenomenological mechanics models can serve as the rational foundation useful to better manufacturing, mechanical characterization and applications of electrospun nanofibers in ever expanding research and industrial areas.

ELECTROSPINNING AND NANOFIBERS—TECHNOLOGICAL DEVELOPMENT AND CHALLENGES

Electrospinning is based on the principle of electrohydrodynamic jetting from a polymer solution or melt [1-5]. This technique has been broadly utilized for fabricating continuous nanofibers and nanowires of natural and synthetic polymers and polymer-derived carbon, ceramics, metals, metal oxides, etc. [7-17]. Schematic setup of conventional needle-based electrospinning is illustrated in figure 1, which is simply made up with an adjustable high-voltage direct current (DC) power supply, a capillary tube, and a conductive fiber collector (usually a metallic plate). During a typical electrospinning process, under the action of electrostatic force (onto the induced charges) on the polymer solution, a droplet forms from the capillary tube and deforms into a Taylor cone [18-20]. When the electrostatic force overcomes the surface tension, a charged thin jet is elongated and accelerated in the electrostatic field. After a variety of jet destabilizations and solvent evaporation (drying), the ultra-thinned jet is finally deposited on the fiber collector to form a nonwoven nanofiber mat. Figure 2 shows a few types of typical nanofibers produced by means of electrospinning technique in our recent research. To date, hundreds of synthetic and natural polymers have been produced by electrospinning [6, 8, 11, 12].

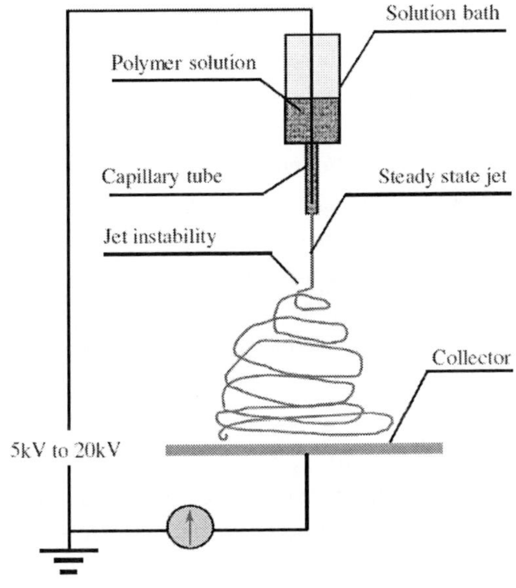

Figure 1. Schematic of electrospinning setup.

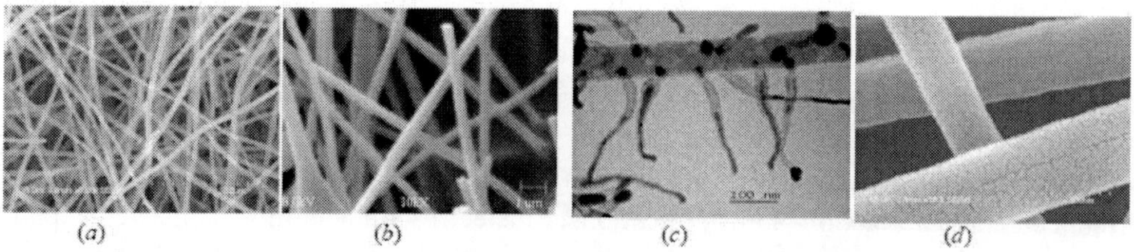

Figure 2. Continuous nanofibers produced by electrospinning: (*a*) polyimide (PI), (*b*) polyacrylonitrile (PAN)-based carbon, (*c*) carbon nanotube (CNT)-grafted carbon [21]. [Copyright (2008) of the Institute of Physics], and (*d*) nano-cracked chromium-coated nanofibers.

Continuous nanofibers collected in electrospinning can be in the form of highly porous nonwoven nanofiber mats or aligned nanofiber films with the aid of specially designed collectors or an auxiliary DC or alterative current (AC) electric field [22]. So far, electrospinning technique and electrospun nanofibers have become a worldwide topic of interest due to their rapidly expanding applications in protective clothing [23], filtration [8, 24-27], templates for synthesizing metallic and polymer nanotubes [8, 28, 29], precursors for fabricating carbon nanofibers [21, 30, 31], and nanofiber composites [32-34]. Furthermore, electrospun nanofibers have also been considered in biomedical engineering and technologies (e.g., medications [35,36], scaffolds for tissue growth [37-42], drug delivery systems [43-46], etc.), dye-sensitized solar cells [47-48], super-hydrophobic surfaces [49, 50], nanofiber sensors [51-55], etc. The near future applications of electrospun nanofibers may also include solar sails, light sails and mirrors in space and nanoelectronics [56], and porous nanofiber electrodes for rechargeable lithium ion batteries [57-59], among others.

Yet, the conventional single-nozzle based electrospinning technique has a very low productivity. One technical resolution of enhancing the nanofiber productivity triggered by the traditional fiber spinning technology is to utilize multiple spinnerets [60] as commercialized in several electrospinning machines in the market (e.g., eSpinner^{TM}). In principle, multi-spinneret based electrospinning can increase the nanofiber productivity in proportion to the number of spinnerets installed; however, severe technical barriers have been experienced such as spinneret clogging due to solvent evaporation near the outlets [3], among others.

Such technical disadvantages in the conventional needle-based electrospinning technique have challenged researchers all over the world to search for more robust and productive electrospinning techniques for continuous, scalable production of nanofibers. Among others, Yarin and Zussman [61] first proposed an experimental prototype of needleless electrospinning as shown in figure 3, in which dozens of jets can be generated simultaneously from a free surface of polymer solution. The seminal contribution of Yarin and Zussman's needleless electrospinning concept is to artificially induce the free-surface destabilization via employing an external magnetic field to a second magnetizable fluid underneath the polymer solution.

Along the vein of free-surface jetting, the research group at Technical University of Liberec in Czech Republic made a technical breakthrough in the needleless electrospinning technology by utilizing free-jetting of a thin-liquid layer formed on a rotating drum [62-65], as shown in figure 4. This technical innovation directly led to the birth of the Nanospider^{TM}, a robust and efficient needleless electrospinning device commercialized successfully by

Elmarco, Inc., the world's first and largest supplier of industrial scale nanofiber production machines.

In recent years, the Nanospider™ needleless electrospinning machines have been installed in over one hundred companies and universities in the U.S. for mass fabrication of continuous nanofibers of various natural and synthetic polymers. It needs to be mentioned that two key technological innovations have been fulfilled through the Nanospider™: (1) thickness of the thin-liquid layer can be tailored through adjusting the drum radius, rotating angular speed, or drum surface properties (e.g., surface roughness, hydrophobicity, etc.); (2) the thin-film thickness is maintained by the rotating drum that makes it possible for robust, mass production of continuous nanofibers. Based on the electrohydrodynamic principle of free jetting from a thin liquid film on a rotating drum as employed in the Nanospider™, new technical features are under further exploration such as mass production of core-shell nanofibers, etc.

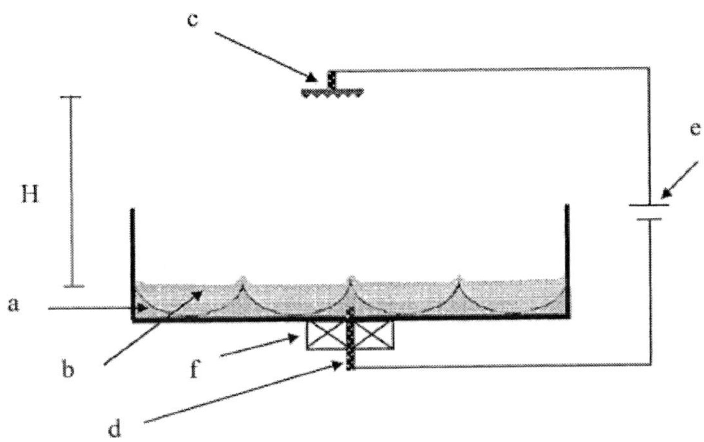

Figure 3. Schematic of the first needleless electrospinning setup proposed by Yarin and Zussman [61]: (*a*) a layer of magnetic liquid, (*b*) a layer of polymer solution, (*c*) counter-electrode located at a distance *H* from the free surface of the polymer solution, (*d*) electrode submerged into the magnetic fluid, (*e*) high voltage source, and (*f*) strong permanent magnet or electromagnet. Copyright (2004) of Elsevier.

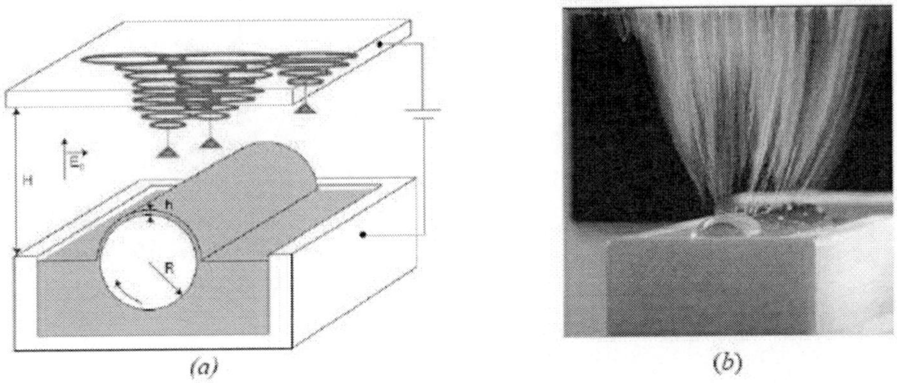

Figure 4. (*a*) Concept of free-surface jetting from thin liquid layer in a high DC electrostatic field [62-65]; (*b*) multiple jets from a rotating drum of a needleless electrospinning machine: Nanospider™ (Elmarco, Inc., Czech, 2010). Copyright (2008) of the American Institute of Physics.

Besides, recent technological development also considers spinning core-shell and hollow nanofibers by means of coelectrospinning [66-68] and emulsion electrospinning [69-72]. The coelectrospinning technique for core-shell nanofibers is based on a double-compartment syringe as shown in figure 5(*a*). During the process, two polymer solutions are supplied simultaneously to form a coaxial Taylor's cone. Upon destabilization of the coaxial Taylor cone in an electrostatic field, a coaxial jet can be formed, stretched, and accelerated underproper condition.

After solvent evaporation (drying), the whipping coaxial jet forms coaxial nanofibers as shown in figure 5(*b*). Compared to conventional single-nozzle based electrospinning, double numbers of material and geometrical parameters (coaxial nozzle) are involved in a coelectrospinning process, which provide more flexible options for controlled coaxial nanofiber fabrication. One superior advantage of the coelectrospinning technique is that it can spin unelectrospinnable materials such as small-molecule monomers, liquids, etc. Furthermore, Yarin and his collaborators [69] recently found that core-shell nanofibers can also be formed by using an emulsion electrospinning technique, which is based still on a conventional single-nozzle based electrospinning setup as shown in figure 6 [69-72]. In their demonstration, the electrospinnable PMMA/DMF and PAN/DMF solutions were first blended and then stored in room temperature without stirring for 12 hours. During this period, the blend decomposed into PMMA/DMF droplets with the diameter of ~100 μm in the PAN/DMF solution.

The resulting emulsion can be electrospun using a standard single-nozzle electrospinning setup (figure 1). During such an electrospinning process, the coaxial Taylor cone can be formed instantaneously at the moment when a PMMA/DMF droplet enters the nozzle exit [72]. Subsequently, the enwrapped PMMA/DMF droplets were deformed, stretched substantially, and finally form the core material after drying. A detailed review was made recently by Yarin [72] on the development of coelectrospinning and emulsion electrospinning techniques and related applications.

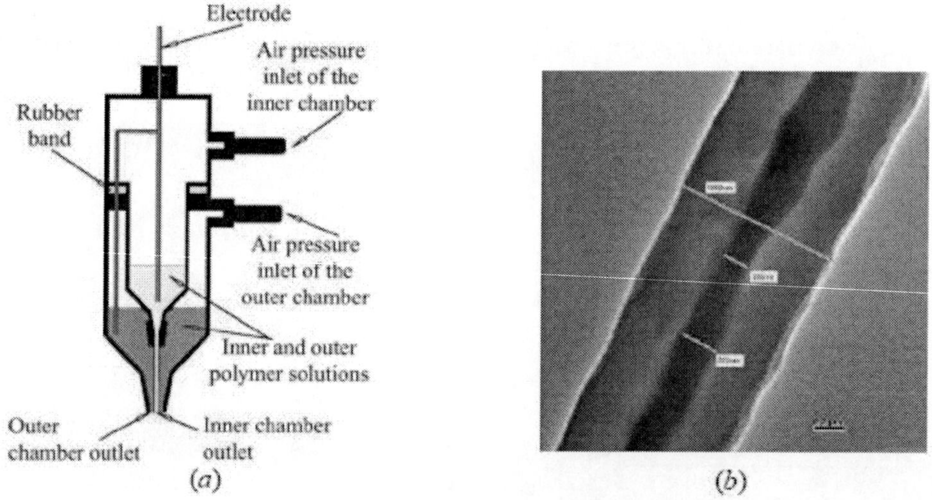

Figure 5. (*a*) Schematic of coelectrospinning setup[66] and (*b*) TEM micrograph of the core-shell structure of unstained samples of coelectrospun poly (ethylene oxide) (PEO, sheath) and poly (dodecylthiophene) (PDT, core) [66]. Copyright (2003) of the Wiley and Sons, Inc.

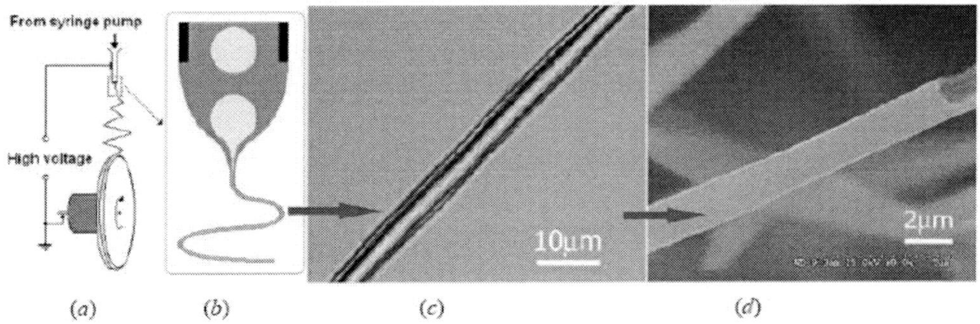

Figure 6. (*a*) Schematic setup of emulsion electrospinning of PMMA/PAN blend in DMF using a single circular nozzle; (*b*) schematic of magnified details of the needle orifice, in which the PMMA/DMF droplets are shown in yellow and the PAN/DMF liquid in blue, the liquids wet the nozzle, and the core-sheath fibers emanating from the Taylor's cone tip are collected on the edge of a rotating wheel; (*c*) SEM micrograph of core-shell fibers produced by this method; and (*d*) SEM micrograph of a hollow fiber obtained through elimination of the core material [69-72]. Copyright (2007) of the American Society of Chemistry.

Nevertheless, there is still lacking the fundamental understanding of the entire process of needleless electrospinning and coelectrospinning such as electrohydrodynamic multi-jet initiation and stretching, solvent evaporation (drying), porous structural formation, and controlled nanofiber collection though several preliminary models have been formulated that were essentially focused on understanding the electrohydrodynamic destabilization of thin liquid films and droplets [62, 73-85]. As a matter of fact, for mass fabrication of nanofibers, the current needleless electrospinning machine (e.g., the Nanospider™) typically needs a substantial period of empirical trial and error in determining the optimal process parameters for a specific polymer-solvent system. Such a process is laborious and usually carried out with the assistance of experienced experts from the machine manufacturers or through special on-site training. Thus, similar to other controllable fabrications, process modeling of electrospinning is extremely important to the fundamental understanding of the electrohydrodynamic phenomenon of electrospinning process and controllable fabrication of nanofibers for wider use of nanofibers.

PROCESS MODELING OF ELECTROSPINNING

As mentioned in the above, the experimental setup for either nozzle-based electrospinning (figure 1), recently developed needleless electrospinning (figure 3), or coelectrospinning (figure 5) is relatively simple and low cost; however the entire electrospinning process is an extremely complicated multi-physics phenomenon coupled with electrohydrodynamics, heat/mass diffusion and transfer, nonlinear rheology, polymer solidification, etc. For noozle-based electrospinning, several theoretical models have been proposed to treat specific one or two individual stages of the process: formation of Taylor cone [18-20], jet initiation, steadily accelerating straight jet, or jet whipping, etc., each of which involves the combined action of electrostatic force, capillary effect, non-newtonian fluid dynamics, heat/mass diffusion and transfer, polymer solidification, and related nonlinear instability at the transition from one stage to the next stage. Historically, Taylor [18] was the pioneer in the study of electrospinning who first established a rigid self-similar theory for the Taylor cone with the half conic angle of 49.3°. The Taylor cone is the envelope cone of the sessile and pendent droplet of the liquid formed at the end of a capillary tube positioned in a steady electrostatic field of not very high amplitude. Spivak and Dzenis [19] provided an asymptotic solution for determining the shape of a weakly conductive droplet in electrostatic field. Yarin [20] advanced Taylor's work [18] by considering the non self-similar solutions to the hyperboloid-shaped liquid droplets, and found that the critical half conic angle of a droplet is 33.5°, which is much closer to the one observed in experiments. Besides, dynamic initiation of the Taylor cone from flat conductive liquids has also been studied within the framework of computational electrohydrodynamics and self-similarity theories [86, 87], among which the half-conic angle of the transient Taylor cone predicted by Reznik et al. [87] is $\leq 30°$, close to their experimental observations. These studies have clarified the physics of droplets before jet initiation in an electrostatic field. Once the jet initiates in the electrostatic field, it first forms a steadily accelerating straight jet as shown in figure 7 till it loses stability to form the whipping jet.

In this stage, several 1D electrohydrodynamic slender jet models have been formulated with different degree of elaboration. To study the asymptotic behavior of a straight jet accelerating in a steady electric field, Kirchenko et al. [88] formulated a 1D conductive jet model in which the liquid was treated as a simple newtonian fluid; this model predicts that the asymptotic jet radius has a power-law decay in the order of $z^{-1/4}$ with the distance z from the spinneret. By extending this model to weakly conductive non-newtonian fluids obeying the Oswald-deWaele power-law, Spivak and Dzenis [89] and Spivak, et al. [90] obtained the power-law asymptotic behavior of the jet radius, in which the index of the power-law decaying function is related to the flow index of the liquids. Hohman et al. [91, 92] formulated an elegant 1D slender jet model of newtonian fluids, in which the additional electric field due to the charged jet is also taken into account; this model is capable of providing the explicit characteristic relations of the jet axisymmetric and whipping instabilities [91-93] and examining the diameter dependency of electrospun nanofibers upon the material and process parameters, validated by experiments [94].

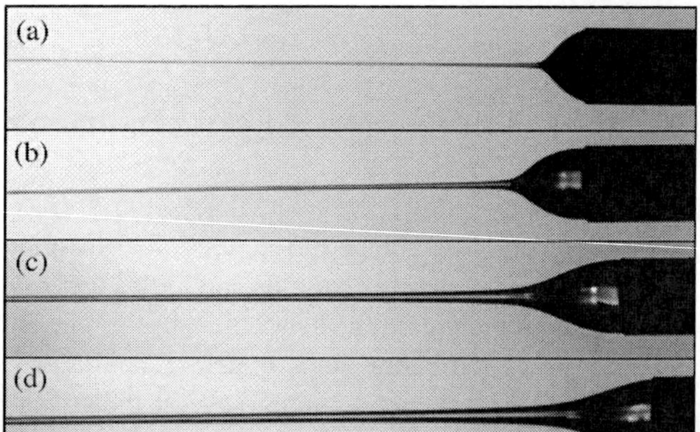

Figure 7. Steadily accelerating straight jets of 2% solutions of polyethylene oxide (PEO) (*MW*=920 kg/mol) in water. Initial jet diameter *D*=45 cm in all cases: (*a*) Flow rate Q = 0.02 ml/min and electric field E = 0.282 kV/cm; (*b*) Q = 0.10 ml/min and E = 0.344 kV/cm; (*c*) Q = 0.50 ml/min and E = 0.533 kV/cm; (*d*) Q = 1.00 ml/min and E = 0.716 [7]. Copyright (2007) of Elsevier.

In addition, Feng [95, 96] refined the 1D jet model by Hohman et al. [91, 92] to eliminate several existing incompatibilities of the boundary conditions at the nozzle tip, and to further tackle the effect of non-newtonian and viscoelastic fluid models on the jetting behavior of an charged jet in an electrostatic field. Yet, to model a realistic electrospinning process, even at the very beginning stages, the liquids of the above models should be matched to the flow in the droplet region at the outlet of the nozzle. In this region, the flow becomes fully two-dimensional (2D axisymmetric) and the slender jet models can be only considered as the first-order approximation [3]. Though a few fully 2D numerical simulations have been devoted to understanding the transition zone between the droplet and the electrically driven jet [97-99], no complete, robust jet model has been available yet to capture the entire electrohydrodynamic phenomenon from the jet initiation to whipping destabilization in a conventional nozzle-based electrospinning process.

Nevertheless, to elucidate the jet bending (whipping) instability in electrospinning, Reneker et al. [100, 101] developed a reduced bead model in which the jet is simply modeled as a chain of charged beads linked through linearly elastic springs and viscous dashpots in series based on a viscoelastic Maxwellian model. This model can reasonably explain the phenomenological jet bending instability, jet conical envelope, drying process, interaction of multiple jets, and study of the effect of process parameters of electrospinning, among others [60, 102-104]. More recently, Dayal and Kyu [105] and Dayal et al. [106] developed a 3D spring-network model coupled with phase separation and solvent evaporation to capture the spatio-temporal evolution of phase separation in a polymer-solution jet during electrospinning. This model can be used to quantitatively explain a few coupled multi-physics phenomena involved in polymer nanofiber formation in electrospinning.

In addition, the above jet models were built on a homogeneous jet. However, the solvent evaporation from the jet surface during electrospinning may lead to inhomogeneity in an instantaneous jet in both the radial and axial directions of dilute polymer jets. Solvent evaporation even becomes very fast in an ultrathin jet. Thus, solvent evaporation plays a crucial role in nanofiber formation by means of electrospinning. Recently, Wu et al. [107] developed a simple 1D axisymmetric jet model to simulate the drying process of an electrospun jet of dilute polymer solution, in which the jet was simplified an ultrathin cylinder of polymer solution.

Phenomenologically, the drying process of a jet suspended in air involves three successive sub-processes [107]: (1) diffusion of solvent molecules from the inside of the jet to its surface, (2) evaporation of solvent molecules from the jet surface, and (3) transport of solvent vapor from the jet surface into the bulk of the atmosphere. In electrospinning, a significant jet shrinking happens due to evaporation of a significant volume of solvent from the dilute solution jet. This model is based on three fundamental assumptions that have been generally used in simulation of drying process of coating systems [107-109]: (1) due to the small jet diameter, the temperature field is considered constant inside the jet; (2) the drying process involving solvent diffusion and transfer is treated as axisymmetric; (3) the effect of residual stresses due to solvent loss on the jet solvent diffusion and transfer is negligible; (4) micro convection due to vapor concentration gradient near the jet is negligible; (5) no chemical reactions take place inside the jet. These assumptions can approximately hold for an ultrathin jet. Thus, for a binary polymer solution used in electrospinning, the governing equation of mass diffusion in an electrospun axisymmetric jet was derived according to the axisymmetric Fick's law of mass diffusion [107]:

$$\frac{\partial C_1}{\partial t} = \frac{1}{r}\frac{\partial}{\partial r}(Dr\frac{\partial C_1}{\partial r}), \tag{1}$$

the initial conditions at the beginning of the drying process are assumed as

$$t = 0: \qquad C_1 = C_{10},\ R_0(0) = R_0, \tag{2}$$

and the boundary conditions at the center and moving surface of the jet are

$$r = 0: \qquad \partial C_1 / \partial r = 0, \tag{3}$$

$$r = R_0(t): \qquad \frac{D}{1-\bar{V}_1 C_1}\frac{\partial C_1}{\partial r} = -\frac{k_g M_1 (P_s - P_\infty)}{RT}, \tag{4}$$

$$r = R_0(t): \qquad \frac{dR_0(t)}{dt} = -\frac{k_g M_1 \bar{V}_1}{RT}(P_s - P_\infty). \tag{5}$$

In the above, C_1 is the transient solvent mass concentration in the polymer solution with its initial value C_{10} as constant over the jet cross-section; R_0 is the initial jet radius; $R(t)$ is the current jet radius; k_g (cm/g) is the coefficient of solvent mass transfer at the jet surface; M_1 is the solvent molar weight; $\bar{V}_1$ is the solvent specific partial volume; T is the absolute temperature near the jet surface; R = 8.3144 J/mol is the universal gas constant; P_∞ is the solvent vapor pressure in the atmosphere far from the jet surface ($P_\infty = 0$ used in the current simulation [107]); P_s is the solvent saturation vapor pressure near the jet surface.

The dependence of P_s upon the solvent concentration near the jet surface can be approached according to the Flory-Huggins equation [110]: $P_s/P_0 = \varphi_1 \exp(\varphi_2 + \chi\varphi_2^2)$, where P_0 is the saturation vapor pressure of the pure solvent, ϕ_1 and ϕ_2 ($\phi_2 = 1-\phi_1$) are the solvent and polymer volume fractions, respectively, and χ is the Flory-Huggins polymer/solvent interaction parameter. χ can be approximated by using Bristow's semi-empirical equation [111, 112]: $\chi = 0.35 + (\delta_1 - \delta_2)^2 \tilde{V}/(RT)$, where $\tilde{V}$ is the solvent molar volume, and δ_1 and δ_2 are the solubility parameters of the solvent and polymer, respectively. In the simulation [107], the value of χ is selected as a constant (i.e., $\chi = 0.45$). The polymer/solvent mutual diffusion coefficient D is described by the Vrentas-Duda free-volume theory [113, 114]: $D = D_1(1-\varphi_1)^2(1-2\chi\varphi_1)$, where D_1 is the solvent self-diffusion coefficient. The above governing partial differential equation (PDE) (1) under the initial and boundary conditions (2-5) can be solved efficiently by using an explicit backward-finite difference method (FDM) [115] based on a moving grid scheme [107].

Wu et al. [107] considered the drying process of a PAN/DMF jet with the initial DMF mass concentration 88%. It needs to be mentioned that the PAN/DMF solution is technologically important as it is commonly used to produce precursors for long carbon nanofibers (CNFs) [116, 117]. In the simulation, the material parameters of the PAN/DMF solution are the follows. The mass densities of DMF (M_w = 73) and PAN (M_w = 150,000) are ρ_{DMF} = 0.944 g/cm^3 and ρ_{PAN} = 1.184 g/cm^3, respectively. The PAN diffusion coefficient in infinite DMF is D_1 = ~1.5×10^{-6} cm. The DMF saturation vapor pressure at 20°C is ~500 Pa. The DMF mass transfer coefficient at the jet surface is k_g = ~0.2 m/s, by analogy with similar solvents [118]. The initial DMF mass ratio of 88% corresponds to C_1 = 0.851. Figures 8 and 9 show the typical numerical results of the drying process of a PAN/DMF jet with varying initial diameter [107]. The numerical results in figure 8 show that solvent evaporation results in inhomogeneity of the solvent concentration in microscopic jets, and such inhomogeneity mitigates with decreasing initial jet radius. Significant jet shrinking due to solvent loss can be also clearly observed.

The numerical results indicate that the composition of the jet could be highly inhomogeneous across the jet due to fast solvent evaporation especially near the jet surface; such inhomogeneity may further influence the rheological properties of an electrospun jet and thus affect the electrohydrodynamic process of the whipping jet, drying duration (figure 9), and resulting nanofiber properties such as the final fiber diameter, fiber morphology [119,120], and mechanical properties, among others.

It needs to be noted that the above 1D drying jet model has oversimplified the electrohydrodynamic effect on the solution diffusion and transfer. During the process, themodel has separated the jet electrohydrodynamics from its mass diffusion and transfer. In reality, such two phenomena are coupled to determine the evolution of the jet diameter.

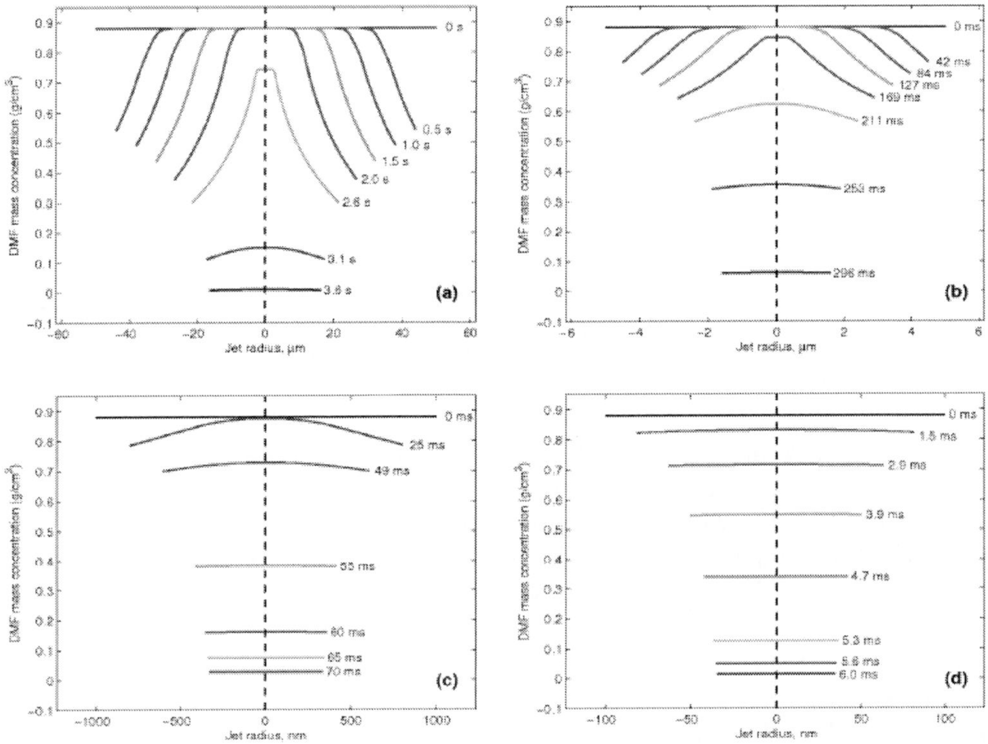

Figure 8. Numerical results of instantaneous solvent (DMF) mass concentration profiles in PAN/DMF solution jets with the initial radii: (*a*) R_0 = 50 μm, (*b*) 5 μm, (*c*) 1000 nm, and (*d*) 100 nm, respectively [107]. Copyright (2011) of the American Institute of Physics.

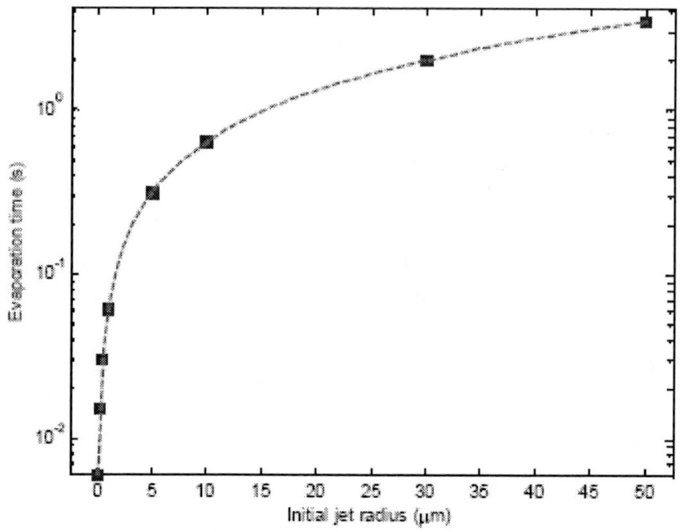

Figure 9. Numerical results of variation of the drying duration with the initial jet radius [107]. Copyright (2011) of the American Institute of Physics.

Therefore, new theoretical effort is still desired in order to gain in-depth understanding of the entire electrospinning process, particularly the various destabilizations, drying process and drying induced nanostructured surface morphology and porous structure formation. These models are expected useful to guide the conventional nozzle-based electrospinning experiments and development of new, innovative models to capture the entire process of needleless electrospinning for production of mass nanofibers [107].

Mechanical Characterization of Electrospun Nanofibers

Similar to other nanostructured and bulk materials, mechanical properties of electrospun nanofibers (e.g., modulus, tensile strength, yielding behavior, etc.) are fundamental to fulfill their targeted structural and functional applications. In the view of end use of nanofibers, controlled electrospinning process is desirable for production of continuous nanofibers carrying with tailorable geometries, stiffness and strength. In addition, the high surface area to volume ratio of nanofibers and their unique manufacturing method may result in the mechanical properties essentially different from those of their bulk counterparts.

Such unique mechanical propertiesmay further influence the deformation, dynamics, stability, adhesion/contact/friction, wetting, and global mechanical response of the resulting nanofibers and nanofiber networks [121-150]. In recent years, cutting-edge micro and nanomechanical characterization techniques [132, 135-150] have been explored for understanding the mechanical response of individual nanofibers produced by electrospinning. Among these, atomic force microscope (AFM) has been widely used for axial modulus and tensile strength measurements of electrospun nanofibers in several groups [136, 138, 139, 141-143]. To date, four AFM-based testing methods have been formulated including AFM-based axial tension test [135, 141], three-point bending test [136, 138, 142], double-beam vibration test [143], and nanoindentation. In an AFM-based axial tension test, the nanofiber segment is fixed between the AFM tip and the substrate (e.g., silicon wafer) by using adhesive.

The tiny tensile force is exerted via the motion of the AFM tip. In contrast, for a micro three-point bending test of nanofibers as shown in figure 10, the nanofiber segment is first clamped across a groove on the substrate using adhesive [136, 138, 142]. The lateral force to induce the deflection is exerted through the AFM tip at the midspan of the fiber segment. In principle, the axial modulus of nanofibers can be also measured by means of nanoindentation similar to that for characterization of thin films. After each test, a force-deformation diagram is obtained, from which the modulus can be extracted. The axial modulus can be also extracted from the resonant frequencies of a pair of microcantilevers linked by the nanofiber segments [143].

In addition, the axial modulus and tensile strength of nanofibers can be also estimated from the average axial modulus and tensile strength of a bundle of nanofibers that can be measured using a microtensile device [142]. Yet, AFM-based testing methods have the technical barriers in determining the ultimate tensile strength and strain to failure of nanofibers. Moreover, idealized experimental conditions such as the clamped ends introduced in three-point bending tests for data reduction may further induce additional deviation of the measurements.

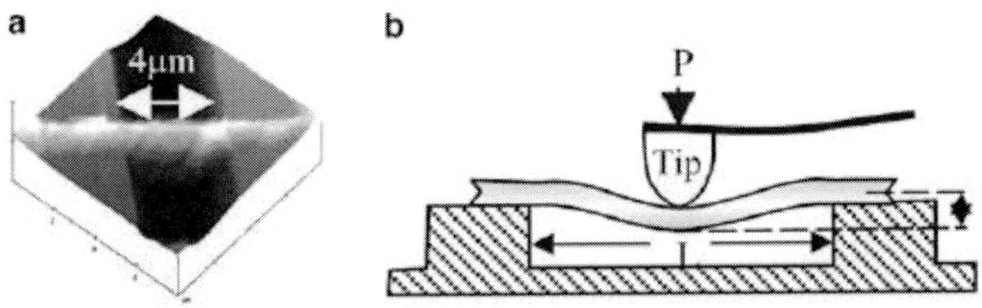

Figure 10. (*a*) AFM contact mode image of a single nanofiber of diameter ~400 nm suspended over an etched groove; (*b*) schematic setup of an AFM-based three-point bending test of electrospun nanofiber [136]. Copyright (2004) of the American Institute of Physics.

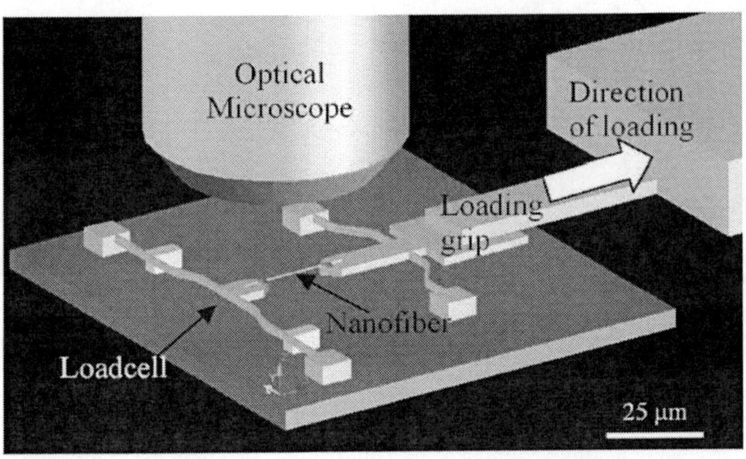

Figure 11. A microtensile testing platform for mechanical characterization of electrospun nanofibers [144,145]. Copyright (2007) of the American Institute of Physics.

Yet, the preferable method for mechanical characterization of nanofibers would be directly testing the individual nanofibers based on micro/nano tensile testers installed with a high-resolution load cell and position controller. Recently commercialized universal micro/nano tensile testers such as the Nano Bionix® and Nano UTM® by Agilent® and microelectromechanical systems (MEMS)-based micro tensile testers (figure 11) have reached the capability of accurately measuring the load-displacement relationship of single nanofibers and nanowires [144-146].

As a result, the full-range stress-strain diagram of a single nanofiber can be determined accurately. For instance, by using a Nano Bionix® machine, Tan et al. [137] have successfully performed the uniaxial tension test of electrospun polycaprolactone (PCL) fibers with the fiber diameters of ~1 μm as shown in figure 12, and they pointed out that the stress-strain diagram of the PCL nanofibers exhibited obvious size effect, i.e., both the ultimate tensile strength and yield stress decrease with increasing fiber diameter while the failure strain increases with the increase of fiber diameter. No noticeable correlation was established between the axial modulus and fiber diameter [137].

Similarly, Chen et al. [132] used a lab-made high-resolution micro tensile station, similar to the Nano UTM® to characterize the modulus and tensile strength of electrospun polyimide [poly(*p*-phenylene biphenyltetra-carboximide), BPDA/PPA] (PI) nanofibers with the diameter of ~300 nm. The experimental results showed that the PI nanofibres had an average ultimate tensile strength of 1.7 ± 0.12 GPa, axial tensile modulus of 76 ± 12 GPa, and ultimate strain of ~3%. These experimental measurements were among the highest ones reported in the literature to date.

In contrast, the precursor nanofibres with the similar fiber diameters and molecular weights carried the average ultimate tensilestrength of 766 ± 41MPa, axial modulus of 13 ± 0.4GPa and ultimate strain of ~43%. In these tension tests, the reliability of the experimental results depends upon the accurate measurement of tensile force, which is in the range from a few millinewtons (μNs) to micronewtons (mNs).

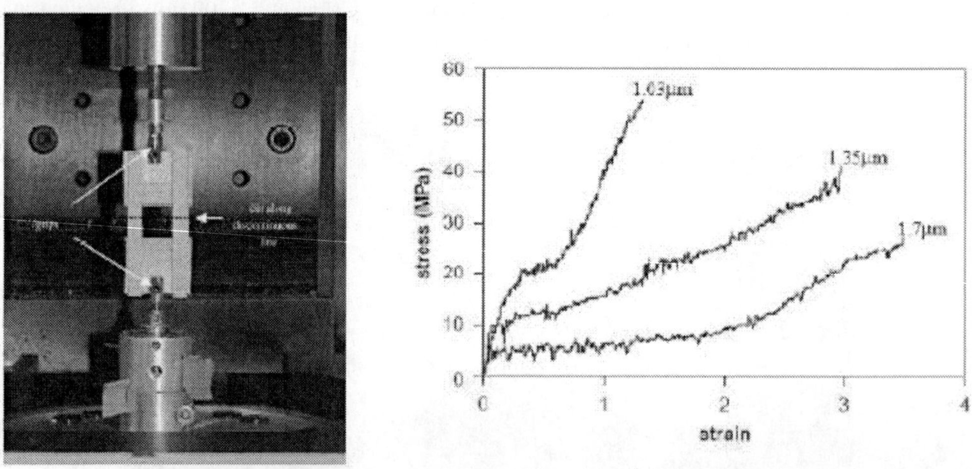

(*a*) (*b*)

Figure 12. (*a*) Microtensile testing setup based on Nano Bionix® in which a single nanofiber is mounted on a cardboard sample-template; (*b*) stress-strain diagrams of three electrospun polycaprolactone (PCL) nanofibers of varying diameters [137]. Copyright (2005) of Elsevier.

In an attempt to accurately capture the tiny tensile forces in a single-nanofiber tension test, recently developed MEMS-based microtensile testers offer an excellent alternative in the view offeasibility and accuracy of the mechanical characterization [144-146], in which the tensile force can be precisely recorded and the loading rate can be controlled extremely lower compared to other micro tensile testers such as Nano Bionix® and Nano UTM®.

To date, increasing experimental results have been accumulated on the mechanical properties of electrospun nanofibers that have greatly enriched the database for in-depth study of the mechanical response of electrospun nanofibers and their dependencies upon the geometric, material, and process parameters. Among others, two unique mechanical behaviors of electrospun polymer nanofibers have been identified experimentally that are essentially different from their large-diameter and bulk counterparts. The first unique discovery is that the axial modulus and ultimate tensile strength of electrospun polymer nanofibers increase abruptly when the fiber diameter is below a certain value [126, 135, 141, 142, 147]. One example is shown in figure 12(b), where the ultimate tensile strength of electrospun PCL nanofibers is nearly doubled when the fiber diameter is decreased from 1.7 μm to 1.03 μm. Ji et al. [151] also discovered that polystyrene (PS) nanofibers quickly increased the axial modulus when their diameters were below ~500 nm; in the case of electrospun poly(2-acrylamido-2-methyl-1-propanesulfonic acid) (PAMPS) nanofibers, they found that such phenomenon was obvious such that the axial modulus increased from ~300 MPa at the diameter of ~110 nm up to ~2.0 GPa when the diameter is ~55 nm. It needs to be mentioned that though the size effect in the mechanical properties of nanomaterials has been widely accepted, it happens at the threshold dimension (i.e., fiber diameter) is much larger than that in metallic nanostructures. Several potential factors may contribute to such size effect, including surface energy [123, 126], chain alignment, and formation of crystalline and semicrystalline structures [140] (figure 13), among others. However, the pure contribution of surface energy is insufficient to interpret such an abrupt increase in the modulus and tensile strength of polymer nanofibers. Besides, the alignment of polymer chains can be also induced in the electrospinning process, in which the electrostatic stretching and constrained polymer chain motion inside the ultrathin jet can contribute the alignment. Nevertheless, recent X-ray diffraction results of electrospun Nylon (polyamide, PA) nanofibers indicated that only a mild monotonous increase can be identified in the crystallinity and orientation of the crystallites inside the electrospun polymer nanofibers with varying fiber diameters [147]. Thus, such a mild increase is also not sufficient to account for the observed large increase in the axial modulus. In addition, the study also pointed out that the average size of crystallites (~4 nm) was largely independent of the fiber diameter. As a result, the crystallinity inside the electrospun PA nanofiberswas nearly constant in the range of fiber diameter under the experimental investigation. To explore the potential physical mechanism of the observed obvious size effect in the tensile modulus of electrospun polymer nanofibers, recently Arinstein et al. [147] proposed the concept of supramolecular structure of amorphous phase in electrospun polymer nanofibers. Such supramolecular structure is made of oriented fragments of polymer chains that may form when the polymer chains are confined inside the polymer nanofibers with the fiber diameter below a certain value. Based on this conceptual model, a preliminary estimate of the transition fiber diameter, below which the polymer nanofibers may exhibit observable size effect in their mechanical properties, is close to those observed in experiments [147]. Yet, this conceptual approach still leaves an outstanding issue: what mechanism governs the likely power-law growth of the axial modulus of electrospun polymer nanofibers with the decrease of fiber diameter below a certain value [126].

Furthermore, surface rippling is another unique phenomenon that was discovered recently in compliant polymer nanofibers (e.g., as-electrospun PAN nanofibers) subjected to axial stretching. Surface rippling or wrinkling can be commonly triggered in compressed compliant surface. Such surface instability in polymer nanofibers was first discovered in MEMS-based micro tension tests of as-electrospun PAN nanofibers [144, 145, 150]. During the tests, periodic ripples suddenly appeared on the fiber surfaces once the axial stretch reach a certain threshold as shown in figure 13 [144, 145, 150]. During the process, the nanofibers did not exhibit obvious softening, and the axial strain to failure (in sense of engineering strain) measured in single-fiber tension tests was up to 60%-130% for as-electrospun PAN nanofibers with the diameter in the range of~300-600 nm. This axial strain is several folds of that measured from PAN microfibers produced via drawing or dry-jet-wet spinning [144, 145]. Such a large axial strain to failure can be attributed to the simultaneous surface rippling of the nanofibers and relevant global softening, which induces the plastic strain much larger than that due to single necking failure in classic polymer microfibers. To explore the physical mechanism of the surface rippling in as-electrospun polymer nanofibers, Naraghi et al. [144, 145] proposed that there exist periodic distribution sites to accommodate the local instabilities. Furthermore, they pointed out that fine circumferential surface cracks were detected on the surface of nanofibers under stretching, and these surface cracks could facilitate the seeking of necking, analogous to fragmentation as evidenced in hard-film-on-soft-substrate systems. Yet, no qualitative and quantitative correlationswere made between the nearly constant wavelength of the periodic surface ripples and the geometrical and material parameters of the nanofibers. In reality, surface defects, voids, and cracks commonly exist on fiber surface, however, subjected to axial stretching, some surface cracking is suppressed while some other ones were enhanced due to the specific instability conditions as treated theoretically by Wu et al. [126] and Wu [127] (See the next section).

Such surface rippling was also observed in pulled-out PAN nanofibers during interlaminar fracture test of polymer composites, in which the electrospun PAN nanofibers were entangled at ply interfaces as nanoreinforcement, as shown in figure 14. In addition, surface ripples were also observed in electrospun polyimide (PI) nanofibers after annealing as shown in figure 15. Figure 15(*a*) shows the image of an as-electrospun PI nanofiber characterized by means of AFM; it can be observed that the as-electrospun PI nanofiber was very smooth. However, when the smooth PI nanofibers were deposited on a silicon wafer in electrospinning andannealed (a process of imidization) at ~400°C in nitrogen environment, surface ripples occurred. Such rippling phenomenon may be related to the axial shrinking during the imdization process.

In principle, formation of surface ripples on the surface of electrospun nanofibers is a phenomenon of surface destabilization that could be correlated to the combined effect of surface tension and nonlinear elasticity of the compliant polymer nanofibers. To date, necking and surface destabilization of classic elastic bars of large diameter subjected to finite axial stretch has been well studied within the framework of nonlinear elasticity in classic solid mechanics [152-154]. In such a study, the effect of surface energy of the elastic bars was very small and therefore ignored due to the large characteristic dimension (diameter) and high elastic modulus of the elastic rods. In contrast, when the fiber diameter is reduced down to nanoscale, the surface tension (surface energy) of the compliant nanofibers may dominate the initiation and propagation of surface rippling. Thus, new theoretical justification is desirable for such a unique surface destabilization phenomenon. Beyond this, due to the unique small characteristic dimensions and very low stiffness of the electrospun polymer nanofibers, systematic theoretical studies are desired to count for the deformation, contact, adhesion, and other mechanical behavior of these ultrathin compliant nanofibers where the effect of surface energy would be taken into account.

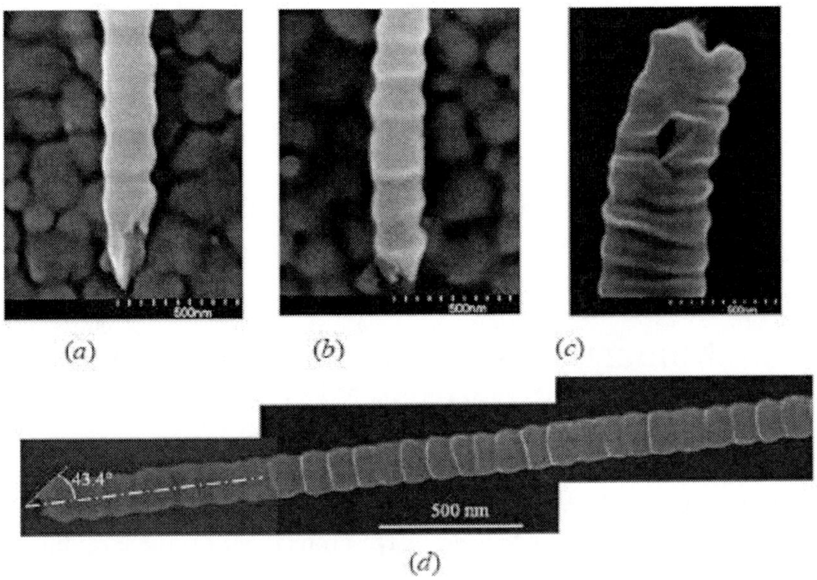

Figure 13. [(*a*) and (b)] SEM micrographs of surface morphology of as-electrospun PAN nanofibers after tensile breakage; the fiber breakage was induced by extrusion of a 45° conic region; (*c*) SEM micrograph of a fiber breakage due to the formation of voids; (*d*). SEM micrograph of ripples formed on the PAN nanofiber surfaces subjected to axial stretching [144]. Copyright (2007) of the American Institute of Physics.

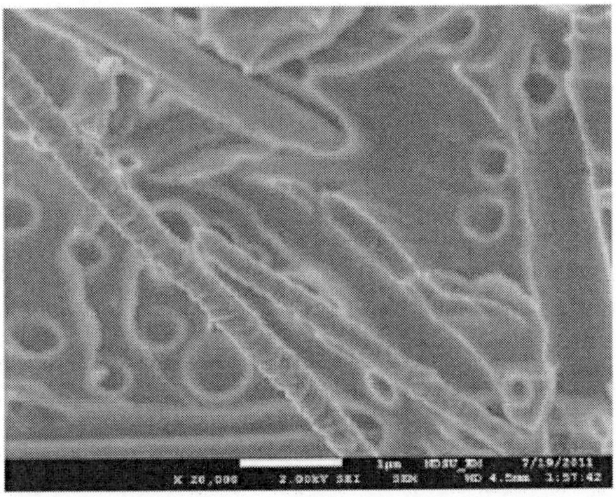

Figure 14. SEM micrograph of the surface morphology of electrospun PAN nanofibers with surface rippling. The PAN nanofibers were inserted at the interfaces of carbon-fiber fabric/epoxy composites and pulled out after mode I interlaminar fracture test.

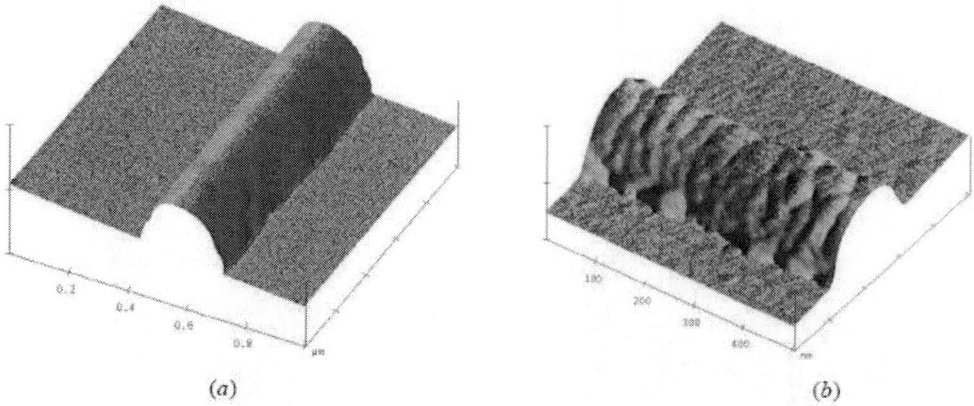

Figure 15. 2 AFM micrographs of surface morphology of (*a*) as-electrospun PI precursor nanofiber and (*b*) PI nanofiber after imdization (fiber diameter: ~250 nm) [126]. Copyright (2008) of the American Physical Society.

MODELING OF MECHANICAL BEHAVIOR OF ELECTROSPUN NANOFIBERS

Phenomenological mechanics models are useful to rationally understand the mechanical behavior of electrospun nanofibers subjected to various loading and environmental effects such as axial tension, contact and adhesion, wave propagation, surface ripping, and wetting, among others. In the recent years, the first author and his co-workers have proposed several continuum mechanics models [121-131] to quantitatively approach the mechanical response of nanofibers by taking into account the effect of surface energy. Due to introduction of the new material parameter of surface energy (surface tension γ: J/m^2) to the nanofiber models, a new intrinsic dimensional parameter $l_0=\gamma/E$ (E: elastic modulus) appears naturally during the modeling process, which can be treated as a criterion whether or not to consider the size effect of the polymer nanofibers. Basically, when the characteristic dimension (i.e., fiber diameter d) is comparable to l_0, the size effect due to the surface energy is appreciable; otherwise, the size effect can be safely ignored, i.e., the surface energy can be ignored in the modeling. It needs to be noted that for polymer nanofibers, especially for those of ultra compliant polymer nanofibers, their moduli are orders lower than those of metallic materials. Therefore, the intrinsic dimensional parameter l_0 of compliant polymer nanofibers is orders larger than those of metallic materials. As a result, compliant polymer nanofibers behave more obvious size effect compared to metallic nanomaterials. In the below, we briefly review several mechanics models relating the mechanical behavior of single nanofibers and fiber networks.

1. Contact and Adhesion of Nanofibers [124]

In electrospinning process, the polymer nanofibers are collected in the form of nonwoven nanofiber mat, in which contact and adhesion between nanofibers is the most common phenomenon.

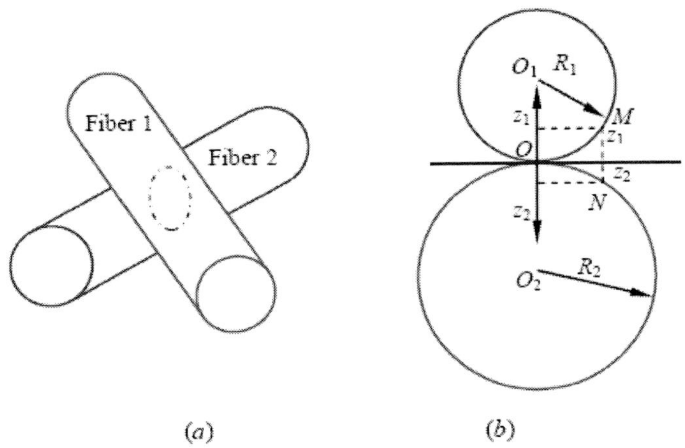

Figure 16. Cylindrical filaments in contact (*a*) contact zone between filaments in contact (*b*) coordinate systems [124]. Copyright (2007) of the Institute of Physics.

The simplest model to count for adhesive contact between nanofibers can be established using the principle of elastic contact between two elastic cylinders positioned in an arbitrary angle and with consideration of the effect of surface adhesion [124]. Such a contact problem can be treated as a special case of the DMT (Derjaguin-Muller-Toporov) theory [155] or the generalized Hertz contact theory [156-159].

Here, let us first consider the classic Hertz contact theory for the contact of two elastic bodies positioned in an arbitrary angle as shown in figure 16. Establish the (x,y)-plane tangent to both the cylindrical surfaces of the cylinder contacting at the origin O. The curvilinear surfaces near the contact point O can be described using two polynomials (ignore the high-order terms):

$$z_1 = A_1x^2 + A_2xy + A_3y^2.$$
$$z_2 = B_1x^2 + B_2xy + B_3y^2. \qquad (6)$$

The distance (gap) between two points at the two curvilinear surfaces such as M and N [see figure 16(*b*)] is

$$z_1 + z_2 = (A_1 + B_1)x^2 + (A_2 + B_2)xy + (A_3 + B_3)y^2, \qquad (7)$$

where A_i (i = 1, 2, and 3) and B_i (i = 1, 2, and 3) are the geometrical parameters that are determined by the surfaces near the contact point above and below. By properly selecting the (x, y)-coordinate system, the cross term xy can be eliminated. Thus, relation (7) can be simplified as

$$z_1 + z_2 = Ax^2 + By^2, \qquad (8)$$

where A and B are coefficients relevant to the principal curvatures of the surfaces in contact and the orientation angle between the two cylinders as

$$A+B=\frac{1}{2}(\frac{1}{R_1}+\frac{1}{R_1^{/}}+\frac{1}{R_2}+\frac{1}{R_2^{/}}),$$
$$B-A=\frac{1}{2}[(\frac{1}{R_1}-\frac{1}{R_1^{/}})^2+(\frac{1}{R_2}-\frac{1}{R_2^{/}})^2+2(\frac{1}{R_1}-\frac{1}{R_1^{/}})(\frac{1}{R_2}-\frac{1}{R_2^{/}})\cos 2\varphi]^{1/2}. \tag{9}$$

In the above, R_1 and $R_1^{/}$ are the principal radii of curvature of the curvilinear surface above; R_2 and $R_2^{/}$ are those of the curvilinear surface below; φ is the orientation angle between the two axes of the cylinders. In the case of two fibers with the equivalent diameter R and fiber orientation angle φ, the coefficients A and $B$are

$$A=\frac{1-\cos\varphi}{2R},\quad B=\frac{1+\cos\varphi}{2R}. \tag{10}$$

Theory of elasticity [157] shows that the sum of the displacements w_1 and w_2 of the points on the curvilinear surfaces in contact above and below is

$$w_1+w_2=(k_1+k_2)\int\frac{p}{r}dA, \tag{11}$$

where pdA is the pressure exerted on an elemental area of the contact surface, r is the distance from this elemental area to the contact center [see figure 16(b)], and k_1 and k_2 are two parameters defined by

$$k_1=\frac{1-\nu_1^2}{\pi E_1},\ k_2=\frac{1-\nu_2^2}{\pi E_2}, \tag{12}$$

where E_i and ν_i (i=1,2) are Young's moduli and Poisson's ratios of the elastic bodies above and below, respectively. The integration (11) must be considered over the entire surface in contact:

$$(k_1+k_2)\int\frac{p}{r}dA=\alpha-Ax^2-By^2, \tag{13}$$

where α the distance reduction between the two elastic bodies subjected to compression, and w_1 and w_2 the displacements of the points on the surfaces in contact above and below, respectively.

Hertz contact theory [156-158] gives the pressure p over the contact surface to satisfy Eq. (13) that is represented by the ordinates of a semi-ellipsoid spanning on the entire contact surface. The peak pressure p_0 is located at the center of the contact zone. By denoting a and b as the semi-axes of the ellipse of the contact zone, the peak pressure can be determined by

$$P=\int pdA=\frac{2}{3}\pi abp_0, \tag{14}$$

and

$$p_0 = \frac{3}{2}\frac{P}{\pi ab}. \tag{15}$$

In the above, P is the compressive force. In the case of polymer nanofibers, P is the resultant adhesive force between two nanofibers in contact that can be estimated by some adhesive theory. The peak pressure is 1.5 times the average pressure on the contact surface, and α is related to the compressive force P and the semi-axes of the ellipse a and b [157]:

$$\alpha = \frac{3P(k_1 + k_2)}{4}\int_0^{\infty}\frac{d\xi}{\sqrt{(a^2+\xi)(b^2+\xi)\xi}}, \tag{16}$$

where the semi-axes a and b are determined as

$$a = m\sqrt[3]{\frac{3\pi}{4}\frac{P(k_1+k_2)}{A+B}}, \quad b = n\sqrt[3]{\frac{3\pi}{4}\frac{P(k_1+k_2)}{A+B}}. \tag{17}$$

In above coefficients m and n are numbers depending only on the ratio (B-A) to (A+B) as tabulated in Table 1 [157].

The compressive force between two contacting nanofibers can be understood as adhesive forces which can be approximated according to Bradley's approach [160]. In this approach, the adhesive force between two unit areas is assumed to follow the long-range Lennard-Jones force:

$$\sigma(z) = \frac{8\Delta\gamma}{3\varepsilon}[(\frac{\varepsilon}{z})^3 - (\frac{\varepsilon}{z})^9]. \tag{18}$$

In the above, ε is a phenomenological distance between two atoms/molecules; z is the distance between two unit areas; and $\Delta\gamma$ is Dupré adhesion energy that is defined by

$$\Delta\gamma = \gamma_1 + \gamma_2 - \gamma_{12}, \tag{19}$$

where γ_1 and γ_2 are respectively the surface energies of the filaments, and γ_{12} is the interface energy. For two equivalentnanofibers in contact, $\Delta\gamma$=2γ. Thus, the adhesive force between two contacting nanofibers can be determined by performing the following integration

$$P = \frac{8\Delta\gamma}{3\varepsilon}\int_A \sigma(z)dA = \frac{8\Delta\gamma}{3\varepsilon}\int_A [(\frac{\varepsilon}{Ax^2+By^2-\alpha+h_0})^3 - (\frac{\varepsilon}{Ax^2+By^2-\alpha+h_0})^9]dA, \tag{20}$$

where the integration domain is over the region out of the contact zone on the tangential plane, i.e., $Ax^2+By^2 \geq \alpha$, h_0 is the minimum gap between surfaces in contact.

Table 1. *m* and *n* values for varying angle φ [157]

θ°	30°	35°	40°	45°	50°	55°	60°	65°	70°	75°	80°	85°	90°
m	2.731	2.397	2.136	1.926	1.754	1.611	1.486	1.378	1.284	1.202	1.128	1.061	1.000
n	0.493	0.530	0.567	0.604	0.641	0.678	0.717	0.759	0.802	0.846	0.893	0.944	1.000

According to Bradley's approach, it assumes $h_0 = \varepsilon$. Substitution of (10) into (20) leads to the compressive force P [124]:

$$P = \frac{4\pi\gamma R}{\sin\varphi}. \tag{21}$$

Consequently, the semi-axes of the elliptical contact zone can be determined by substituting (10) and (21) into (17) as

$$a = m\sqrt[3]{\frac{6\pi(1-\nu^2)}{\sin\varphi}\frac{R^2\gamma}{E}}, \quad b = n\sqrt[3]{\frac{6\pi(1-\nu^2)}{\sin\varphi}\frac{R^2\gamma}{E}}, \tag{22}$$

and m and n are two functions with respect to φ, respectively, as tabulated in Table 1 [157]. The distance reduction of the nanofibers α due to the distributed adhesive forces over the two curvilinear surfaces can be determined by substituting (21) and (22) into (13):

$$\alpha = \frac{6\pi(1-\nu^2)}{\sin\varphi}\frac{\gamma R}{E}\int_0^{\infty}\frac{d\xi}{\sqrt{(a^2+\xi)(b^2+\xi)\xi}}. \tag{23}$$

The above relations (20-23) govern the adhesive contact in polymer nanofibers. As an example, the area of the contact zone can be determined by relation (22) as

$$A_c = \pi ab = \pi mn\left[\frac{6\pi(1-\nu^2)}{\sin\varphi}\frac{R^2\gamma}{E}\right]^{2/3}. \tag{24}$$

When the two fibers are cross-positioned (orthogonal), the area of the contact zone reaches its minimum value as [124]

$$(A_c)_{\min} = 6^{2/3}\pi^{5/3}(1-\nu^2)^{2/3}\left(\frac{R^2\gamma}{E}\right)^{2/3}. \tag{25}$$

Figure 17 shows the numerical results of the area of contact zone with varying fiber diameters and fiber surface tension [124]. In the numerical process, the surface tensions of the nanofibers are selected as γ=0.05 N/m and γ = 0.1 N/m, respectively. The Young's modulus of the fibers is given as E=20 GPa. Two families of polymer nanofibers are used. The first family has the fiber radii of 2μm, 5μm, 10μm, 15μm, and 20μm; and the second has the fiber radii of 100nm, 200nm, 500nm, 1000nm, and 1500nm.

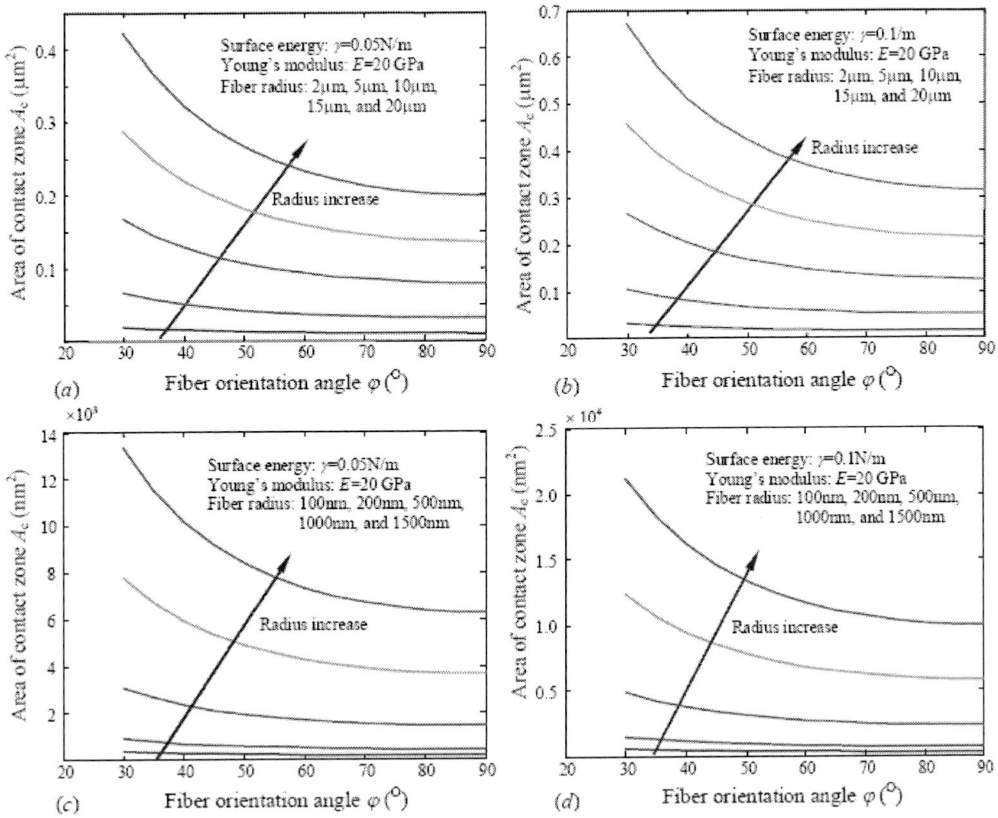

Figure 17. Variation of the area of contact zone vs. the angle between fibers (*a*) $\gamma = 0.05$N/m and (*b*) $\gamma = 0.1$N/m for fiber radii of 2μm, 5μm, 10μm, 15μm, and 20μm, respectively, and (*c*) $\gamma = 0.05$N/m and (*d*) $\gamma = 0.1$N/m for fiber radii of 100nm, 200nm, 500nm, 1μm, and 1.5μm, respectively [124]. Copyright (2007) of the Institute of Physics.

It can be observed from figure 17 that at a given polymer nanofiber radius, the contact are decreases rapidly with increasing angle φ between two fibers due to the rapid decrease of adhesive force with growing angle φ [124].

The above model can be used for determining the contact force, contact stress, and area of contact zone. The results given by the above model can be consideredfor exploring the mechanisms of nanofiber friction, sliding, and compression hysteresis in nanofibrous materials subjected to external loading. These parameters are expected important when taking into account the mechanical response of ultrathin soft nanofibers, where the adhesive force-induced contact will result in significant deformation and stresses in the nanofibers.

2. Collapse of Adhesive Nanofibers [125]

In electrospinning, the nanofibers are commonly collected as porous nonwoven nanofiber mat, where a majority of fibers are suspended. Besides, in many nanofiber-related applications such as nanofiber sensors and transducers, the nanofibers will work as free-standing member. Thus, adhesion between the nanofiber and surroundings (e.g., other nanofibers, substrates, etc.) may significantly influence the performance of the nanofiber.

Herein, we define the phenomenon that due to external perturb, a nanofiber is stuck to another nanofiber or substrate as nanofiber *collapse*. Figure 18 shows a typical example of nanofiber collapse due to surface adhesion. Consider a simple case of two arbitrarily positioned nanofibers collapsing due to surface adhesion as shown in figure 19. To simplify the process, the two fibers are assumed to carry the same fiber diameter, length and surface properties.

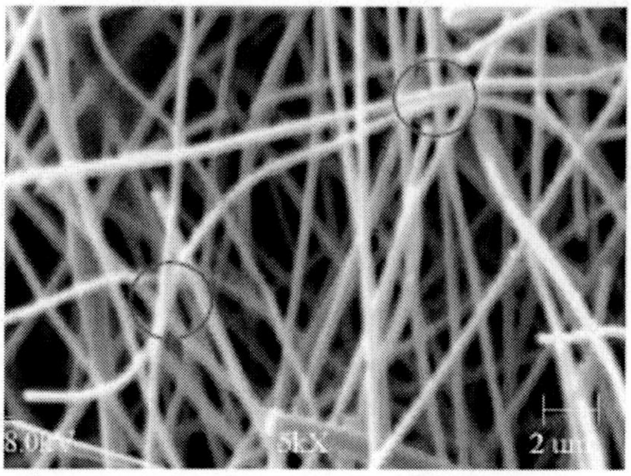

Figure 18. Adhesion between nanofibers within a PAN nanofiber network (Nanofiber diameter: ~300 nm, circles indicate the adhesion zones) [125]. Copyright (2007) of the Institute of Physics.

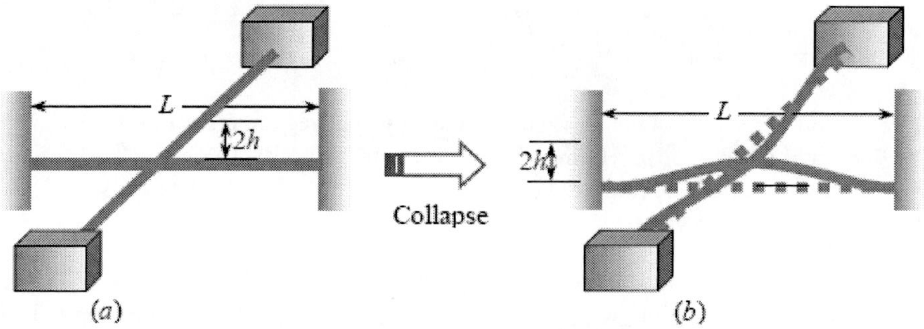

Figure 19. Adhesion between two nanofiber segmentspositioned in an arbitrary angle [125]. Copyright (2007) of the Institute of Physics.

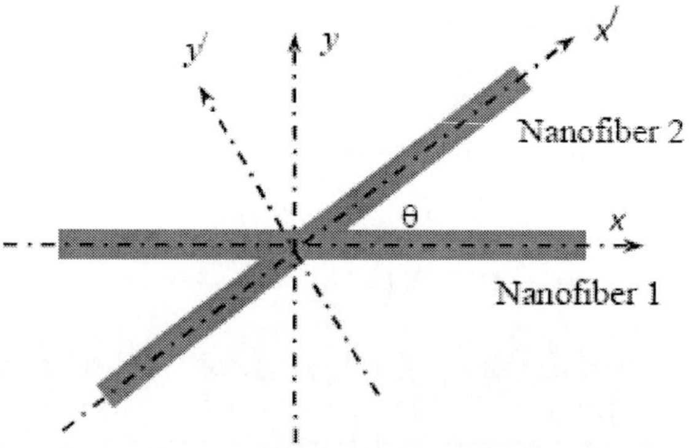

Figure 20. Geometries of two fibers positioned at an arbitrary angle θ.

Consider two uniform fiber segments positioned in two parallel horizontal planes with a distance h. The spatial angle between the fiber axes is denoted as θ. Simple relationships can be established between two coordinate systems attached to the fiber axes as shown in figure 20:

$$\begin{cases} x' = x\cos\theta + y\sin\theta, \\ y' = -x\sin\theta + y\cos\theta. \end{cases} \tag{26}$$

After fiber collapse due to surface adhesion as shown in figure 19(b), the asymptotic distance between two points on the surfaces of deflected fiber segments near the contact point is [125]

$$\begin{aligned} z &= 3h[(\frac{2x}{L})^2 - \frac{2}{3}(\frac{2x}{L})^3] + \frac{y^2}{2r} + 3h[(\frac{2x'}{L})^2 - \frac{2}{3}(\frac{2x'}{L})^3] + \frac{y'^2}{2r} + h_0 \\ &\approx \frac{12h}{L^2}x^2 + \frac{y^2}{2r} + \frac{12h}{L^2}x'^2 + \frac{y'^2}{2r} + h_0 \\ &\approx \frac{12h}{L^2}[(1+\cos^2\theta)x^2 + \sin 2\theta xy + \sin^2\theta y^2] \\ &+ \frac{1}{2r}[\sin^2\theta x^2 - \sin 2\theta xy + (1+\cos^2\theta)y^2] + h_0, (\text{for } x/L \ll 1 \text{ and } x'/L \ll 1) \end{aligned} \tag{27}$$

where local coordinates have been used. By using Bradley's approach [160], integrating relation (18) over the area out of the contact zone yieldsthe adhesive force between the two fibers:

$$\begin{aligned} P &= \int_\Gamma \sigma(z)dA = \frac{8\Delta\gamma}{3\varepsilon}\int_{-\infty}^{+\infty}\int_{-\infty}^{+\infty}[(\frac{\varepsilon}{z})^3 - (\frac{\varepsilon}{z})^9]dxdy, \\ &= \pi\Delta\gamma L/\sqrt{D}, \end{aligned} \tag{28}$$

where D is the determinant of a positive definite 2×2 matrix relating the fiber aspect ratio L/r, fiber distance h_c, and angle between fibers θ, i.e.

$$D = \begin{vmatrix} (1+\cos^2\theta)\frac{12h}{L} + \sin^2\theta\frac{L}{2r} & \sin\theta\cos\theta(\frac{12h}{L} + \frac{L}{2r}) \\ \sin\theta\cos\theta(\frac{12h}{L} + \frac{L}{2r}) & \sin^2\theta\frac{12h}{L} + (1+\cos^2\theta)\frac{L}{2r} \end{vmatrix}. \tag{29}$$

Substitution of (28) into the deflection formula of the nanofiber segment fixed at two ends and with a midspan deflection h: $P = 48\pi Er^4h/L^3$ leads to the critical condition for nanofiber collapse as

$$\frac{h_c}{L} = \frac{1}{48\sqrt{D}}\frac{\Delta\gamma}{Er}(\frac{L}{r})^3. \tag{30}$$

With Dupré adhesion energy $\Delta\gamma=2\gamma$ for two equivalent nanofiber segments, relation (30) can be rewritten as

$$\frac{h_c}{L}=\frac{1}{24\sqrt{D}}\frac{\gamma}{Er}(\frac{L}{r})^3. \tag{31}$$

Furthermore, in the special cases of $\theta = 0^\circ$ and $\theta = 90^\circ$, the corresponding collapse conditions are respectively

$$(\frac{h_c}{L})^{3/2}=\frac{1}{48\sqrt{6}}\frac{\gamma}{Er}(\frac{L}{r})^{5/2}, \tag{32}$$

and

$$\frac{h_c}{L}=\frac{1}{12}(\frac{L}{r})^4\frac{\gamma}{Er}. \tag{33}$$

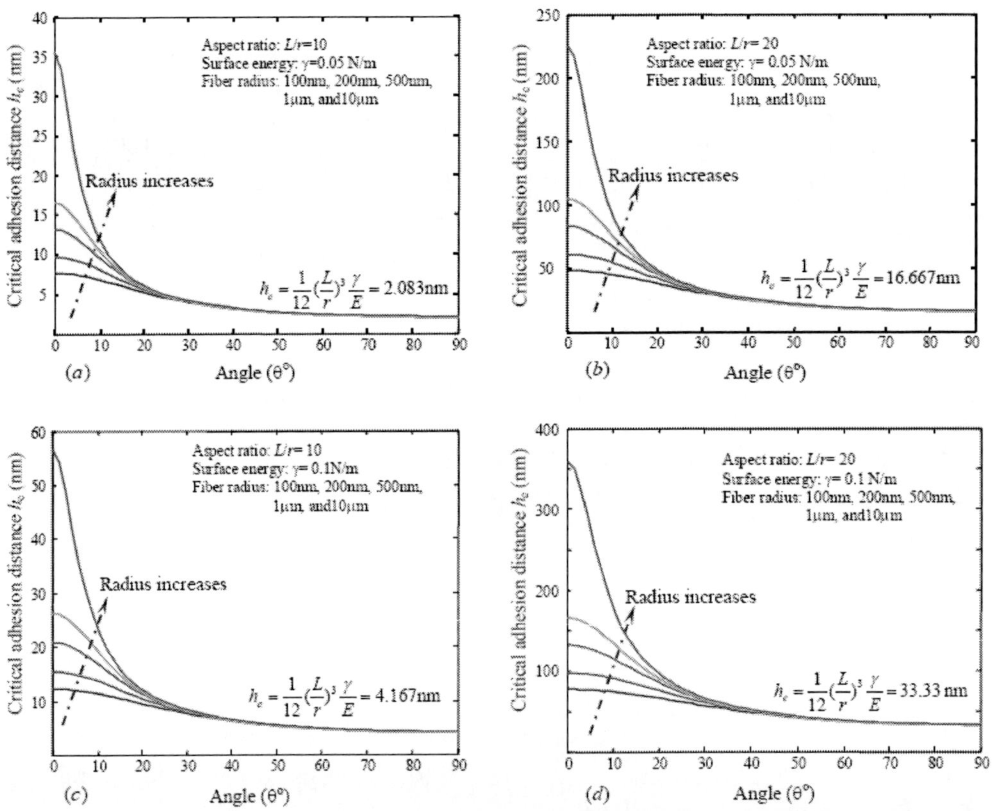

Figure 21. Variation of the critical collapse distance h_c between fiber segments vs. the nanofiber orientation angle θ at varying fiber radius r and aspect ratio L/r: (*a*) $L/r = 10$, $\gamma = 0.05$ N/m; (*b*) $L/r = 20$, $\gamma = 0.05$ N/m; (*c*) $L/r = 10$, $\gamma = 0.1$ N/m; and (*d*) $L/r = 20$, $\gamma = 0.1$ N/m [125]. Copyright (2007) of the Institute of Physics.

The general collapse criterion (31) involves the intrinsic length $l_0 = \gamma/E$, thus the following conclusion can be drawn that size effect does exist to govern the nanofiber collapse. For the convenience of discussion, relation (31) can be further cast into

$$\frac{h_c}{r} = \frac{1}{24\sqrt{D}}\frac{\gamma}{Er}\left(\frac{L}{r}\right)^4. \tag{34}$$

Numerical results on the dependency of the critical collapse distance upon the surface energy (surface tension) γ, fiber radius r are plotted in figure 21. During the numerical process, the surface energies of the fiberare selected as $\gamma = 0.05$ N/m and $\gamma = 0.1$ N/m, respectively; the Young's modulus is chosen as $E = 2$ GPa [125]. The numerical results shown in figure 21 clearly indicate thesize effect. Given a fiber pair, the critical collapse distance h_c decreases rapidly with increasing angle θ; meanwhile the critical collapse distance decreases with increasing surface energy. In addition, given an aspect ratio of the fiber segment L/r, at a small angle θ, h_c increases with the increase of fiber radius; however, at a relatively large angle θ, h_c tends to a constant, i.e., $(1/12)(L/r)^3\gamma/E$. This constant depends only upon the aspect ratio L/r and the intrinsic length $\gamma/E = 0.025$ nm for $\gamma = 0.05$ N/m, and $\gamma/E = 0.05$ nm for $\gamma = 0.1$ N/m. This parameter is expected to be useful for collapse analysis and reliable design of nanofiber networks as well as nanofiber devices. In addition, at a given fiber radius, h_c grows abruptly with the increase of L/r. Such effect can be interpreted in the sense of variation of the bending stiffness of a fiber segment which rapidly decreases with increasing fiber segment length based on a reciprocal cubic law.

3. Axial Deformation, Wave Propagation and Surface Rippling in Polymer Nanofibers [122, 123,126]

As discussed in the above, continuous nanofibers produced by electrospinning represent a new class of 1D nanomaterials. Their continuity, high surface area to volume ratio, and tailorable diameter (1-2,000 nm), surface morphology and microstructures make electrospun nanofibers numerous innovative applications in nanocomposites, filtration, protective clothing, biomedical engineering and technologies, nanodevices, etc. [7-17]. The mechanical behaviors of nanofibers are fundamental to their broad applications. In this section, simple 1D nonlinear continuum mechanics models [122, 123, 126] developed recently by the first author and his co-workers are reviewed; these models have taken into account the combined effect of surface energy and nonlinear elasticity in the mechanical responses (e.g., deformation, wave propagation, and surface rippling, etc.) of compliant polymer nanofibers. These models are capable of interpreting a series of size-related mechanical phenomena as observed in polymer nanofibers, and can also predict new surface stability conditions useful to nanofiber fabrication.

3.1. Static Tensile Deformation Subjected to Axial Stretching [123]

Consider a compliant polymer nanofiber as a circular cylinder at stretch-free state. Two configurations are adopted to describe the motion of a material point of the fiber, i.e., the stretch-free configuration (with surface tension ignored) and the current configuration subjected to a finite axial stretch. The corresponding coordinates of the material point are denoted by (R, Θ, Z) and (r, θ, z), respectively. Subjected to uniform axial stretching, the axisymmetric deformation can be described as:

$$r = \lambda_1 R \ (0 \le R \le R_0), \qquad \theta = \Theta \ (0 \le \Theta \le 2\pi),$$
$$z = \lambda_3 Z \ (-\infty \le Z \le +\infty), \tag{35}$$

where R_0 is the initial radius of the stretch-free nanofiber (with surface-tension ignored), and λ_1 and λ_3 are the transverse and longitudinal stretches, respectively.

The constitutive law of the polymer nanofibers is modeled to follow that of isotropically, incompressible hyperelastic Mooney-Rivlin solid [161], which can be expressed as the Cauchy stress tensor **T** vs. the left Cauchy-Green tensor **B** corresponding to deformation (35):

$$\mathrm{T} = -p\mathrm{I} + 2c_1\mathrm{B} - 2c_2\mathrm{B}^{-1}, \tag{36}$$

where p is the hydrostatic pressure, and c_1 and c_2 are two material constants. The stress equilibrium equations are:

$$\partial T_{rr} / \partial r + (T_{rr} - T_{\theta\theta}) / r = 0, \tag{37}$$

$$\partial T_{\theta\theta} / \partial \theta = 0, \tag{38}$$

$$\partial T_{zz} / \partial z = 0. \tag{39}$$

Two traction boundary conditions (BCs) on the fiber surface and along the axial direction are evoked:

$$T_{rr} = -\gamma / r_0 \ \ (r = r_0), \tag{40}$$

$$P = 2\pi \int_0^{r_0} r T_{zz} dr + 2\pi r_0 \gamma. \tag{41}$$

In the above, γ (N/m) is the surface energy of the amorphous polymer fiber which is assumed being independent of the current fiber radius r_0 or the applied axial stretch λ_3, and P is the axial tensile force. Solving (37-39) under BCs in (40) and (41) yields [123]

$$P = 2\pi R_0^2 (\lambda_3 - \lambda_3^{-2})(c_1 - c_2 \lambda_3^{-1}) + \pi R_0 \gamma \lambda_3^{-1/2}, \tag{42}$$

which shows the surface/size effect in the axial deformation.

3.2. Longitudinal Wave Propagation in Pre-Stretched Nanofibers [126]

When considering longitudinal wave propagation in a pre-stretched polymer nanofiber, the dynamic motion is treated as a small disturbance superimposed on the axisymmetric deformation of the nanofiber given in (35). The perturbed state is described by coordinates ($\tilde{r}$, $\tilde{\theta}$, $\tilde{z}$). Thus, for a small disturbance, coordinates of a material point in the perturbed configuration can be described as

$$\tilde{r} = [\lambda_1 + f(Z,t)]R, \qquad \tilde{\theta} = \Theta, \qquad \tilde{z} = \lambda_3 Z + g(Z,t), \tag{43}$$

where $f(Z, t)$ and $g(Z, t)$ are two small time-dependent disturbance functions satisfying BCs in (40) and (41).

The governing wave propagation equation can be derived by using the principle of least action (Hamilton's law):

$$\delta \int_{t_1}^{t_2} L dt = 0, \tag{44}$$

where

$$L = T - \Pi. \tag{45}$$

In the above, t_1 and t_2 are two arbitrary starting and ending times, L is the modified lagrangian, and T and Π are respectively the kinetic energy and potential energy of the nanofiber segment under consideration which are defined as [126]

$$T = 1/4\pi R_0^4 \int_{L_0} \rho \dot{f}^2 dZ + 1/2\pi R_0^2 \int_{L_0} \rho \dot{g}^2 dZ, \tag{46}$$

$$\Pi = 2\pi \int_L \int_0^{R_0} eRdRdZ + 2\pi R_0 \gamma \int_{L_0} (\lambda_1 + f)(\lambda_3 + g_z) dZ - P\int_{L_0} (\lambda_3 + g_z - 1) dZ, \tag{47}$$

in which $\dot{f}$ and are respectively the transverse deformation rate (strain rate) and the axial translational speed of a material point, and ρ is the mass density, e is the strain energy density of incompressible hyperelastic Mooney-Rivlin solid defined by [161]

$$e = c_1(I_1 - 3) + c_2(I_2 - 3), \tag{48}$$

with I_1 and I_2 respectively the first and second invariants of the left Cauchy-Green tensor of perturbed deformations given in (43). The three terms in (47) are the contributions due to elastic strain energy, surface energy, and mechanical work done by the axial force P, respectively. In addition, the integration with respect to Z runs over the entire length L_0 of the fiber segment under consideration.

After several variational operations and eliminating the higher order terms based on small perturbation assumption, the necessary condition for an extremum of functional in (44) leads to a linearized longitudinal wave propagation equation:

$$\begin{aligned} &R_0^2\rho\ddot{g}_{ZZ} - 8\lambda_3^3\rho\ddot{g} - 2(c_1 + c_2\lambda_3^{-1})R_0^2 g_{ZZZZ} \\ &+ 8(4c_1 + 2c_1\lambda_3^3 + 6c_2\lambda_3^{-1} - \gamma/(2R_0)\lambda_3^{1.5})g_{ZZ} = 0. \end{aligned} \tag{49}$$

To determine the wave dispersion relation, assume the travelling longitudinal wave to carry the mode:

$$g(Z,t) = A_0 \exp[ik(Z-ct)], \tag{50}$$

where A_0 is the complex amplitude of the wave disturbance, k is the wave number, and c is the wave phase speed. Substituting (50) into (49) results into the wave dispersion relation:

$$Ak^2c^2 + Bc^2 - Ck^2 - D = 0, \tag{51}$$

where

$$A = R_0^2\rho,\ B = 8\lambda_3^3\rho,\ C = 2(c_1 + c_2\lambda_3^{-1})R_0^2, \tag{52}$$

$$D = 8[4c_1 + 2c_1\lambda_3^3 + 6c_2\lambda_3^{-1} - \gamma/(2R_0)\lambda_3^{1.5}]. \tag{53}$$

By introducing the intrinsic length l_0 [123] and reference wave phase speed c_0 of the polymer material:

$$l_0 = \gamma/c_1,\ c_0 = \sqrt{c_1/\rho}, \tag{54}$$

Relation (51) can be recast as

$$(\frac{c}{c_0})^2 = \frac{2[1+(c_2/c_1)\lambda_3^{-1}]k_0^2 + 8[4+2\lambda_3^3 + 6(c_2/c_1)\lambda_3^{-1} - 0.5(l_0/R_0)\lambda_3^{1.5}]}{k_0^2 + 8\lambda_3^3}. \tag{55}$$

In the above, if ignoring the transverse motion of the nanofiber in (49), the wave speed can be determined from (55) as

$$(c/c_0)^2 = 4\lambda_3^{-3} + 2 + 6(c_2/c_1)\lambda_3^{-4} - 0.5(l_0/R_0)\lambda_3^{-1.5}. \tag{56}$$

In addition, in the limiting case of wave propagation in a stretch-free nanofiber (λ_3=1), i.e., an infinitesimal deformation, relation (56) can be reducedinto

$$c = \sqrt{E_T/\rho - \gamma/(2\rho R_0)}, \tag{57}$$

where E_T is the tangential modulus of hyperelastic Mooney-Rivlin solid at $\lambda_3 = 1$:

$$E_T = 6(c_1 + c_2). \tag{58}$$

Relation (57) covers the case of longitudinal waves propagating in linearly elastic nanowires (with Poisson's ratio $\upsilon = 0.5$) [122].

3.3. Surface Rippling of Nanofibers Subjected to Axial Pre-Stretch [126]

The governing equation of longitudinal wave propagation (49) can be utilized for the study of surface rippling phenomenon as observed recently in compliant as-electrospun PAN nanofibers subjected to axial stretching [126, 144, 145, 150]. Governing equation of surface rippling in polymer nanofibers can be obtained by eliminating the two inertia terms in Eq. (49) such that

$$Cg_{ZZZZ} - Dg_{ZZ} = 0, \tag{59}$$

where C and D are defined in (52) and (53). The ripple wave number k can be determined by assuming the rippling modes in the form:

$$g(Z) = B_0 \exp(ikZ), \tag{60}$$

where B_0 is the complex amplitude of wave disturbance similar to that in (50). Substitutionof (60) into (59) results in the ripple wave number:

$$k^2 = -D/C. \tag{61}$$

For physically meaningful rippling mode, k must be a positive number. This condition implies that surface rippling happens only under the condition $D < 0$.

Based on relation (53), the critical condition to trigger surface rippling in compliant nanofibers can be determined as

$$R_0 < (l_0/4)/[2\lambda_3^{-1.5} + \lambda_3^{1.5} + 3(c_2/c_1)\lambda_3^{-2.5}]. \tag{62}$$

Relation (62) can be further used to determine an important geometrical parameter in nanofiber fabrication, i.e., the minimum critical fiber radius $(R_C)_{min}$, below which surface rippling can even happen spontaneously without any pre-stretch (i.e., $\lambda_3 = 1$). By setting $\lambda_3=1$ in relation (62), it leads to

$$(R_C)_{min} = \gamma/(2E_T) = (\gamma/12)/(c_1 + c_2). \tag{63}$$

For typical rubbery polymer nanofibers, relation (63) determines that $(R_C)_{min} \approx 10\text{-}40$ nm. In principle, below $(R_C)_{min}$, compliant nanofibers cannot exist due to their spontaneous surface destabilization (rippling). This indicates that it is physically impossible to fabricate compliant nanofibers with the fiber radius below $(R_C)_{min}$. Detailed discussions on surface rippling in polymer nanofibers are available elsewhere [126]. In summary, the above 1D nonlinear continuum mechanics models can be utilized to clarify the dependencies of axial deformation, longitudinal wave dispersion, and surface rippling condition in compliant polymer nanofibers upon the surface energy, nonlinear elastic properties, and fiber diameter. These results can be considered as the theoretical basis of manufacturing, static and dynamic characterization of compliant nanowires/nanofibers, nanofiber device design, etc.

3.4. Hydroelastic Response of Nanofibers [129]

In Section 3.3, 1D continuum mechanics models have been discussed to understand the mechanical behavior of nanofibers. In reality, a significant number of polymers work in aqueous environments such as human body wound dressing, in-plants, liquid filters, etc. In such cases, moisture and liquids may alter the mechanical behavior of these polymers, especially in the case of ultrathin fibers where the capillary force would be significant. Here introduce our recent study on the axial mechanical response of an ultrathin polymer fiber wetted with droplets and subjected to axial stretch as shown in figure 22 [129]. The polymer fiber is modeled as an incompressible, isotropic, hyperelastic Mooney-Rivlin solid [161] that can be expressed in terms of Cauchy stress tensor $\mathbf{T}$ vs. the left Cauchy-Green tensor $\mathbf{B}$ as shown in Eq. (36). Similar to those discussed in Section 3.3, the nanofiber is treated as an ultrathin circular cylinder of initial radius R_0. Two configurations are introduced to describe the deformation of the fiber, i.e., undisturbed stretch-free (without droplet and fiber surface stress) with coordinates (R, Φ, Z) and current configuration (after deformation under the capillary force and stretch) with coordinates (r, φ, z), respectively. We only consider the deformation of the fiber within the droplet and avoid the complicated phenomena near the wetting fronts.

The governing equations of the fiber are the same as Eqs. (37-39), while two traction BCs involving capillary effect are triggered. At the fiber surface, both interface tension and capillary effect yield uniform radial compression:

$$T_{rr} = -\gamma_{LS} / r_0 - p_c + 2\gamma_{LV} \sin\theta / L. \tag{64}$$

In the above, the first term is the compressive stress due to the specific interface energy γ_{LS} (N/m) which is assumed to be independent of the radius and the fiber axial stretch λ_3; and r_0 is the current radius; the second term is the capillary pressure p_c due to the droplet that can be determined by solving the Young-Laplace equation; the third term is the capillary forces at the two wetting fronts by averaging over the wetting length. In the axial direction, force equilibrium reads [129]

$$P = 2\pi \int_0^{r_0} r T_{zz} dr + 2\pi r_0 \gamma_{LV} \cos\theta, \tag{65}$$

where P is the axial tensile force required to maintain the axial equilibrium. Solving Eqs. (37-39) under BCs (64) and (65) with the aid of the available solution of a droplet wetting a fiber

[162,163] yields the relationship between the engineering stress σ of the fiber within the droplet and the axial stretch λ_3 [129]:

$$\sigma = \frac{P}{\pi R_0^2} = 2(\lambda_3 - \frac{1}{\lambda_3^2})(c_1 + \frac{c_2}{\lambda_3}) + \frac{2\gamma_{LV}\cos\theta - \gamma_{LV}}{\sqrt{\lambda_3}R_0} - \frac{p_c}{\lambda_3} + \frac{2\gamma_{LV}\sin\theta}{\lambda_3 L}. \tag{66}$$

In the above, p_c and L are respectively the capillary pressure and wetting length, which can be determined by solving the relevant Young-Laplace equation [162,163]. Numerical results of detailed scaling properties of relation (66) are plotted in figure 23 [129].

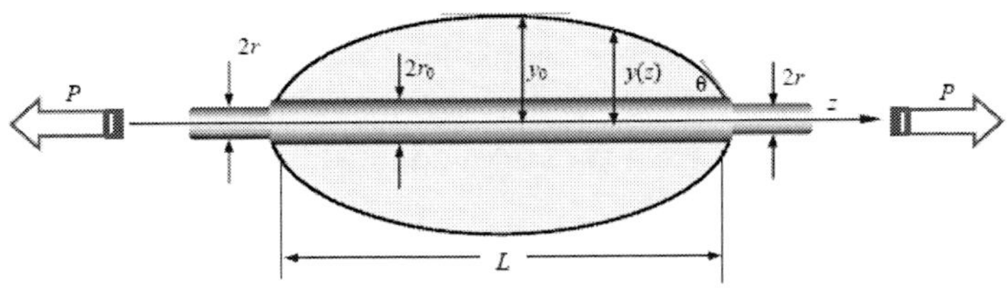

Figure 22. Geometries of a barrel-shaped droplet wetting on a thin compliant fiber subjected to axial stretch. In the diagram, r is the current fiber radius out of the droplet, and r_0 is the fiber radius within the droplet [129]. Copyright (2010) of the American Institute of Physics.

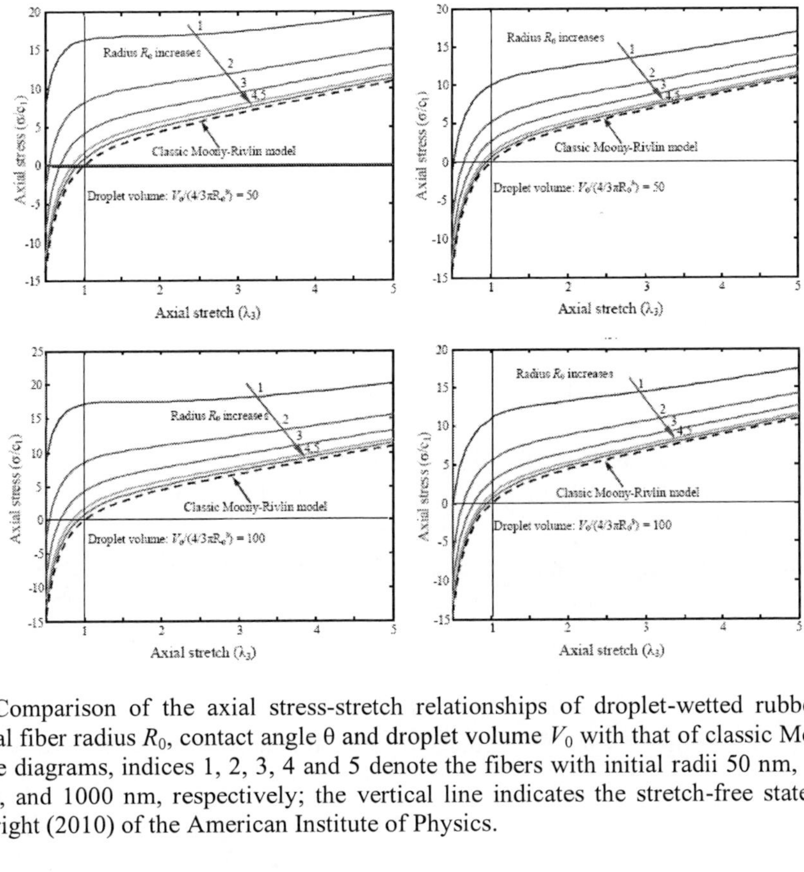

Figure 23. Comparison of the axial stress-stretch relationships of droplet-wetted rubbery fibers at varying initial fiber radius R_0, contact angle θ and droplet volume V_0 with that of classic Mooney-Rivlin model. In the diagrams, indices 1, 2, 3, 4 and 5 denote the fibers with initial radii 50 nm, 100 nm, 200 nm, 500 nm, and 1000 nm, respectively; the vertical line indicates the stretch-free state, i.e. $\lambda_3 = 1$ [129]. Copyright (2010) of the American Institute of Physics.

CONCLUDING REMARKS

In summary, in this chapter detailed review has been given to the production, mechanical characterization and mechanics modeling of continuous nanofibers by means of the novel, low-cost, top-down electrospinning technique. Several recently developed experimental techniques for nanofiber fabrication and mechanical characterization of continuous nanofibers are reviewed; the unique mechanical responses of electrospun nanofibers such as the size effect in the tensile mechanical behavior have been discussed. Several novel 1D continuum mechanics models have been discussed which can be utilized to correlate the experimental observations of electrospun nanofibers including the size effect in mechanical properties, wave propagation, surface rippling, adhesion and contact, surface wetting, and hydroelastic response, among others. These mechanics models are deemed useful to provide novel fundamental understanding of the principles for manufacturing and characterization of continuous electrospun nanofibers for a variety of applications.

ACKNOWLEDGMENT

Partial support of the research by the NSF, U.S. DOE EPSCoR-SUNRISE, ND NASA EPSCoR, ND Soybean Council, NDSU Development Foundation, and the Ashland Inc. (OH) is gratefully appreciated.

REFERENCES

[1] J. Doshi and D. H. Reneker, *J. Electrostatics* 35, 151 (1995).

[2] D. H. Reneker and I. Chun, *Nanotechnology* 7, 216 (1996).

[3] D. H. Reneker, A. L. Yarin, E. Zussman, and H. Xu, *Adv. Appl. Mech.* 41, 43 (2007).

[4] D. H. Reneker and A. L. Yarin, *Polymer* 49, 2387 (2008).

[5] Y. Dzenis, *Science* 304, 1917 (2004).

[6] D. Li and Y. N. Xia, *Adv. Mater.* (Weinheim, Ger.) 16, 1151 (2004).

[7] G. C. Rutledge and S. V. Fridrikh, *Adv. Drug Delivery Rev.* 59, 1384 (2007).

[8] Z. M. Huang, Y. Z. Zhang, M. Kotaki, and S. Ramakrishna, *Compos. Sci. Tech.* 63, 2223 (2003).

[9] A. Greiner and J. H. Wendorff, *Angew. Chem. Int. Ed.* 46, 5670 (2007).

[10] R. Ramaseshan, S. Sundarrajan, R. Jose, and S. Ramakrishna, *J. App. Phys.* 102, 111101 (2007).

[11] S. Ramakrishna, K. Fujihara, W. E. Teo, T. C. Lim, and Z. Ma, *An Introduction to Electrospinning and Nanofibers*, World Scientific, Singapore (2005).

[12] A. L. Andrady, *Science and Technology of Polymer Nanofibers*, Jon Wiley and Son (2008).

[13] Z. L. Wang (ed.), *Nanowires and Nanobelts: Materials, Properties and Devices*, volume 2: Nanowires and Nanobelts of Functional Materials, Springer, New York (2003).

[14] S. Guceri, Y. G. Gogotsi, and V. Kuznetsov (ed.), *Nanoengineered Nanofibrous Materials*, Kluwer Academic Publishers, Boston (2003).
[15] G. Cao, *Nanostructures and Nanomaterials: Synthesis, Properties and Applications*, Imperial College Press, London, U.K. (2004).
[16] I. S. Chronakis, *J. Mater. Processing. Tech.* 167, 283 (2005).
[17] A. Baji, Y. W. Mai, S. C. Wong, M. Abtahi, and P. Chen, *Compos. Sci. Tech.* 70, 703 (2010).
[18] G. Taylor, *Proc. Roy. Soc. Lond. A* 313, 453 (1969).
[19] A. F. Spivak and Y. A. Dzenis, *J. Appl. Mech.* 66, 1026 (1999).
[20] A. L. Yarin, S. Koombhongse, and D. H. Reneker, *J. Appl. Phys.* 89, 3018 (2001).
[21] C. Lai, Q. Guo, X. F. Wu, D. H. Reneker, and H. Hou, *Nanotechnology* 19, 195303 (2008).
[22] W. E. Teo and S. Ramakrishna, *Nanotechnology* 17, R89 (2006).
[23] P. Gibson, H. Schreuder-Gibson, and D. Rivin, *Colloids Surfaces A –Physicochem. Eng. Asp.* 187, 469 (2001).
[24] R. Gopal, S. Kaur, Z. W. Ma, C. Chan, S. Ramakrishna, and T. Matsuura, *J. Membrane Sci.* 281, 581 (2006).
[25] R. S. Barhate and S. Ramakrishna, *J. Membrane Sci.* 296, 1 (2007).
[26] B. Maze, H. V. Tafreshi, Q. Wang, and B. Pourdeyhimi, *J. Aerosol Sci.*38, 550 (2007).
[27] Y. M. Sang, Q. B. Gu, T. C. Sun, F. S. Li, and C. Z. Liang, *J. Hazardous M.* 153, 860 (2008).
[28] M. Bognitzki, H. Q. Hou, M. Ishaque, T. Frese, M. Hellwig, C. Schwarte, A. Schaper, J. H. Wendorff, and A. Greiner, *Adv. Mater.* 12, 637 (2000).
[29] W. X. Liu, M. Graham, E. A. Evans, and D. H. Reneker, *J. Mater. Res.* 17, 3206 (2002).
[30] J. Sutasinpromprae, S. Jitjaicham, M. Nithitanakul, C. Meechaisue, and P. Supaphol, *Polym. Int.* 55, 825 (2006).
[31] K. S. Yang, D. D. Edie, D. Y. Lim, Y. M. Kim, and Y. O. Choi, *Carbon* 41, 2039 (2003).
[32] J. S. Kim and D. H. Reneker, *Polym. Compos.* 20, 124 (1999).
[33] Y. Dzenis, *Science* 319, 419 (2008).
[34] X. F. Wu, *Fracture of Advanced Composites with Nanostructured Interfaces: Fabrication, Characterization and Modeling*, VDM Publishing House, Germany (2009).
[35] D. Smith, D. H. Reneker, W. Kataphinan, and A. Dabney, *US Patent* 6821479 (2001).
[36] D. Smith and D. H. Reneker, *US Patent* 6753454 (2004).
[37] W. J. Li, C. T. Laurencin, E. J. Caterson, R. S. Tuan, and F. K. Ko, *J. Biomed. Mater. Res.* 60, 613 (2002).
[38] J. A. Matthews, G. E. Wnek, D. G. Simpson, and G. L. Bowlin, *Biomacromolecules* 3, 232 (2002).
[39] C. Burger, B. S. Hsiao, and B. Chu, *Ann. Rev. Mater. Res.* 36, 333 (2006).
[40] Q. P. Pham, U. Sharma, and A. G. Mikos, *Tissue Eng.* 12, 1197 (2006).
[41] C. P. Barnes, S. A. Sell, E. D. Boland, D. G. Simpson, and G. L. Bowlin, *Adv. Drug Delivery Rev.* 59, 1413 (2007).
[42] J. Xie, X. Li, and Y. Xia, *Nature Mater.* 29, 1775 (2008).

[43] E. R. Kenawy, G. L. Bowlin, K. Mansfield, J. Layman, D. G. Simpson, E. H. Sanders, and G. E. Wnek, *J. Contr. Release* 81, 57 (2002).
[44] S. Y. Chew, Y. Wen, Y. Dzenis, and K. W. Leong, *Curr. Pharm. Design* 12, 4751(2006).
[45] D. Liang, B. S. Hsiao, and B. Chu, *Adv. Drug Delivery Rev.* 59, 1392 (2007).
[46] S. G. Kumbar, S. P. Nukavarapu, R. James, M. V. Hogan, and C. T. Laurencin, *Recent Patents Biomed. Eng.* 1, 68 (2008).
[47] K. Onozuka, B. Ding, Y. Tsuge, T. Naka, M. Yamazaki, S. Sugi, S. Ohno, M. Yoshikawa, and S. Shiratori, *Nanotechnology* 17, 1026 (2006).
[48] M. Y. Song, D. K. Kim, K. J. Ihn, S. M. Jo, and D. Y. Kim, *Nanotechnology* 15,1861 (2004).
[49] L. Jiang, Y. Zhao, and J. Zhai, *Ange. Chem. Int. Ed.* 43, 4438 (2004).
[50] Y. Miyauchi, B. Ding, and S. Shiratori,*Nanotechnology* 17, 5151 (2006).
[51] X. Y. Wang, C. Drew, S. H. Lee, K. J. Senecal, J. Kumar, and L. A. Samuelson, *J. Macromolecular Sci.* A39, 1251 (2002).
[52] B. Ding, J. H. Kim, Y. Miyazaki, and S. M. Shiratori, *Sensors and Actuators B-Chem.* 101, 373 (2004).
[53] D. Aussawasathien, J. H. Dong, and L. Dai, *Synthetic Metals* 154, 37 (2005).
[54] M. K. Tiwari, A. L. Yarin, and C. M. Megaridis, *J. Appl. Phys.* 103, 044305 (2008).
[55] B. Ding, M. Wang, J. Y. Yu, and G. Sun, *Sensors* 9, 1609 (2009).
[56] I. D. Norris, M. M. Shaker, F. K. Ko and A. G. MacDiarmid, *Synth. Metals* 114, 109 (2000).
[57] M. J. Laudenslager, R. H. Scheffler, and W. M. Sigmund, *Pure Appl. Chem.* 82, 2137 (2010).
[58] D. L. Schulz, J. Hoey, J. Smith, A. Elangovan, X. Wu, I. Akhatov, S. Payne, J. Moore, P. Boudjouk, L. Pederson, J. Xiao, and J. G. Zhang, *Electrochem. Solid-State Lett.* 13, A143 (2010).
[59] X. Zhang, L. Ji, O. Toprakci, Y. Liang, and M. Alcoutlabi, *Polym. Rev.* 51, 239 (2011).
[60] S. A. Theron, A. L. Yarin, E. Zussman, and E. Kroll, *Polymer* 46, 2889 (2005).
[61] A. L. Yarin and E. Zussman, *Polymer* 45, 2977 (2004).
[62] A. Sarkar and D. Lukas, *6th Int. Conf.-TEXSCI*, June 5-7, Czech Republic (2007).
[63] D. Lukas, A. Sarkar, and P. Pokorny, *J. Appl. Phys.* 103, 084309 (2008).
[64] D. Lukas, A. Sarkar, L. Martinova, K. Vodsed'alkova, D. Lubasova, J. Chaloupek, P. Pokorny, P. Mikes, J. Chvojka, and M. Komarek, *Textile Progress* 41, 59 (2009).
[65] S. Petrick and M. Maly, *Production of Nozzleless Electrospinning Nanofiber Technology* (Presentation), Liberc, Czech Republic (2009).
[66] Z. Sun, E. Zussman, A. Yarin, and J. H. Wendorff, and A. Greiner, *Adv. Mater.* 15, 1929 (2003).
[67] D. Li and Y. N. Xia, *Nano Lett.* 4, 933 (2004).
[68] D. Li, A. Babel, S. A. Jenekhe, and Y. N. Xia, *Adv. Mater.* 16, 2062 (2004).
[69] A. V. Bazilevsky, A. L. Yarin, and C. M. Megaridis, *Langmuir* 23, 2311 (2007).
[70] Y. Dror, W. Salalha, R. Avrahami, E. Zussman, A. L. Yarin, R. Dersch, A. Greiner, and J. H. Wendorff, *Small* 3 1064 (2007).
[71] A. L. Yarin, E. Zussman, J. H. Wendorff, and A. Greiner, *J. Mater. Chem.* 17, 2585 (2007).
[72] A. L. Yarin, *Polym. Adv. Tech.* 22, 310 (2011).

[73] T. Miloh, B. Spivak, and A. L. Yarin, *J. Appl. Phys.*106, 114910 (2009).
[74] X. F. Wu and Y. A. Dzenis, *J. Phys. D. Appl. Phys.* 38, 2848 (2005).
[75] L. D. Landau and E. M. Lifshitz, *Fluid Mechanics* (2nd Ed.), Oxford: Butterworth-Heinemann (1987).
[76] E. Schaffer, T. Thurn-Albrecht, T. P. Russell, and U. Steiner, *Nature* 403, 874 (2000).
[77] E. Schaffer, T. Thurn-Albrecht, T. P. Russell, and U. Steiner, *Europhys. Lett.* 53, 518 (2001).
[78] L. F. Pease and W. B. Russel, *J. Non-Newtonian Fluid Mech.* 102, 233 (2002).
[79] L. Wu and S. Y. Chou, *Appl. Phys. Lett.* 82, 3200 (2003).
[80] L. Wu and S. Y. Chou, *J. Non-Newtonian Fluid Mech.* 125, 91 (2005).
[81] R. Verma, A. Sharma, K. Kargupta, and J. Bhaumik, *Langmuir* 21, 3710 (2005).
[82] N. Wu, L. F. Pease, and W. B. Russel, *Langmuir* 21, 12290 (2005).
[83] N. Wu, L. F. III Pease, and W. B. Russel, *Adv. Functional Mater.* 16, 1992 (2006).
[84] D. Kim and W. Lu, *Phys. Rev. E* 73, 035206 (2006).
[85] S. Srivastava, P. D. S. Reddy, C. Wang, D. Bandyopadhyay, and A. Sharma, *J. Chem. Phys.* 132, 174703 (2010).
[86] V. G. Suvorov and E. A. Litvinov, *J. Phys. D: Appl. Phys.* 33, 1245 (2000).
[87] S. N. Reznik, A. L. Yarin, A. Theron, and E. Zussman, *J. Fluid Mech.* 516, 349 (2004).
[88] V. N. Kirichenko, I. V. Petryanov-Sokolov, N. N. Suprun, and A. A. Shutov, *Sov. Phys. Dokl.* 31, 611 (1986).
[89] A. F. Spivak and Y. A. Dzenis, *Appl. Phys. Lett.* 73, 3067 (1998).
[90] A. F. Spivak, Y. A. Dzenis, and D. H. Reneker, *Mech. Res. Comm.* 27, 37 (2000).
[91] M. M. Hohman, M. Shin, G. Rutledge, and M. P. Brenner, *Phys. Fluids* 13, 2201 (2001).
[92] M. M. Hohman, M. Shin, G. Rutledge, and M. P. Brenner, *Phys. Fluids* 13, 2221 (2001).
[93] Y. M. Shin, M. M. Hohman, M. P. Brenner, and G. C. Rutledge, *Appl. Phys. Lett.* 78, 1149 (2001).
[94] S. V. Fridrikh, J. H. Yu, M. P. Brenner, and G. C. Rutledge, *Phys. Rev. Lett.* 90, 144502 (2003).
[95] J. J. Feng, *Phys. Fluids* 14, 3912 (2002).
[96] J. J. Feng, *J. Non-Newtonian Fluid Mech.* 116, 55 (2003).
[97] I. Hayati, *Colloids Surface* 65, 77 (1992).
[98] F. J. Higuera, *J. Fluid Mech.* 484, 303 (2003).
[99] F. Yan, B. Farouk, and F. Ko, *Aerosol. Sci.* 34, 99 (2003).
[100] D. H. Reneker, A. L. Yarin, H. Fong, and S. Koombhongse, *J. Appl. Phys.* 87, 4531 (2000).
[101] A. L. Yarin, S. Koombhongse, and D. H. Reneker, *J. Appl. Phys.* 89, 3018 (2001).
[102] S. A. Theron, E. Zussman, and A. L. Yarin, *Polymer* 45, 2017 (2004).
[103] C. J. Thompson, G. G. Chase, A. L. Yarin, and D. H. Reneker, *Polymer* 48, 6913 (2007).
[104] Z. P. Zhou, X. F. Wu, X.Q. Gao, L. Jiang, Y. Zhao, and H. Fong, *J. Phys. D: Appl. Phys.* 44, 435401 (2011).
[105] P. Dayal and T. Kyu, *Phys. Fluids* 19, 107106 (2007).
[106] P. Dayal, J. Liu, S. Kumar, and T. Kyu, *Macromolecules* 40, 7689 (2007).
[107] X. F. Wu, Y. Salkovskiy, and Y. A. Dzenis, *Appl. Phys. Lett.,* 98, 223108 (2011).

[108] R. Saure, G. R. Wagner, and E. U. Schlunder, *Surface Coatings Tech.* 99 253 (1998).

[109] R. Saure, G. R. Wagner, and E. U. Schlunder, *Surface Coatings Tech.* 99, 257 (1998).

[110] P. J. Flory, *Principles of Polymer Chemistry*, (Cornell University Press, Ithaca, 1953).

[111] G. M. Bristow and W. F. Watson, *Trans. Farady Soc.* 54, 1731 (1958).

[112] S. J. Farlow, *Partial Differential Equations for Scientists and Engineers*, Dover, New York (1993).

[113] J. L. Duda, J. S. Vrentas, S. T. Ju, and H. T. Liu, *AIChE J.* 28, 279(1982).

[114] J. M. Zielinski and J. L. Duda, *AIChE J.* 38, 405 (1992).

[115] J. Crank, *The Mathematics of Diffusion*, Oxford University Press, Oxford (1975).

[116] Y. A. Dzenis and Y. Wen, *Mater. Res. Soc. Symp.-Proc.* 702, 173 (2002).

[117] D. H. Reneker and H. Fong (editors), *Polymeric Nanofibers*, Washington, DC: American Chemical Society (2005).

[118] F. P. Incropera and D. P. DeWitt, *Fundamentals of Heat and Mass Transfer* (4th ed.), John Wiley and Sons, New York (1996).

[119] S. Koombhongse, W. Liu, and D. H. Reneker, *J. Polym. Sci. B: Polym. Phys.* 39, 2598 (2001).

[120] M. Bognitzki, W. Czado, T. Frese, A. Schaper, M. Hellwig, M. Steinhart, A. Greiner, and J. H. Wendorff, *Adv. Mater.* 13, 70 (2001).

[121] X. F. Wu and Y. A. Dzenis, *J. Appl. Phys.* 98, 093501 (2005).

[122] X. F. Wu and Y. A. Dzenis, *J. Appl. Phys.* 100, 124318 (2006).

[123] X. F. Wu and Y. A. Dzenis, *J. Appl. Phys.* 102, 044306 (2007).

[124] X. F. Wu and Y. A. Dzenis, *J. Phys. D: Appl. Phys.* 40, 4276 (2007).

[125] X. F. Wu and Y. A. Dzenis, *Nanotechnology* 18, 285702 (2007).

[126] X. F. Wu, Y.Y. Kostogorova-Beller, A.V. Goponenko, H.Q. Hou, and Y.A. Dzenis, *Phys. Rev. E* 78, 061804 (2008).

[127] X. F. Wu, *J. Appl. Phys.* 107, 013509 (2010).

[128] X. F. Wu, A. Bedarkar, and K.A. Vaynberg, *J. Colloid Interface Sci.* 341, 326 (2010).

[129] X. F. Wu, A. Bedarkar, and I. S. Akhatov, *J. Appl. Phys.* 108, 083518(2010).

[130] A. Bedarkar and X. F. Wu, *J. Appl. Phys.* 106, 113527 (2009).

[131] A. Bedarkar, X. F. Wu, and A. Vaynberg, *Appl. Surface Sci.* 256, 7260 (2010).

[132] F. Chen, X. W. Peng, T. T. Li, S. L. Chen, X. F. Wu, D. H. Reneker, and H. Q. Hou, *J. Phys. D: Appl. Phys.* 41, 025308 (2008).

[133] E. Zussman, D. Rittel, and A. L. Yarin, *Appl. Phys. Lett.* 82, 3958(2003).

[134] C. Y. Cheng, J. Chen, F. Chen, P. Hu, X. F. Wu, D. H. Reneker, and H. Q. Hou, *J. Appl. Polym. Sci.* 116, 1581 (2010).

[135] E. P. S. Tan and C. T. Lim, *Rev. Sci. Instrum.* 75, 2581 (2004).

[136] E. P. S. Tan and C. T. Lim. *Appl. Phys. Lett.* 84, 1603 (2004).

[137] E. P. S. Tan, S. Y. Ng, and C. T. Lim, *Biomater.* 26, 1453 (2005).

[138] E. P. S. Tan , C. N. Goh, C. H. Sow, and C. T. Lim, *Appl. Phys. Lett.* 86, 073115 (2005).

[139] E. P. S. Tan and C. T. Lim, *Compos. Sci. Tech.* 66, 1102 (2006).

[140] C. T. Lim, E. P. S. Tan, and S. Y. Ng, *Appl. Phys. Lett.*92, 141908 (2008).

[141] E. Zussman, M. Burman, A. L. Yarin, R. Khalfin, and Y. Cohen, *J. Polym. Sci. B: Polym. Phys.* 44, 1482 (2006).

[142] M. K. Shin, S. I. Kim, S. J. Kim, S. K. Kim, H. Lee, and G. M. Spinks, *Appl. Phys. Lett.* 89, 231929 (2006).

[143] P. A. Yuya, Y. K. Wen, J. A. Turner, Y. A. Dzenis, and Z. Li, *Appl. Phys. Lett.* 90, 111909 (2007).
[144] M. Naraghi, I. Chasiotis, H. Kahn, Y. K. Wen, and Y. Dzenis, *Appl. Phys. Lett.* 91, 151901 (2007).
[145] M. Naraghi, I. Chasiotis, H. Kahn, Y. Wen, and Y. Dzenis, *Rev. Sci. Instrum.* 78, 085108 (2007).
[146] B. A. Samuel, M. A. Haque, B. Yi, R. Rajagopalan, and H. C. Foley, *Nanotechnology* 18, 115704 (2007).
[147] A. Arinstein, M. Burman, O. Gendelman, and E. Zussman, *Nature Nanotechnology* 2, 59 (2007).
[148] D. Edmondson, N. Bhattarai, S. Jana, A. Kim, and M. Q. Zhang, *Appl. Phys. Lett.* 94, 103101 (2009).
[149] Y. S. Zhang, E. P. S. Tan, C. H. Sow, and C. T. Lim, *Nanomechanical Characterization of One-Dimensional Nanostructures*, pp. 105, in *Micro and Nano Mechanical Testing of Materials and Structures* (edited by F. Q. Yang and J. C. M. Li), Springer (2008).
[150] M. Naraghi and I. Chasiotis, *Mechanics of PAN Nanofibers* in *Major Accomplishments in Composite Materials and Sandwich Structures: An Anthology of ONR Sponsored Research* (edited by I. M. Daniel), pp. 757, Springer (2009).
[151] Y. Ji, B. Q. Li, S. R. Ge, J. C. Sokolov, and M. H. Rafailovich, *Langmuir* 22, 1321 (2006).
[152] B. D. Coleman, *Arch. Ration. Mech. Anal.* 83, 115 (1983).
[153] H. H. Dai and Z. Cai, *Acta Mech.* 139, 201 (2000).
[154] Y. B. Fu and R. W. Ogden, *Nonlinear Elasticity: Theory andApplications,* Cambridge University Press, Cambridge, U. K. (2001).
[155] B. V. Derjaguin, V. M. Muller, and V. P. Toporov, *J. Colloid Interface Sci.* 53, 314 (1975).
[156] K. L. Johnson, *Contact Mechanics*, Cambridge: Cambridge University Press (1985).
[157] S. Timoshenko and J. N. Goodier, *Theory of Elasticity* (2nd Ed.), New York: McGraw-Hill (1965).
[158] L. D. Landau and E. M. Lifshitz, *Theory of Elasticity* (3rd Ed.), London: Butterworth-Heinemann (1986).
[159] D. Maugis, *Contact, Adhesion and Rupture of Elastic Solids*, Heidelberg: Springer-Verlag (2000).
[160] R. S. Bradley, *Philo. Mag.*13, 853 (1932).
[161] S. C. Cowin and S. B. Doty, *Tissue Mechanics*, Springer, New York (2007).
[162] B. J. Carroll, *J. Colloid Interface Sci.* 57, 488 (1976).
[163] X. F. Wu, Y.A. Dzenis, *Acta Mech.* 185, 215(2006).

NANOCOMPOSITES

In: Nanostructures
Editor: Yu Dong

ISBN: 978-1-62618-081-9

Chapter 9

POLYMER-LAYERED SILICATE NANOCOMPOSITES: FABRICATION AND PROPERTIES

Hazizan Md Akil[1,2,*] ***and Mohd Fadli Ahmad Rasyid***[1]

[1]School of Material and Mineral Resources Engineering, Universiti Sains Malaysia, Pulau Pinang, Malaysia

[2]Cluster for Polymer Composites (CPC), Engineering and Technology Research Platform, Universiti Sains Malaysia, Engineering Campus, Pulau Pinang, Malaysia

ABSTRACT

The growth in research on polymer-layered silicate nanocomposites (PLSN) has increased tremendously over the past two decades. The reasons for such growth are the remarkable improvements in nanocomposites at a relatively low filler content and better dispersion via intercalation and exfoliation mechanisms as compared to conventional composites. Among the properties that are reported to have increased are strength and modulus under tensile and flexural loading conditions. Apart from these improvements, other properties are also reported to have increased such as thermal stability, fire resistance and gas impermeability. Montmorillonite (MMT) and saponite are among the most commonly used clays in producing PLSN due to their ion-exchange properties and swelling behaviour. However, there are also other types of clays that have equal potential such as Smectite, Vermiculite and Muscovite. These clays have a 2:1 layered structure and they are different in terms of their layer charge density, layer charge site, and charge distribution. Muscovite, in particular, has the highest layer charge density and homogeneous charge distribution. In addition, it has a high aspect ratio, a well-defined crystal structure and is much cheaper than MMT. The general formula of Muscovite is KAl2(AlSi3)O10(OH)2. It has a monoclinic structure with a very high layer charge density which is close to 1.0 equivalent per O10(OH)2.

In this study, Muscovite has been chemically treated to produce organically modified Muscovite (O-Muscovite) using a cation exchange treatment in order to improve the performance of the PLSN. The nanocomposite was produced using the melt compounding technique with Muscovite and O-Muscovite as the fillers at various filler

* E-mail: hazizan@eng.usm.my.

concentrations and with different types of compatibilizers. PLSN was characterized using mechanical and thermal characterization techniques. The dispersion of clay in the polymer matrix was evaluated using TEM analysis.

INTRODUCTION

Polymer-layered silicate nanocomposites (PLSN) are a new class of composite materials composed of hybrid organic polymers and inorganic materials [1-3]. They are classified as a third type of nanocomposites in which one dimension of the dispersed particles is in the nanometre range.

Over the past decade, PLSN have become increasingly popular due to remarkable improvements in terms of engineering properties when compared to pure polymers and conventional composites (macros and nanocomposites) such as strength and modulus under tensile and flexural loading conditions [1, 4-6], thermal stability [7-9], gas impermeability[10-11] and flammability [12-13] that are achievable at low filler loadings of typically less than 5 wt%.

The key to this improvement is the ability of the layered silicate to disperse individually and the high aspect ratio of the silicate platelets within the polymer matrix.

Conventionally, micros and nanocomposites are filled with fillers in the form of particles (e.g. silica), fibres (e.g. glass fibres) or plate-shaped particles (e.g. saponite), in order to improve their properties and reduce the cost. Consequently, these fillers have resulted in increased weight and brittleness [14-15].

Toyota, one of the primary pioneers in PLSN, has begun a detailed study on PLSN and has promoted many other researches in the composites industries. Through their Toyota Central Research and Development Co. Inc. (TCRD), they have developed a nylon-6 layered silicate filled with montmorillonite (MMT) produced by in situ polymerization and which shows an improvement in terms of thermal and mechanical properties[16]. They have also reported on various other types of polymer–clay hybrid nanocomposites based on epoxy resin and polystyrene, acrylic polymers, rubber, and polyimides formed using a similar approach. In addition, Vaia et al. [17] discovered a method to melt mixed polymers with layered silicates without using an organic solvent and where the intercalation of the polymer chains into the galleries of the layered silicate could occur spontaneously on heating a mixture of polymer and silicate clay powder above the polymer glass transition or melting temperature.

Based on these two major findings, PLSN is now being used in many areas, especially in packaging, automotive, and in building and construction. These three application areas are the driving force behind the production of PLSN. In applications in the automotive field especially, Toyota and General Motors have produced plastic parts for engine compartments, body side mouldings and assist step components which have contributed to weight reduction and production costs [18]. The addition of a 5 wt% layered silicate particle increases the strength, rigidity and heat resistance as much as a 30 wt% of conventional fillers would do.

However, there are two major problems to rendering layered silicates with polymer matrices. Firstly, the layered silicates are not easily dispersed in the polymers due to their preferred face-to-face stacking in agglomerated tactoids, and secondly, the tactoids cannot be dispersed into the discrete monolayers due to the intrinsic incompatibility of the hydrophilic layered silicates with the hydrophobic polymers [19]. In order to overcome this problem, the hydrophilic silicate surface needs to be modified into an organophilic silicate to make it compatible to the organic polymer matrices in achieving their exfoliation. Generally, this modification can be done by an ion exchange treatment with a cationic surfactant, including primary, secondary, tertiary and quaternary alkyl ammonium or alkyl phosphonium. The commonly used alkyl ammonium modifications include octadecyltrimethylammonium, dioctadecyldimethylammonium and dimethyldioctadecylammonium [20-21]. The cationic surfactant is used to lower the surface energy of the silicate surface and to improve wetting with the polymer matrix, thus producing better interaction between the two phases and better engineering properties.

On the other hand, layered silicates are clay minerals which are built from two structural units. The simplest ones are the 1:1 structures (example: kaolinite), where a silica tetrahedral sheet is fused to an aluminium octahedron and shares the oxygen atoms.

Montmorillonite (MMT) [21-23] and saponite [9, 24-25] are among the most commonly used layered silicates in producing PLSN due to their ion-exchange properties and swelling behaviour.

However, there are also other types of clays which have attracted much less attention but which are equally potent, such as Vermiculite and Muscovite. These clays belong to the 2:1 layered structure, which is the same crystallite structure as MMT but has a different layer charge density, layer charge site, and charge distribution. Muscovite, in particular, has the highest layer charge density and homogeneous charge distribution. In addition, it has a higher aspect ratio than MMT.

In fact, research on muscovite as a filler in PLSN has not yet been explored extensively. This is because muscovite does not swell in water and the ion exchange treatment for muscovite is not easily performed. It will be interesting to investigate whether a muscovite layered silicate nanocomposite can be obtained.

EXPERIMENTAL PROCEDURES

Materials

A polypropylene (PP), denoted as TITANPRO PX617, with a melt flow index (MFI) of 1.7g/10 min and density of 0.9g/cm^3 and provided by Titan PP Polymers (M) Sdn. Bhd. (Malaysia), was used as a matrix. The compatibilizers, PP-g-MA and PP-POSS, containing 8-10 wt% of maleic anhydride (MA) and 10 wt% of methyl polyhedral oligomeric silsesquioxane (POSS) respectively, were purchased from Sigma Aldrich. Cetyltrimethylammonium bromide (CTAB) was supplied by Sigma Aldrich. The muscovite clay was the product of Bidor Mineral (M) Sdn. Bhd. with a cation exchange capacity (CEC) of 82meq/100 g.

Ion Exchange Treatment of Clays

Muscovite powder (10 g) and $LiNO_3$ powder (170 g) were mechanically mixed. After the mixing was completed, the mixture was heated in a furnace at 300 ^{0}C for 12 hours. The resulting product was washed with deionized water and filtered, and then dried at 110 ^{0}C. The $LiNO_3$ treatment was repeated seven times. The product was named Li-muscovite. Consequently, 3.0 g of cetyltrimethylammonium bromide (CTAB) was dissolved in 50mL of deionized water, and then 0.5 g of Li-muscovite was mixed with the CTAB solution.

The mixture was then placed in a Teflon-lined stainless-steel autoclave in a hydrothermal reactor (100 mL) and heated at 170°C for 12 hours. After the reaction, the surfactant-muscovite (organoclay) product was filtered and then washed three times with ethanol before being dried in a vacuum oven at room temperature.

Preparation of PLSN

Polypropylene nanocomposites, containing 5 wt% of organo muscovite, were fabricated via the melt compounding method. The total mixing time was 8 minutes. First, the polypropylene and compatibilizer were mixed for 2 minutes at 180 °C and 50 rpm using an internal mixer known as the Thermo Haake Polydrive R600 which was fitted with cam blades. Then the organoclay was added to the mixture and allowed to mix for 6 minutes at 180 °C and 50 rpm. The polypropylene granulates and the organo muscovite were then dried prior to being blended in a vacuum oven for 24hours at 80^{0}C in order to remove the moisture.

Characterization

Wide Angle X-Ray Analysis (WAXD)

The presence of the PP structural phases in the nanocomposites were analysed through the Wide Angle X-ray Diffraction (WAXD) on a Bruker diffractometer with a Cu target of k = 1.5405 A° at a generator voltage of 35 kV and a generator current of 30 mA. The experiments were performed in the range of $2\Theta = 1\text{-}10^o$ with a scanning rate of 2°/min. The interlayer distance of organoclay in the composite was calculated from the (001) peak by using Bragg's equation of $2d\sin\Theta = n\lambda$.

Transmission Electron Microscopy (TEM)

The nanostructure of the nanocomposites and the dispersion of the muscovite were observed by using High-resolution Transmission electronic microscopy (HRTEM, LIBRA® 120) at an acceleration voltage of 120 kV and equipped with energy dispersive X-ray spectroscopy (EDX). Before the observation, the sample was maintained at 60 ^{0}C for 12 hours.

TEM analyses were performed on previously dispersed muscovite organoclays that were deposited in a 200-mesh copper Holey carbon grid and then dried before observation. The nanocomposites in the form of 100 nm microtome sections were obtained with a Power Tome PC microtome equipped with a FC-4E cryo-chamber at -90 ^{0}C and analysed by TEM. The samples were taken from the middle of the tensile bar.

Mechanical Properties

The mechanical properties of the nanocomposite samples were tested using an Instron Universal Model 5533 according to the ASTM D638 for tensile tests and the ASTM D790 for flexural tests, and the notched Izod impact test was performed using an impact pendulum tester (Zwick Model 5101) with a 7.5 J hammer according to the ASTM D256. All the tests were performed at room temperature and five replicated specimens from each compatibilizer were prepared for the tests.

Thermal Properties

The Perkin Elmer Pyris-1 series of differential scanning calorimetry (DSC) was used to obtain the crystallization and melting behaviour of the nanocomposites. The samples were heated from 30^0C to 200^0C at a heating rate of 10^0C/min. under a nitrogen flow of 30 ml/min. The samples were kept at this temperature for 5 minutes before cooling down, in order to ensure that the materials were melted uniformly and also to eliminate the previous thermal history. Subsequently, the samples were cooled to 40^0C at a cooling rate of 10^0C/min. and the corresponding melting temperature (Tm), crystallization temperature (Tc) and degree of crystallinity were recorded from the heating and cooling curves. The thermal stability was studied from the TGA curves employing the Perkin Elmer TGA 7 equipment. Samples of 5 mg each were heated from 30^0C to 600^0C C at a heating rate of 108^0C/min in an atmosphere of nitrogen and the corresponding weight loss was recorded.

RESULTS AND DISCUSSION

X-Ray Diffraction (XRD)

The X-ray diffractogram of neat clay (Muscovite), organoclay (0-Muscovite) and PP layered silicate nanocomposites are shown in Figure 1 and the basal spacing values are summarized in Table 1. From the diffractogram, the basal spacing can be established and the values of the spacing can be compared. In general, muscovite is not as easily treated using the ion exchange method as compared to montmorillonite. Muscovite has a stronger electrostatic force holding the aluminosilicate layers and the interlayer cations (K^+) together. Therefore, the $LiNO_3$ treatment was carried out to replace the interlayer cations in the muscovite with the Li+ by melting the $LiNO_3$. In the $LiNO_3$ treatments, the spacing of (001) in plane $d_{(001)}$ of the muscovite was enlarged, providing the possibility for organic intercalation to take place. During the ion exchange treatment, a long alkyl chain (CTAB) was expected to replace the potassium ion and result in the expansion of the silicate layer, and that would ease the penetration of the polymer molecules in between the layers. This characteristic would offer a good distribution of filler particles and an agglomeration effect could be avoided or reduced. This homogenization effect would influence the behaviour of the composite itself in terms of its mechanical and thermal properties. The increase in basal spacing can be seen in the position of the peaks in the X-Ray diffractogram results as shown in Figure 1.

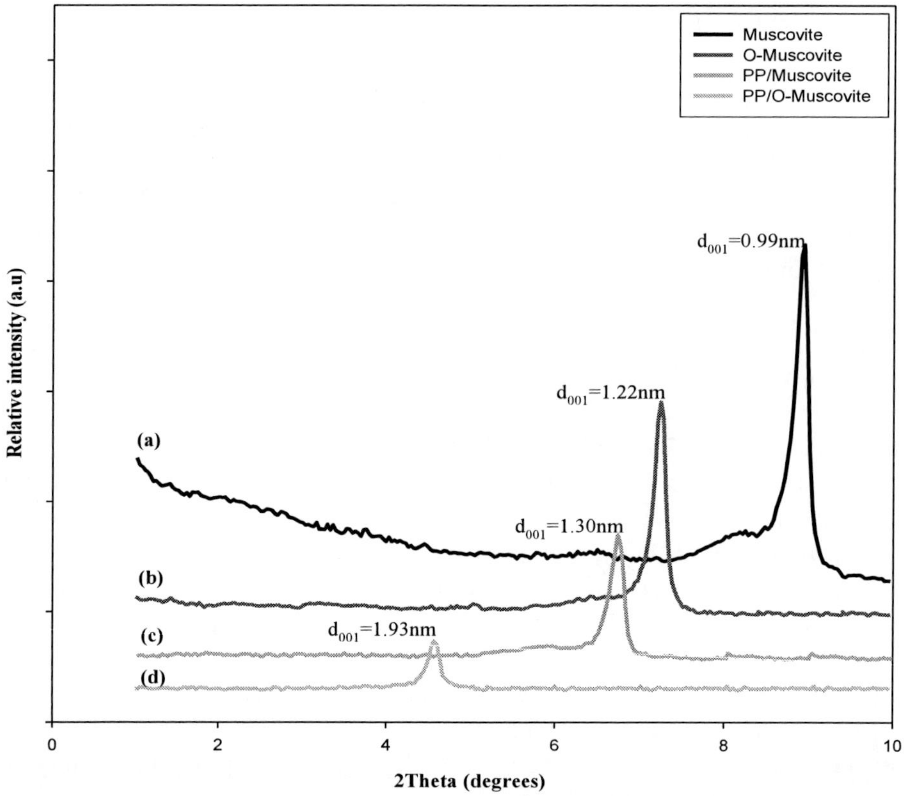

Figure 1. XRD Diffraction pattern of (a) original muscovite, (b) O-Muscovite, (c) PP/muscovite, (d) PP/O-muscovite.

As can be seen from Figure 1, the (001) plane diffraction peak of the neat clay appears around $2\theta = 8.94^0$, corresponding to the interlayer distance or basal spacing (d) of 0.99 nm calculated from Bragg's law, while the (001) diffraction peak of the organo clay appears around $2\theta = 7.25$, $d_{(001)} = 1.22$ nm. The basal spacing of clay was obviously increased after the treatment with $LiNO_3$ and CTAB corresponding to 23% of increment and confirms the occurrence of surfactant molecule intercalation between the muscovite clay layers. Therefore, it reveals that the ion exchange treatment was successfully done in order to introduce and intercalate long carbon chain alkyls into the interlayers of clay. The diffraction peak of the O-Muscovite/PP Layered Silicate Nanocomposites shifted towards lower angles corresponding to a basal spacing of $d_{(001)}$ =1.93 nm with the diffraction peak appearing around $2\theta = 4.57^0$. The shift of the silicate reflection can be interpreted as the PP chains having been intercalated into the galleries of muscovite. Figure 2 shows that the addition of a compatibilizer increased the basal spacing as the d_{001} peak shifted to lower angles and confirmed the intercalation of nanolayered clay particles as reported by other researchers [26-27]. This indicates that the organo muscovite formed a nanocomposite with polypropylene as a result of the increase in the basal spacing caused by the intercalation of polymer molecules. A shift to lower angles in the peak represents the formation of an intercalated structure [28]. The PP and the functionalized PP intercalation between clay platelets were the factors that contributed to this increment.

Table 1. The basal spacing of Muscovite and Muscovite nanocomposites

Sample	d (001) peak (nm)
Muscovite	0.99
O-Muscovite	1.22
PP/muscovite	1.30
PP/O-muscovite	1.93

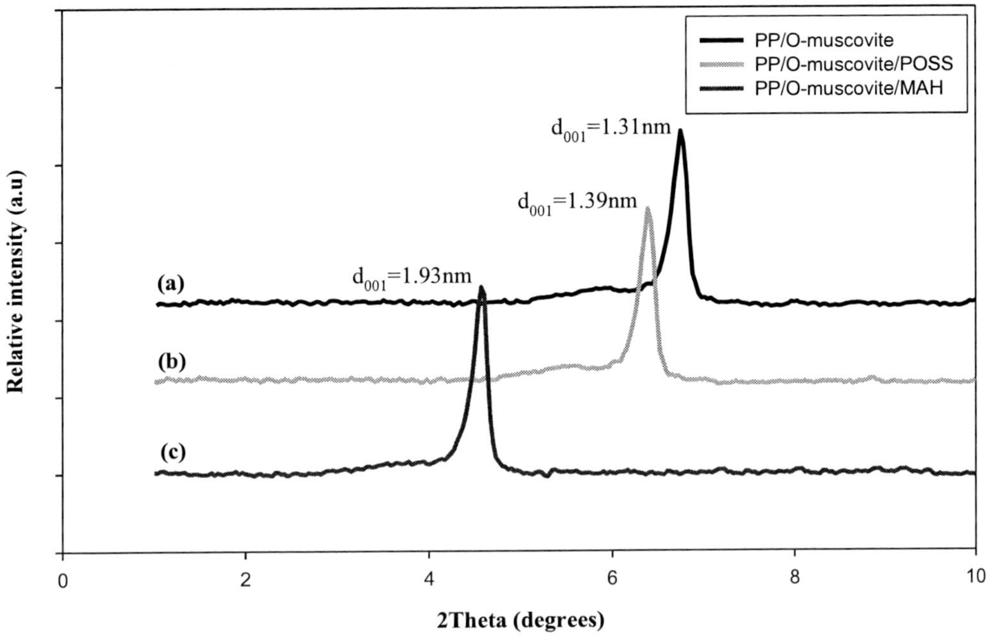

Figure 2. XRD Diffraction pattern of (a) PP/O-muscovite, (b) PP/O-muscovite/POSS with 5 wt% of Muscovite and 5 wt% compatibilizer, (c) PP/O-muscovite/MAH with 5 wt% of Muscovite and 5 wt% compatibilizer.

The nanocomposites obtained with the MAH functionalized PP showed the highest increment compared to the POSS functionalized PP. When using the MAH functionalized PP, the basal spacing moved to 1.93nm and to 1.39 nm for the POSS functionalized PP.

Transmission Electron Microscopy (TEM)

Transmission electron microscopy (TEM) allows for the direct imaging of the filler dispersion in a much smaller region than what is probed by XRD. The dispersion of the organoclays within the PP matrix was further corroborated by the TEM images. The TEM images obtained for the PP layered silicate nanocomposites at 5 wt% filler loading are shown in Figures 3 and 4 respectively. The dark line represents the intersection of the silicate layers while the white background corresponds to the PP matrix. The organically modified muscovite nanocomposites showed good dispersion and orientation of the organo layers, whereas the neat muscovite nanocomposites contained larger tactoids as can be seen in Figure 3(a)-(b).

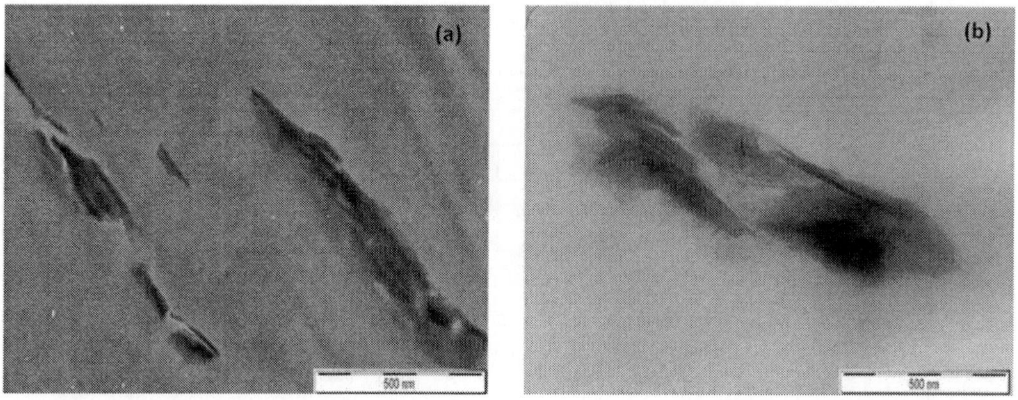

Figure 3. TEM morphological structures of 5 wt% filler loading (a) PP/muscovite; (b) PP/O-muscovite nanocomposite.

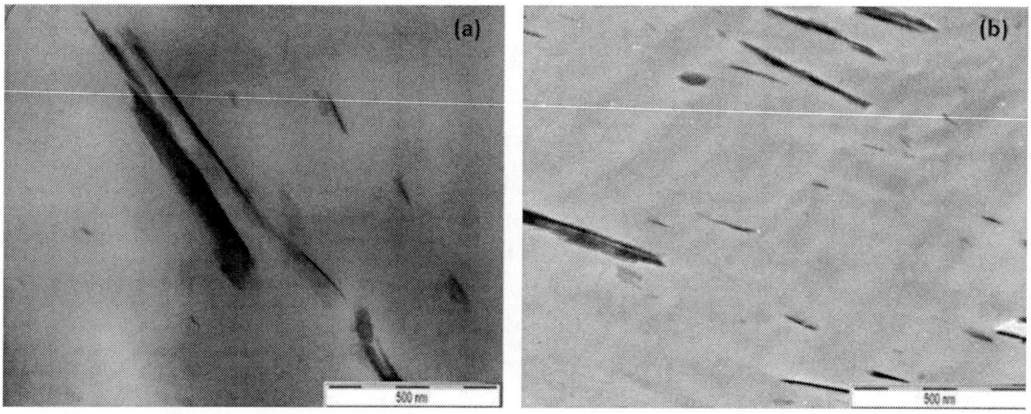

Figure 4. TEM morphological structures of 5 wt% filler loading (a) PP/O-muscovite/POSS nanocomposite; (b) PP/O-muscovite/MAH nanocomposite.

It can be concluded from the homogenously dispersed organo layers in the PP matrix and the XRD results that the O-Muscovite/ PP layered silicate nanocomposites formed an intercalated structure. Furthermore, the PP-g-MAH compatibilized PP nanocomposites (Figure 4(a)) showed a higher degree of disordered structures and exfoliated layers than the PP-POSS compatibilized PP nanocomposites (Figure 4(b)).

It can also be observed that large aggregates were present and that the clay was not well dispersed in the PP-POSS nanocomposites. The MAH nanocomposite underwent a better intercalation and dispersion than the PP-POSS nanocomposite and the non-compatibilized nanocomposite.

Mechanical Properties

Figures 5 and 6 show the variations of the flexural properties of the nanocomposites as a function of the filler loading. The addition of organically modified clays (O-Muscovite) into the composite structure caused an increase in the mechanical properties with optimum results at 5 wt%, wherein the flexural strength increased by 4% and the flexural modulus increased by 20%, respectively as compared to the virgin matrix.

This small increase in the flexural strength was partly due to the occurrence of agglomeration at higher filler loadings, which reduced the interfacial bonding between the clay particles and the polymer matrix, and thus there was no further increase of strength to higher levels as can be seen in Figure 8.

The increase in modulus was attributed to the high stiffness of the nanolayers due to the intensive interaction between the silicate layers with the chains of PP and the organic modification which improved the interfacial adhesion between the O-muscovite and the PP resin matrix [29]. In the light of the above discussion, it is suggested that the presence of the organoclay must have had a positive effect on the mechanical properties of the PP/organoclay nanocomposites.

Figure 7. shows the variation of the flexural strength and modulus of the organo muscovite/ PP layered silicate as a function of different compatibilizers. From the figure, it is clear that the addition of the PP-g-MAH and PP-POSS increased the tensile strength, but not significantly. The PP-g-MAH compatibilized PP nanocomposites have better flexural properties as compared to the PP-POSS compatibilized PP and the non-compatibilized nanocomposite.

This is because the PP-g-MAH compatibilized PP nanocomposites showed a homogenous morphology and supported the compatibility improvement between PP and the organoclay. Furthermore, MAH has a greater polarity than POSS. POSS exists as a cage-type structure and is non-polar. MAH is a rigid, five-membered ring with a permanent dipole moment [30]. Due to this effect, MAH shows better flexural properties and is more favourable compared to POSS.

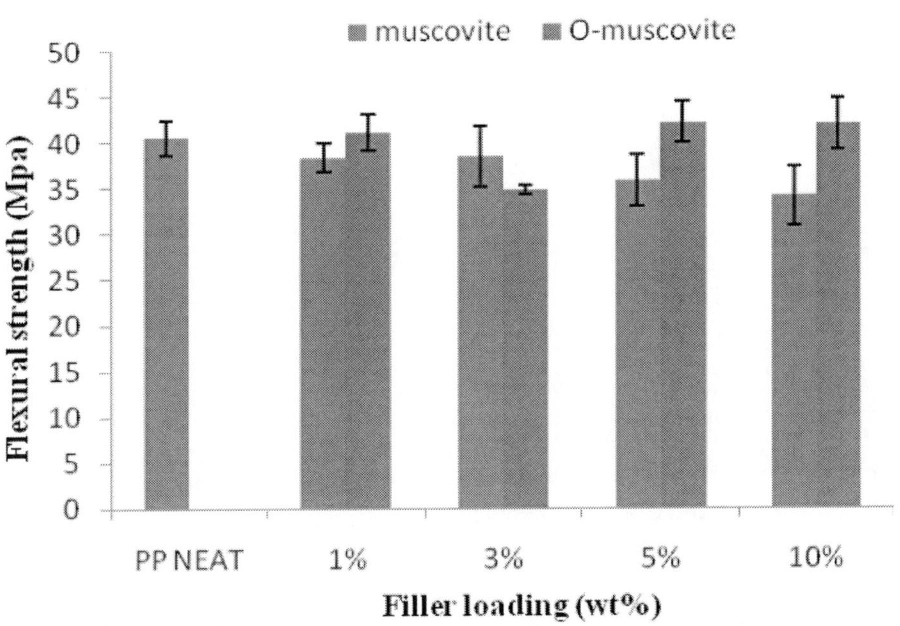

Figure 5. Variation of flexural strength (MPa) of organo muscovite/ PP layered silicate nanocomposite with a function of filler loading.

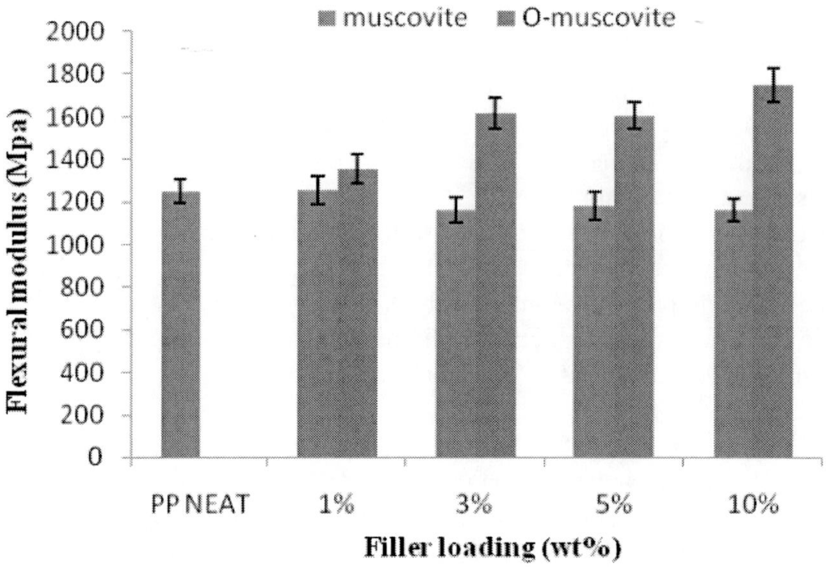

Figure 6. Variation of flexural modulus (MPa) of organo muscovite/ PP layered silicate nanocomposite as a function of filler loading.

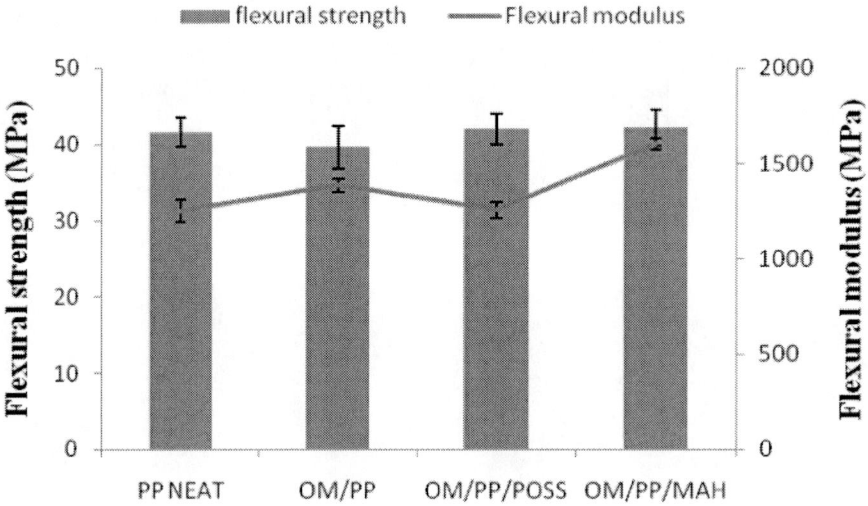

Figure 7. Variation of flexural strength (MPa) and tensile modulus (MPa) of organo muscovite/ PP layered silicate nanocomposite with 5 wt% filler content.

The results of the notched Izod impact strength measurement for the PP layered silicate nanocomposite are presented in Figures 9 and 10. The impact strength was reduced with the addition of muscovite at 1wt% filler loading. This reduction is attributed to the poor adhesion between the clay particles and the matrix as a result of incompatibility when the rigid filler is added. However, when the muscovite content is higher than 1wt%, the impact strength shows an increment up to an optimum value of 28 J/m for the PP/O-muscovite nanocomposite, corresponding to 5 wt% of the muscovite.

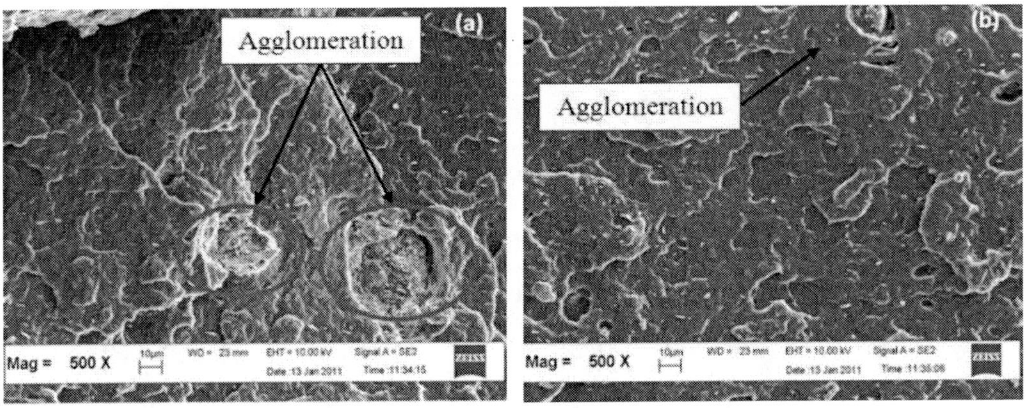

Figure 8. SEM micrographs of 5 wt% filled organoclay /PP nanocomposites: (a) muscovite/PP, (b) O-muscovite/PP.

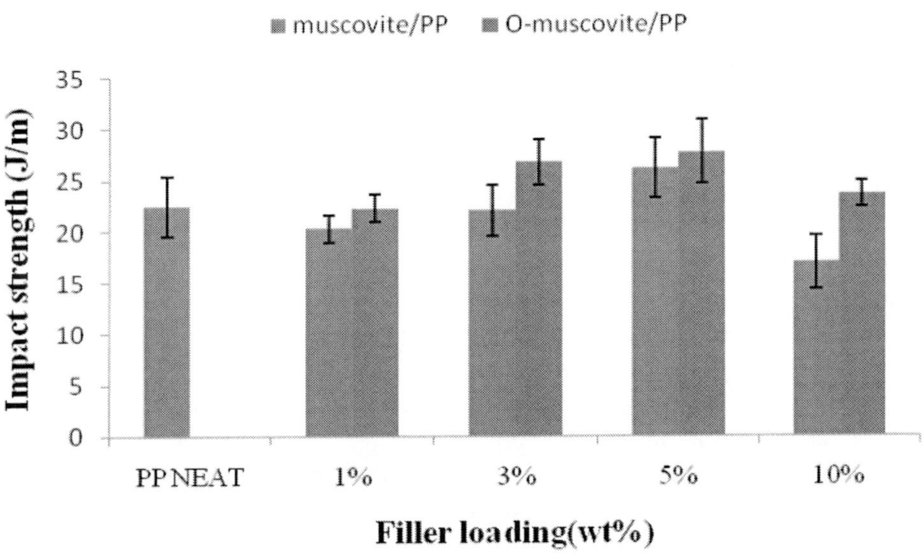

Figure 9. Variation of impact strength (J/m) of organo muscovite/PP layered silicate nanocomposite with 5 wt% filler content.

This increment almost reached 23% compared to the neat PP, while beyond 5 wt%, the value decreased with an increase in the muscovite content. The decrease in the impact strength is attributed to the agglomeration that occurred due to the non-uniform dispersion of clay in the polymer matrix. In addition, the impact strength was reduced for the non-compatibilized nanocomposite systems. This is attributed to the agglomeration that occurred due to the non-uniform dispersion of clay in the polymer matrix. In all the nanocomposite specimens, the maximum impact strength was observed in the PP-g-MAH compatibilized PP nanocomposites. The increment almost reached 23% compared to the neat PP as shown in Figure 9. This is due to the good adhesion between the clay particles and the PP matrix as a result of the compatibility effect when a compatibilizer was added. In addition, it was also probably due to the interdiffusion of the polymer matrix through the interlayer galleries of the muscovite.

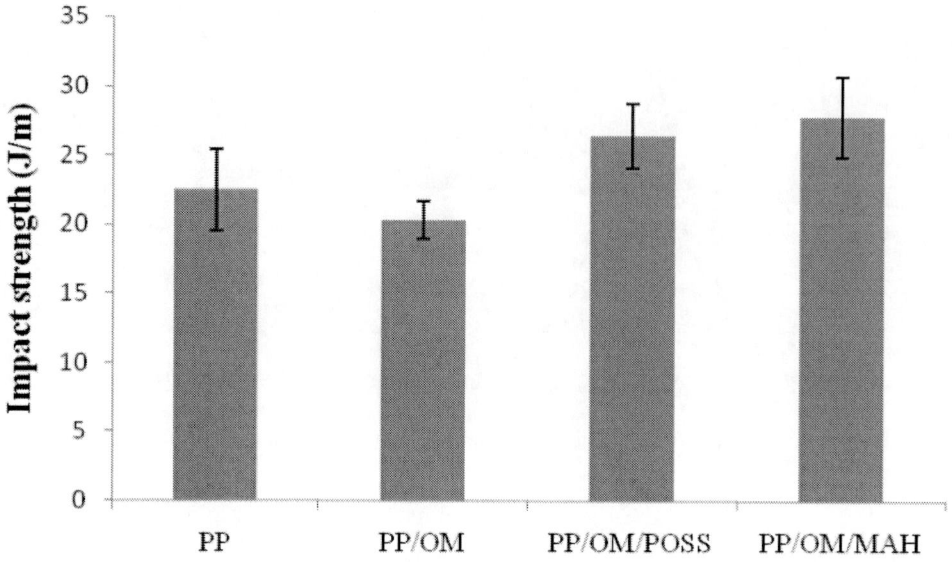

Figure 10. Variation of impact strength (J/m) of organo muscovite/PP layered silicate nanocomposite with 5 wt% filler content.

Thermal Properties

The effect of filler loading on the thermal properties of Muscovite/PP Layered Silicate Nanocomposites in connection with their crystallization behaviour was studied using DSC. The properties derived from these curves are given in Table 1.

As can be seen, the melting point (T_m) of the nanocomposites showed a reduction as compared to the neat PP because the increased interactions between the clay particles and the PP matrix affected the mobility of the chains [31].

The decrease in the degree of crystallinity (X_c) with the clay loading is attributed to the presence of the clay and the strong interaction between PP and the clay, which limited the motion of the polymer chains. Thus, the polymer chains could not move easily into the crystalline phase.

Table 1. DSC results of PP and Muscovite/PP Layered Silicate Nanocomposite

Specimens	Filler loading (wt%)	Tc(^{0}C)	Tm(^{0}C)	ΔHm (J/g)	Xc (%)
PP	0	109	168	88	42.1
PP/muscovite	1	116	166	85	40.7
PP/O-muscovite	1	113	164	80	38.3
PP/muscovite	3	113	165	79.5	38.0
PP/O-muscovite	3	114	164	77	36.8
PP/muscovite	5	111	164	76.5	36.6
PP/O-muscovite	5	114	163	76.2	36.5
PP/muscovite	10	115	164	73	34.9
PP/O-muscovite	10	114	163	72	34.4

Table 2. DSC results of non-compatibilized and compatibilized PP/O-muscovite nanocomposite

Specimens	Filler loading (wt%)	Tc (^{0}C)	Tm (^{0}C)	ΔHm (J/g)	Xc (%)
PP	0	109	168	88	42.1
PP/O-muscovite	5	112	166	79	37.8
PP/O-muscovite/MAH	5	114	163	76.2	36.5
PP/ O-muscovite/POSS	5	111	162	78	37.3

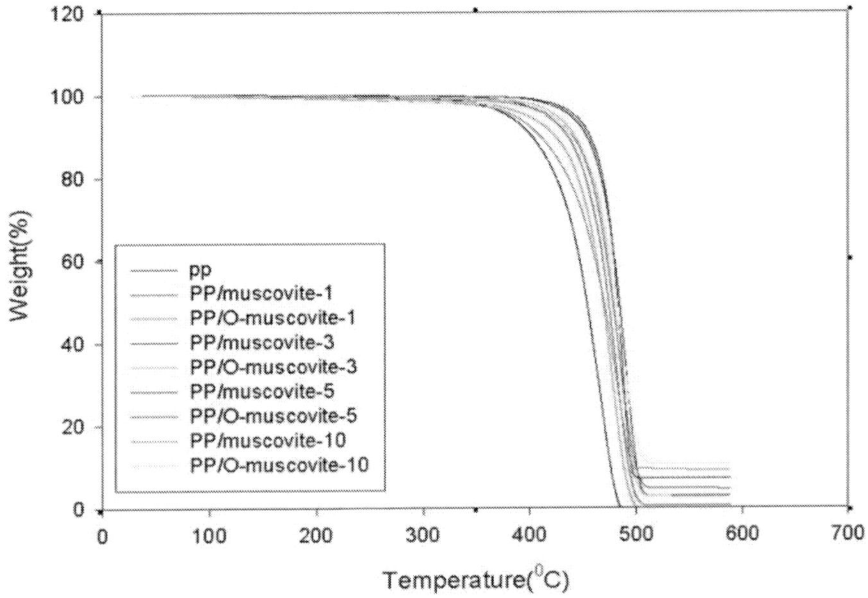

Figure 10. TGA thermograph of PP nanocomposite as a function of filler loading.

The crystallization temperature (T_c) of PP was 109^{0}C and the addition of clay into the PP matrix resulted in an increase in T_c, which is explained by the fact that the clay layers acted as efficient nucleating agents for the crystallization of PP.The DSC results for the non-compatibilized and compatibilized PP nanocomposites at 5 wt% filler loadings are summarized in Table 2. The DSC thermograms of the PP/O-muscovite nanocomposites revealed that the melting temperature (T_m) was slightly reduced with the addition of the O-muscovite and the compatibilizer.

However, the crystallization temperature (T_c) increased with the incorporation of the PP-g-MAH and the PP-POSS compatibilizers. Such a nucleation effect was often reported for organoclays and compatibilizers. The crystallinity of the nanocomposite increased slightly with the addition of the compatibilizer and the highest increment was recorded in the PP-g-MAH compatibilized PP nanocomposites. The thermal stability of PP and the PP/clay nanocomposites were studied by TGA analysis as shown in Figure 10. It could be clearly observed that the thermal stability of the base polymer increased with the filler loading from the initial decomposition temperature of 465^{0}C to 489^{0}C with a final weight loss of 1.2% and 2.7%, respectively.

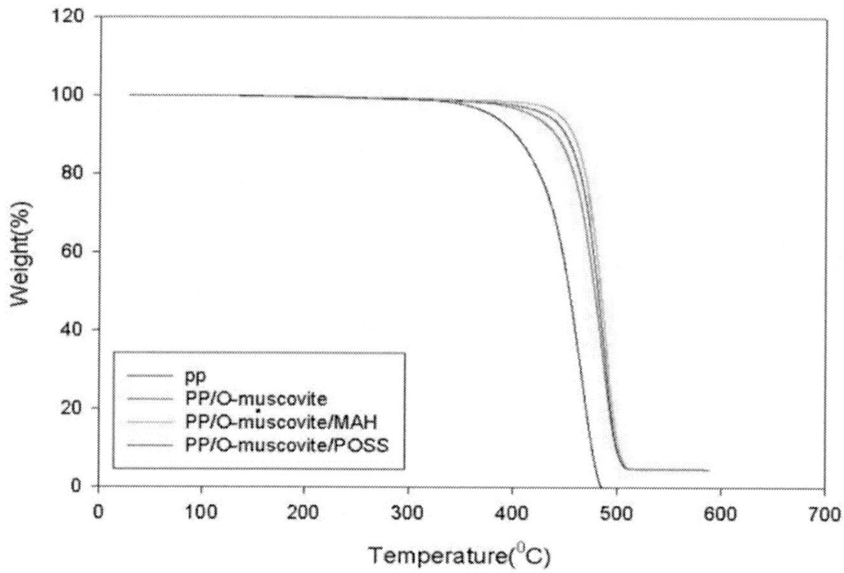

Figure 11. TGA thermograph of PP nanocomposite as a function compatibilizer.

The increased thermal stability of the nanocomposites is attributed to the organic/inorganic interaction between the polymers and the nanoclays [32]. By comparing the thermal stability and weight loss of the nanocomposites in the presence of PP-g-MAH, it was observed that the weight loss at the final decomposition temperature was higher for the non-compatibilized than for the compatibilized nanocomposites as shown in Figure 11. The enhanced thermal stability is attributed to the strong interaction of the base polymers with the clay surface through the chemical linkage between the compatibilizer and the muscovite, which intervened between the surface polarity of the clay and the clay–polymer interface. The organoclay delayed the volatization of the products generated at the temperature of the carbon–carbon bond scission of the polymer matrix.

Conclusion

The PP/muscovite nanocomposites were successfully prepared by employing the melt compounding technique at various filler concentrations and with different types of compatibilizers. The muscovite was modified using the cation exchange treatment to produce O-muscovite. The WAXD patterns and TEM images showed that the cation exchange treatment of clay can improve the clay dispersion in the PP matrix in the presence of a compatibilizer.

The PP/muscovite nanocomposites showed an increase in their mechanical properties with the addition of organically modified clays (O-Muscovite) into the composite structure. The PP-g-MAH compatibilizer showed better mechanical properties compared to PP-POSS. The addition of a compatibilizer has a nucleating effect at higher temperatures due to the crystallization which takes place upon cooling. The TGA thermographs showed a higher thermal stability in the case of the PP/O-muscovite nanocomposites which was further increased with the use of a compatibilizer.

References

[1] E. P. Giannelis, *Advanced Materials.*8, 29-35 (1996).

[2] M. Ogawa and K. Kuroda, *Bulletin of the Chemical Society of Japan.* 70, 2593-2618 (1997).

[3] A. Okada and A. Usuki, *Materials Science and Engineering: C.* 3, 109-115 (1995).

[4] J. Li, M. T. Ton-That, W. Leelapornpisit and L. Utracki, *Polymer Engineering and Science.* 47 (2007).

[5] E. P. Giannelis, *Applied Organometallic Chemistry.* 12, 675-680(1998).

[6] T. Fornes, P. Yoon, D. Hunter, H. Keskkula and D. Paul, *Polymer.* 43, 5915-5933 (2002).

[7] J. Zhang and C. A. Wilkie, *Polymer.* 47, 5736-5743 (2006).

[8] P. C. LeBaron, Z. Wang and T. J. Pinnavaia, *Applied Clay Science.* 15, 11-29 (1999).

[9] M. Alexandre and P. Dubois, *Materials Science and Engineering R: Reports.* 28, 1-63 (2000).

[10] J. W. Gilman, A. B. Morgan, R. Harris Jr, E. Manias, E. P. Giannelis and M. Wuthenow, *Proc. New Advances in Flame Retardant Technology. Proceedings. Fire Retardant Chemicals Association, Lancaster, PA, USA,* pp. 9–22, (1999).

[11] S. Ahmadi, Y. Huang and W. Li, *Journal of Materials Science.* 39, 1919-1925 (2004).

[12] J. W. Gilman, T. Kashiwagi and J. D. Lichtenhan, *SAMPE Journal(USA).* 33, 40-46 (1997).

[13] J. W. Gilman, T. Kashiwagi, J. E. T. Brown, S. Lomakin, E. P. Giannelis and E. Manias, *Materials and Process Affordability: Keys to the Future.* 43, 1053-1066 (1998).

[14] H. Fischer, *Materials Science and Engineering: C.* vol. 23, 763-772 (2003).

[15] G. Lagaly, *Applied Clay Science.* 15, 1-9 (1999).

[16] M. Okamoto and S. S. Ray, *Encyclopedia of nanoscience and nanotechnology.* vol. 8, 791-843 (2004).

[17] R. A. Vaia, H. Ishii and E. P. Giannelis, *Chemistry of materials.* 5, 1694-1696 (1993).

[18] S. N. Bhattacharya, M. R. Kamal and R. K. Gupta, *Polymeric nanocomposites: theory and practice*, Hanser Gardner Pubns, Ohio (2008).

[19] S. Pavlidou and C. Papaspyrides, *Progress in polymer science.* 33, 1119-1198 (2008).

[20] M. A. Osman and J. E. P. Rupp, *Macromolecular Rapid Communications,* 26, 880-884 (2005).

[21] V. Mittal, *Journal of Thermoplastic Composite Materials.* 20, 575 (2007).

[22] T. Mandalia and F. Bergaya, *Journal of Physics and Chemistry of Solids.* 67, 836-845 (2006).

[23] G. Jimenez, N. Ogata, H. Kawai and T. Ogihara, *Journal of Applied Polymer Science,* 64, 2211-2220 (1997).

[24] R. K. Bharadwaj, *Macromolecules,* 34, 9189-9192 (2001).

[25] Y. Kojima, A. Usuki, M. Kawasumi, A. Okada, T. Kurauchi and O. Kamigaito, *Journal of Polymer Science Part A: Polymer Chemistry.* 31, 983-986 (1993).

[26] A. Martínez-Gómez, E. Pérez and C. Álvarez, *Polymer,* 50, 1447-1455 (2009).

[27] J. K. Mishra, K. J. Hwang and C. S. Ha, *Polymer.* 46, 1995-2002 (2005).

[28] S. Parija, S. Nayak, S. Verma and S. Tripathy, *Polymer Composites,* 25, 646-652 (2004).

[29] E. S. A. Rashid, M. F. A. Rasyid, H. M. Akil, K. Ariffin and C. C. Kooi, *Applied Clay Science.* 52, 295-300 (2011).

[30] García–Martínez J., Laguna O., Areso S., Collar E, *Journal of Applied Polymer Science,* 81, 625-636 (2001).

[31] P. Maiti, P. H. Nam, M. Okamoto, N. Hasegawa and A. Usuki, *Macromolecules,* 35, 2042-2049 (2002).

[32] J. G. Doh and I. Cho, *Polymer Bulletin,* 41, 511-518 (1998).

In: Nanostructures
Editor: Yu Dong

ISBN: 978-1-62618-081-9

Chapter 10

HOLISTIC APPROACH AND DEVELOPMENT ON POLYPROPYLENE (PP) / CLAY NANOCOMPOSITES FROM PROCESSING, MATERIAL CHARACTERIZATION TO NUMERICAL MODELING

***Yu Dong*[1*] *and Debes Bhattacharyya*[2]**

[1]Department of Mechanical Engineering, Curtin University, Perth, Australia
[2]Center for Advanced Composite Materials, Department of Mechanical Engineering, The University of Auckland, Auckland, New Zealand

ABSTRACT

Since polypropylene (PP)/clay nanocomposites were first commercialized for manufacturing seat backs of 2004 Honda Acura TL cars, extensive research work was conducted to focus on improvements of functional and mechanical properties of PP/clay nanocomposites via melt compounding, mainly based on "trial and errors" methods. Despite generally superior functional properties, mechanical properties of such nanocomposites are quite inconsistent which are very hard to be satisfied with the real applications of end users, and thus inevitably make adverse impact on relevant theoretical and numerical predictions. As expected, the sophisticated development of manufacturing methods and sound knowledge of individual effects of processing parameters play a key role on material performances.

This chapter elaborates a holistic and systematic study on the determination of optimal material formulations using Taguchi design of experiments (DoE) in melt compounding process, material characterization via X-ray diffraction (XRD) analysis, scanning electron microscopy (SEM), transmission electron microscopy (TEM), mechanical testing, differential scanning calorimetry (DSC) and dynamic mechanical thermal analysis (DMTA), as well as numerical modeling using representative volume elements (RVEs) and image-based extended object-oriented finite element (OOF) analysis. The clay type and clay content, compatiblizer content and PP type were investigated specifically to obtain the combination of material processing factors for

* Corresponding Author's Email Address: Y.Dong@curtin.edu.au

maximizing (not simultaneously) the tensile/flexural moduli and tensile/flexural/impact strengths of fabricated nanocomposites. PP/clay nanocomposites with good individual mechanical properties were then synthesized in a globally sub-optimal factorial combination for the subsequent material characterization. Two-dimensional 3×3 array RVE models were employed to predict the elastic moduli of PP/clay nanocomposites in relation to the clay content, clay aspect ratio and particle dispersion patterns. Alternatively, OOF modeling technique was used to map the real micro/nanostructures of clay platelets and predict the overall material behavior. The numerical results from both methods were compared with the experimental data and available composites theoretical models. Very good agreement was shown to coexist, establishing the viability of proposed numerical approaches, especially with the OOF method based on the imaging analysis of captured real morphological structures.

INTRODUCTION

The successful utilization of layered silicates as one of most popular nanoparticles in polymer nanocomposites (e.g. nylon 66/clay nanocomposites) offers the great insight to researchers and engineers on the future material merits of light weight, cost-effectiveness and multifunctional properties such as high stiffness and strength, excellent thermal and barrier properties at a relatively low clay content (≤ 5 wt%). Depending on the interactions and structures of constituents, polymer/clay nanocomposites can be classified as "phase separated or immiscible", "intercalated" and "exfoliated" structures, illustrated in Figure 1. Immiscible composites (or so-called "microcomposites") are produced with quite minor property enhancements if the polymer molecular chains are unable to intercalate clay platelets and thus increase their interlayer spacing. Beyond this type of classical composites, when polymer chains are inserted into stacks of layered clays with the interlayer expansion, a well-ordered intercalated sandwich structure takes place, which is built up with repeatable polymeric and silicate layers.

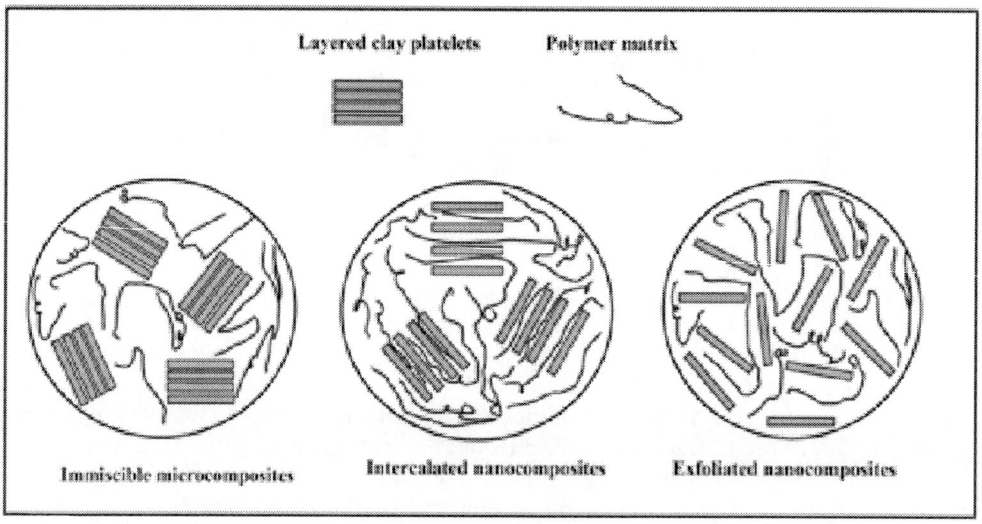

Figure 1. Morphological structures of polymer/clay nanocomposites [1].

In the exfoliated structure, individual clay platelets are completely peeled apart or delaminated and homogeneously dispersed into the continuous polymer matrix. PP/clay nanocomposites with low cost are in a market-driven demand since PP, as one of the favorite polyolefin materials, possesses the great potential due to its low density, high thermal stability, excellent processibility as well as wide usages in automotive and packaging industries. However, its low modulus relative to engineering polymers (e.g. nylon 6) might limit the material selection to meet the requirements of end users. This drawback can be alleviated by the addition of effectively dispersed clays. PP/clay nanocomposites, prepared initially by either conventional intensive shear mixing or twin screw extrusion, and then by the secondary injection molding or compression molding processes, have been successfully produced with partially intercalated/exfoliated structures [2-6]. The major problem encountered is due to the low interactions and incompatibility between the non-polar hydrophobic PP and hydrophilic clay to obstruct the effective intercalation at the early phase. The effective approach to increase their compatibility is to mix the PP matrix with the coupling agents or compatibilizers such as maleic anhydride grafted PP (MAPP) oligomer in order to facilitate the penetration of PP chains into clay platelets whose interlayer spacing is enlarged accordingly. Despite the lack of experimentally complete clay exfoliation till now, the improvements of stiffness and strength for PP/clay nanocomposites at a low clay content are still attractive from the industrially economical point of view. Previous work [7-10] performed by other researchers demonstrated the inconsistencies of modulus and strength increases possibly due to the complex influence of the material selection and the melt processing methods.

The early investigation conducted by Hasegawa et al. [4] at Toyota Central R & D laboratories concentrated on the preparation of PP-clay hybrids in presence of a typical MAPP oligomer using a twin screw extrusion process at 200-230°C. The increase of weight ratio between MAPP and clay was found to achieve more uniform clay dispersion. The hybrid tensile moduli appeared to be enhanced by the maximum value of 30% with increasing the MAPP content up to 21.6 wt% (clay content: 7.2 wt%). Unfortunately, the tensile strengths remained comparable to that of neat PP, which was further confirmed by both Gicía-López et al. [7] and Dong et al. [8]. Liu et al. [5] used co-interaction organophilic clays with a much larger interlayer spacing, which were modified by alkyl ammonium, to prepare PP/clay nanocomposites via grafting-melt compounding. Partially intercalated/exfoliated clay structure was observed with the increases of tensile strength and tensile modulus up to 27 and 42%, respectively in 7 wt% clay filled PP nanocomposites when compared with those of neat PP. The additional clays did not change the crystal structure but accelerated the crystallization process considerably. Modesti et al. [10] found that the optimized nanocomposite processing condition could be accomplished at a higher screw speed and lower barrier temperature profile, leading to a high degree of intercalation/exfoliation with the strong compatibility of MAPP between clays and PP matrix. In such a processing condition, the significant improvement of mechanical properties was obtained with the maximum increases of tensile modulus about 130%, flexural modulus about 35% and impact strength about 50% in 5 wt% filled PP nanocomposites (MAPP content: 5 wt%). However, Gicía-López et al. [7] detected that impact strengths of PP/clay nanocomposites worsened as opposed to that of neat PP and continuously decreased in excess of MAPP. To improve the clay exfoliation level, Lee et al. [9] introduced an effective in-situ ultrasonic technique that resulted in an increase of clay interlayer spacing in all PP nanocomposites. In particular, the ultrasonic processing became more effective in dispersing and exfoliating clay in low molecular weight PP. The resulting flexural modulus was enhanced by over 43 and 54% for nanocomposites containing low and medium molecular weight PPs, respectively; whereas almost no modulus change occurred in high molecular weight PP.

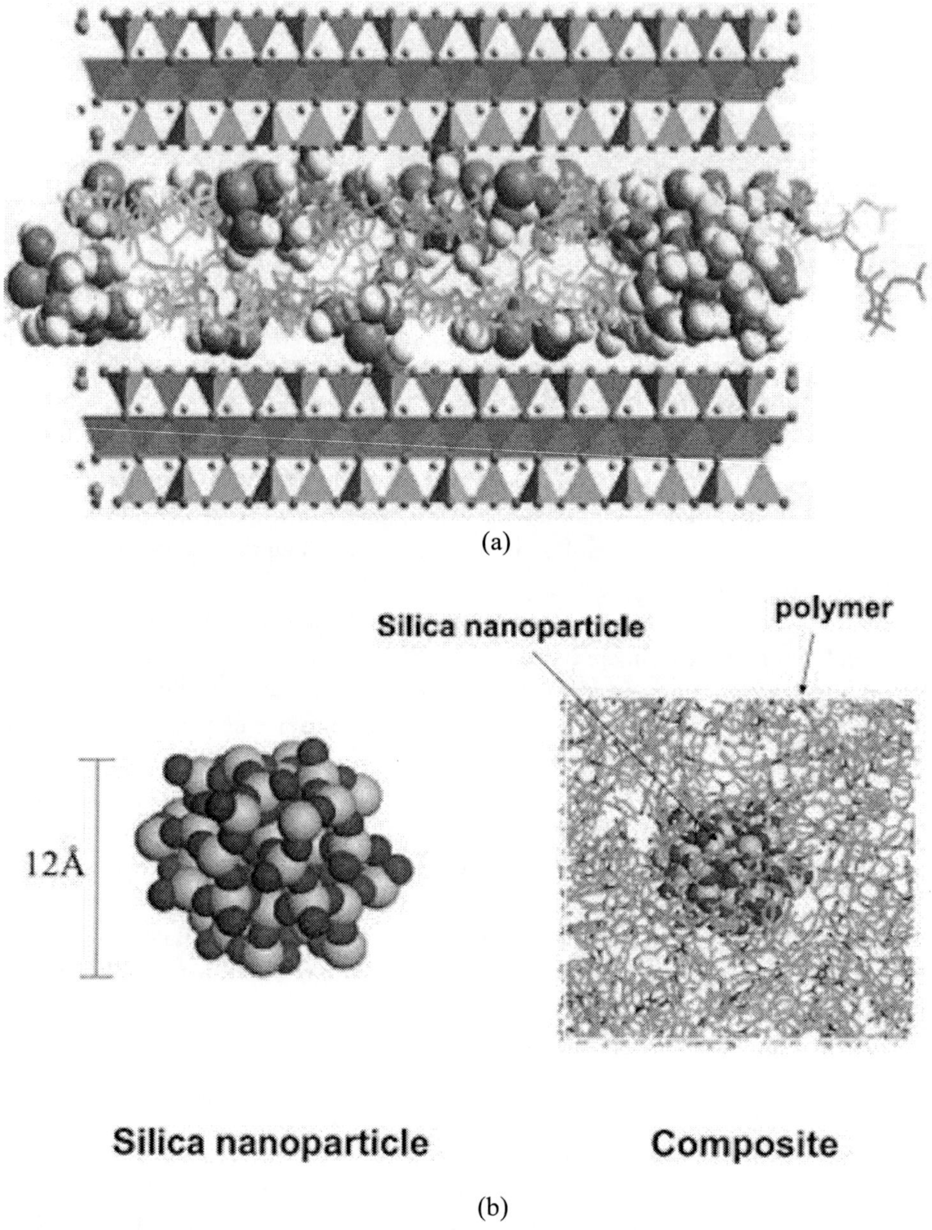

Figure 2. MD simulation models of polymer/clay nanocomposites [17, 18]:
(a) intercalated polyethylene oxide (PEO)/ montmorillonite (MMT) nanocomposites and (b) polyimide/silica nanocomposites.

These contradictory results necessitate a systematic study in the formulation and fabrication of such nanocomposites in place of the "trial and error" experimental method in the common polymer blending approach. On the other hand, despite tremendous opportunities offered by experimentally dispersing nanofillers into the continuous polymer matrix in a nanocomposite system, an effective multi-scale numerical approach to predict the overall material behavior of nanocomposites from the complex constituents and morphological structures is still at an infant stage. Three well-developed approaches of composites theoretical models, finite element analysis (FEA) and molecular dynamic (MD) simulation have been implemented for the modulus prediction of nanocomposites.

The popular Halpin-Tsai and Mori-Tanaka models [11, 12] are very simple and rough closed-form analytical solutions to originally predict the material properties of short-fiber composites. Some success has been achieved by using these two models to estimate the elastic modulus [13-16], thermo-mechanical properties [13] and the reinforcement efficiency of polymer/clay nanocomposites [16]. Alternatively, MD simulation offers a far more accurate model development of thermodynamics and kinetics of formation and estimates the fundamental physical properties at a macroscopic level by means of statistical mechanics, which is based on the discrete atom or molecule as the basic unit. This sophisticated computational technique can bring MD models down to the nanoscaled level to predict their relevant molecular interactions [17-19], interlayer spacing [20-22], binding energy [20-23] and elastic properties [18]. Typical examples of MD simulation models of nanocomposites are demonstrated in Figure 2, which consider the molecular/atomic structures of both clay silicates and polymer matrices for the predictions of interlayer spacing and binding energies. Nevertheless, this approach solely focuses on the local molecular/atomic interactions as well as hierarchical structures and behavior. As a consequence, scaling up the integrated lengths or huge time scales appears to be formidable when globally characterizing the nanocomposite structures. When the large computational time and memory are of great concern, a representative volume element (RVE) domain, also known as "unit cell", with the periodic conditions, can be carefully employed to capture the sufficient information on the simplified repeatable micro/nanostructures, and then determine the corresponding effective microscopic properties of a composite system such as elastic modulus and Poisson's ratio, etc.

A typical periodic RVE model for polymer/clay nanocomposites, as illustrated in Figure 3, underwent the well-established random Monte-Carlo algorithm, which has been extensively applied in modeling the mechanical behavior of short-fiber composites [24,25], whisker and platelet filled polymer [25, 26], and more recently nanocomposites [15, 25, 27]. Furthermore, the effects of clay platelet aspect ratio, clay dispersion patterns, clay content and constituent properties on the stiffness of polymer/clay nanocomposites can be plausibly simulated using RVE models to better understand the intercalated/exfoliated structure-property relationship [28, 29]. Nonetheless, for simplicity either well-aligned RVE [28, 29] or randomly distributed RVE with the Monte Carlo method [15] used in micro/nanostructures of composite materials cannot conform to the actual complex and highly heterogeneous morphological structures of nanocomposites. As a result, the object-oriented finite element (OOF) analysis [30-32] enables to compensate for the lack of a fine detailed structural interpretation in dealing with a small-scaled and disordered heterogeneous material system. OOF analysis is considered as a novel numerical approach based on mapping the computational grids from incorporated real morphological SEM/TEM images as seen from Figure 4. The primary benefit of using OOF technique lies in the combination of geometric data (e.g. particle size, shape, spatial position and actual orientation, etc.) in the real form of micro/nanostructures and fundamental material parameters (e.g. elastic modulus, Poisson's ratio and coefficient of thermal expansion of constitutive phases) to understand the overall material behavior. Although OOF is temporarily limited to the calculations of elasticity and thermal conductivity in 2-D microstructures, very promising numerical results have been obtained in analyzing the stress transfer [33], fracture mechanisms [34-36], crack propagations [37], residual thermal stresses [38, 39] and thermal conductivities [40] of heterogeneous materials. More recently, OOF modeling has also been employed in predicting the elastic moduli of nanocomposites, in reasonable agreement with experimental data and theoretical analyses of Halpin-Tsai and Lewis-Nielsen models [41-43]. This book chapter aims to holistically develop a unique systematic approach for determining the preferred material formulations and manufacturing procedures using Taguchi design of experiments (DoE), experimentally characterizing the nanocomposite materials and developing numerical simulations based on both RVE and OOF models in comparison to the conventional composites theories.

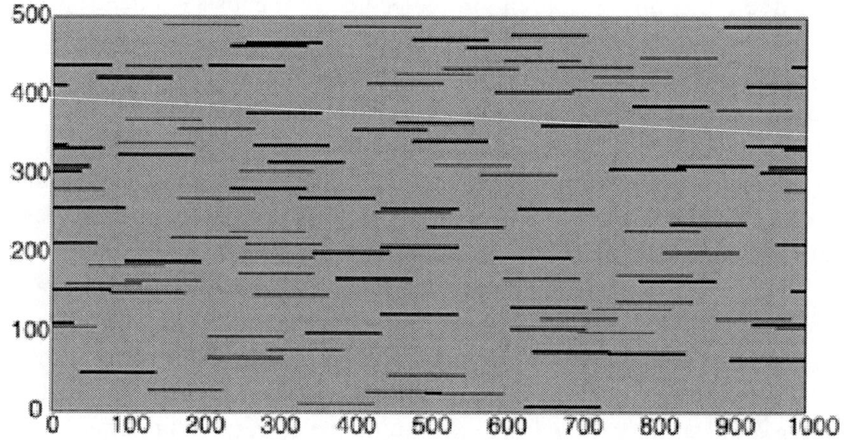

(a)

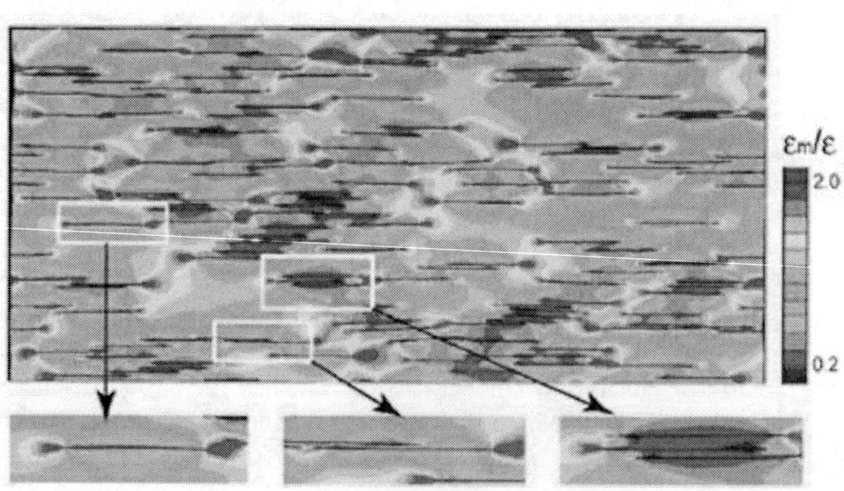

(b)

Figure 3. RVE model for micromechanical simulation of nylon 6/clay nanocomposites [15]: (a) clay particle distribution with periodic boundary conditions and (b) axial strain contour.

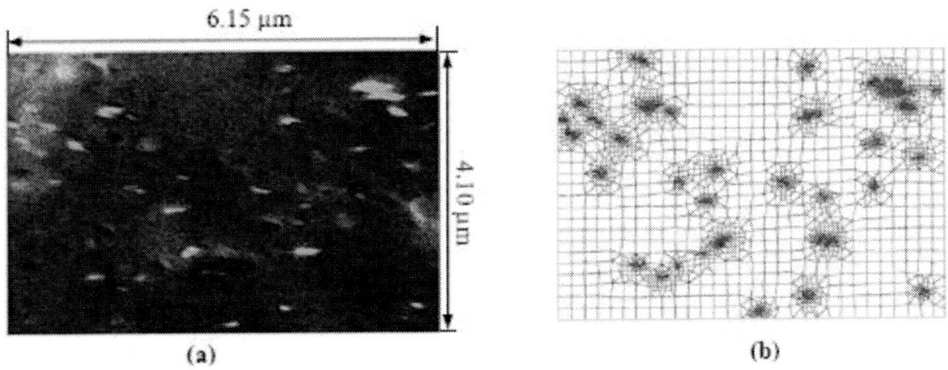

Figure 4. A typical micro/nanostructure based OOF model of PP/clay nanocomposites: (a) SEM micrograph and (b) finite element mesh [8].

The clay distribution effects on the deformation mechanisms were particularly evaluated by OOF modeling under linear elastic conditions. Ultimately, the processing-micro/nanostructure-property relationship of tailored PP/clay nanocomposites could be better understood under this comprehensive investigation.

EXPERIMENTAL WORK

Material Features and Formulations

Three grades of commercially available polypropylene (PP) with the various melt flow indices (MFI*), referred to as PP-Co M710 (MFI= 0.6 g/10 min), PP-Hom Y130 (MFI=4.0 g/10 min) and PP-Hom H380F (MFI=25 g/10 min), were provided by Clariant (New Zealand) Ltd. PP-Co M710 is a high impact PP copolymer with ethylene content of 6 wt%. Maleic anhydride grafted PP (MAPP) Exxelor™ PO1020 (MA content: 0.5-1 wt%, MFI= ~ 430 g/10 min) was selected as the compatibilizer from ExxonMobil Chemical (Germany). The layered silicates used in this study were three types of NANOLIN™ organomodified clay denoted as DK1N, DK2 and DK4, which were obtained from Zhejiang Fenghong Clay Chemicals Co., Ltd, China. They were modified by different levels of cation exchange reaction with octadecylammonium salt [44] in order to broaden their interlayer spacings (d_{001}=2.29, 2.25 and 3.56 nm, respectively).

Table 1. Four factors and three levels used in L_9 DoE layout [47]

Factor	Level		
	1	2	3
A: Clay type	DK1N	DK2	DK4
B: Clay content (wt%)	3	5	10
C: MAPP content (wt%)	5	10	20
D: PP type	PP-Co M710	PP-Hom Y130	PP-Hom H380F

* Melt flow conditions 2.16 kg/230°C according to ASTM D 1238.

Table 2. L_9 DoE layout for the fabrication of PP/clay nanocomposites [47]

Trial number	Symbol	Clay type	Clay content (wt%)	MAPP content (wt%)	PP type
1	RR1: DK1N/MAPP/Y130 (5/10/85)*	DK1N	5	10	PP-Hom Y130
2	RR2: DK4/MAPP/Y130 (3/20/77)	DK4	3	20	PP-Hom Y130
3	RR3: DK2/MAPP/Y130 (10/5/85)	DK2	10	5	PP-Hom Y130
4	RR4: DK1N/MAPP/M710 (3/5/92)	DK1N	3	5	PP-Co M710
5	RR5: DK2/MAPP/H380F (3/10/87)	DK2	3	10	PP-Hom H380F
6	RR6: DK2/MAPP/M710 (5/20/75)	DK2	5	20	PP-Co M710
7	RR7: DK4/MAPP/M710 (10/10/80)	DK4	10	10	PP-Co M710
8	RR8: K1N/MAPP/H380F (10/20/70)	DK1N	10	20	PP-Hom H380F
9	RR9: DK4/MAPP/H380F (5/5/90)	DK4	5	5	PP-Hom H380F

Note: * Values in parentheses (e.g. 5/10/85) indicate the contents (wt%) of clay, MAPP and PP in PP/clay nanocomposites.

The initial material formulations were statistically determined by using well-known Taguchi design of experiments (DoE) [45, 46], which could easily detect the most significant factors for maximizing the mechanical properties of PP/clay nanocomposites and work out their ultimate preferred formulations in a limited number of trials. Taking this DoE response into consideration, a "larger-the-better" characteristic formula was employed in expression of

$$S/N = -10\log\left(\frac{1}{n}\sum_{i=1}^{n}\frac{1}{y_i^2}\right) \tag{1}$$

where S/N is the signal-to-noise ratio, n is the number of material samples in each trial and y is the measured response value, namely the normalized moduli or strengths of nanocomposites (over those of corresponding neat PP). A four-factor and three-level Taguchi DoE was performed with the main effects shown in Table 1. This resulted in a detailed L_9

DoE layout illustrated in Table 2 in the limited nine-trial runs. For simplicity, the factorial interactions were totally neglected with the assumption of all four independent factors.

Melt Processing

PP/clay nanocomposites were synthesized by melt compounding using a co-rotating intermesh twin screw extruder Brabender DSE 20 (screw diameter D_0 = 20 mm, L_0:D_0 = 40:1 where L_0 is the length of a screw). The four-zone temperature profile was set at 185, 195, 200 and 210°C from the feeding zone to heating zones with the die temperature of over 203°C. The high shear screw configuration designed for this work is depicted in Figure 5, which consists of a series of conveying elements (SE), wide kneading blocks (KBW), reverse elements (SE L) and wide reverse kneading blocks (KBW L). Conveying elements with different pitches and lengths control the material covey rate and wide kneading blocks apply to enhance the mixing and achieve high shear intensity; whereas reverse screw elements and reverse kneading blocks generate the back pressure to retard the melt flow for the further mixing. Instead of using conventional hand mixing and hopper feeding, a twice direct compounding method was introduced with its detailed process illustrated in Figure 6.

The first step was to feed PP and MAPP pellets into the high speed extruder at a screw speed of 200 rpm and a feed rate of 3.0 kg/h for all PP grades. For a good control of material feed rate, WOYWOD Plasticolor 1000 and 2200 pellet feeders, as shown in Figure 7(a), were utilized to feed the desirable contents of MAPP and PP, respectively. The clay powders were then downstream fed into the fully melted PP/MAPP mixture to produce the initial nanocomposite batches. A SCHENCK AccuRate agitational powder feeder Model 300 was operated in clay feeding with the modified attachment of an Ideal electric marker to the feeder nozzle to vibrantly break up clay aggregates and improve the clay dispersability, Figures 7(b) and (c). The corresponding feeder parameters and feed rates for individual PP, MAPP and clay powders were initially calibrated. Feeder speed parameters were subsequently rectified in the confirmation tests to achieve desirable material contents. A constant agitation setting for the powder feeder was set in the entire extrusion process.

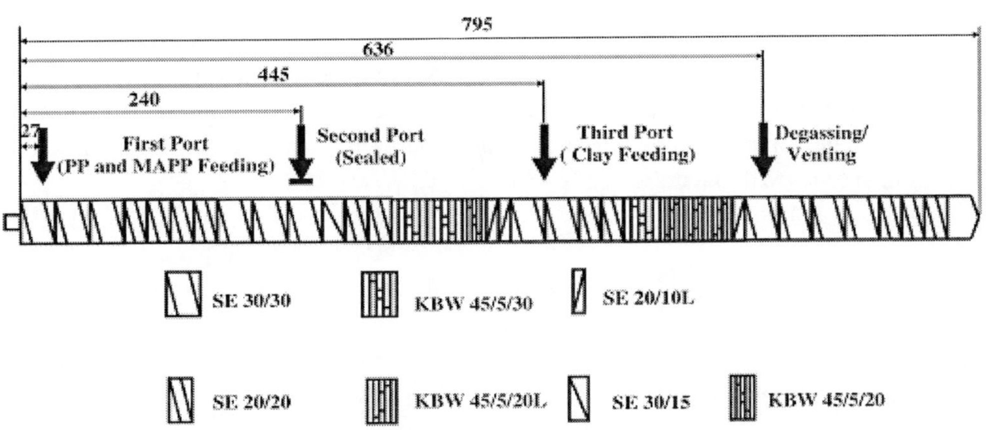

Figure 5. A high shear screw configuration used in twice direct compounding process [47].

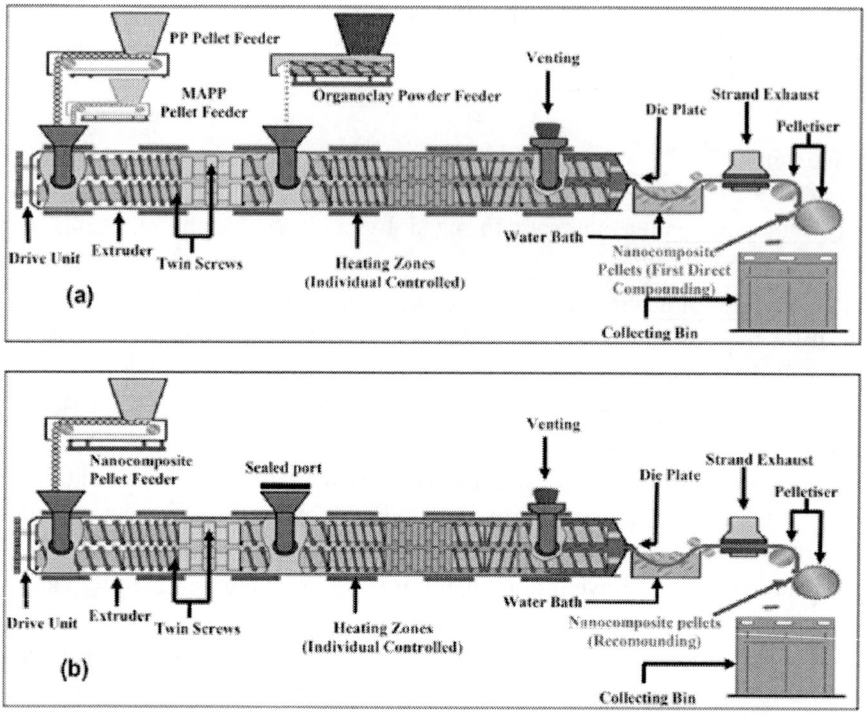

Figure 6. Schematic diagram of twice direct compounding process: (a) first direct compounding and (b) recompounding [47].

At the second step, initially prepared nanocomposite pellets were recompounded at a relatively low screw speed of 100 rpm with the same feed rate using a Plasticolor 2200 pellet feeder, which is believed to further promote the effective clay dispersion by extending the residence time of material mixtures. All raw materials and later collected nanocomposite pellets were dried in a vacuum oven at 90 °C for more than 16 h. Material testing samples were prepared on a BOY 50A injection molder with a temperature profile of 195, 200, 210 and 190 °C from the first, second, third feeding zones to the nozzle. The die temperature remained at 25 °C and the injection pressure was about 60-80 bars. The produced samples were immediately sealed in plastic bags and then placed in a silica-gel contained vacuum desiccator for moisture absorption for over 24 h. As a reference material, neat PP was only subjected to the first direct compounding and then injection molding in the same processing conditions as nanocomposite counterparts. This ensures that the thermal degradation of PP could diminish to a great extent.

Material Characterization Techniques

Tensile samples of PP/clay nanocomposites were examined using a Bruker D8 ADVANCE X-ray diffractometer at 40 kV and 40 mA with CuK_{α} X-ray beam (λ=0.15406 nm). The XRD patterns were scanned in a small angle range of 2θ=2-10° to investigate the clay intercalation level and dispersability, and at wider angles of 2θ=2-30° to detect PP crystalline structures in nanocomposites.

(a)

(b) (c)

Figure 7. Feeder setup for melt compounding of PP/clay nanocomposites: (a) polymer pellet feeders Plasticolor 1000 (A) and 2200 (B), (b) SCHENCK AccuRate powder feeder model 300 (in cross section view) and (c) Ideal electric marker (C) as a vibrator.

The interlayer spacing (d_{001}) was determined from the 2θ position of clay (001) basal reflection peak based on Bragg's law ($n\lambda=2\ d_{001}\sin\theta$). The same criterion applies to (002) and (003) peaks with the corresponding interlayer spacings of d_{002} and d_{003}. All the XRD patterns were smoothed by using a peak fitting analysis based on the Gaussian distribution function. The fractured surface morphology of PP/clay nanocomposites were analyzed via a Philips XL30S Field Emission Gun scanning electron microscope at 5.0 kV. The tensile sample of nanocomposites were cryogenically fractured parallel to the melt flow direction to produce nice brittle fracture surfaces further subjected to platinum sputter coating. To graphically evaluate the intercalated/exfoliated clay structures, TEM characterization was also employed on cryosectioned ultrathin nanocomposite samples (nominal thickness: 70 nm) with the aid of a Philips CM12 transmission electron microscope at an accelerating voltage of 120 kV. These samples were cryogenically microtomed in the longitudinal tensile direction (i.e. similar to the melt flow direction) at -80ºC with a Hitachi S-4700 ultramicrotome and were subsequently collected on 300 mesh copper grids. DSC analysis was performed to study the melting and crystallization behavior of PP/clay nanocomposites with a DSC-Q1000 TA instrument. The DSC samples of about 10 mg were first equilibrated at -60 ºC and then heated up to 220 ºC at 10 ºC/min. Isothermal condition was maintained for 5 mins in the following step to eliminate the material thermal history. Furthermore, the samples were cooled down to -60 ºC at the same rate of 10 ºC/min. A second scan was conducted under the same thermal conditions. The final crystallization and melting curves were chosen from the first cooling and second heating scans, respectively. The melting temperature (T_m), crystallization temperature (T_c) and degree of crystallinity (X_c) of PP matrix in PP/clay nanocomposites were determined in DSC measurement. X_c was calculated based on the equation shown below

$$X_c(\%) = \frac{\Delta H_m}{(1-W_f)\Delta H_f^0} \times 100 \qquad (1)$$

where ΔH_m (J/g) is the heat of fusion of PP matrix, W_f is the weight fraction of clay particles and is the heat of fusion of pure crystalline PP in which ΔH_f^0 =209 J/g [48].

The dynamic mechanical thermal analysis (DMTA) was conducted using a Rheometric Scientific DMTA V analyzer in a single cantilever mode at 0.01% strain with an oscillatory frequency of 1 Hz. The DMTA has the temperature range from -60 to 150°C at the temperature sweep of 5°C/min in the nitrogen atmosphere. The DMTA samples with the nominal dimensions of 27×10×2.5 mm were cut from the injection molded strip-like samples. At least two samples were used at each measurement for the test reproducibility. The glass transition temperature (T_g) was determined based on the peak of tanδ as a sensitive indicator of transitions.

Mechanical Testing

Mechanical properties were determined by using tensile, flexural and impact tests according to ASTM D638, D790 and D6110, respectively. Tensile testing was carried out on a universal tensile machine Instron 1185 at room temperature. The chord tensile modulus (0.05-0.25% strains) with an extensometer (gauge length: 50 mm) at a crosshead speed of 5 mm/min and the tensile yield strength at 50 mm/min were measured accordingly. The flexural testing was undertaken using a laboratory-scaled three-point bending rig on the Instron 1185 with the chord flexural modulus (0.05-0.25% strains) and flexural strength (calculated from 5% strain) determined at a speed of 1.22 mm/min. The Charpy impact strength was obtained by breaking notched injection molded impact samples using a CEAST Resil 25 pendulum impact testing machine. The reported results are based on the average data of over five samples with calculated standard deviations.

COMPOSITES THEORETICAL MODELS

Composites theoretical models usually follows the fundamental assumption including (i) linear elastic material behavior of fillers and polymer matrix; (ii) axisymmetric fillers in identical shape and size that are characterized by material parameters such as aspect ratio and shape parameter and (ii) well-bonded filler-polymer interfacial area with the phenomenological negligibility of interfacial slip, filler-polymer debonding and matrix cracking.

Most of these models are implemented to predict the tensile moduli of two-phase polymer composites with the reinforcements of unidirectionally well aligned (or randomly oriented) fiber-like, flake-like, disk-like platelet fillers. Conventional composites theories used in Halpin-Tsai model [13, 15, 28, 49], Mori-Tanaka model [15, 16, 27], Ji's model [50-52], Voigt upper bound and Reuss lower bound [16] recently applied to the modeling work of polymer/clay nanocomposites with satisfactory results. As a consequence, a detailed review of composites theoretical models in mathematical forms is elaborated in the following sections for the further reference.

Halpin-Tsai Model

Halpin-Tsai model [11] represents one of most popular composites theories to predict the tensile moduli of unidirectional composites as a function of filler aspect ratio. It can deal with a variety of reinforcement geometries for discontinuous reinforcements such as fiber-like or flake-like fillers to benefit the fiber composites industry. The longitudinal and transverse moduli E_{11} and E_{22} of a composite material in Halpin-Tsai model are expressed in a general form

$$\frac{E_c}{E_m} = \frac{1+\zeta\eta\phi_f}{1-\eta\phi_f} \tag{2}$$

$$\eta = \frac{\frac{E_f}{E_m}-1}{\frac{E_f}{E_m}+\zeta} \tag{3}$$

where E_c, E_f, E_m are Young's moduli of composites, fillers and polymer matrix, respectively. ζ is a shape parameter that depends on the filler geometry and loading direction. ζ=2(L/d) for fibers or ζ=2(L/t) for disk-like platelets when the longitudinal elastic modulus E_{11} is calculated. Since transverse elastic modulus E_{22} is relatively insensitive to fiber aspect ratio, ζ=2 is used in this case as an approximation. L, d, t, are referred to as the length, diameter and thickness of dispersed fillers, respectively.

Modified Halpin-Tsai Model

A modulus reduction factor (MRF) is often introduced in equation (2) as a matter of fact that 2-D disk-like clay platelets contribute less to the unidirectional reinforcement as opposed to 1-D fiber-like fillers. Such modified Halpin-Tsai model is thus written as

$$\frac{E_c}{E_m} = \frac{1+\zeta(MRF)\eta\phi_f}{1-\eta\phi_f} \tag{4}$$

For randomly dispersed short fibers, MRF is generally between 0.167 and 1 [53]. In a special case of platelet fillers, MRF=0.66 [49] was found to well predict the tensile moduli of rubber/clay nanocomposites in a wide range of clay volume fractions.

Voigt Upper Bound and Reuss Lower Bound

When $\zeta \to \infty$, the Halpin-Tsai model equations reach the upper bound known as Voigt rule of mixtures (ROM) [54] in which fibers and polymer matrix is subjected to an isostrain condition with the same uniform strains. The Voigt upper bound is derived as

$$E_c = \phi_f E_f + (1-\phi_f)E_m \tag{5}$$

On the contrary, if $\zeta \to 0$, the lower bound is converged from Halpin-Tsai equations under isostress condition with equal stresses for fibers and polymer matrix. This lower bound is called Reuss inverse rule of mixture (IROM) [55] or simplified as Reuss lower bound in a form of

$$\frac{1}{E_c} = \frac{\phi_f}{E_f} + \frac{(1-\phi_f)}{E_m} \tag{6}$$

Voigt upper bound and Reuss lower bound offers initial simple limitations of elasticity of multi-phase composites due to the only consideration of filler volume fraction in equations (5) and (6) whilst filler shape, filler geometry, filler orientation and spatial arrangement are completely ignored in such theoretical calculations.

Modified Rule of Mixture (MROM)

Similar to modified Halpin-Tsai model, MRF can also be incorporated in equation (5) to develop a modified rule of mixture (MROM) [56] with flake-like fillers shown as

$$E_c = \phi_f E_f (MRF) + (1-\phi_f)E_m \tag{7}$$

There are two major different forms to express MRF based on Riley's rule [57] and Padawer and Beecher's rule [58], respectively:

Riley form for MRF:

$$MRF = 1 - \frac{\ln(u+1)}{u} \tag{8}$$

Padawer and Beecher form for MRF:

$$MRF = 1 - \frac{\tanh u}{u} \tag{9}$$

with

$$u = \frac{1}{\alpha}\sqrt{\frac{\varphi_f G_m}{E_f(1-\varphi_f)}} \tag{10}$$

where is the inverse aspect ratio of dispersed fillers, and G_m is the shear modulus of the polymer matrix.

Mori-Tanaka Model

For a non-dilution composite system, Mori-Tanaka model [12, 59] was developed to predict the elastic moduli of composites with aligned spheroidal fillers reinforced into isotropic matrix.

It is derived from the principle of Eshelby's inclusion model to predict the elastic stress field in and around an ellipsoidal particle within an infinite matrix [60]. The particle interactions are normally not explicitly considered since the average matrix strain applies to all particles by assumption. The longitudinal and transverse elastic moduli with better prediction for relatively high-aspect-ratio fillers are given by

$$\frac{E_{11}}{E_m} = \frac{A_0}{A_0 + \phi_f(A_1 + 2\nu_m A_2)} \tag{11}$$

$$\frac{E_c}{E_m} = \frac{E_{22}}{E_m} = \frac{E_{33}}{E_m} = \frac{2A_0}{2A_0 + \phi_f[-2\nu_m A_3 + (1-\nu_m)A_4 + (1+\nu_m)A_5 A_0]} \tag{12}$$

where ν_m is the Poisson's ratio of polymer matrix, and parameters A_0-A_5 are matrix/filler properties and Eshelby tensor components [60], which depend on aspect ratios of particles and dimensionless elastic constants of the matrix. E_{11}, E_{22} and E_{33} are composite moduli in the three orthogonal axes 1, 2 and 3 under applied loads, as illustrated in Figure 8. Accordingly, the elastic modulus of composites on the major plane 2-3 of disk-like platelets E_c can be determined from equation (12). $E_{//}$ and $E_{\perp}$ are the composite moduli in parallel (longitudinal) and perpendicular (transverse) directions to the major axis of fillers.

Hui-Shia Model

Hui-Shia model can be used for the prediction of composite tensile modulus with unidirectionally aligned platelet reinforcements in a simple hypothesis of perfect interfacial bonding between polymer matrix and platelets. Its mathematical equations are formulated as

Longitudinal elastic modulus (E_{11})

$$\frac{E_c}{E_m}\frac{E_{11}}{E_m} = \frac{1}{1-\frac{\phi_f}{4}\left[\frac{1}{\xi}+\frac{3}{\xi+\Lambda}\right]} \tag{13}$$

Transverse elastic modulus (E_{22})

$$\frac{E_{22}}{E_m} = \frac{1}{1-\left(\frac{\phi_f}{\xi}\right)} \tag{14}$$

with

$$\xi = \phi_f + \frac{E_m}{E_f - E_m} + 3(1-\phi_f)\left[\frac{(1-g)\alpha^2 - \frac{g}{2}}{\alpha^2 - 1}\right] \tag{15}$$

$$g = \frac{\pi}{2}\alpha \tag{16}$$

$$\Lambda = (1-\phi_f)\left[\frac{3(\alpha^2+0.25)g - 2\alpha^2}{\alpha^2 - 1}\right] \tag{17}$$

where for disk-like platelets ($\alpha \le 0.1$).

Laminate Model

In the real morphology of polymer/clay nanocomposites, some degrees of misalignment and random orientation of clay particles are quite evident, which, however, do not coincide with the simple assumption of unidirectionally aligned fillers in conventional composites theories. In the case of completely random orientation in all three orthogonal directions, the approximation equations for fiber and platelet reinforced composites moduli $E_{\text{ran-3D}}$ based on the laminate theory [61, 62] are expressed as

$$E_{ran-3D}^{fibres} = 0.184\,E_{//} + 0.816\,E_{\perp} \tag{18}$$

$$E_{ran-3D}^{platelets} = 0.49\,E_{//} + 0.51\,E_{\perp} \tag{19}$$

Laminate model reveals a good analytical approach for the modulus prediction of polymer nanocomposites with a dominantly random-oriented clay dispersion pattern which is one step closer to the actual micro/nanostructures.

Ji's Model: Modified Takayanagi Model

Takayanagi model focuses on a variety of heterogeneous materials with two viscoelastic constituent phases based on simple mechanical models, consisting of partly connected elements either in series or in parallel, to determine their viscoelastic properties [63].

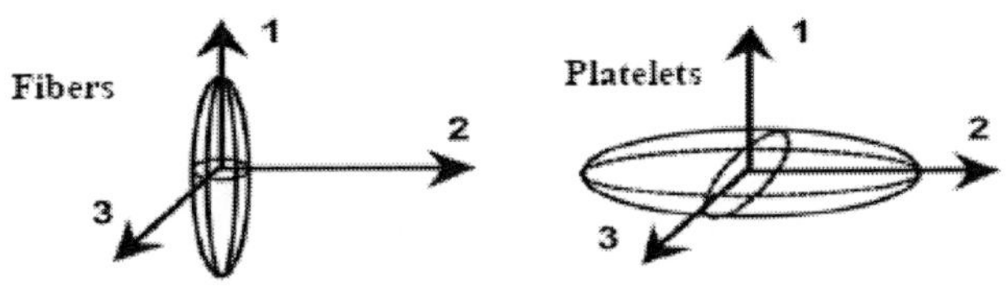

Applied load direction	Major axis/plane		Mori-Tanaka composite modulus	
	Fibers	Platelets	Fibers	Platelets
1			$E_{11} = E_{//}$	$E_{11} = E_{\perp}$
2	Axis 1	Plane 2-3	$E_{22} = E_{\perp}$	$E_{22} = E_{//}$
3			$E_{33} = E_{\perp}$	$E_{33} = E_{//}$

Figure 8. Physical representations and coordinate systems for composite stiffness calculations with the reinforcements of fibers and disk-like platelets in Mori-Tanaka equations [13].

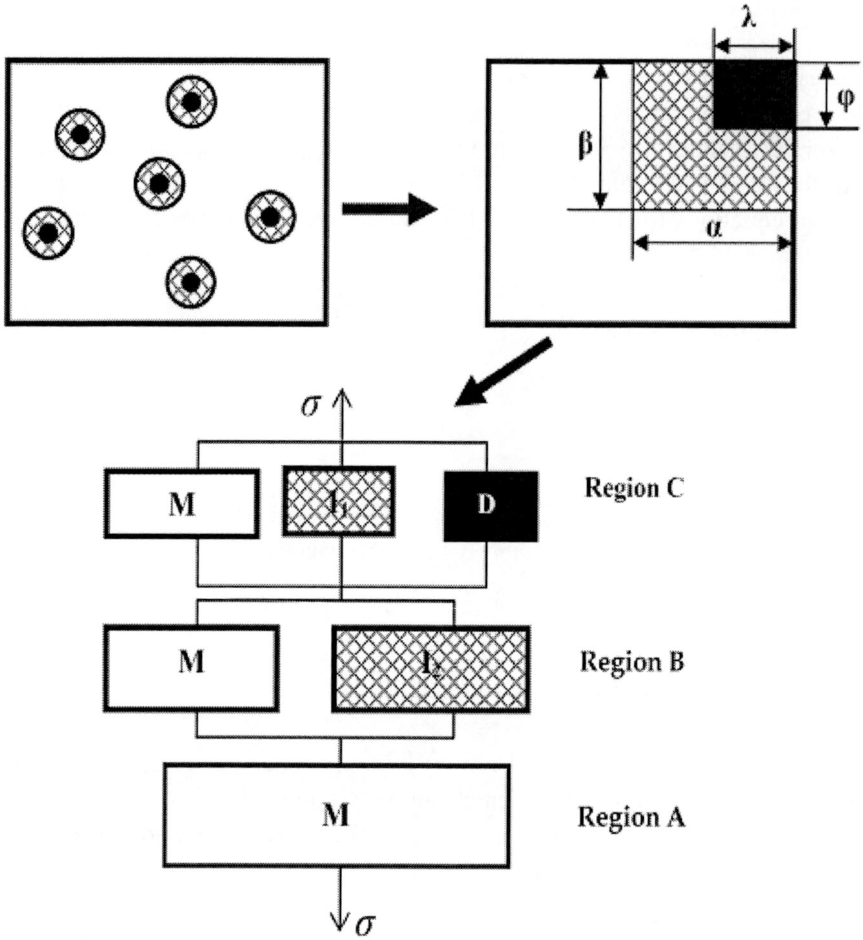

Figure 9. Three-phase Takayanagi model with parallel and series arrangement [64].

When the constrained interfacial region (or "interphase") is incorporated around clay particles in a nanocomposite system, a three-phase modified Takayanagi model, also called Ji's model [64] is normally employed in which three phases are connected to one another in series and parallel, as depicted in Figure 9. In a composite unit cell or RVE, the three phases of dispersed phase "D", interface "I" and matrix "M" are equivalently integrated into individual constituent phases in geometry. λ, φ and α, β represent the dimensions in the transverse and longitudinal directions for dispersed phase and interface, respectively. The tensile modulus of polymer nanocomposites with randomly oriented platelet reinforcements is given by

$$E_c = \left[\left[\frac{1-\sqrt{[2(\tau/t_0)+1]\phi_f}}{E_m} \right] + \frac{\sqrt{[2(\tau/t_0)+1]\phi_f}-\sqrt{\phi_f}}{\{1-\sqrt{[2(\tau/t_0)+1]\phi_f}\}E_m+\sqrt{[2(\tau/t_0)+1]\phi_f}(k-1)E_m/\ln k} + \frac{\sqrt{\phi_f}}{\{1-\sqrt{[2(\tau/t_0)+1]\phi_f}\}E_m+\{\sqrt{[2(\tau/t_0)+1]\phi_f}-\sqrt{\phi_f}\}(k+1)E_m/2+\sqrt{\phi_f}E_f} \right]^{-1} \quad (20)$$

where is the thickness of interfacial region, and t_0 is the platelet thickness. k represents the modulus ratio of interface neighboring on the platelet surface. It is assumed that the interphase modulus decreases linearly along the normal direction from kE_m on the filler interface to E_m on the matrix interface.

RVE Numerical Simulation

2-D RVE model was developed on a continuum scale by implementing the commercial finite element package ANSYS. As a simplest case, clay platelets are assumed to be unidirectionally well-aligned and homogeneously dispersed into polymer matrix. Clay aspect ratio, clay content, platelet filler dispersion pattern and the interphase properties between clay platelets and polymer matrix are the four major modeling parameters chosen in this RVE simulation.

In particular, the aspect ratios have been fixed to be 10 and 100 with the clay particle thickness of t =10 and 1 nm, respectively, which are two idealized benchmarks of full intercalation and exfoliation for a nanocomposite system accordingly [16]. With a further hypothesis of periodic particle structures in RVE models, three typical RVE patterns have been investigated, namely “stack”, “staggering” and “randomly distributed” patterns, Figure 10. a and b are the length and width of a single RVE domain (a= 200 nm as a constant for simplicity). ΔL is the offset between clay platelets in the stack and c is the lateral distance between two clay platelets. Mathematically $c=b$ for the stack rows as shown in Figure 10(a). The geometric parameter b and staggering factor f_s [29] for the platelet overlap can be derived as

$$b = \frac{\rho_f}{\rho_m}\left[\frac{Lt}{a}\left(\frac{1}{W_f}-1\right)\right] + \frac{Lt}{a} \approx \frac{\rho_f}{\rho_m}\left[\frac{Lt}{a}\left(\frac{1}{W_f}-1\right)\right] \tag{21}$$

and

$$f_s = \frac{2\Delta L}{b} \tag{22}$$

In the simplified equation (21), ρ_f and ρ_m are referred to as the densities of clay particles and PP matrix. The term of Lt/a in equation (21) is regarded as a trivial part and thus is ignored provided that geometric parameters of RVE models are determined in approximately 95% confidence level.

The interphase properties have been varied in range of $0.5E_m$, E_m and $2E_m$ for the interphase modulus E_i, and of $0.5t$, t and $2t$ for the interphase thickness t_i. The constituent properties used in the RVE simulation are elaborated in details elsewhere [1]. The boundary conditions have been set in such a manner of zero X and Y displacements on the left boundary (i.e. fully constrained) and respectively applied equivalent strain of 0.05 and 0.25% on the right boundary of a RVE model, Figure 10(a).

The resulting average stress is calculated from the sum of the nodal forces implemented on the right boundary. The elastic modulus for a RVE model material obeys the Hooke's law in the following equation

$$E_c = \frac{\sigma}{\varepsilon_0} = \frac{\sum_{i=1}^{n} F_i}{\varepsilon_0 A_0} = \frac{\sum_{i=1}^{n} F_i}{\varepsilon_0 (1 \cdot 3b)} = \frac{\sum_{i=1}^{n} F_i}{3b\varepsilon_0} \tag{23}$$

in which is the applied strain, and is the resulting axial stress. F_i is the nodal force at node i, and A_0 is the cross section area of the established 3×3 array RVE domain based on the assumption of plane stress and unit thickness.

OOF MODELING

The significantly enhanced mechanical properties of PP/clay nanocomposites like the elastic modulus can be achieved by using a small amount of well-dispersed clay particles as rigid nanofillers. An alternative modeling approach based on the object-oriented finite element (OOF) code OOF 2.0.4 [31], developed by the National Institute of Standards and Technology (NIST), was also employed to predict elastic moduli of nanocomposites at different clay contents.

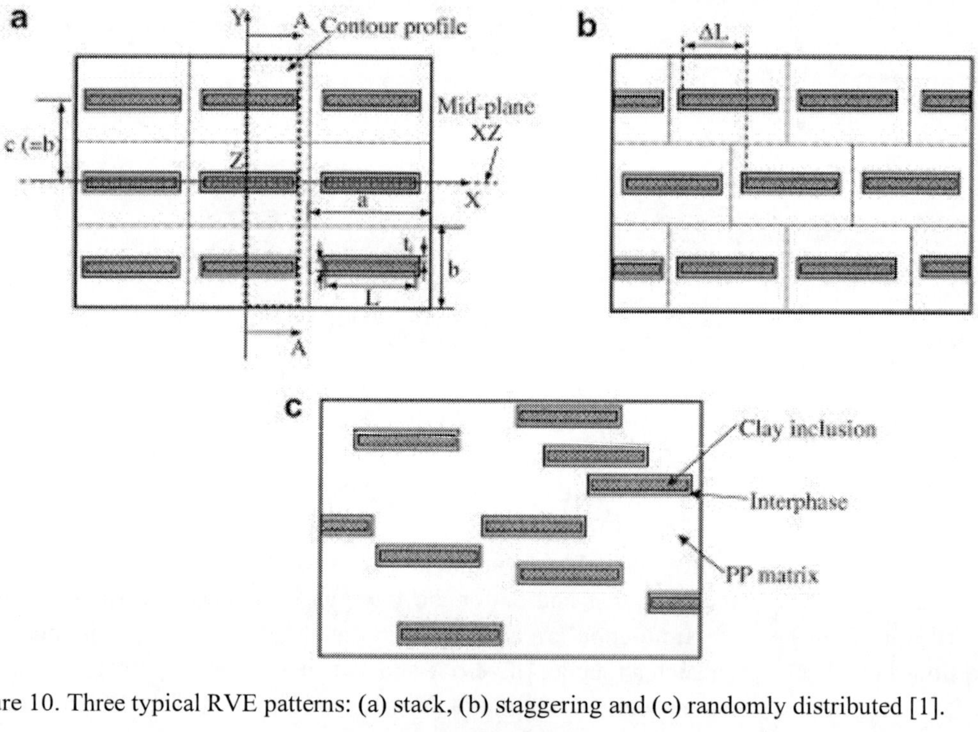

Figure 10. Three typical RVE patterns: (a) stack, (b) staggering and (c) randomly distributed [1].

The Fedora Red Hat Core 4 Linux system was selected to install OOF 2.0.4 on a Dell INSPIRON™ 8600 laptop (based on the setup of Windows XP and Linux dual boot system with 1.6 GHz processors and 2 GB RAM). As illustrated in Figure 11(a), a representative region of interest was chosen from a typical TEM micrograph for 5 wt% filled PP/clay nanocomposites. The constituent properties of clay and PP matrix were subsequently assigned after the image segmentation using a pixel selection tool, Figure 11(b), then followed by the generation of 2-D finite element mesh from the OOF skeleton to display detailed micro/nanostructures of clay platelets in actual size, shape, orientation and spatial position with an adaptive mesh refinement scheme, Figure 11 (c). It is hypothesized that both clay particles and PP matrix behave as isotropic linearly elastic materials with the perfect interfacial bounding between these two constituents. The material properties of each constituent used in OOF modeling are listed in [65]. The simple case of "ideally isotropic homogenized particles" [13, 15, 66, 67] for the intercalated clay platelets is only considered for the calculations of elastic modulus and Poisson's ratio of clay particles since the intercalation effect can be quite dominant in the real morphology of PP/clay nanocomposites. The boundary conditions were defined as such that both the x and y displacements U_x and U_y on the left boundary were set to zero with full constraints whilst U_x on the right boundary was equivalent to 0.05 and 0.25% elastic strains, respectively.

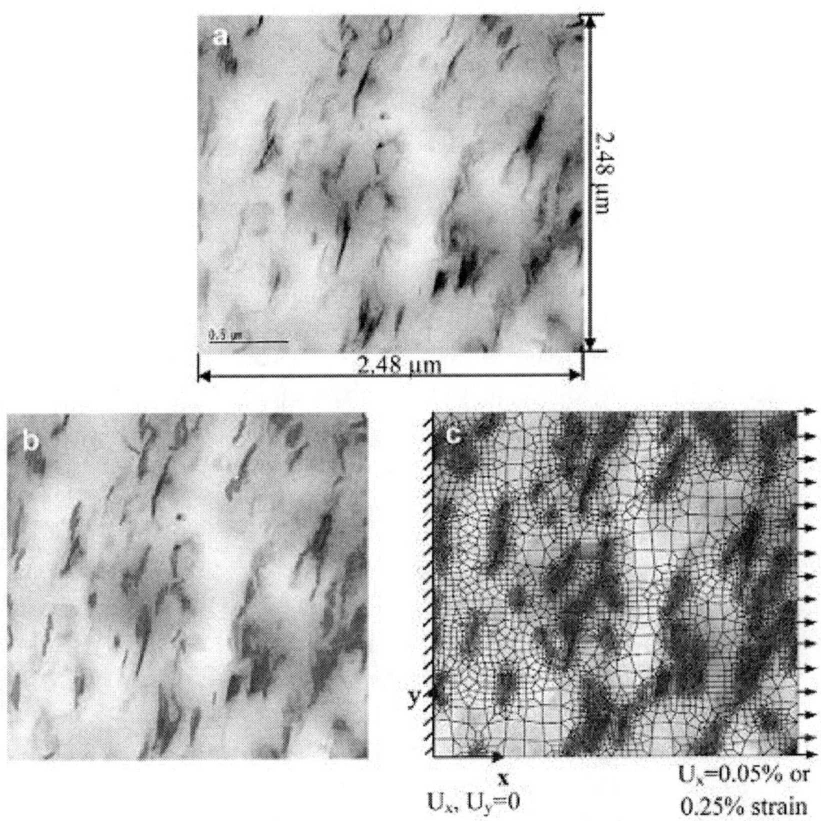

Figure 11. OOF modeling steps based on 5 wt% filled PP/clay nanocomposites: (a) original representative region of interest from a TEM image, (b) image segmentation from pixel selection and (c) finite element mesh with the boundary conditions [65].

The effective in-plane elastic modulus of nanocomposites E_{eff} was estimated from the numerically calculated nodal reactions (i.e. the sum of the boundary forces) acting on the right boundary, resulting from the in-plane tension of OOF models. E_{eff} is then expressed as

$$E_{eff} = \frac{L_x \sum_{i=1}^{n} F_i}{L_y \Delta L} \tag{24}$$

where F_i is the reaction in the ith node in the x direction. n is the set number of boundary nodes under the nonzero displacement condition. ΔL is the prescribed in-plane displacement, and L_x and L_y are the horizontal and vertical dimensions of an OOF modeling domain along the x and y directions, respectively. The assumptions of unit thickness and plane stress were made to simplify the OOF modeling process.

RESULTS AND DISCUSSION

Significant Factors and Preferred Formulations in DoE

Taguchi DoE method reveals the statistical analysis of *S/N* ratios with conceptual approach, consisting of the effect-graphing and visual identifications of significant factors in the material processing of PP/clay nanocomposites. In addition, the Pareto analysis of variance (ANOVA) technique [45] further enables to examine corresponding determined mechanical properties [47] in each of formulated nanocomposites, listed in Table 2. The general criterion to detect the significant factors in this DoE study was based upon the derived cumulative contribution percentage of about 90% along with the technical and economic considerations for non-significant factors. The Pareto ANOVA diagrams for each of mechanical parameters are displayed in Figure 12. As clearly seen, PP type (factor D) and clay content (factor B) were found to be two most significant factors to improve tensile moduli of nanocomposites with the contribution percentages of 60 and 33%, respectively; whereas clay type (factor A) and MAPP content (factor C) are quite trivial factors with the sum of contribution percentages becoming less than 10%, Figure 12 (a). The similar Pareto ANOVA diagram was also shown in Figure 12(b) for the enhanced tensile strengths, except for the addition of clay type as the third significant factor. As for the flexural properties, the increases of flexural modulus and strength are ascribed to the strong influences of PP type (contribution percentages: 53 and 77%) and clay content (contribution percentage: 43 and 21%) accordingly, Figures 12(c) and (d). On the other hand, impact strengths tend to be affected by PP type, MAPP content and clay type with the respective contribution percentages of 67, 20 and 8%. In comparison, MAPP content as the second most significant factor impacts more greatly on the impact properties as opposed to the tensile and flexural properties. Since the "larger-the-better" characteristics in Taguchi DoE method mean that the higher the sum of *S/N* ratio, the better the response is for the factorial effect, the best combination of factors to enhance each of mechanical parameters are determined from the sum of *S/N* ratio diagrams in Figure 13, along with the respective compositions and estimates of error variance presented in Table 3. This DoE study is solely based on the maximization of individual mechanical properties. Considering the difficulty in achieving the global property enhancement of PP/clay nanocomposites, a good balance has to be established to increase tensile/flexural properties and simultaneously reduce the sacrificed impact strength. As depicted in Figure 13(a), the best combination of significant factors to obtain the highest value for the sum of S/N ratio, in response to the improved tensile modulus, is at level 3 of clay content (i.e. 10 wt% clays) and level 3 of PP type (i.e. PP-Hom H380F). Hence the preferred factor-level condition becomes $A_3B_3C_3D_3$, in which both clay type (DK4 clays) and MAPP content (20% MAPP), as two non-significant factors, remain at the same level 3.

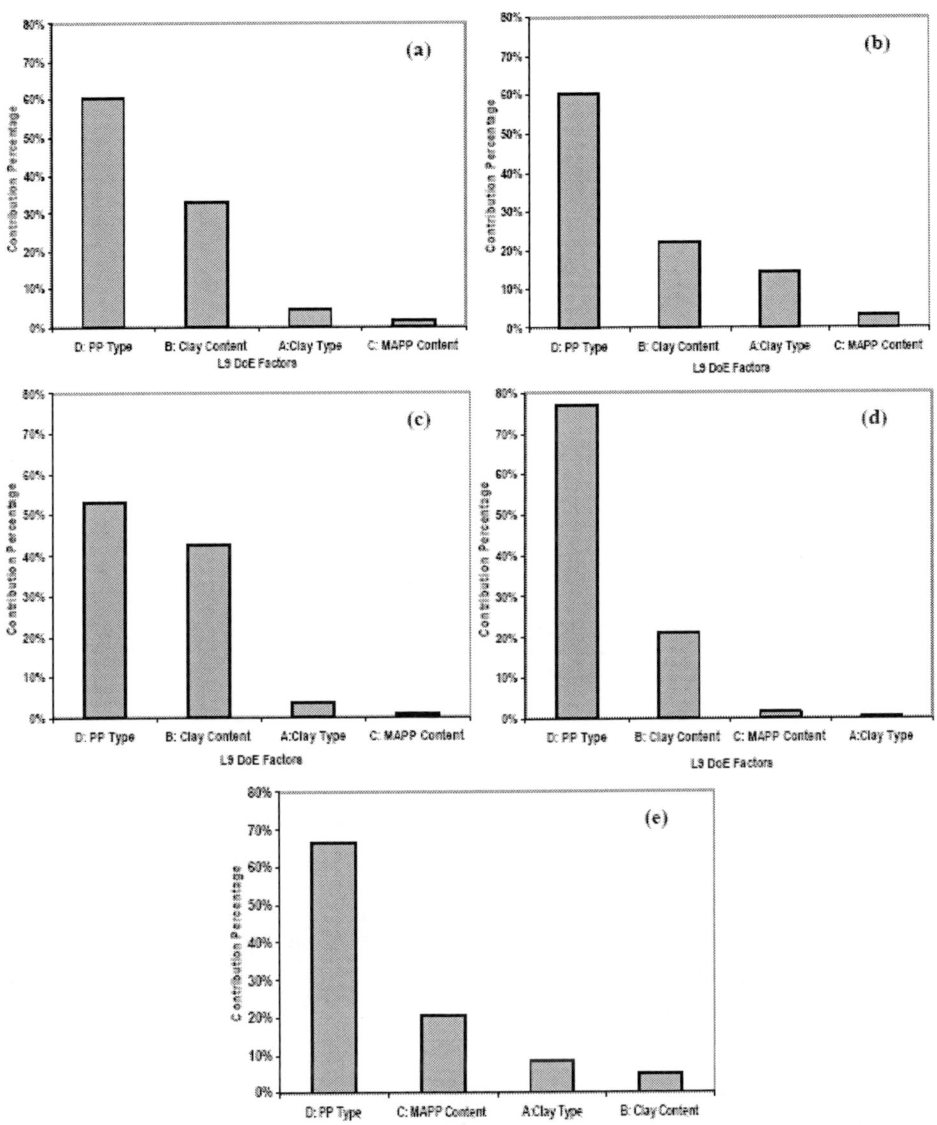

Figure 12. Pareto ANOVA diagrams for the enhancement of mechanical parameters of PP/clay nanocomposites: (a) tensile modulus, (b) tensile strength at yield, (c) flexural modulus, (d) flexural strength and (e) Charpy impact strength [47].

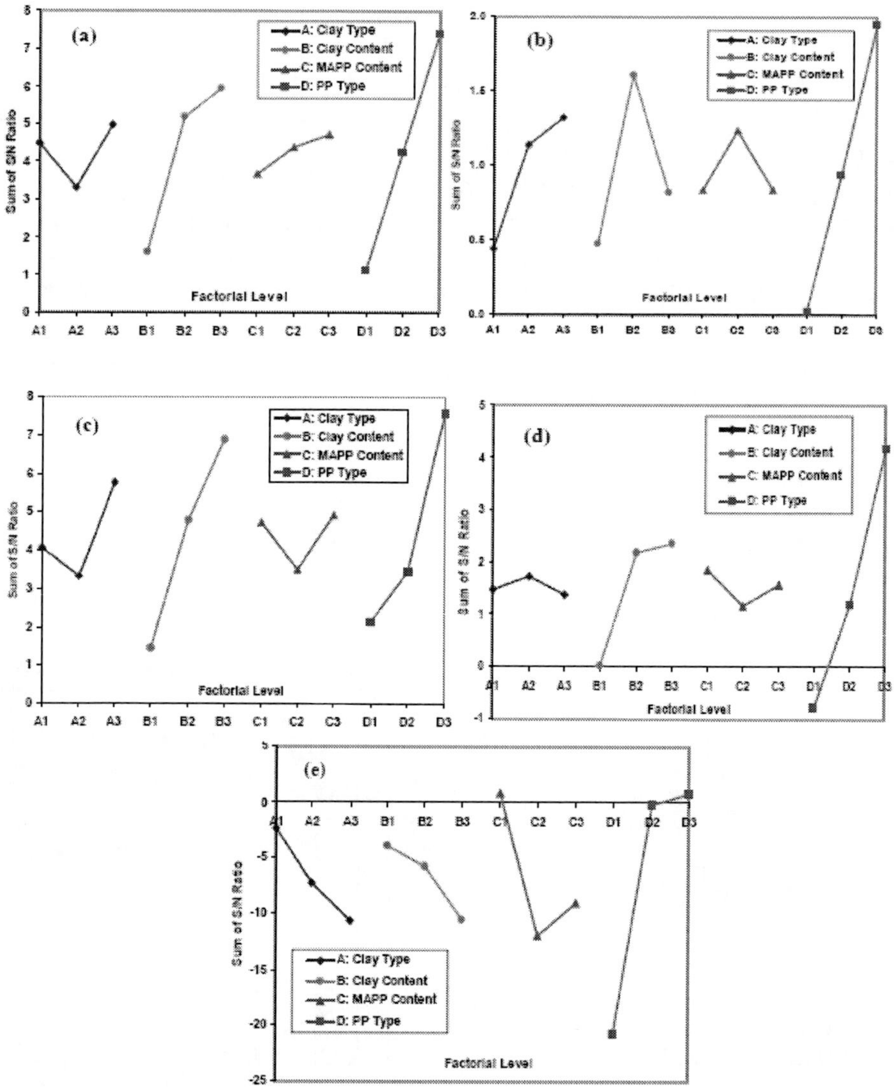

Figure 13. Sum of *S/N* ratio diagrams for the enhancement of mechanical parameters of PP/clay nanocomposites: (a) tensile modulus, (b) tensile strength at yield, (c) flexural modulus, (d) flexural strength and (e) Charpy impact strength [47].

For the maximization of the tensile strength, $A_3B_2C_2D_3$ (i.e. 5 wt% DK4 clays, PP-Hom H380F and 10 wt% MAPP) was detected as the corresponding best combination, Figure 13(b). In a similar manner, the best combinations to enhance the flexural modulus and strength are $A_3B_3C_3D_3$ (i.e. 10 wt% DK4 clays, PP-Hom H380F and 20 wt% MAPP) and $A_2B_3C_1D_3$ (i.e. 10 wt% DK2 clays, PP-Hom H380F and 5 wt% MAPP), respectively, Figures 13(c) and (d). Conversely, for the response of maximized impact strength, the majority of the calculated sums of S/N ratios appear to be negative at each factor level, Figure 13(e), which suggests that the additional clays in the presence of MAPP tend to deteriorate the impact properties, especially for PP-Co M710 based nanocomposites. Thus the best combination is $A_1B_1C_1D_3$ (i.e. 3 wt% DK1N clays, PP-Hom H380F and 5 wt% MAPP).

Table 3. Summary of preferred formulations to improve mechanical properties of PP/clay nanocomposites [47]

Larger-the-better L_9 DoE response (normalized)		Preferred factorial level combination	Composition (wt%)	Estimate of error variance
Tensile properties	Tensile modulus	$A_3B_3C_3D_3$	DK4/MAPP/H3 80F (10/20/70)	0.35
	Tensile strength	$A_3B_2C_2D_3$	DK4/MAPP/H3 80F (5/10/85)	0.04
Flexural properties	Flexural modulus	$A_3B_3C_3D_3$	DK4/MAPP/H3 80F (10/20/70)	0.31
	Flexural strength	$A_2B_3C_1D_3$	DK2/MAPP/H3 80F (10/5/85)	0.05
Impact properties	Impact strength	$A_1B_1C_1D_3$	DK1N/MAPP/ H380F (3/5/92)	9.70

Table 4 Optimized material formulations of PP/clay nanocomposites

Sample batch symbol	Composition (wt%)		
	DK4 clay	MAPP	PP-Hom H380F
Opt1	3	6	91
Opt2	5	10	85
Opt3	8	16	76
Opt4	10	20	70

As a matter of fact, PP type and clay content demonstrate overall significant effects on the mechanical property enhancement, particularly with PP-Hom H380F as the suitable PP grade for the best material performance. In view of global enhancement of mechanical properties, the final optimized nanocomposite formulations were determined from selected PP-Home H380F, MAPP and DK4 clays with 3, 5, 8 and 10 wt% clay contents (weight ratio of MAPP: clay, WR=2:1), as presented in Table 4, subjected to the identical sample preparation conditions.

Clay Dispersion and Intercalation Effect

The intercalation effect with respect to the clay dispersion was investigated via XRD analysis on optimized PP/clay nanocomposites after the DoE study. Their XRD spectra are plotted in Figure 14, along with the reported interlayer spacing values (d_{001}, d_{002} and d_{003}) shown in Table 5.

Three distinct (001), (002) and (003) reflection peaks occur at 2.48, 4.88 and 7.36° for raw DK4 clay powders, which correspond to 3.56, 1.81 and 1.20 nm, respectively. Regardless of clay and MAPP contents used, (001), (002) and (003) peaks for optimized nanocomposites remain in the close positions to those of DK4 clays. Quite differently, the peak intensities are reduced accordingly with the lowest level for Opt1, which is most likely due to the breakdown mechanism of clay tactoids into finer particles in the melt compounding process. The first peaks are mildly shifted to lower diffraction angles when the clay content decreases, Figure 14. The corresponding interlayer spacings of nanocomposites are reported to increase from 3.62 (Opt4) to 3.73 nm (Opt1) relative to 3.56 nm for DK4 clays, resulting from the limited intercalation.

The second and third reflection peaks are very similar despite the broadened XRD peaks with decreasing the clay content, which implies that clay tactoids break down into much thinner stacks of layered structures with the possibility of exfoliation.

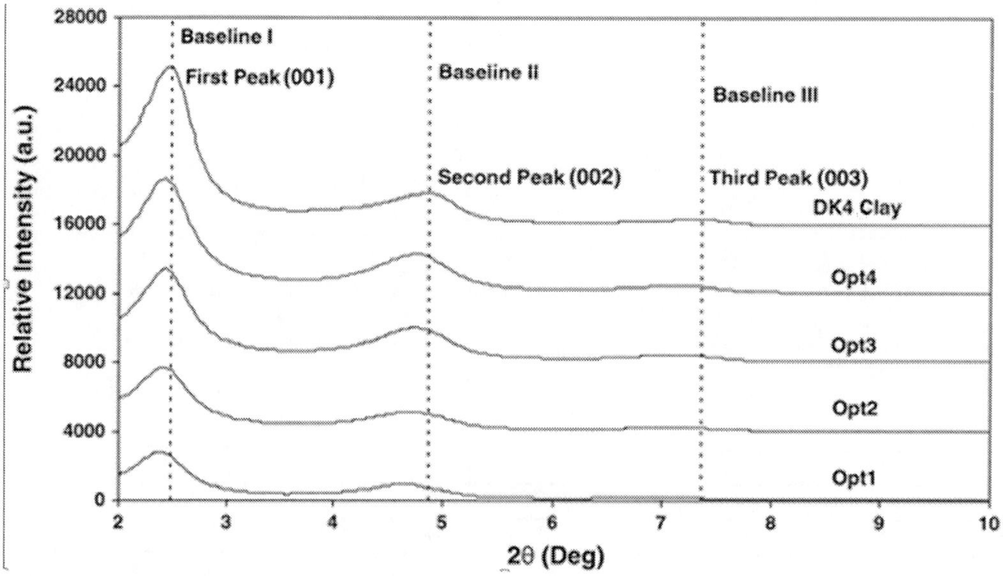

Figure 14. XRD spectra of DK4 clays and optimised PP/clay nanocomposites [68].

Table 5. XRD features of DK4 clays and optimized PP/clay nanocomposites [68]

Material type	2θ (1st peak)	d_{001} (nm) (1st peak)	2θ (2nd peak)	d_{002} (nm) (2nd peak)	2θ (3rd peak)	d_{003} (nm) (3rd peak)
DK4 clays	2.48°	3.56	4.88°	1.81	7.36°	1.20
Opt1	2.37°	3.73	4.64°	1.90	7.04°	1.25
Opt2	2.39°	3.70	4.69°	1.88	7.06°	1.25
Opt3	2.40°	3.68	4.78°	1.85	7.18°	1.23
Opt4	2.44°	3.62	4.78°	1.85	7.26°	1.22

SEM Analysis

Observed from the fracture surfaces of optimized nanocomposites, a good clay dispersion with locally existing clay tactoids in size of less than 2 μm is manifested, especially in Opt1 and Opt2 filled with less than 5 wt% clays, Figures 15(a) and (b). Nevertheless, by increasing the clay contents in range of 8-10 wt%, a great amount of clay tactoids do not appear prevalently, possibly due to the dual favorable effects of compatibilizer MAPP and organomodified DK4 clay with relatively large interlayer spacing, Figures 15(c) and (d). Much weaker van der Waals forces caused by the large interlayer spacing (d_{001}=3.56 nm) facilitate the break-down effect of clay tactoids when sufficient MAPP molecules migrate between the clay interlayers to achieve the good compatibility. As a consequence, by means of intensive mechanical shearing stress in the twin screw extrusion, the majority of large DK4 clay particles are separated into tiny clay platelets in size of 500 nm to 1 μm.

TEM Characterization

Despite a mixed morphological structure with combined intercalated stacks and exfoliated platelets, the overall good clay dispersion has been found at the low clay contents of 3 and 5 wt% in good accordance with corresponding SEM results, Figures. 16(a) and (b). Apart from the overall arrow-indicating clay major orientations, clay platelets in the lateral dimension of 200–500 nm still exist in skewed and curved structures with some extents of localized random orientation, which is believed to be caused by the core effect of injection molded nanocomposite samples [13].

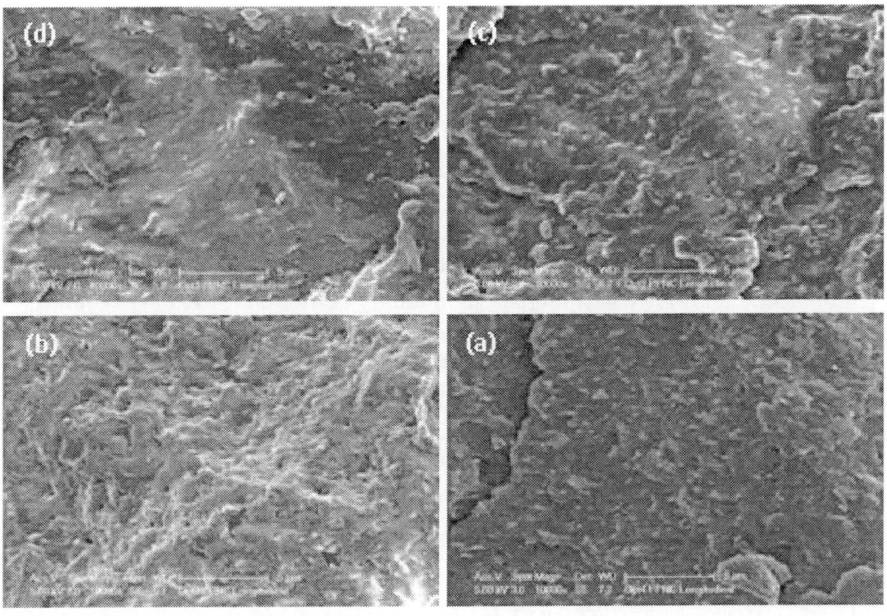

Figure 15. SEM micrographs of optimized PP/clay nanocomposites at ×10,000 magnification: (a) Opt1, (b) Opt2, (c) Opt3 and (d) Opt4. The scale bars represent 5μm and the arrows indicate clay tactoids [68].

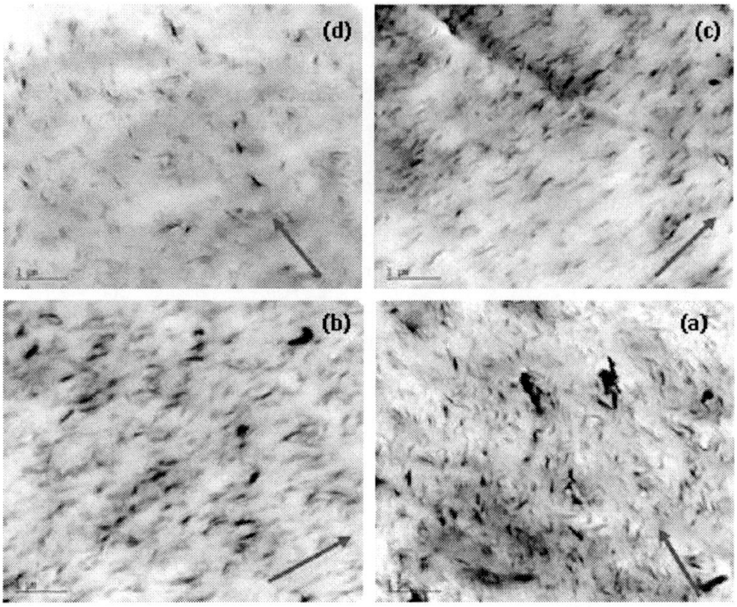

Figure 16. TEM micrographs of optimized PP/clay nanocomposites at ×15,000 magnification: (a) Opt1 (b) Opt2, (c) Opt3 and (d) Opt4. The scale bars represent 1μm and the arrows indicate clay major orientation directions [68].

At the higher clay contents of 8–10 wt%, the clay dispersion has been greatly improved and becomes reasonably uniform despite some typical clay tactoids, Figures 16(c) and (d). It is convincing that the presence of MAPP benefits the size reduction of clay agglomerates due to its effective compatibilizer role.

Mechanical Properties

Mechanical properties of finally optimized formulation are depicted in Figure 17 as opposed to those of neat PP-Hom H380F. The uses of clay particles and MAPP as the compatibilizer within the PP matrix have been proven to yield the overall property enhancements. Both tensile and flexural moduli demonstrate the better improvements in a relatively monotonic manner when increasing the clay content. At the clay content of 10 wt%, the tensile and flexural moduli have been enhanced by 41% and 61%, respectively. The strengths also increase up to 16% (tensile) and 22% (flexural), but beyond the threshold of 3 wt% clays, they become more or less level-off regardless of the clay content. The better tensile and flexural properties could arise from a higher level of intercalated/exfoliated clay structures and more active interfacial interactions, particularly for the low clay content of less than 5 wt%. The strength level-off/retention effect at high clay contents (8–10wt%) might be interpreted by the proportionally increased MAPP content (relative to the clay content level) in the optimized formulation to promote the clay dispersion and delamination in compensation for the negative effect of clay agglomeration that is often the reason for the strength reduction. Additionally, the impact strength has shown a 50% improvement at 3 wt% clay inclusions, and subsequently declines to the value of neat PP for 10 wt% filled nanocomposites.

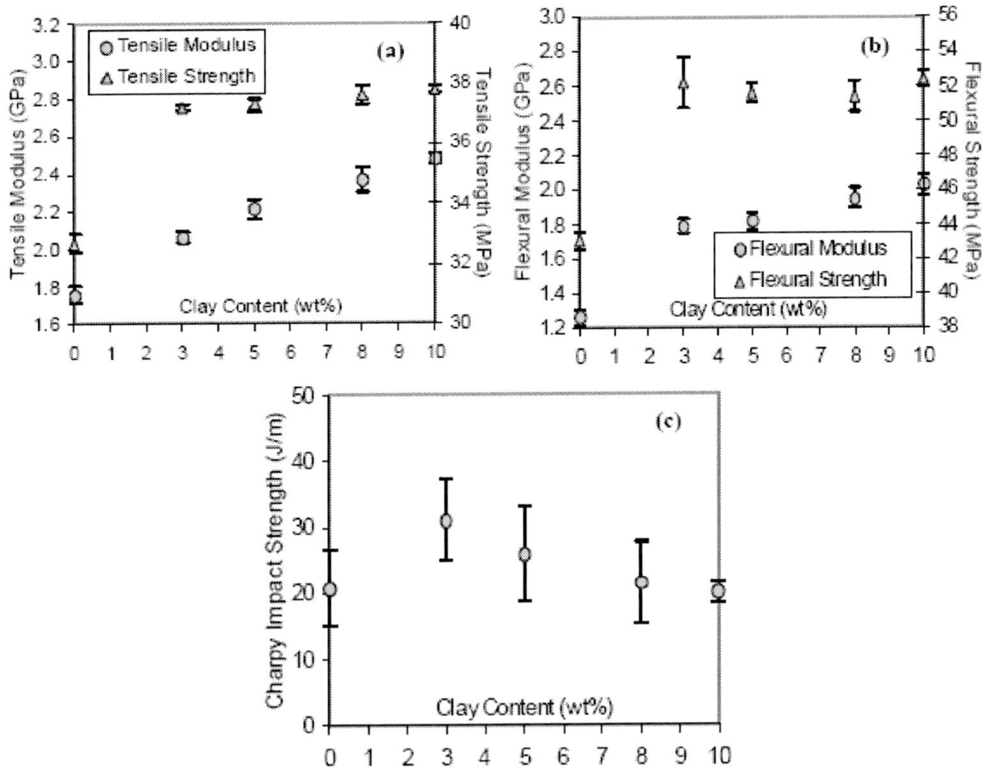

Figure 17. Mechanical properties of optimized PP/clay nanocomposites: (a) tensile properties, (b) flexural properties and (c) Charpy impact strength [69].

The higher impact strength at a low clay content possibly benefits from more uniform clay dispersion leading to the better intercalation/exfoliation level. The reduction of impact strength often occurs at a high clay content because of the induced material brittleness from the proportional amount increases of clays and MAPP. In general polymer blending, MAPP can naturally generate more brittle characteristic than neat PP itself.

PP Crystalline Structures Using Wide-Angle XRD (WAXD)

The crystalline structures of neat PP-Hom H380F, MAPP and corresponding optimized nanocomposites were detected in WAXD spectra, Figure 18. Obviously, five typical characteristic peaks of most common α-PP take place in all materials at 2θ angles of about 14.0, 16.8, 18.6, 21.2 and 21.9°, corresponding to the lattice planes of (110), (040), (130), (111) and (131), respectively [71, 72].

Apart from the different relative peak intensities, the same number of nanocomposite WAXD peaks relative to those of neat PP suggests that the additional clays do not greatly affect the crystalline structures of PP matrix in PP/clay nanocomposites. In the other aspect, the intensities of α (040) have presented overall significant alterations and all nanocomposites possess more considerably enhanced α (040) intensities when compared to those of neat PP-Hom H380F and MAPP.

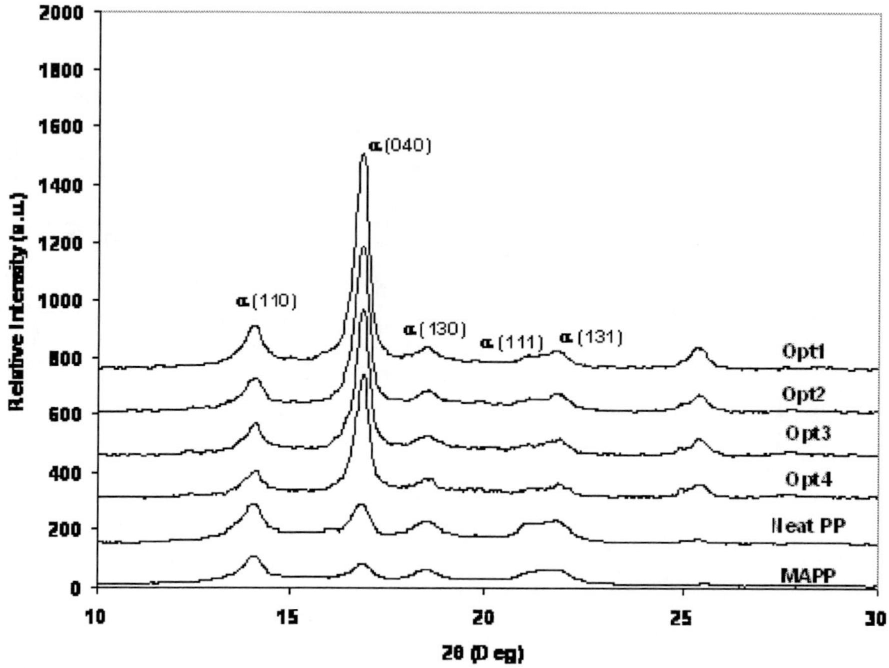

Figure 18. WAXD spectra of neat PP-Hom H380F, MAPP and optimized PP/clay nanocomposites. The curves are shifted vertically for clarity [70].

As detailed elsewhere [73, 74], the formation of α-PP, as a most stable crystalline form in melt-crystallization process, can make great contributions to the enhancement of tensile modulus and higher yield stress, which coincides well with previously determined tensile properties of optimized nanocomposites. The α (040) intensity is relatively reduced with increasing the clay content, which could be attributed to the low molecular weight MAPP proportionally increased in amount for the optimized formulations; whereas MAPP reveals little influence on the transformation of crystalline structures.

DSC Data

The melting and crystallization behavior of optimized nanocomposites were characterized by obtaining DSC thermal analysis results depicted in Figure 19, along with the determined DSC parameters and calculated degree of crystallinity (X_c) listed in Table 6. All nanocomposites show melting temperatures (T_m) at about 164°C irrespective of clay and MAPP contents. This thermal behavior is very similar to those of neat PP-Hom H380F and MAPP, which indicates that the additions of clays and MAPP have very insignificant impact on the melting behavior of PP/clay nanocomposites. The crystallization temperatures (T_c) of all nanocomposites increase up to 118°C regardless of the clay content, as opposed to 113 °C for PP-Hom H380F and 112 °C for MAPP accordingly. Such increase is expected due to the good balance between the positive clay nucleating effect and detrimental influence of the excessive amount of MAPP that migrates into the PP bulk matrix and further hinders the nucleation.

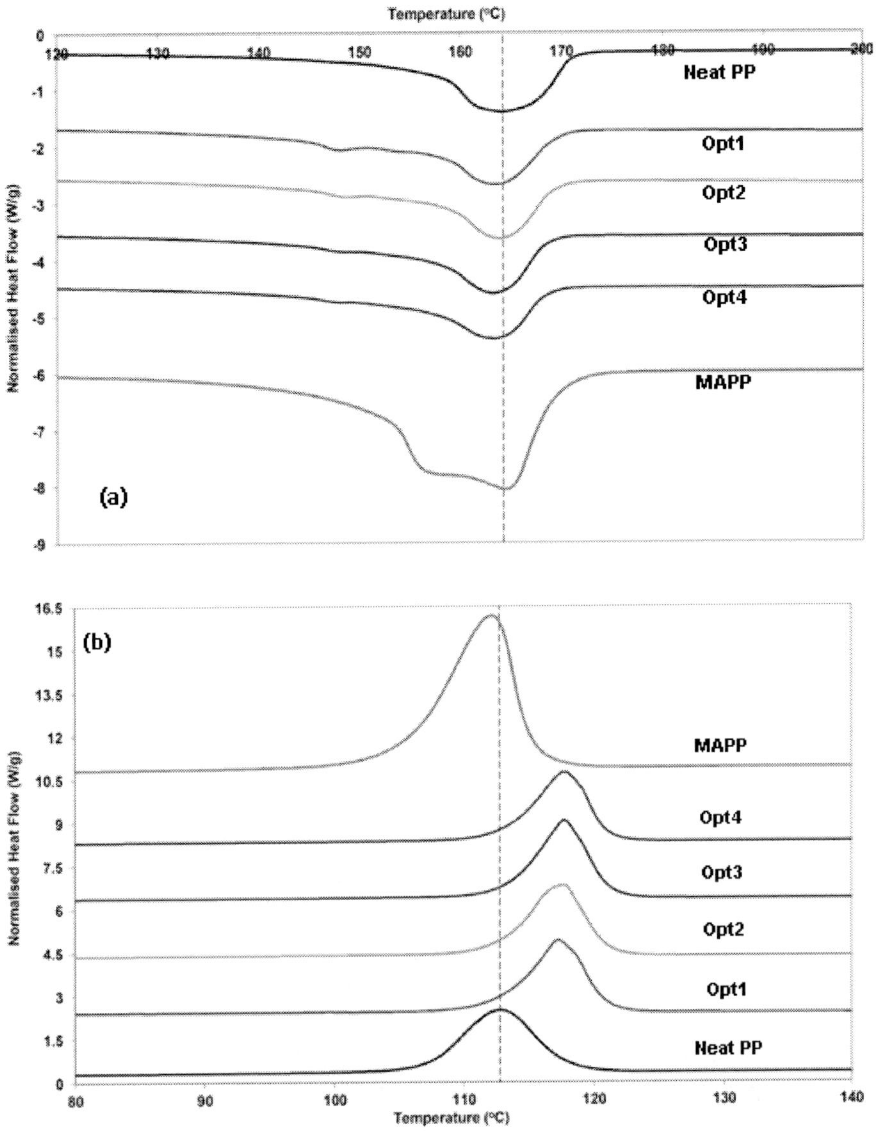

Figure 19. DSC curves of neat PP-Hom H380F, MAPP and optimized PP/clay nanocomposites: (a) heating scan and (b) cooling scan. The curves are shifted vertically for clarity [70].

More nucleating cores can be generated with the aid of better clay dispersion, along with more active interfacial interactions, in order to facilitate the crystallization process and achieve better nucleating effect and higher T_c [2]. The higher T_{ic} shown in such nanocomposites also signifies a higher extent of segregation of dispersed clay platelets possibly around the spherulite/interspherulite boundaries, leading to the enhanced intercalation [75]. The X_c of optimized nanocomposites slightly increases up to 46.3% for Opt3 relative to 41.8% for neat PP, again resulting from the positive nucleating effect of dispersion clays on crystallinity. Notwithstanding that MAPP has been found to be a high crystalline PP with the X_c value of 92.3%, it has surprisingly very minor effect on the crystallinity level of all nanocomposite counterparts.

Table 6. DSC characteristic parameters and level of crystallinity of PP/clay nanocomposites [70]

Material type	Clay /MAPP content (wt%)	T_m (°C)	T_c (°C)	ΔH_m (J/g)	X_c (%)
Neat PP	0/0	164.2	112.8	87.32	41.8
MAPP	0/100	164.3	112.0	192.90	92.3
Opt1	3/6	163.3	117.3	92.86	45.8
Opt2	5/10	163.9	117.7	89.06	44.9
Opt3	8/16	163.1	117.8	88.96	46.3
Opt4	10/20	163.3	117.8	82.79	44.

Thermo-Mechanical Results

The glass transition temperature (T_g) as well as the reinforcement effect, owing to the additional clays and MAPP, are under investigation in DMTA analysis. The dynamic elastic modulus E' and loss tangent tanδ as a temperature function for optimized PP/clay nanocomposites are displayed in Figure 20. The typical values of E' at five different temperatures in range from -30 to 135°C, and the T_g values are presented in Table 7. Evidently, two relaxations appear in neat PP-Hom H380F and nanocomposites, which are identified as α and β peaks. α peak of PP crystalline regions is known as the shoulder-like weak peak close to 100 °C whilst β peak presents the glass-rubber relaxation of PP amorphous portion, which should be generally assigned to T_g [48]. Seen from Figure 20(b) and Table 7, much higher T_g becomes manifested in most nanocomposites as compared to that of neat PP (T_g=14.7 °C).

Table 7. Dynamic elastic moduli E' of optimized PP/clay nanocomposites at different temperatures and corresponding glass transition temperatures (T_g) [70]

	Dynamic elastic modulus* E' (GPa)					
	-30°C	5°C	25°C	90°C	135°C	T_g (°C)
Neat PP	2.10	1.82	1.38	0.41	0.16	14.7
Opt1	2.23 (1.06)	1.93 (1.06)	1.49 (1.08)	0.48 (1.18)	0.23 (1.42)	18.9
Opt2	2.40 (1.14)	2.10 (1.16)	1.60 (1.16)	0.48 (1.18)	0.21 (1.29)	17.7
Opt3	2.76 (1.31)	2.45 (1.35)	1.85 (1.35)	0.52 (1.27)	0.20 (1.23)	18.8
Opt4	2.95 (1.40)	2.61 (1.44)	1.97 (1.43)	0.53 (1.30)	0.20 (1.24)	21.2

* The values in parentheses are the relative dynamic elastic moduli of nanocomposites to those of neat PP-Hom H380F.

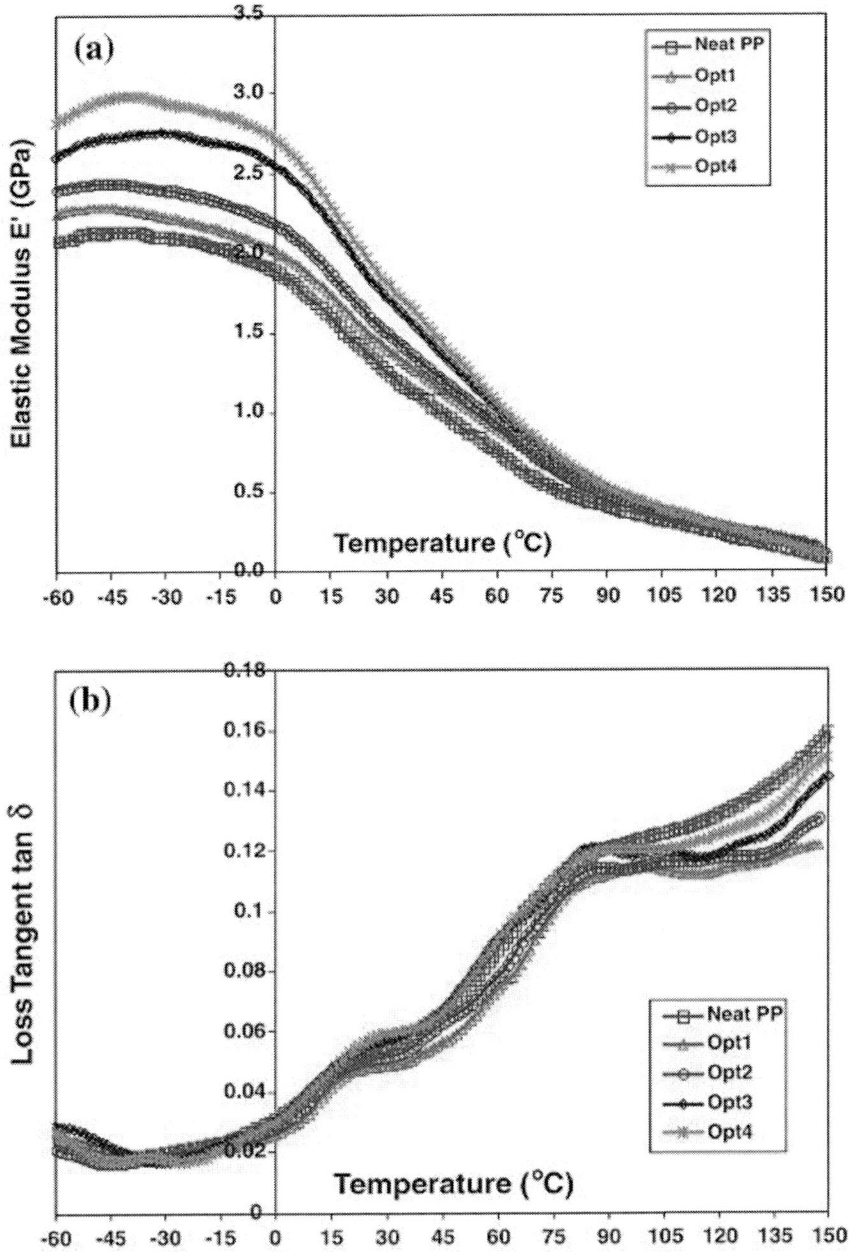

Figure 20. DMTA curves for neat PP-Hom H380F and optimized PP/clay nanocomposites: (a) dynamics elastic modulus E' and (b) loss tangent $\tan\delta$ [70].

When the clay content increases from 5 to 10 wt%, a further enhanced T_g up to the maximum 21°C for Opt4 is obtained, which might be again explained by the major benefits from the overall uniform clay dispersion and active interfacial interactions between PP molecular chains and intercalated clay platelets. Additionally, more intercalated clay platelets can offer higher mobility restrictions to PP chains, thus resulting in the shift of T_g to a higher temperature. With respect to the reinforcement effect, all the nanocomposites demonstrate higher E' than neat PP as expected, shown in Figure 20(a) and Table 7. When increasing the clay content, E' values are enhanced monotonically between -30 and 90 °C with the maximum increase of 44% for Opt4. In particular, at 135 °C within the rubbery plateau, the reinforcement effect has not diminished but instead reveals relatively high level at low clay contents (3-5 wt%) with the maximum increase of 42% for Opt1.

RVE Simulation Work

The clay dispersion effect on the overall mechanical behavior of nanocomposites is numerically evaluated by means of three aforementioned RVE patterns on the elastic modulus with W_f=5 wt% and t_i=2t exhibited in Figure 21. The randomly distributed pattern resembles the real morphological structures of nanocomposites except that the well-aligned clay structures and constant clay aspect ratios are assumed in this RVE simulation. For exfoliated nanocomposites, the relative modulus E_c/E_m is given in the following order

$$\left(\frac{E_c}{E_m}\right)_{STK} < \left(\frac{E_c}{E_m}\right)_{RAD} < \left(\frac{E_c}{E_m}\right)_{STG(fs=1)} \tag{25}$$

For intercalated nanocomposites, a totally different relationship is established in the form of

$$\left(\frac{E_c}{E_m}\right)_{STK} \approx \left(\frac{E_c}{E_m}\right)_{RAD} \approx \left(\frac{E_c}{E_m}\right)_{STG(fs=1)} \tag{26}$$

Observed from equation (26), the clay dispersion patterns have very minor influence on relative elastic moduli of intercalated nanocomposites. As a result, since clay particles tend to prevalently form the intercalated structures in PP/clay nanocomposites, only the stack pattern is implemented in the subsequent RVE modeling for simplicity. In view of the stress transfer between clay platelets and PP matrix, the axial stress and strain distributions of cross section A-A (see Figure 10(a)) are depicted for exfoliated and intercalated RVE models (W_f= 5 wt %, t_i=2t), respectively in Figures 22 and 23, which are both confirmed in axisymmetric characteristics along the mid-plane XZ. For exfoliated RVE models, much higher axial stresses occur in the clay inclusions and clay/interphase regions with the maximum stress of over 48 times greater than those around PP matrix by the effective transfer of interfacial shear stresses, Figure 22 (a). Nevertheless, the axial stress distributions have made very little differences especially when the interphase modulus is equal to or higher than the matrix modulus. Axial strains increase remarkably in PP matrix, but oppositely become much smaller around neighboring areas of clays and interphases when decreasing the interphase modulus, Figure 22(b). This numerical finding reflects the "anti-straining" phenomenon due to the constraints of PP matrix deformation by individually exfoliated rigid clay platelets. Intercalated RVE models, as shown in Figure 23(a), undergo the changes of maximum stress between clay particles and PP matrix from 7.4 to 3.8 with the reduced interphase modulus from 2E_m to 0.5E_m. By contrast, the stress transfer in exfoliated nanocomposites is far more efficient, which in turn leads to the higher modulus and strength improvements.

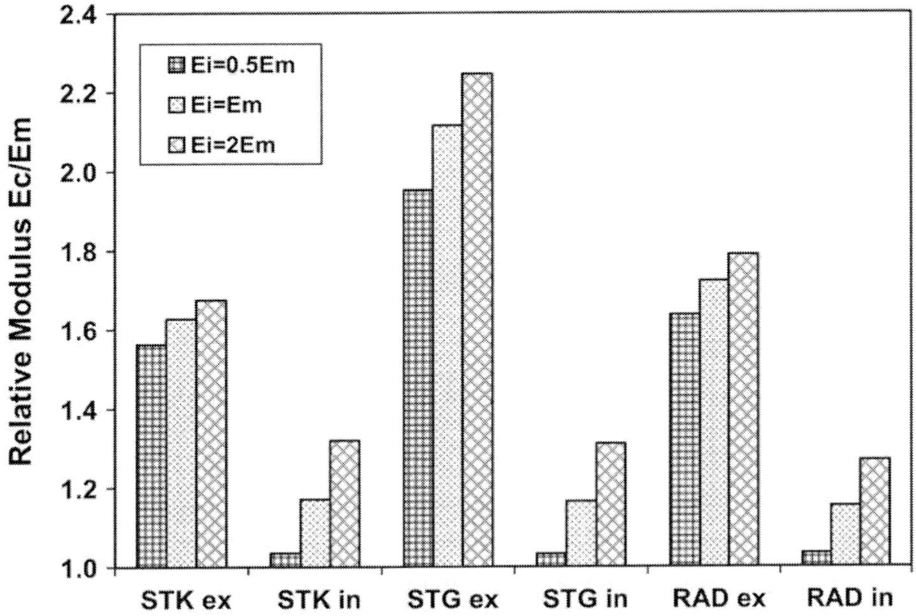

Figure 21. Effect of RVE patterns on the relative moduli of exfoliated and intercalated nanocomposites (W_f = 5 wt %, t_i = 2t). Symbols of ''STK'', 'STG'' and ''RAD'' are denoted as the stack, staggering (f_s = 1) and randomly distributed patterns, respectively. ''ex'' and ''in'' represent the exfoliated and intercalated RVE models [1].

The axial strains for intercalated nanocomposites appear to be independent of the interphase modulus owing to very similar strain distributions in Figure 23(b). In particular, the maximum axial strain of PP matrix in an intercalated RVE model is approximately 4 times higher than that in an exfoliated counterpart. Since intercalated RVE models are subjected to the greater deformation in large PP matrix areas and relatively small axial stresses in clay inclusions, the modulus enhancement is expectedly inferior to that in an exfoliated structure. The comparisons between RVE simulation results using the stack pattern (W_f = 5 wt%, t_i = 2t), experimental data of optimized nanocomposites from Figure 17(a), and composites theoretical models are demonstrated in Figures 24 and 25 in terms of intercalated and exfoliated structures. A narrow band of numerical results is evident amongst all exfoliated RVE models with L/t=100 by varying the interphase modulus in spite of slightly greater modulus enhancement of nanocomposites arising from the higher interphase modulus, Figure 24. Modified Halpin-Tsai model (MRF=0.66) and Hui-Shia model present relatively close predictions with the numerical results when compared to Halpin-Tsai model and modified rule of mixture (MRF=0.66) that might be considered as the upper bounds of exfoliated nanocomposites. The overprediction of all theoretical models on the relative modulus E_c/E_m infers that conventional micromechanical models for particulate composites might not be applicable to the stiffness prediction of well exfoliated nanocomposites. In recognition of manufacturing difficulty and practical unavailability of fully exfoliated PP/clay nanocomposites, the numerical approach still shows a generally-known trend of enhanced relative modulus with increasing the clay content, which can be taken as the typical reference for the further investigation.

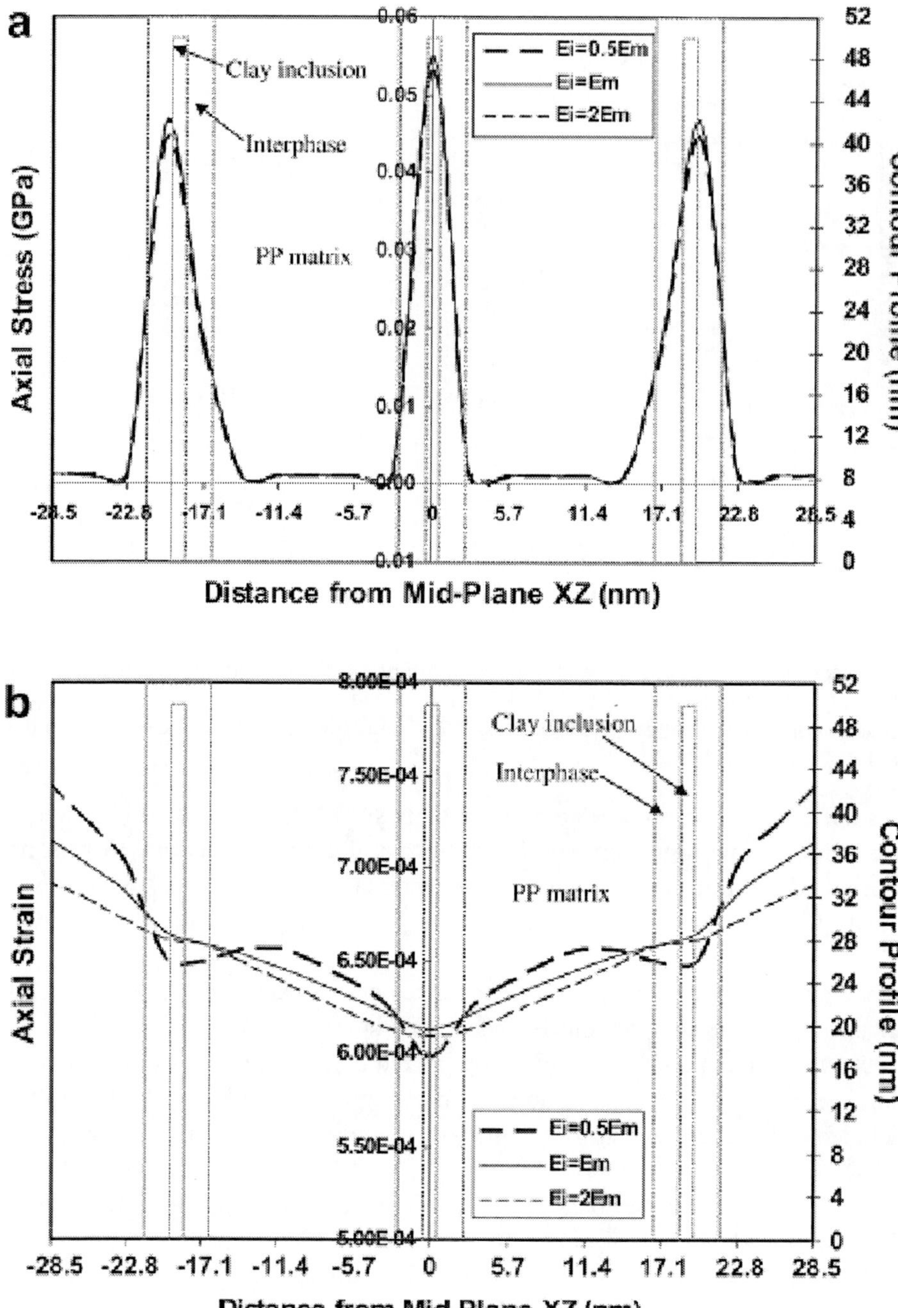

Figure 22. Axial stress (a) and axial strain (b) distributions of exfoliated RVE models along the mid-plane XZ in the stack pattern ($W_f = 5$ wt%, $t_i = 2t$) [1].

On the other hand, the greatest divergence from experimental data appears in the numerical results of intercalated nanocomposites (L/t=10) with the interphase modulus E_i=$2E_m$, giving the highest modulus enhancement and best fitting to Halpin-Tsai model, Figure 25.

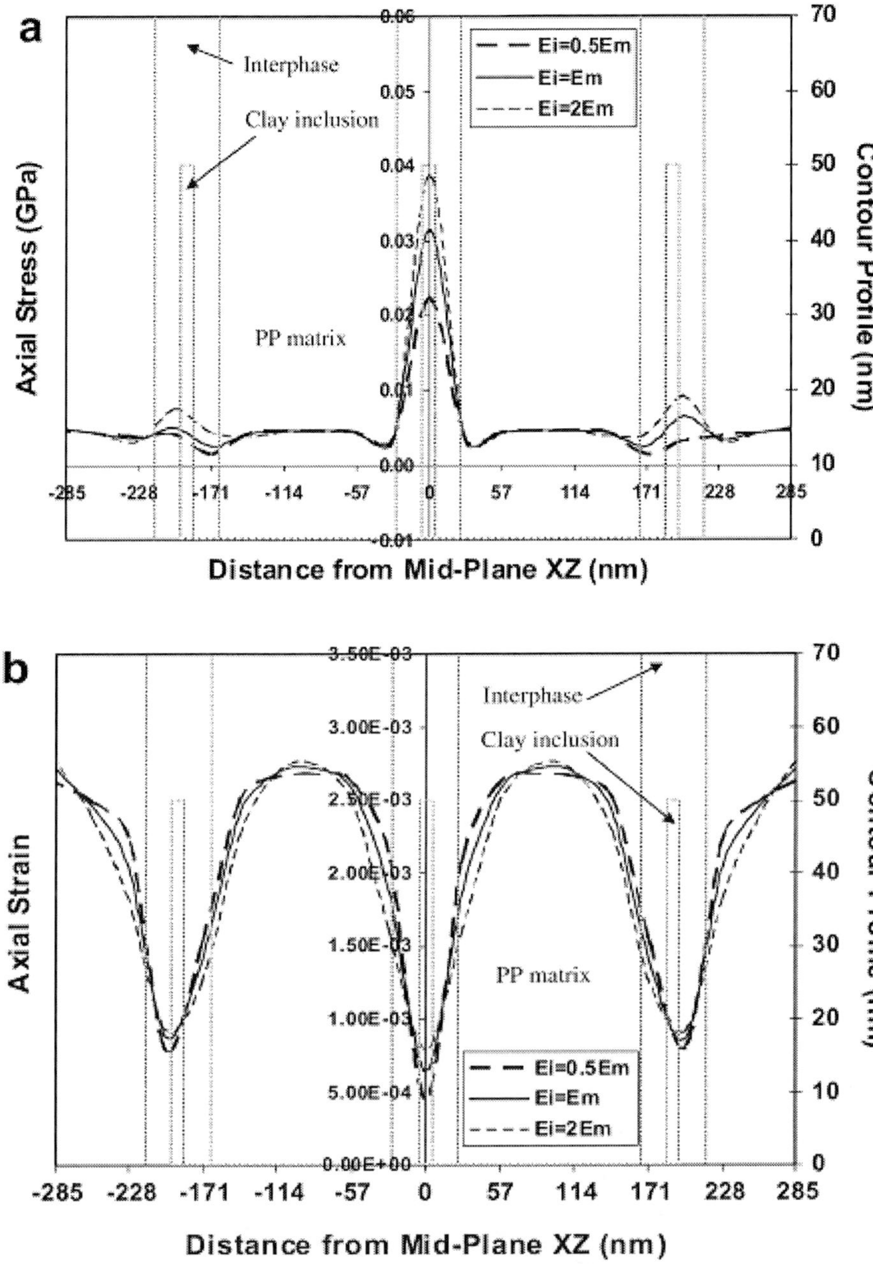

Figure 23. Axial stress (a) and axial strain (b) distributions of intercalated RVE models along the mid-plane XZ in the stack pattern (W_f = 5 wt%, t_i = $2t$) [1].

The intercalated RVE models with E_i=0.5E_m yield the least modulus enhancement of less than 5%; whereas those models with E_i=E_m attain the best agreement with Hui-Shia model. The experimental data become more consistent with those estimated by the modified Halpin-Tsai model (MRF=0.66), which tends to occur at higher clay content of over 8 wt%. The underestimation of experimental data by RVE models with E_i=E_m might be caused by the only hypothesis of complete stacks of layered intercalated structures in RVE modeling.

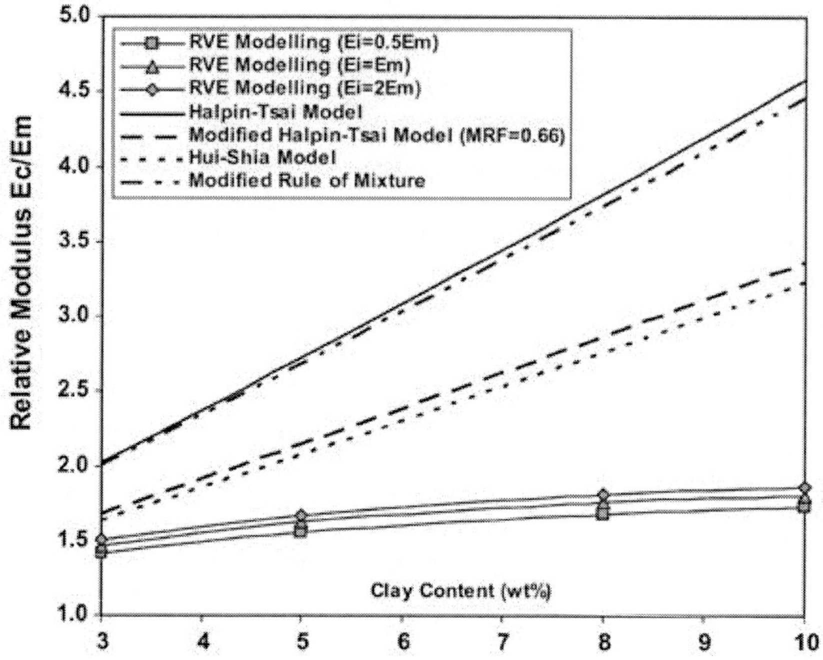

Figure 24. Comparisons of relative moduli of exfoliated nanocomposites between RVE modeling results (L/t=100 and t_i=$2t$) and composites theoretical models [1].

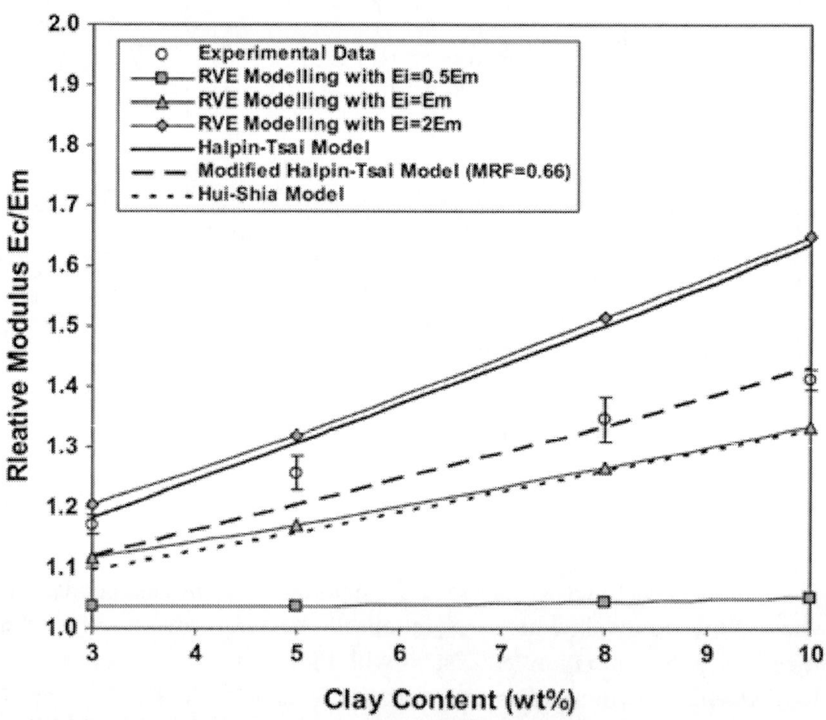

Figure 25. Comparisons of relative moduli of intercalated nanocomposites between RVE modeling results (L/t=10 and t_i=$2t$), experimental data and composites theoretical models [1].

However, the previous TEM results, demonstrated in Figure 16, reveal a mix of intercalated and exfoliated clay dispersion in the real nanocomposite morphology though admittedly the intercalation is more dominant. Consequently, the further promoted elastic modulus by a certain exfoliation level with relative large clay aspect ratios is negligible in the complete intercalated RVE models. When taking the experimental data as the benchmark, the interphase modulus E_i could fall between E_m and $2E_m$ (but much closer to E_m). This claim is totally different from the general assumption of E_i=$2E_m$ employed in modeling nylon 6/clay nanocomposites [29]. The additional low molecular weight MAPP as the compatibilizer leads to the reduction of interphase modulus since their short molecular chains are often highly mobilized around clay platelet surfaces and interlay areas [28]. It is also worth mentioning that an excessive amount of MAPP due to the negative matrix plasticization effect [47, 68, 70] tends to further deteriorate the interphase modulus, thus damaging the modulus enhancement mechanism. Though the experimental measurement of interphase modulus cannot be achieved easily, the use of parametric study in RVE modeling still brings an important qualitative evaluation of interphase properties.

OOF Modeling Prediction

As a matter of fact, the clay orientation is the other important factors to significantly influence the elastic moduli of nanocomposites besides clay aspect ratio and shape parameter [13, 76]. To reflect the nanofiller orientation effect, OOF modeling results are compared with the experimental data of optimized PP/clay nanocomposites and typical Halpin-Tsai and Hui-Shia models in either unidirectional platelet alignment or 3-D random platelet orientation, as depicted in Figure 26. A linearly increasing trend for elastic modulus was clearly captured by modeling results when the clay content increases, which are in good accordance with experimental data but present a slight overpredictive tendency. The plausible explanation for such discrepancy can be related to the missing constituent properties of MAPP as the compatibilizer and interphases, and the artefacts and phase distinction issues in the imaging analysis during the pre-processing. The other possible reason could be due to the simple assumption of perfect interfacial bonding whilst polyolefin nanocomposites such as PP/clay nanocomposites are actually in lack of strong adhesion with weak van der Waals forces employed in the clay interlayers [77]. Alternatively, in the presence of prevalent intercalated platelets within PP matrix, the experimental data are also fitted with composites theoretical models at various clay aspect ratios (L/t=5-80). The unidirectional alignment state is based upon the theoretical calculations of unmodified Halpin-Tsai model and Hui-Shia model in equations (2) and (13), respectively; whereas the 3-D random orientation state is represented by Halpin-Tsai laminate hybrid and Hui-Shia laminate hybrid models derived from equations (2), (13), (14) and (19). At a fixed clay aspect ratio, each curve for the random orientation, as shown in Figures 26(b) and (d), is situated below corresponding one for the unidirectional alignment, Figures 26(a) and (c). It is manifested that in conventional composites theory, deviations from unidirectional reinforcement inevitably result in the sizable reduction of composites stiffness and such reduction level arises from the filler shape [13]. This modulus reduction phenomenon on account of different filler orientation states appears to be more pronounced for Halpin-Tsai laminate hybrid model as opposed to the unmodified Halpin-Tsai model by dealing with discontinuous fiber-liker fillers, Figures 26(a) and (b).

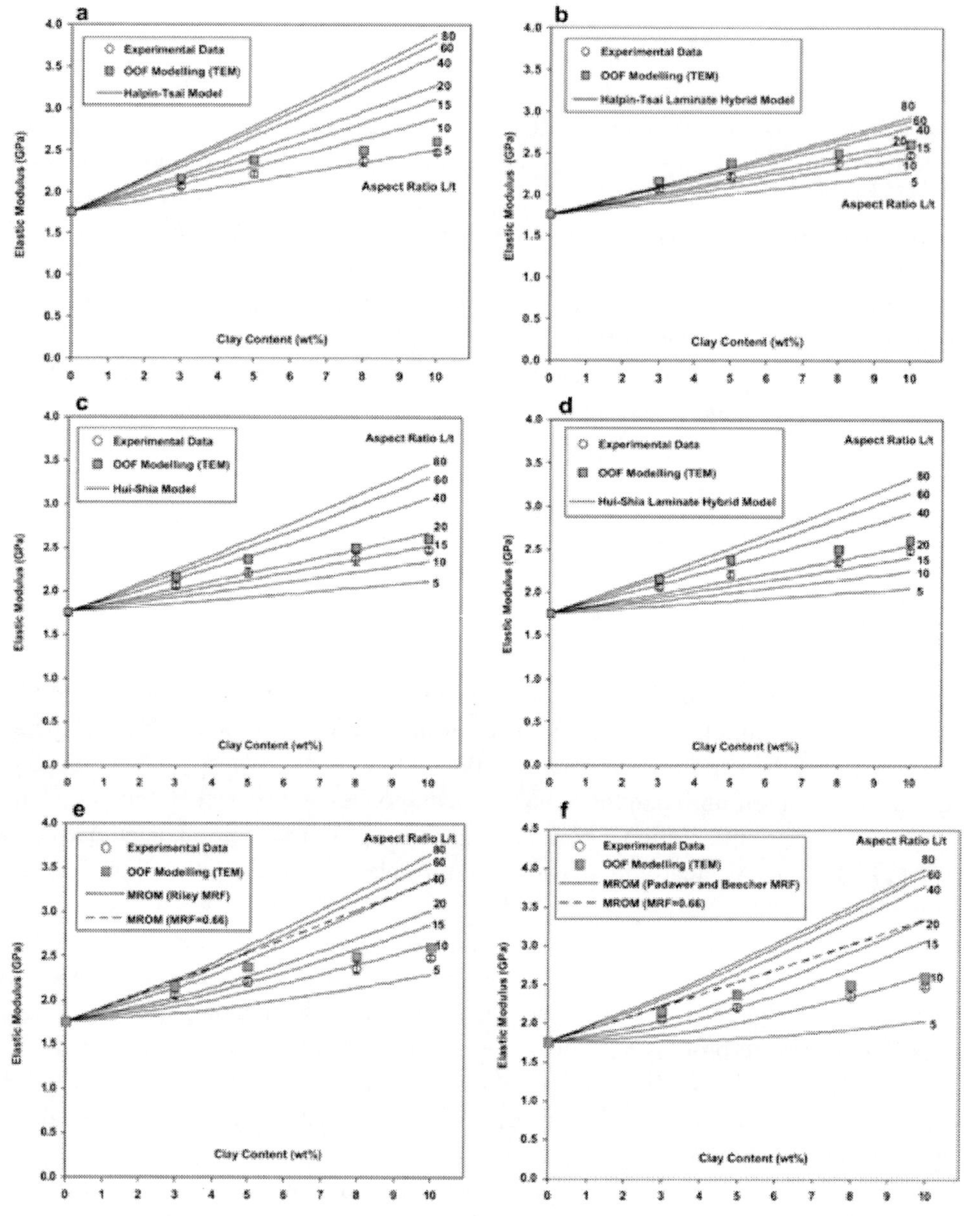

Figure 26. Comparisons of experimental data, OOF modeling (TEM) results and composites theoretical models for optimized PP/clay nanocomposites including (a) Halpin-Tsai model, (b) Halpin-Tsai laminate hybrid model, (c) Hui-shia model and (d) Hui-Shia laminate hybrid model, (e) modified rule of mixture (MROM) with Riley form of MRF and MRF=0.66 and (f) modified rule of mixture (MROM) with Padawer and Beecher form of MRF and MRF=0.66 [65].

Nonetheless, due to the disk-like platelet filler geometry with the biaxial reinforcement effect, Hui-Shia laminate hybrid model also reveals less modulus reduction when compared with unmodified Hui-Shia model, Figures 26(c) and (d). As a result, when a nanocomposite system requires an effective multi-directional reinforcement, clay platelets as nanofillers are more favorable owing to its insensitivity to the orientation states relative to fiber-like fillers. Additionally, the involvement of modulus reduction factor (MRF) in modified rule of mixture

(MROM) does not show the good correlation with the experimental data as evidenced by great deviations from both Riley form and Padawer and Beecher form in MROM, Figures 26(e) and (f). Beyond the aspect ratio of 10 as the intercalation threshold [16], consistently higher elastic moduli have been found for Padawer and Beecher form relative to Riley form. The MROM model with MRF=0.66 overestimates experimental data and the extent of discrepancy becomes more significant at the higher clay contents from 8-10 wt%. The clay dispersion effect for optimized PP/clay nanocomposites has been visually presented from the uniaxial stress contours (σ_{xx}) under 0.25% elastic strain, Figure 27. Despite the lower tensile stresses detected in the PP matrix-rich region, higher stresses are often visible around the interfacial area between clay particles and PP matrix, which suggests the effective load transfer by the interface from the matrix to the fillers in PP/clay nanocomposites. Moreover, increasing the clay content from 3 to 10 wt% means that more frequent particle-particle contact happening due to the increased number of clay particles, and thus the decreased interparticle spacing, leading to the stress enhancement on PP matrix. Such stress intensification effect is quite prevalent at the sharp edges of clay particles where stress concentration sites occur and clay platelets are especially tailored to be well-aligned with the loading direction.

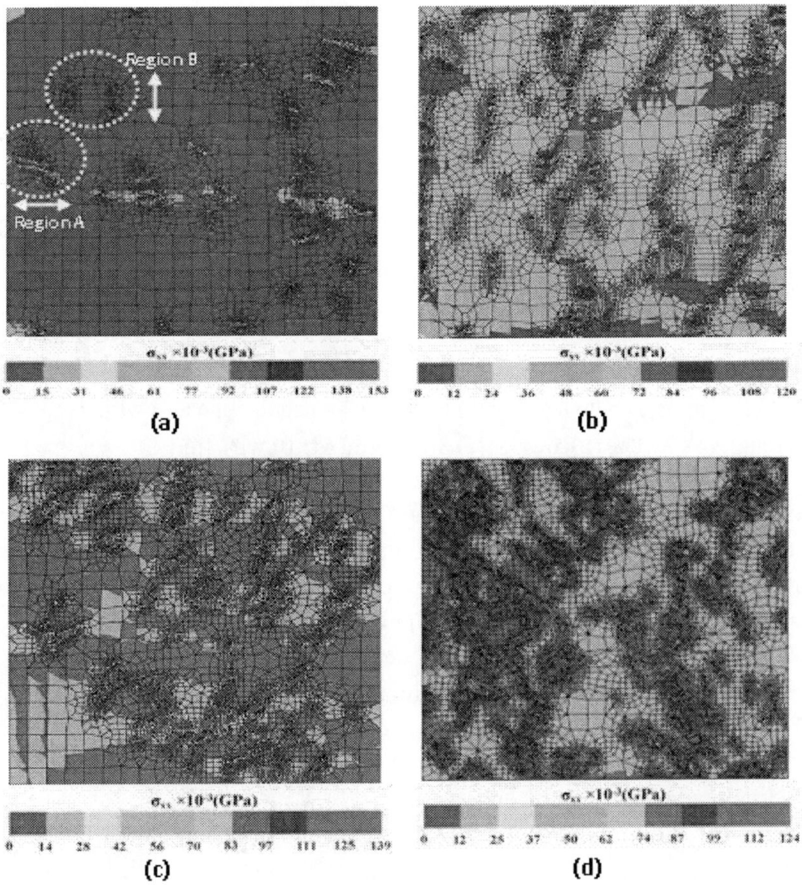

Figure 27. Uniaxial stress (σ_{xx}) contours of optimised PP/clay nanocomposites with various clay contents (U_x=0.25% strain): (a) 3 wt%, (b) 5 wt%, (c) 8 wt% and (d) 10 wt% [65].

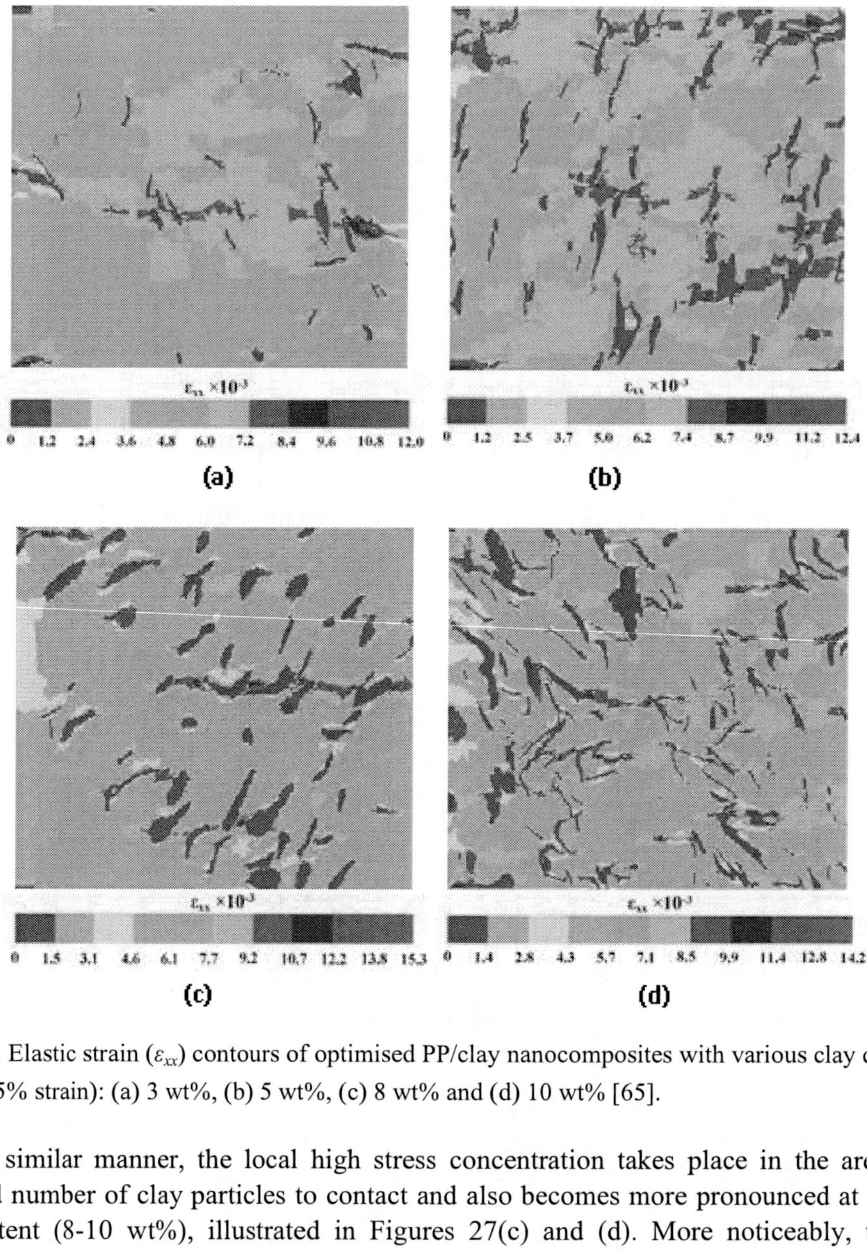

Figure 28. Elastic strain (ε_{xx}) contours of optimised PP/clay nanocomposites with various clay contents (U_x =0.25% strain): (a) 3 wt%, (b) 5 wt%, (c) 8 wt% and (d) 10 wt% [65].

In a similar manner, the local high stress concentration takes place in the areas with increased number of clay particles to contact and also becomes more pronounced at a higher clay content (8-10 wt%), illustrated in Figures 27(c) and (d). More noticeably, the clay orientation state can substantially influence the local stresses in a unidirectional tensile mode. For instance, at Region A in Figure 27(a), better aligned clay particles along the loading direction give higher stresses in contrast to those almost perpendicular to the applied load in Region B. This result confirms an important standpoint in the general composites theory that the preferred filler orientation contributes to the variations of composite modulus and strength in the longitudinal and transverse directions.

Therefore, a good alignment of high-aspect-ratio clay particles within PP matrix is quite vital for the effective stress transfer in a nanocomposite system, which practically diminishes in the real nanocomposite morphology with different extents of random clay orientation. The occurrences of nanocomposite damage, crack and failure could be induced by such high local stress fields around the clay particles, especially for the higher clay contents with clay particle clustering and contact increments.

The corresponding elastic strain contours (ε_{xx}) for the local deformation mechanism are exhibited in Figure 28. Clay particles as rigid fillers can often possess very excellent deformation/straining resistance owing to their high elastic modulus. As a consequence, elastic strains for all clay particles appear to be relatively small when compared to those of PP matrix irrespective of the clay contents in such nanocomposites. PP matrix regions adjacent to clay particles indicate the higher strains parallel to the loading direction. In spite of existing intercalated clay structures, large particle size and close particle contact are quite crucial in the enhancement of deformation level within PP matrix in the proximity of clay particles, where the weak zones are easily developed towards the mechanical failure. Contrarily, less elastic deformation occurs in the particle-free PP matrix areas, and by increasing the clay content, elastic strain distributions become more inhomogeneous. OOF modeling results apparently demonstrate that poor mechanical properties of PP/clay nanocomposites are often ascribed to the particle agglomeration with higher clay contents. They also elucidate the importance of proper nanocomposite processing in order to achieve the uniform clay dispersion with the minimal effect of stress concentrations prone to the nanocomposite structural failure.

CONCLUSION

A holistic approach in a systematic manner has been successfully implemented to statistically determine the optimized material formulations of PP/clay nanocomposites and establish their processing-structure-property relationship using the comprehensive material characterization and numerical modeling. Taguchi design of experiments has been found to be very effective in the appropriate material selection and suitability of manufacturing conditions. The real morphology of PP/clay nanocomposites is confirmed to be in the mixed form of intercalated and exfoliated structures despite the dominant intercalation effect, as characterized by XRD, SEM and TEM. In addition, the remarkable reinforcement effect of additional clay particles is well presented in mechanical and thermo-mechanical properties of PP/clay nanocomposites.

The α-PP formation in the PP crystalline structure, detected by WAXD, is believed to be the main contributor for the enhancement of nanocomposite mechanical properties. Also observed in DSC results, clay particles behave as the nucleating agents to accelerate the PP crystallization process. The novel OOF modeling technique offers the greater insight on the valid predictions for elastic moduli of PP/clay nanocomposites when compared to RVE modeling counterpart and composites theoretical models. Such unique advantages of OOF modeling lie in the clear understanding of effective stress transfer and deformation mechanism of clay particles on the basis of mapping the real micro/nanostructures. In summary, the homogeneous clay dispersion still plays a vital part in nanocomposite fabrication and processing in view of experimental and numerical approaches used in this study. After all, clay agglomerates eventually could generate the localized stress concentrations with the crack initiation and structural failure of nanocomposites.

Acknowledgements

The authors sincerely acknowledge research financial supports from the Tertiary Education Commission (TEC), New Zealand to Dr. Yu Dong through the Bright Future Top Achiever Doctoral Scholarship and the Foundation for Research, Science and Technology (FRST funding # UOAX 0406), New Zealand.

References

[1] Y. Dong and D. Bhattacharyya, *Comput. Mater. Sci.* 49, 1 (2010).
[2] S. G. Lei, "Formulation and mechanical properties of polypropylene nanocomposites", M.ASc thesis, Concordia University, Department of Mechanical and Industrial Engineering (2003).
[3] W. Lertwimolnun and B. Vergnes, *Polymer* 46, 3462 (2005).
[4] N. Hasegawa, M. Kawasumi, M. Kato, A. Usuki and A. Okada, *J. Appl. Polym. Sci.* 67, 87 (1998).
[5] X.H. Liu and Q.J. Wu, *Polymer* 42, 10013 (2001).
[6] Y. Wang, F. B. Cheng and K. C. Wu, *J. Appli. Polym. Sci.* 93, 100 (2004).
[7] D. Gicía-López, O. Picazo, J. C. Merino and J. M. Pastor, *Eur. Polym. J.* 39, 945 (2003).
[8] Y. Dong, D. Bhattacharyya and P. J. Hunter, *Compos. Sci. Technol.* 68, 2864 (2008).
[9] E. C. Lee, D. F. Mielewski and R. J. Baird, *Polym. Eng. Sci.* 44, 1773 (2004).
[10] M. Modesti, A. Lorenzetti, D. Bon and S. Besco, *Polymer* 46, 10237 (2005).
[11] J.C. Halpin and J.L. Kardos, *Polym. Eng. Sci.* 16, 344 (1976).
[12] T. Mori and K. Tanaka, *Acta. Metall.* 21, 571 (1973).
[13] T.D. Fornes and D.R. Paul, *Polymer* 44, 4993 (2003).
[14] D.A. Brune and J. Bicerano, *Polymer* 43, 369 (2002).
[15] N. Sheng, M.C. Boyce, D.M. Parks, G.C. Rutledge, J.I. Abes and R.E. Cohen, *Polymer* 45, 487 (2004).
[16] J.J. Luo and I.M. Daniel, *Compos. Sci. Technol.* 63, 1607 (2003).
[17] E. Hackett, E. Manias and E.P. Giannelis, *Chem. Mater.* 12, 2161 (2000).
[18] G.M. Odegard, T.C. Clancy and T.S. Gates, *Polymer* 46, 553 (2005).
[19] K.S. Katti, D. Sikdar, D.R. Katti, P. Ghosh and D. Verma, *Polymer* 47, 403 (2006).
[20] M. Fermeglia and S. *Pricl, Prog. Org. Coat.* 58, 187 (2007).
[21] R. Toth, A. Coslanich, M. Ferrone, M. Fermeglia, S. Pricl, S. Miertus and E. Chiellini, *Polymer* 45, 8075 (2004).
[22] M. Fermeglia, M. Ferrone and S. Pricl, *Mol. Simul.* 30, 289 (2004).
[23] R. Toth, M. Ferrone, S. Miertus, E. Chiellini, M. Fermeglia and S. Pricl, *Biomacromolecules* 7, 1714 (2006).
[24] P. J. Hine, H. R. Lusti and A. A. Gusev, *Compos. Sci. Technol.* 62, 1445 (2002).
[25] H. R. Lusti, "Property predictions for short fiber and platelet filled materials by finite element calculations", PhD thesis, Swiss Federal Institute of Technology (ETH) Zurich, Institute of Polymers, Department of Materials (2003).
[26] A. Gusev, *Macromolecules* 34, 3081 (2001).

[27] K. Hbaieb, Q. X. Wang, Y. H. J. Chia and B. Cotterell, *Polymer* 48, 901 (2007).
[28] L. J. Zhu and K. A. Narh, *J. Polym. Sci. Pt. B-Polym. Phys.* 42, 2391 (2004).
[29] R. S. Fertig III and M. R. Garnich, *Compos. Sci. Technol.* 64, 2577 (2004).
[30] S. A. Langer, E. R. Fuller Jr. and W. C. Carter, *Comput. Sci. Eng.* 3, 15 (2001).
[31] S. A. Langer, A. C. E. Reid, S. I. Haan and R. E. Garcia, The OOF2 manual: Revision 3.7 for OOF2 Version 2.0.4, the National Institute of Standards and Technology (NIST), USA. [Online]: http://www.ctcms.nist.gov/~langer/oof2man/index.html
[32] N. Chawla, B. V. Patel, M. Koopman, K. K. Chawla, R. Saha, B. R. Patterson, E. R. Fuller and S. A. Langer, *Mater. Charact.* 49, 395 (2003).
[33] C. H. Hsueh, E. R. Fuller Jr., S. A. Langer and W. C. Carter, *Mater. Sci. Eng. A-Struct. Mater. Prop. Microstruct. Process.* 268, 1 (1999).
[34] M. H. Zimmermann, D. M. Baskin, K. T. Faber, E. R. Fuller Jr., A. J. Allen and D. T. Keane, *Acta Mater.* 49, 3231 (2001).
[35] V. Cannillo, G.C. Pellacani, C. Leonelli, A.R. Boccaccini, *Compos. Pt. A-Appl. Sci. Manuf.* 34, 43 (2003).
[36] V. Cannilo, C. Leonelli, T. Manfredini, M. Montorsi, P. Veronesi, E.J. Minay and A. R. Boccaccini, *Compos. Sci. Technol.* 65, 1276 (2005).
[37] V. Cannillo, T. Manfredini, M. Montorsi, C. Siligardi and A. Sola, *J. Eur. Ceram. Soc.* 26, 3067 (2006).
[38] V. R. Vedula, S. J. Glass, D. M. Saylor, G. S. Rohrer, W. C. Carter, S. A. Langer and E. R. Fuller, *J. Am. Ceram. Soc.* 84, 2947 (2001).
[39] V. Cannillo, C. Leonelli and A. R. Boccaccini, *Mater. Sci. Eng. A-Struct. Mater. Prop. Microstruct. Process.* 323, 246 (2002).
[40] Z. Wang, A. Kulkarni, S. Deshpande, T. Nakamura and H. Herman, *Acta. Mater.* 51, 5319 (2003).
[41] V. Cannillo, F. Bondioli, L. Lusvarghi, M. Montorsi, M. Avella, M. E. Errico and M. Malinconico, *Compos. Sci. Technol.* 66, 1030 (2006).
[42] M. Avella, F. Bondioli, V. Cannillo, M. E. Errico, A. M. Ferrari, B. Focher, M. Malinconico, T. Manfredini and M. Montorsi, *Mater. Sci. Technol.* 20, 1340 (2004).
[43] F. Bondioli, V. Cannillo, E. Fabbri and M. Messori, *J. Appli. Polym. Sci.* 97, 2382 (2005).
[44] X. Kornmann, H. Lindberg and L. A. Berglund, *Polymer* 42, 1303 (2001).
[45] S. H. Park, *Robust design and analysis for quality engineering*, Chapman and Hall, London (1996).
[46] R. H. Lochner and J. E. Matar, *Design for quality: an introduction to the best of Taguchi and western methods of statistical experimental design*, Quality Resources and ASQC Quality Press, Milwaukee, WI (1990).
[47] Y. Dong and D. Bhattacharyya, *Compos. Pt. A-Appl. Sci. Manuf.* 39, 1177 (2008).
[48] J. E. Mark, *Physical properties of polymer handbook*, AIP Press, New York (1996).
[49] Y. P. Wu, Q. X. Jia, D. S. Yu and L. Q. Zhang, *Polym. Testing* 23, 903 (2004).
[50] X. L. Ji, J. K. Jing, W. Jiang and B. Z. Jiang, *Polym. Eng. Sci.* 42, 983 (2002).
[51] L. Cauvin, D. Kondo. M. Brieu and N. Bhatnagar, *Polym. Testing* 29, 245 (2010).
[52] G. T. Wang, H. Y. Liu, Z. Z. Yu and Y. W. Mai, *J. Reinf. Plast. Compos.* 28, 1625 (2009).
[53] W. H. Bowyer and M. G. Bader, *J. Mater. Sci.* 7, 1315 (1972).
[54] W. Voigt, *Ann. Phys.* 38, 573 (1889).

[55] Reuss, *ZAMM* 9, 49 (1929).
[56] Y. Hui and D. Shia, *Polym. Eng. Sci.* 38, 774 (1998).
[57] V. R. Riley, *Interaction effects in polymer composites*, in: *Polymer conference series*, University of Utah, Salt Lake City,1970.
[58] G. E. Padawer and N. Beecher, *Polym. Eng. Sci.* 10, 185 (1970).
[59] G. P. Tandon and G. J. Weng, Polym. Compos. 5, 327 (1984).
[60] J. D. Eshelby, *Proc. R. Soc. London A* 241, 376 (1957).
[61] M. van Es, J. Turnhout and E. Van der Giessen, *Comparing polymer-clay nanocomposites with conventional composites using composite madelling*, in: S. Al-Malaika, A. W. Golovoy (editors), *Specialty polymer additives: principles and applications*, Blackwell Science, Oxford (2001).
[62] D. Hull and T. W. Clyne, *An introduction to composite materials*, 2nd ed., Cambridge University Press, New York (1996).
[63] R. A. Dickie, *J. Appli. Polym. Sci.* 17, 45 (1973).
[64] X. L. Ji, J. K. Jing, W. Jiang and B. Z. Jiang, *Polym. Eng. Sci.* 42, 983 (2002).
[65] Y. Dong and D. Bhattacharyya, *Polymer* 51, 816 (2010).
[66] Y. Dong, "Multi-scale effects on deformation mechanisms of polymer nanocomposites: experimental characterisation and numerical study", PhD thesis, the University of Auckland, Department of Mechanical Engineering (2008).
[67] D.H. Kim, P.D. Fasulo, W.R. Rodgers and D.R. Paul, *Polymer* 48, 5308 (2007).
[68] Y. Dong and D. Bhattacharyya, *J. Mater. Sci.* 47, 3900 (2012).
[69] Y. Dong and D. Bhattacharyya, "Optimisation of manufacturing parameters and morphology-based numerical analysis of nanocomposites", Conference proceedings of "Processing and Fabrication of Advanced Materials-XVII", the Seventeenth International Symposium on Processing and Fabrication of Advanced Materials (PFAM XVII), 15-17, Dec. 2008 (Indian Institute of technology, Delhi, India) pp559-579, N. Bhatnagar, T. S. Srivatsan (editors), I. K. International Publishing House, New Delhi, India (2009).
[70] Y. Dong and D. Bhattacharyya, *J. Mater. Sci.* 47, 4127 (2012).
[71] L. Zhao, J. Li, S. Guo and Q. Du, *Polymer* 47, 2460 (2006).
[72] Q. Zheng, Y. Shangguan, S. Yan, Y. Song, M. Peng and Q. Zhang, *Polymer* 46, 3163 (2005).
[73] S. C. Tjong, J. S. Shen and R. K. Y. Li, Polym. Eng. Sci. 36, 100 (1996).
[74] J. Karger-Kocsis and J. Varga, J. Appli. Polym. Sci. 62, 291 (1996).
[75] P. Maiti, P. H. Nam, M. Okamoto, N. Hasegawa and A. Usuki, *Macromolecules* 35, 2042 (2002).
[76] X. Li, H. Gao, W. A. Scrivens, D. Fei, V. Thakur, M. A. Sutton, A. P. Reynolds and M. L. Myrick, *Nanotechnology* 16, 2020 (2005).
[77] V. J. Mittal, *Thermoplast. Compos. Mater.* 20, 575 (2007).

ABOUT THE EDITOR

Yu Dong received a B.E. in Mechanical Engineering and Automation from Wuhan University of Science and Technology (WUST), China in 2000, M.E. (1st Class Honours) and Ph.D. in Mechanical Engineering from the University of Auckland, New Zealand in 2002 and 2008, respectively. Currently he is a Senior Lecturer in Department of Mechanical Engineering, Curtin Univerisity, Perth, Australia. Dr. Dong has widespread research interests including polymer nanocomposites, electrospun nanofibres, micromechanical modelling, nanomaterial processing and characterisation and statistical design of experiments. He is a Member of Editorial Boards for *International Journal of Polymers and Technologies, Journal of Modern Applied Science* and *Journal of Applied Physics Research*. He also serves as a selected reviewer for *Journal of Nanoscience and Nanotechnology Letters*, *Journal of Materials Science and Engineering A*, *Journal of composite materials*, *Journal of Composites Part B* and *Chemical Engineering Journal*. Dr. Dong is actively affiliated to several global engineering professional bodies as members of American Society of Mechanical Engineers (ASME), Society of Plastic Engineers (SPE), USA, Institute of Nanotechnology (IoN), UK and Australasian Association for Engineering Education (AAEE) as well as a key researcher for the ATN-ISTA Australia-China NanoNetwork and Australian Research Network for Advanced Materials (ARNAM). To recognise his great contribution in polymer/clay nanocomposites, he was appointed as a Member of Organising Committee for the Nineteenth International Symposium on Processing and Fabrication of Advanced Materials (PFAM XIX), Auckland, New Zealand, January 14-17, 2011. He is a recipient of Best paper Silver Award at the 5th Asian-Australasian Conference on Composite Materials (ACCM-5) in Hong Kong, 2006, excellent mentorship for Curtin BHP Billiton Award Winning Team in Sustainable Design and Appropriate Technology Themes in EWB Challenges, Australia in 2009 and Curtin Highest RPI Points in Science & Engineering for ECR Publications in 2011.

INDEX

A

B

C

D

E

F

G

H

I

K

L

M

N

Q

R

S

T

U